Handbook of Applied Mycology

Volume 5:
Mycotoxins in Ecological Systems

Handbook of Applied Mycology

Series Editor
Dilip K. Arora

Banaras Hindu University
Varanasi, India

Volume 1 Soil and Plants, edited by *Dilip K. Arora, Bharat Rai, K. G. Mukerji*, and *Guy R. Knudsen*

Volume 2 Humans, Animals, and Insects, edited by *Dilip K. Arora, Libero Ajello*, and *K. G. Mukerji*

Volume 3 Foods and Feeds, edited by *Dilip K. Arora, K. G. Mukerji*, and *Elmer H. Marth*

Volume 4 Fungal Biotechnology, edited by *Dilip K. Arora, Richard P. Elander, and K. G. Mukerji*

Volume 5 Mycotoxins in Ecological Systems, edited by *Deepak Bhatnagar, Eivind B. Lillehoj, and Dilip K. Arora*

Handbook of Applied Mycology

Volume 5:
Mycotoxins in Ecological Systems

edited by

Deepak Bhatnagar

U.S. Department of Agriculture
New Orleans, Louisiana

Eivind B. Lillehoj

Lillehoj and Associates
Kimballton, Iowa

Dilip K. Arora

Centre of Advanced Study in Botany
Banaras Hindu University
Varanasi, India

MARCEL DEKKER, INC. New York • Basel • Hong Kong

Library of Congress Cataloging-in-Publication Data

Mycotoxins in ecological systems / edited by Deepak
Bhatnagar, Eivind B. Lillehoj, Dilip K. Arora.
p. cm. -- (Handbook of applied mycology ; v. 5.)
Includes bibliographical references and index.
ISBN 0-8247-8551-7 (acid free paper)
1. Mycotoxins. 2. Aflatoxins. 3. Fungi--Ecology.
I. Bhatnagar, Deepak. II. Lillehoj, E.B. III. Arora,
D.K. IV. Series.
QP632.M9M94 1992
615.9'5292--dc20 91-39242
CIP

This book is printed on acid-free paper.

MARCEL DEKKER, INC.
270 Madison Avenue, New York, New York 10016

Current printing (last digit):
10 9 8 7 6 5 4 3 2 1

PRINTED IN THE UNITED STATES OF AMERICA

HANDBOOK INTRODUCTION

Applied mycology is one of the most stimulating and rapidly evolving areas of the biological sciences. It encompasses many facets of agricultural, industrial, pharmacological, medical, and food sciences, all of which include organisms or processes that are subject to biotechnological manipulation. However, published information on the practical significance of fungi is fragmented in many specialized publications. There is an obvious need for a comprehensive treatment of the basic principles, methods, and applications of mycology as an applied, integrated, and multidisciplinary subject. The potential usefulness of such a handbook is suggested by the practical significance of mycological research, by the uniqueness and increasing coherence of the field's conceptual directions, and by the international scope of the subject. These five volumes of the *Handbook of Applied Mycology* attempt to fulfill this need. These volumes present and collate the major aspects of applied mycology and are designed as a modern standard reference for the work in this field.

In developing these volumes we decided that the treatment of the subject should be comprehensive but concise enough to enable completion within a set of five volumes published closely together rather than a continuous encyclopedic series taking several years to complete. However, this does not preclude the possibility of publishing supplementary volumes, as appropriate, to communicate new information. Further, we wanted to develop a treatise that would offer a detailed overview of applied aspects of mycology in terms of its theoretical, methodological, and empirical contributions. By emphasizing the interdisciplinary aspects of the subject, we hope that the handbook will highlight recent prospective linkages among diverse research paradigms.

Since the information presented in these volumes is expected to serve students as well as veterans in the field, both basic and advanced specialized materials are included. In addition to addressing researchers in mycology, the handbook will prove useful to a broad spectrum of professional groups, including biotechnologists, medical scientists, molecular biologists, food scientists, pharmacologists, soil microbiologists, microbial ecologists, and plant pathologists. Each chapter is intended to provide a balanced view of the current state of knowledge within an expanding field of interest. The selection of the chapters and overall organization of the book were guided by several insightful colleagues from around the world. These volumes were authored by a large team of international experts. Every chapter within the five volumes has been edited by a distinguished group of scientists. As a result, nine editors and over 200 authors representing 20 countries have contributed to this project.

Because of the multidisciplinary nature of the subject, judgment had to be made about the relative importance and significance of the hundreds of potential topics related to applied mycology. Also, a balance had to be struck between comprehensive coverage and intellectual appeal of the field. Readers should find this work useful for in-depth information and as a route to additional information contained in its critical discussions and wide range of references provided. Each of the five volumes is intended to be self-contained. Therefore, some degree of duplication of materials, especially basic principles, is inevitable.

No work of this magnitude can be accomplished without the support and contributions of many individuals. The planning and eventual production of this handbook were a collaborative effort. I am deeply indebted to the group of coeditors and many colleagues who have assisted me throughout all stages of the project. I also appreciate the dedication and hard work of the chapter authors, who cooperated persistently in the production of these volumes. I acknowledge with gratitude the support of M. V. Wiese, N. W. Schaad, S. Stotzsky, J. S. Singh, Bharat Rai, R. S. Upadhyay, V. S. Jaiswal, B. Thompson, Jon Olson, A. K. Pandey, Rejeev Gaur, and Shusma Gupta, whose interest and cooperation have been invaluable to me. I am grateful to the University of Idaho for providing assistance in the form of staff support and facilities, and the Banaras Hindu University, India, for granting leave during the period I edited these volumes. I appreciate the encouraging support provided by Ms. Sandra Beberman of Marcel Dekker, Inc., at all stages of production of this book. My gratitude is expressed to my teacher, J. L. Lockwood, and to my parents, whose example has always been my guiding principle. Finally, to Meenakshi, Sidhartha, and Gautam for their patience, encouragement, and willingness to forgo other interests so that the handbook might be completed, I am grateful.

Dilip K. Arora

PREFACE

In the postantibiotic era, mycotoxin studies have emerged as the tour de force of applied mycology. Previous volumes in the current series have described numerous aspects of mycotoxins and their implications in varied contexts. An overview of the information shows that mycotoxins are particularly important in agriculture since their presence exerts a negative economic impact on relevant commerce as well as severe health problems after exposure to contaminated food and feed. During the antibiotic era, the study of microbial secondary metabolites focused on the implication of the substances in human medicine but the ecological nature of secondary metabolites remained largely a mystery. The microbial-toxin era introduced a new and fascinating aspect of applied mycology, namely, *agro-ecological aspects of mycotoxins*.

Contemporary mycotoxicology has emerged as a single discipline from the background of a variety of investigative pursuits. Although initial considerations of mycotoxins appeared to have the qualitites of a Tower of Babel, threads of common intradisciplinary interest began to develop as the discipline matured. A critical conceptual problem paralleled the quandary of the antibiotic period, i.e., the fundamental mystery of the biological logic for production of secondary metabolites.

The initial diversity of approaches in mycotoxin research provided diffuse goals but also yielded interactive symbioses with profound implications. For example, initial considerations of an evolutionary explanation for fungal acquisition of genetic information for synthesis of complex, carcinogenic/mutagenic hydrocarbons such as aflatoxin from innocuous polyketides provided an extraordinary enigma for microbiologists. At the same time, oncologists were carefully elucidating the structural and functional aspects of chemical carcinogens with particular emphasis on their lock-and-key properties in target genomes. The investigations yielded unequivocal evidence for the unique fit of carcinogens at specific DNA sites. Integration of the two concepts provided the basis for an exciting new departure in understanding the importance of coevolution of pertinent genetic information for synthesis of aflatoxin in the context of effecting specific modifications of target genomes.

Elucidation of the mycotoxin phenomenon clearly requires an integrated approach. The current volume has been commissioned to identify ecological implications of mycotoxins through consideration of three aspects of the toxins: (1) genetic and evolutionary origin of the information required by fungi for synthesis of toxic metabolites, (2) specific aspects of mycotoxin-mediated, acute cytotoxicity in target tissues, and (3) unique modulation of host-cell genomes by chronic exposure to toxic fungal metabolites. A unified consideration of the seemingly disparate elements is developed to produce a structure for new, relevant hypotheses of the biology of synthesis of toxic secondary metabolites.

Scientists who have contributed to the volume have presented a panoply of exciting information. The expositions provide an unusual cross-section of scientific information that is a genuine testament to the creative and audacious nature of mycotoxicologists. Students of biology will be rewarded by careful perusal of the presentations and fascinated by their extraordinary extrapolations. The editors want to thank, in addition to the authors, other scientists who provided critical comments and suggestions and the editorial staff of Marcel Dekker, Inc., in particular Andrew Berin, for careful review of the material; their efforts have vastly improved the expositions.

Deepak Bhatnagar
Eivind B. Lillehoj
Dilip K. Arora

CONTENTS

CONTRIBUTORS

HERMAN AUTRUP* Fibiger Institute, Copenhagen, Denmark

JUDITH L. AUTRUP Laboratory of Environmental Carcinogenesis, Fibiger Institute, Copenhagen, Denmark

GEORGE S. BAILEY Department of Food Science and Technology, Oregon State University, Corvallis, Oregon

MARIAN N. BEREMAND† Mycotoxin Research Unit, Agricultural Research Service, National Center for Agricultural Utilization Research, U.S. Department of Agriculture, Peoria, Illinois

J. W. BENNETT Department of Cell and Molecular Biology, Tulane University, New Orleans, Louisiana

DEEPAK BHATNAGAR Southern Regional Research Center, Agricultural Research Service, U.S. Department of Agriculture, New Orleans, Louisiana

KRISHNA S. BILGRAMI Department of Botany, Bhagalpur University, Bhagalpur, India

FUN S. CHU Department of Food Microbiology and Toxicology, University of Wisconsin—Madison, Madison, Wisconsin

THOMAS E. CLEVELAND Southern Regional Research Center, Agricultural Research Service, U.S. Department of Agriculture, New Orleans, Louisiana

---

Present affiliations:
*Institute of Environmental and Occupational Medicine, University of Aarhus, Aarhus, Denmark.
†Department of Plant Pathology and Microbiology, Texas A&M University, College Station, Texas.

RODERICK H. DASHWOOD[*] Oregon State University, Corvallis, Oregon

ANNE E. DESJARDINS Mycotoxin Research Unit, Agricultural Research Service, National Center for Agricultural Utilization Research, U.S. Department of Agriculture, Peoria, Illinois

PATRICK F. DOWD Mycotoxin Research Unit, Agricultural Research Service, National Center for Agricultural Utilization Research, U.S. Department of Agriculture, Peoria, Illinois

DAVID L. EATON Department of Environmental Health and Institute for Environmental Studies, University of Washington, Seattle, Washington

KENNETH C. EHRLICH Southern Regional Research Center, Agricultural Research Service, U.S. Department of Agriculture, New Orleans, Louisiana

DOUGLAS E. GOEGER Department of Preventive Medicine and Community Health, Division of Human Nutrition, University of Texas Medical Branch, Galveston, Texas

J. D. HENDRICKS Oregon State University, Corvallis, Oregon

SHUBHA KALE Department of Cell and Molecular Biology, Tulane University, New Orleans, Louisiana

NANCY P. KELLER[†] Southern Regional Research Center, Agricultural Research Service, U.S. Department of Agriculture, New Orleans, Louisiana

EIVIND B. LILLEHOJ Lillehoj and Associates, Kimballton, Iowa

PATRICIA M. LOVELAND Department of Food Science and Technology, Oregon State University, Corvallis, Oregon

KOENRAAD MARIËN[‡] Oregon State University, Corvallis, Oregon

SUSAN P. McCORMICK Mycotoxin Research Unit, Agricultural Research Service, National Center for Agricultural Utilization Research, U.S. Department of Agriculture, Peoria, Illinois

HIDEKI MORI Department of Pathology, Gifu University School of Medicine, Gifu, Japan

HOWARD S. RAMSDELL[§] Department of Environmental Health, University of Washington, Seattle, Washington

---

Present affiliations:

[*]Department of Environmental Biochemistry, University of Hawaii, Honolulu, Hawaii.

[†]Department of Plant Pathology and Microbiology, Texas A&M University, College Station, Texas.

[‡]Department of Health, State of Washington, Olympia, Washington.

[§]Department of Environmental Health, Colorado State University, Fort Collins, Colorado.

RONALD T. RILEY Toxicology and Mycotoxins Research Unit, Richard B. Russell Agricultural Research Center, Agricultural Research Service, U.S. Department of Agriculture, Athens, Georgia

KAUSHAL K. SINHA Department of Botany, Bhagalpur University, Bhagalpur, India

FREDRIK CARL STØRMER Department of Environmental Medicine, National Institute of Public Health, Oslo, Norway

NEIL W. WIDSTROM Insect Biology and Population Management Research Laboratory, Agricultural Research Service, U.S. Department of Agriculture, Tifton, Georgia

DAVID E. WILLIAMS Department of Food Science and Technology, Oregon State University, Corvallis, Oregon

1

AFLATOXIN: GENETIC MOBILIZATION AGENT

EIVIND B. LILLEHOJ *Lillehoj and Associates, Kimballton, Iowa*

I. INTRODUCTION

The 1990s issue in a new era—the age of environmental concerns. A bitter lesson was learned in the post—World War II period that involved extraordinary variation in perceptions that ranged from limitless, unfettered economic development to the horrors of environmental degradation. Positive aspects of Borlaug's green revolution faded into the disaster of overpopulation and attendant starvation in the third world. Technologically advanced societies were confronted with numerous negative aspects of relentless increases in demands for natural resources including land for agricultural production. In addition to "domestication" of vast, natural ecosystems, developed agricultural systems were subjected to accelerated intensive practices; a critical product of the activities was drastic reduction in the number of species and biological diversity. Clearly, future agriculture will require evaluation in not only economic but also environmental terms.

II. AFLATOXIN, A SECONDARY FUNGAL METABOLITE

A. Historical Perspective

Although there is a broad international consensus on the need to protect the earth's environment, specific examples of unambiguous influence of environmental deterioration on human health are limited. However, the discoveries of the occurrence of naturally occurring toxins in agricultural commodities during the past 30 years provide an extraordinary model with direct influence on human health. Aflatoxin is a secondary metabolite of a fungus, *Aspergillus flavus*, that was discovered in agricultural commodities in the 1960s [1]. Secondary metabolites are substances produced by living cells that are not essential for the growth of the producing organism [2, 3]. The value of secondary metabolites to the producing organisms has aroused spirited discussion [4, 5]. An overview of the relevant literature demonstrates: (1) characterization of hundreds of secondary metabolites with complex structure [6], and (2) specificity of chemical structure—biological response interactions in numerous systems [1, 7]. The observations provide a compelling basis for identification of secondary metabolites as genetically determined substances that confer a distinct advantage on the producing microbes.

B. Aflatoxin Toxicity—Initial Observations

Aflatoxin was originally identified in peanut meal as a fungal toxin that killed turkeys [1]. Subsequent studies identified acute toxic responses in animals to doses ranging from 0.3—0.6 mg aflatoxin per kilogram body weight in rabbits and ducklings to 10 mg/kg in hamsters and rats [1]. Although acute aflatoxin toxicity represented a real problem for agriculture, the most important aspect of the toxicity was the discovery that animals exposed to the toxin exhibited increased tumor formation, particularly in the liver. Extended feeding periods at low levels (15 ppb) produced a high incidence of liver tumors in test animals [1]. The discovery of the naturally occurring carcinogen in food commodities took place during the period when the pertinent scientific community considered chemical carcinogenesis from the empirical perspective with little understanding of causality. Initial observations on aflatoxn-mediated toxicity in microbes showed inhibition of growth and formation of aberrant cell structures [8]; an association was discovered between aflatoxin and the inhibition of nucleic acid synthesis [9]. The studies provided preliminary evidence of aflatoxin as an inhibitor of genetic replication. Subsequent research identified aflatoxin-induced removal of repressor control of enzyme synthesis in bacterial cells [10].

A number of studies on the effects of aflatoxin in tissue and organ cultures provided further preliminary information. In parallel with observations in microbes, the toxin blocked nucleic acid synthesis in mammalian cells [11]. Malignant transformations in tissue culture cells were produced by aflatoxin [12], and subsequent investigations identified specific aspects of these transformations [13].

III. AFLATOXIN PRODUCTION IN THE AGROECOSYSTEM

Initial observations on aflatoxin contamination of agricultural commodities assumed that the toxin-producing fungus was an opportunistic saprophyte with no parasitic capability [1]. Subsequently, *A. flavus* strains were found capable of infecting developing peanuts, cottonseed, and corn kernels and contaminating the seed with aflatoxin before harvest [14]. No evidence existed for the common occurrence of aflatoxin in crops prior to the 1960s. Although some scientists believed that the fungus and associated toxin had occurred in prior periods, a larger consensus began to develop that linked the presence of the opportunistic, toxin-producing fungus with the intensive agricultural practices introduced in the post—World War II period [5]. The concept paralleled developments in related areas of agriculture that were identifying the limitations of intensive agriculture [15].

Since its inception, agriculture has introduced disequilibrium into stable ecological communities. The magnitude of the deviation from equilibrium is a product of the intensity of relevant practices. During the past 50 years, an extraordinary increase in intensive practices has been introduced that can be characterized by: (1) high technology plants with elevated productivity, (2) wide use of industrially derived fertilizers and pesticides, (3) increased mechanization, and (4) continuous cropping of limited species with restricted genetic variation. The aggregate effect of intensive agriculture is extraordinary homogenization of the biota within agroecosystems and related marked shifts in the attendant pest species. Harper [16] has identified the dilemma: "Pest epidemics can often be ascribed to some activity of man that has brought a floristically diverse ecosystem closer to monoculture...."

Intensive agricultural practices appear to be responsible for the creation of unique agroeconiches that under certain conditions select toxin-producing fungi by unknown mechanisms (Fig. 1). Since a major component of contemporary crop production systems is based on intensive practices, it appears unlikely that large-scale, effective alternatives will be introduced soon. Therefore, the scientific community has assumed a responsibility for elucidating the observed phenomena and providing the

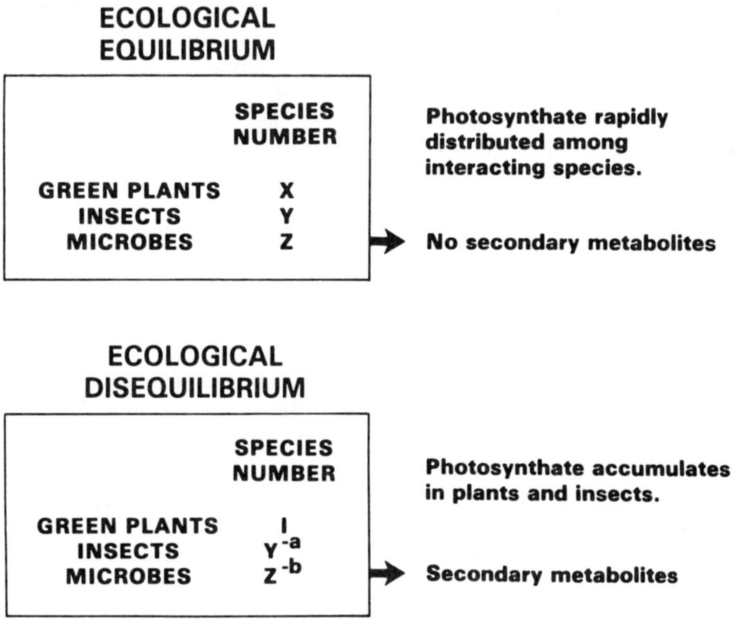

FIGURE 1. Secondary metabolite production and ecosystem interactions.

basis for future control of toxin-producing microbes. Characterization of the evolutionary basis for production of toxins and their effects in target cells appears to be one of the basic challenges of further research. Two major aspects of deciphering the mystery of ecological implications of fungal toxins such as aflatoxin are: (1) mechanisms involved in synthesis of the toxin, and (2) molecular aspects of the toxic response. Recent work has identified some of the enzymes that are responsible for synthesis of aflatoxin [17]. Further work is being carried out to identify the genes involved in elaboration of the enzymes. The current pathway includes: polyketide → norsolorinic acid → averantin → averufanin → averufin → versiconal hemiacetal acetate → versicolorin A → sterigmatocystin → *O*-methylsterigmatocystin → aflatoxin B_1. Examination of the toxic properties of the intermediates in the aflatoxin pathway and the final toxin demonstrate a broad pattern of increased toxicity from the initial pathway compound (norsolorinic acid) to the most toxic substance, the final product, aflatoxin B_1 [18]. A hypothesis has been postulated that provides an evolutionary explanation for the aflatoxin biosynthetic pathway [18].

IV. AFLATOXIN GENOTOXICITY

A. Introduction

A comprehensive understanding of the implications of microbial toxins in agroecosystems requires fundamental knowledge of the specific effects of toxins on intracellular targets. Although initial observations of aflatoxin toxicity were based on empirical evidence of acute responses, the carcinogenic properties of the substance directed work toward elucidation of the molecular events in cellular transformations. The interest in aflatoxin became part of the larger scientific interest in events associated with the initiation of carcinogenic processes. Preliminary studies of aflatoxin and other chemical carcinogens provided compelling evidence for an effect of toxins on cellular nucleic acid and associated genomes and genetic processes.

B. Microbes

A great deal of scientific interest has been devoted to the fascinating phenomenon of bacterial virus (phage) infections and the subsequent regulation of lysis of host cells or incorporation of the virus into the host chromosome (lysogenic state). A complex regulatory system has evolved to maintain the lysogenic state through appropriate repressors [19]. The phage incorporated into the host chromosome (prophage) can be maintained through many generations. Under certain conditions, the repressor system is compromised, the insertion process reversed, the prophage released, and the lytic state established. It has been known for some time that certain compounds that damage DNA can convert the lysogenic state to the lytic state [20]. Lillehoj and Ciegler [21] identified the aflatoxin-induced transition of a lysogenic bacterial strain to the lytic state. The studies provided direct evidence for a genetics-based role of aflatoxin as a cellular toxin that resembled the activity of ultraviolet (UV) radiation and other radiomimetic compounds.

In addition to induction of the lytic state, UV radiation and radiomimetic agents produced mutations and other genetic modifications [22]; the effects were linked to errors in DNA repair. Mutations were associated with absence of *rec A* and *lex A* genes and inducible, error-free repair. Certain types of DNA damage produced by carcinogenic compounds were attributed to error-prone (SOS) repair [23, 24]. Aflatoxin was found to initiate SOS repair in *Escherichia coli* [25]. A detailed examination of the effect of aflatoxin on the viral–bacterial interaction showed that the toxin induced λ phage excision from the chromosome by cleavage of the repressor that involved *rec A* activation [26]. Bacterial cells deficient in DNA repair and recombination enzymes exhibited enhanced sensitivity to aflatoxin [27, 28].

The increased awareness of problems associated with environmentally based mutagenesis and carcinogenesis was distinctly increased by the introduction of a sensitive and simple bacterial test for chemical mutagens by Ames and his colleagues [29, 30]. The test was based on development of histidine-requiring mutants of *Salmonella typhimurium*. Reversion to histidine independence by specific strains identified either base-pair or frameshift mutations. Aflatoxin was found to be an extremely effective mutagen in the Ames test [31–33]; the toxin was also an effective mutagen in insects [34, 35].

C. Tissue Culture

Studies in tissue culture elucidated a clear relationship between the presence of aflatoxin and dose-dependent transformations of nominal tissue cultures to tumor cells [36]. Further characterization of the genotoxicity of aflatoxin in cultured cells was carried out in systems designed to identify the unscheduled DNA synthesis (UDS) associated with DNA repair. A key investigation demonstrated that aflatoxin-mediated excision repair in human cells was linked to the presence of an activating microsomal system [24]. Subsequently, the induction of DNA repair in liver of intact animals by aflatoxin was identified [37].

In addition to aflatoxin-based DNA repair in tissue culture cells, the tests provided information on the effect of certain transforming viruses. Treatment of hamster embryo cells with carcinogens elevated transformation by subsequent exposure to a simian virus SA7 [38]. Aflatoxin increased SA7 transformations when it was added prior to the virus but inhibited viral activity when added after viral adsorption to cells. The studies suggested that DNA repair provided additional sites for virus integration into host chromosomes. Sarasin and Hanawalt [39] linked DNA damage and SOS repair with elevated viral activity. Increased viral development in aflatoxin-treated cells was also related to aflatoxin-mediated diminution of interferon in monkey kidney cells [40]. Studies in chick embryo cells found an aflatoxin-mediated, dose-dependent increase in UDS that was associated with induction of mixed function oxidase (MFO) activity [41].

V. AFLATOXIN–DNA BINDING

A. Introduction

Preliminary studies of aflatoxin toxicity demonstrated genetic effects that appeared to follow an initial binding of the toxin to nucleic acid. Intensive consideration of the nature of carcinogen binding to DNA identified requirements for planar structures to allow for fits into the grooves of the DNA helix [1, 42]. Detailed structure–function relationships defined two unique regions (K and L) of carcinogens that appeared to be required for initiation of the oncogenic process [43, 44]. Examination of aflatoxin demonstrated a K region at the unsaturated carbons of the terminal furan and a clearly identified L region [45]. The studies related the region of pertinent reactivity of the toxin on the unsaturated carbon atoms of the terminal furan. Subsequent improvement in quantum calculations identified a unique bay region in aflatoxin and other chemical carcinogens [45–47]. The bay region appeared to indicate a location that supported epoxide formations; the observations provided the basis for the requisite bay region in a polycyclic aromatic hydrocarbon at the site for metabolic activation from an inert potential carcinogen to an active proximate carcinogen. An overview of the initial studies on aflatoxin–DNA binding provided evidence of a structural basis for the toxin to insert into a DNA groove with subsequent activation by MFO enzymes to activate the compound into an active mutagen or carcinogen.

B. Cytochrome P-450 Activation of Aflatoxin

Early mixed function oxidase activities of tissues were subsequently associated with a unique hemoprotein called cytochrome P-450 [48]. Many drugs, xenobiotics, and endogenous cellular metabolites are chemically modified by processes associated with cytochrome P-450 in microsomes of the endoplasmic reticulum [49, 50] (Fig. 2). Reducing potential for the cytochrome P-450 reactions is mediated by reduced nicotinamide adenine dinucleotide phosphate (Fig. 3). The oxidized hemoprotein binds a substrate and undergoes a single-electron reduction with formation of a molecular oxygen compex. The complex provides the energy for activation of the substrate molecule with formation of electrophilic intermediates that can react with nucleophilic sites in nucleic acid. An epoxide across the carbon atoms of the terminal furan is considered the active intermediate of the cytochrome P-450–mediated oxidation with the formation of a subsequent number of stable derivatives including aflatoxin M_1, P_1, Q_1, and $B_{2}a$ (Fig. 3). In the presence of appropriate agents such as glucuronic acid, sulfates, and glutathione, the hydroxylated derivatives can form complexes that are excreted [51]. Although the cytochrome P-450 system functions to detoxify xenobiotics such as aflatoxin, it also produces the active DNA-binding intermediate. Karenlampi [52] demonstrated an unequivocal relationship between cellular P-450 and aflatoxin toxicity.

Unraveling the mystery of control of the flow of events in the cytochrome P-450 complex between detoxification and activation of the DNA-binding intermediate represents a truly exceptional challenge. Although some pertinent information is available, the basic functions involved in the process are currently ambiguous.

Structural facets of biologically active compounds provide an important influence on the cytochrome P-450 process. The fit of the substrate molecule into the microsomal aggregate defines the carbon atoms available for oxidation. A number of microsomal activators and inhibitors have been

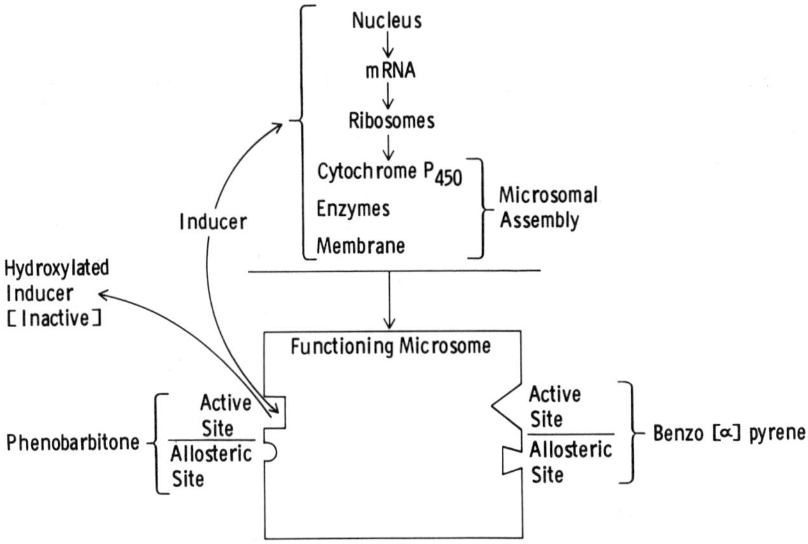

FIGURE 2. Microsomal cytochrome P-450 complex.

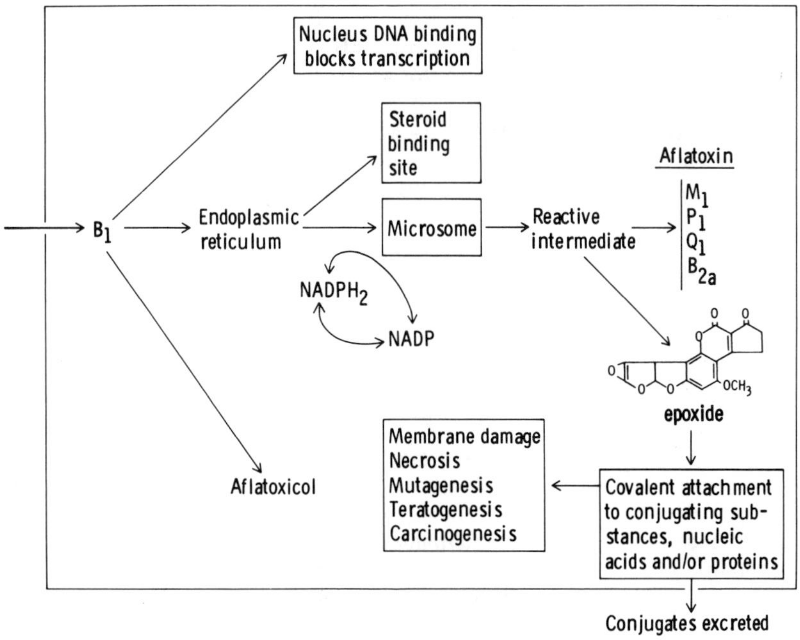

FIGURE 3. Intracellular metabolism of aflatoxin.

characterized [48, 53]. Substances with an impact on transcription/translation or enzyme assembly of the microsomal aggregate can modify the catalytic specificity of the unit. Rates of microsomal activities can also be modified by allosteric type modifications (Fig. 2). For example, steroid hormones could function as allosteric modulators of microsomal activity or competitors for xenobiotic reaction sites. A family of isozymes of cytochrome P-450 have been identified in recent years with associated variation in function [54, 55]. A genetic basis in mice for variation in P-450 specificity has been found responsible for production of either aflatoxin-DNA conjugate or the aflatoxin M_1 derivative [56]. Varied patterns of specificity have also been observed with purified cytochrome P-450 and aflatoxin preparations in: (1) covalent DNA attachments, (2) *S. typhimurium* mutation/reversion, and (3) induction of bacterial SOS error-prone repair [57–60].

C. Aflatoxin-DNA Adduct

A number of exciting studies have been carried out in recent years to identify the exact location of the cytochrome P-450–mediated covalent attachment to DNA. In a perceptive study, mild acid hydrolysis of DNA from aflatoxin-treated liver yielded 8,9-dihydro-8-(N^7-guanyl)-9-hydroxy-aflatoxin B_1 (AFB_1-N^7-Gua) (Fig. 4). A minor product of the acid hydrolysis was identified as 8,9-dihydro-8-(N^5-formyl-2',5',6'-triamino-4'-oxo-N^5-pyrimidyl)-9-hydroxyaflatoxin B_1 (AFB_1-formamidopyrimidine). In a series of the most important experiments of contemporary mycotoxicology, scientists at MIT and the University of York elucidated many of the mysteries associated with the aflatoxin–DNA binding phenomenon [61–63]. In vitro studies provided evidence that the AFB_1-N^7-Gua was rapidly removed by hydrolysis from DNA relative to AFB_1-formamidopyrimidine [64, 65].

Exposure of bacterial cells to low levels of aflatoxin for 35 min yielded 15–22 AFB_1 adducts in the genome [66]. Approximately 70% of the covalently bound toxin was identified as AFB_1-N^7-Gua. Studies of animals

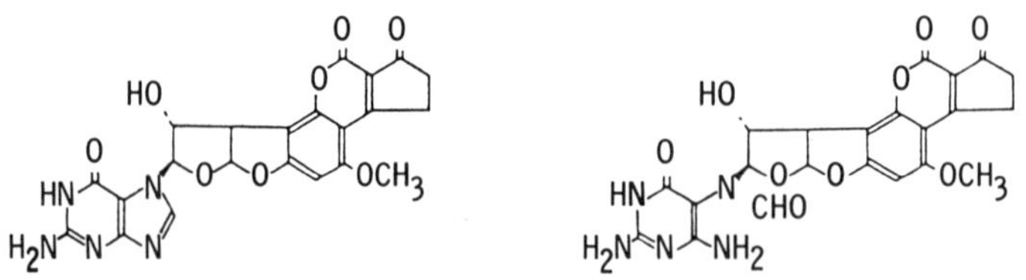

8,9-Dihydro-8-(N^7-guanyl)-9-hydroxy-aflatoxin B_1

8,9-Dihydro-8-(N^5-formyl-2',5',6'-triamino-4'-oxo-N^5-pyrimidyl)-9-hydroxy-aflatoxin B_1

FIGURE 4. Structure of aflatoxin-guanine adduct.

exposed to aflatoxin also showed that the major DNA conjugate was AFB_1-N^7-Gua with adduct presence ranging from 2 to 90 residues per 10^6 DNA bases [61]. Preferential binding studies of the target guanine (G) residues within the double-stranded DNA helical structure have identified single guanine residues flanked by adenine (A) and thymine (T) as poor targets, whereas guanine was a good target when flanked by cytosine (C), as in CCG [67, 68].

A summary of current information on binding of aflatoxin to DNA provides compelling evidence for the event as a key component in the mutagenic/carcinogenic process. Binding involves a number of sequential steps: (1) intercalation of the toxin into the helical structure of DNA, (2) oxidation of the unsaturated (C-8, C-9) carbon atoms of the terminal furan by cytochrome P-450 to form an electrophilic intermediate, and (3) nucleophilic attack of C-8 of aflatoxin on N-7 of guanine and formation of a covalent bond in an AFB_1-N^7-Gua interaction. A number of cellular responses are elicited by the DNA damage imposed by formation of the aflatoxin adduct that culminates in one of two general processes: (1) AFB_1-N^7-Gua adduct is excised from the intact DNA helix as a single-strand product and subsequent elicitation of error-free and error-prone (SOS) repair, or (2) AFB_1-N^7-Gua adduct is not excised but is converted to AFB_1-formamidopyrimidine, which causes errors in subsequent transcription [42, 69–71].

D. Excretion of Aflatoxin-DNA Adduct

A single 25-μg dose of aflatoxin produced a single AFB_1-N^7-Gua adduct per 50,000 bases in rat liver after 2 h of incorporation into DNA of approximately 1% of toxin [61]. The toxin-DNA adduct is rapidly excised and excreted; between 30 and 40% of the adduct is removed within the first 24 h after treatment [72]. The discovery of rapid, dose-dependent excretion of the AFB_1-N^7-Gua adduct provided a unique opportunity for nonintrusive assessment of actual aflatoxin-mediated damage to the genetic system of exposed animals and humans. New detection technology based on monoclonal antibody for the adduct provides an effective basis for broad assessment of genotoxicity in populations at risk [73, 74].

VI. AFLATOXIN-BASED INCREASE IN GENOMIC VARIATION

A. Introduction

The body of accumulated knowledge on mutagenesis and carcinogenesis identifies a clear association between events associated with DNA damage and attendant variation in a target genome. Microbiologists have known for some time that induction of stable mutations in microbial populations can be acquired only under conditions of elevated toxicity. There appears to be a link between broad acute toxicity of certain agents and irreversible genetic changes. A number of fascinating discoveries during the past few decades have elucidated mobile genetic elements in the biota with extraordinary hereditary implications. Assessment of environmental influences on variation in genetic systems within an agroeconiche represents a contemporary research challenge.

B. Plasmids

Initial characterization of antibiotic resistance in bacteria identified mobile extrachromosomal genetic elements (plasmids) that were capable of autonomous intracellular replication [19, 75]. In addition to R factors in antibiotic resistance, plasmids were linked to sources of genetic information for bacterial fertility through conjugation (F factors). Subsequent research has characterized a number of additional plasmid-based traits including restriction endonucleases, DNA repair enzymes, aromatic hydrocarbon degradation enzymes, and bacteriocin/hemolysin production [19]. Nominal plasmids are circular DNA that exhibit the ability for gene transfer and incorporation into host cell chromosomes (episomes) [19]. Plasmid-linked genes generally code for proteins involved in the propagation and transfer of the plasmid and a second group of enzymes such as catalysts involved in denaturation of antibiotics. Plasmid genes resemble bacterial phages in their capacity for intracellular replication and insertion into the host chromosome. The plasmid facility for rapid movement within sensitive populations provides a unique mechanism for rapid propagation and distribution of genetic information. The veritable explosion of current interest in genetic engineering is based primarily on the innate capacity of plasmids for acquisition of "foreign" or recombinant DNA. The importance of environmental genotoxins such as aflatoxin as inducing agents for mediating rapid exchanges among interacting species cannot be overestimated. The introduction of a plasmid with associated information for enhanced error-prone DNA repair dramatically increased sensitivity to aflatoxin in the Ames test [76]. Levels of 0.1 μg of the toxin produced 2000–2200 revertants in the plasmid-containing bacteria compared with 0–80 revertants in plasmid-free strains. The work suggested that aflatoxin-induced nicks in host DNA and plasmid-based, error-prone repair enymzes produced mutations. Subsequently, a family of plasmids were discovered that confer protection against DNA damage through error-prone repair [77]. Clearly, environmental agents that impose damage on cellular DNA would select for plasmids with error-prone repair enzymes that could increase mutation rates.

C. Transposons

While microbiologists were intensively engaged in characterization of plasmids, maize geneticists were elucidating another set of mobile genetic elements. McClintock and her colleagues made a number of discoveries that provided the basis for a new group of genetic regulators [78, 79]. The control components possessed elements of mobility and the ability to insert into genes (transposition) with attendant regulatory control. Insertions appeared to be expressed in developmental patterns that mediated mutations in pertinent genes. Subsequently, similar elements were identified in bacteria and were named transposons [19]. Transposon mobility was linked to defined DNA segments that were bounded at both ends with specific nucleotide sequences. However, transposition differs from traditional recombination in not being insertion-dependent on relationship of sequences in the donor and recipient sites [19]. Transposons provide a unique facility for sponsoring rearrangements of genetic material within a genome. For example, it appears likely that transposons have been actively involved in the evolution of some bacterial plasmids through reactions involving recognition of transposon sequences that facilitate gene transfer between plasmids and bacterial chromosomes [19].

D. Retroviruses/Retrotransposons

Retroviruses are RNA viruses that are propagated by conversion into duplex DNA through the enzymatic action of reverse transcriptase. Since their discovery, the viruses have gradually assumed greater scientific interest because of their role in general biological activities. The retroviral replicative intermediate, the provirus, resembles many other types of mobile genetic elements such as transposons [80]. Contemporary recognition of the importance of reverse transcription has largely developed from comparisons of nucleotide sequences. The retroviral provirus has terminal inverted repeat sequences that are embedded in terminal direct repeated sequences. The provirus is integrated into the host chromosome at many different sites in small direct repeated sequences. Therefore, the nucleotide sequences at the ends of the provirus are constant whereas the sequences adjoining the ends are different [19, 80]. Genetic elements that have been introduced via reverse transcription can be determined by identification of diagnostic nucleotide sequences. Retroviruses are widespread in nature and have been detected in a broad range of animals; pertinent research has focused on leukemia viruses in chickens and recently on the AIDS virus in humans [19, 81]. Some retroviruses have become incorporated into host chromosomes and are transmitted via germ line cells to progeny. More than 10% of the human genome appears to be integrated DNA versions of RNA molecules [80].

Genotoxic agents appear to mobilize genetic elements through DNA binding and subsequent repair processes [82]. Endogenous retroviruses in mouse embryo cells were initiated to an active, extrachromosomal state by aflatoxin [83]. Dose-response studies showed that virus expression was elicited by 0.067 μM aflatoxin. Initial consideration of reverse transcription focused on retroviral propagation. However, subsequent studies of the enzyme have expanded to the importance of the catalyst [84]. Hepatitis B virus, a DNA virus, was found to have an RNA intermediate that used reverse transcriptase in nominal replications of the virus [85, 86]. The term *retrotransposon* was created to define homology between transposons and retroviruses in reverse transcriptase–mediated reactions; they would include retroviruses with no known extracellular phase [84].

RNA particles (copia) with retroviruslike properties have been found in cells of *Drosophila melanogaster* [87, 88]. In an intriguing investigation, the effect of aflatoxin on transposable elements was examined [89]. The toxin was topically applied to larvae, and eye color in adults was later evaluated. Three unstable alleles were used to monitor toxin effect.

Results of the studies demonstrated aflatoxin-mediated somatic changes at the three loci.

Transposable elements in yeast cells (Ty elements) have been linked to increased mutation levels [90]. Exposure of pertinent yeast cells to aflatoxin produced mitotic segregants and gene conversions to tryptophan prototrophy [91]. The observations provided support for the premise of aflatoxin specificity in initiation of mutagenic events that may, in the instance of Ty, involve elaboration of the mobile genetic elements.

E. Oncogenes

Oncogenes code for products that have the ability to transform eukaryotic cells into tumor cells [19, 80]. Unique oncogene sequences have been identified in transforming viruses and in cellular genomes. The capacity of retroviruses to incorporate genes into host chromosomes initiated interest in retroviral oncogenes and their chromosomal counterparts, protoncogenes [19]. Studies in rats have identified aflatoxin-mediated activation of oncogenes [92]; *ras* oncogenes, a group of highly conserved genes, were activated by the toxin. Single-base mutations induced by carcinogens appear to be related to oncogene activation [93]. A subsequent investigation verified aflatoxin activation of *ras* oncogenes and provided evidence for gene activation through point mutations of the 12th codon that involved a G-C to A-T transition [94]. Specific regions of the oncogene were preferentially bound by aflatoxin; pertinent DNA fragments were characterized by restriction endonuclease tests [95]. Toxin binding occurred predominantly in fragments with high levels of G-C and unique flanking regions. A 3.1 *Hind*III restriction product with a Nar 1, an endonuclease reaction site of GCCGGC, fragment were detected. The Nar location has been considered diagnostic for the 12th amino acid glycine (GGC) in the *ras* oncogene. Elevated toxin affinity for the second guanine in a CGGC tetranucleotide was also identified. In related studies, aflatoxin binding to guanine was linked to interference in G-C pairing [96, 97]. The direct interference was not unequivocally associated with the initiation of oncogene activity. However, the observations provided compelling evidence for specificity in the aflatoxin–oncogene binding phenomenon.

F. Humans and Other Primates

A body of evidence has accumulated over the past 20 years that clearly identifies the presence of genotoxins in the environment. Basic motivation for relevant research projects emanated from a general concern for the eposure of humans to naturally occurring mutagens and carcinogens. The exceptional discoveries in molecular genetics in the past 10 years have provided an entirely new basis for consideration of genotoxins. Results of aflatoxin studies provide a basis for assessment of the impact of genotoxins on human health.

Initial studies identified broad occurrence of aflatoxin in food and/or feed in certain regions; exposure of monkeys to the toxin was associated with viral disease and induction of liver tumors. [1, 98–100]. The presence of aflatoxin in food has also been linked to elevated occurrence of hepatocellular carcinoma (HCC) [101]. Additional studies have extended observations that associate aflatoxin-contaminated food with human disease [102].

In Africa, kwashiorkor, a common disease of malnourished children, has been linked to ingestion of aflatoxin [103]. The studies implicated the inability of tissues in afflicted children to process or transport the toxin. In a subsequent investigation of kwashiorkor in Sudan and Ghana, elevated levels of aflatoxin derivatives were detected and were linked to liver malfunction [104, 105]. Examination of the interaction between diet and aflatoxin toxicity demonstrated a link between protein deprivation and reduced cytochrome P-450; restricted MFO activity would restrict the ability to detoxify the toxin [106]. Studies in Africa and China provide a clear association between malnutrition and aflatoxin detoxification in humans. Malnutrition is often produced by extended drought periods. The absence of moisture of crops has encouraged the planting of drought-resistant crops such as peanuts. Unfortunately, elevated temperatures and drought stress enhance the development of *A. flavus* in peanuts and thus increase aflatoxin contamination of the commodity [107].

Studies of humans infected with hepatitis B virus (HBV) have established an association between the virus and HCC with routine occurrence in areas with enhanced incidence and levels of aflatoxin in food [108, 109]. Since the HBV has been identified as a retroviral type of virus, exposure to a foodborne carcinogen such as aflatoxin in induction of a provirus in humans represents a fascinating research challenge. Retroviral sequences have been identified in primate and human chromosomes [81, 110, 111]. Comparisons of type C retroviruses in human and monkey genomes have been utilized in consideration of evolutionary relationships [112]. A study of the human genome elucidated a provirus and a class of retroviral sequences that demonstrated homology with HTLV-1, an infectious human virus [113]. The results provide unequivocal evidence for the retroviral-type sequences in the human genomes, but the elaboration of infectious viruses from the genome remains uncertain.

New technologies have provided tools for the determination of aflatoxin-DNA adducts such as AFB_1-N^7-Gua in urine [114, 115]. Aflatoxin was detected in food and urine of people in a region of China with elevated occurrence in HCC [116]. A good correlation was found between aflatoxin levels in food from a household and excretion of the toxin by residents. The average toxin level was 119 ppb in 253 samples with a projected daily intake of 58 μg/person, and the daily excretion rate was 0.93 μg/person. In an area of Kenya with elevated levels of HCC and foodborne aflatoxin, the toxin-DNA adduct was routinely detected in human urine [117].

Animals exposed to aflatoxin exhibit diminution in immunocompetence, with specific inhibition of immunoresponse in cell-mediated systems [118, 119]. Lymphocyte studies have demonstrated aflatoxin-mediated damage to chromosomes [120, 121].

Aflatoxin inhibits the production of natural killer cells involved in the immune surveillance for elmination of tumor cells. The compromised immune system has been linked to aflatoxin exposure and subsequent viral infections [122].

VII. CONCLUSIONS

Scientific discoveries and the incredible advances in technology during the past few years have introduced a new age of environmental concerns with particular emphasis on human health. A specific area of interest includes the occurrence of genotoxic agents in foodstuffs. A significant scientific effort has been devoted to identification of factors that contribute to the elaboration of genotoxic carcinogens such as aflatoxin. An overview of the relevant observations suggests that the infection of developing crops by aflatoxin-producing fungi and toxin contamination of developing seed is a contemporary phenomenon that can be attributed to intensive, high technology agricultural practices. Therefore, elevated levels of aflatoxin in foods is a new, ecologically based problem that requires imaginative responses. Appropriate reaction to hazards associated with the toxin requires detailed understanding of the cellular responses in exposed humans with emphasis on damage to genetic systems.

Traditional consideration of the importance of mutagens and carcinogens such as aflatoxin focused on interactions with target DNA and subsequent relevant changes in genetic expression. The relatively recent discoveries of the widespread occurrence of mobile genetic elements with regulatory influence on genetic control dramatically expand our understanding of genetic systems. The presence of proviruses and retroviral elements in the human genome and their activation by certain chemical compounds such as aflatoxin provide an irresistible motivation for reassessment of the influence of environmental genotoxins on health. Clearly, old mysteries such as association between mutagens and viruses and between carcinogens and oncogenes will be resolved in the new conceptual context. Genotoxins appear to initiate active viral forms from chromosomal proviruses; simultaneous toxin-mediated immunosuppression provides an opportunity for establishment of unique retroviral-type disorders. Genotoxic compounds also activate other mobile genetic elements such as transposons that exhibit the potential for profound effects on host genomes. The extraordinary respository of "nonfunctional" genetic sequences provides for an essentially limitless array of genetic combinations that can be elicited under specific conditions [123]. It has been proposed that mutation of endogenous genes to the deregulated, tumorigenic state provides an evolutionary fitness trait that selects for a particular clone of cells at the expense of the host organism [123]. Destabilization of a genome could inherently initiate expression of oncogenes that in the broader, evolutionary context reflects increased "adaptiveness" potential.

An overview of the information on genotoxins and their effects raises important questions on cause—effect relationships. Dobzhansky [124] provided a guide for dealing with the complexity of interactive, biological problems: "In biology nothing makes sense except in the light of evolution." It is apparent that genotoxins provide a mechanism for genome destabilization and attendant potential for genetic rearrangements. Genotoxic agents appear to have evolved as ecological signals for periods of disequilibrium that will require adaptiveness among the interacting species. During periods of biological equilibrium, elaboration of genotoxins within an econiche would be suppressed and the genomes of the attendant biota would be relatively stable. It appears likely that at the extremes of contemporary technology in agriculture, profound species disequilibrium has occurred through homogenization of the biota [18]. Lederberg [123] provides an exemplary analysis of the dilemma associated with understanding biological equilibria:

> We can hardly be sure that such an equilibrium for earth even includes the human species even as we contrive to eliminate some of the others. Our propensity for technological sophistication harnessed to intraspecies competition adds a further dimension of hazard.

REFERENCES

1. Detroy, R. W., Lillehoj, E. B., and Ciegler, A. Aflatoxin and related compounds. In *Microbial Toxins*, Vol. 6 (A. Ciegler, S. Kadis, and S. Ajl, Eds.), Academic, New York, 1971, p. 3.
2. Aharanowitz, Y., and Demain, A. L. Thoughts on secondary metabolism. *Biotechnol. Bioeng.*, *22(Suppl. 1)*:5, 1980.
3. Bu'Lock, J. Mycotoxins as secondary metabolites. In *The Biosynthesis of Mycotoxins* (P. S. Steyn, Ed.), Academic, New York, 1980, p. 1.
4. Waksman, S. Antibiotics. *Biol. Rev.*, *23*:452, 1948.
5. Lillehoj, E. B. Evolutionary basis and ecological role of toxic microbial secondary metabolites. *J. Theor. Biol.*, *97*:325, 1982.
6. Turner, W. B., and Aldrich, D. C. *Fungal Metabolites*, Vol. II, Academic, New York, 1983, p. 1.
7. Corcoran, J. W., and Hahn, F. E. *Antibiotics*, Vol. III. *Mechanisms of Action of Antimicrobial and Antitumor Agents*, Springer-Verlag, New York, 1975, p. 1.
8. Lillehoj, E. B., and Ciegler, A. Aflatoxin B_1 binding and toxic effects on *Bacillus megaterium*. *J. Gen. Microbiol.*, *54*:185, 1968.
9. Lillehoj, E. B., and Ciegler, A. Inhibition of deoxyribonucleic acid synthesis in *Flavobacterium aurantiacum*. *J. Bacteriol.*, *94*:787, 1967.
10. Lillehoj, E. B., and Ciegler, A. Aflatoxin B_1 effect on enzyme biosynthesis in *Bacillus cereus* and *Bacillus licheniformis*. *Can. J. Microbiol.*, *16*:1059, 1970.
11. Engelbrecht, J. C., and Altenkirk, B. Comparison of some biological effects of sterigmatocystin and aflatoxin analogues on primary cell culture. *J. Natl. Cancer Inst.*, *48*:1647, 1972.
12. Toyoshima, K., Hiasa, Y., Ito, N., and Tsubura, Y. *In vitro* malignant transformations of cells derived from rat livers by means of aflatoxin B_1. *Gann*, *61*:55, 1970.
13. Boreiko, C. J., Ragan, D. L., Abernathy, D. J., and Frazelle, J. H. Initiation of C3H/10T1/2 cell transformation by *N*-methyl-*N'*-nitro-*N*-nitrosoguanidine and aflatoxin B_1. *Carcinogenesis*, *3*:391, 1982.
14. Lillehoj, E. G. The aflatoxin in maize problem: the historical perspective. In *Aflatoxin in Maize: A Proceedings of the Workshop* (M. S. Zuber, E. B. Lillehoj, and B. L. Renfro, Eds.), CIMMYT, Mexico, D. F., 1987, p. 14.

15. Batra, S. W. T. Biological control in agroecosystems. *Science, 215*: 134, 1982.
16. Harper, J. L. *Population Biology of Plants*, Academic, New York, 1977, p. 1.
17. Bhatnagar, D., Cleveland, T. E., and Lillehoj, E. B. Enzymes in aflatoxin B_1 biosynthesis: strategies for identifying pertinent genes. *Mycopathology, 107*:75, 1989.
18. Lillehoj, E. B. Aflatoxin, an ecologically-elicited, genetic activation signal. In *Mycotoxins and Animal Foods* (J. E. Smith and R. S. Henderson, Eds.), CRC Press, Boca Raton, Fla., 1991.
19. Lewin, B. J. *Genes*, Vol. III, Wiley, New York, 1987, p. 1.
20. Lwoff, A. Lysogeny. *Bacteriol. Rev., 17*:269, 1953.
21. Lillehoj, E. B., and Ciegler, A. Aflatoxin induction of lysogenic bacteria. *Appl. Microbiol., 20*:782, 1970.
22. Witkin, E. M. Ultraviolet mutagenesis and inducible DNA repair in *Escherichia coli. Bacteriol. Rev., 40*:869, 1976.
23. Sarasin, A., Goze, A., Devoret, R., and Moule, Y. Induced reactivation of UV-damaged phage in *E. coli* K12 host cells treated with aflatoxin B_1 metabolites. *Mutat. Res., 42*:205, 1977.
24. Sarasin, A., Smith, C. A., and Hanawalt, P. C. Repair of DNA in human cells after treatment with activated aflatoxin B_1. *Cancer Res., 37*:1786, 1977.
25. Quillardet, P., Huisman, O., D'ari, R., and Hofnung, M. SOS chromotest, a direct assay of induction of an SOS function in *Escherichia coli* K12 to measure genotoxicity. *Proc. Natl. Acad. Sci. USA, 79*:5971, 1982.
26. Moreau, P. L., Fanica, M., and Devoret, R. Cleavage of repressor and induction of an SOS function in *Escherichia coli. Carcinogenesis, 1*:837, 1980.
27. Moreau, P. O., Bailone, A., and Devoret, R. Prophage lambda induction in *Escherichia coli* K12 *emv*A *uvr*B: a highly sensitive test for potential carcinogenesis. *Proc. Natl. Acad. Sci. USA, 73*:3700, 1976.
28. Ueno, Y., and Kubota, K. DNA-attacking ability of carcinogenic mycotoxins in recombination-deficient mutant cells of *Bacillus subtilis. Cancer Res., 36*:445, 1976.
29. Ames, B. N. The detection of chemical mutagens with enteric bacteria. In *Chemical Mutagens—Their Principles and Methods for Detection* (A. Hollaender, Ed.), Plenum, New York, 1971, p. 267.
30. Maron, D. M., and Ames, B. N. Revised methods for the *Salmonella* mutagenicity test. In *Handbook of Mutagenicity Test Procedures*, 2nd ed. (B. J. Kilbey, W. Nichols, and C. C. Ramel, Eds.), Elsevier, Amsterdam, 1984, p. 94.
31. Ueno, Y., Kubota, K., Ito, T., and Nakamura, Y. Metabolism of carcinogenic mycotoxins in *Salmonella typhimurium. Cancer Res., 38*:536, 1978.
32. Ueno, Y., Tashiro, R., Hawaikawa, K., and Emi, Y. Metabolism of aflatoxins and their toxicity. In *Problems of Threshold in Chemical Mutagenesis* (Y. Tazima, Ed.), Environmental Mutagen Society of Japan, Tokyo, 1984, p. 61.

33. Wong, J. J., Singh, R., and Hsieh, D. P. H. Mutagenicity of fungal metabolites related to aflatoxin in biosynthesis. *Mutat. Res., 44*: 447, 1977.
34. Abdou, R. F., Megalla, S. E., and Azab, S. G. Mutagenic effects of aflatoxin B-1 and G-1 on the Egyptian cotton leaf-worm, *Spodoptera littoralis* (Boisd.). *Mycopathology, 88*:23, 1984.
35. Belitsky, G. A., Khovanova, E. M., Budunova, I. V., and Sharuptis, H. G. Mycotoxin induction of somatic mosaicism in *Drosophila* and DNA repair in mammalian liver cell culture. *Cell Biol. Toxicol., 1*: 133, 1985.
36. Cortesi, E., Sasffinotti, U., Donovan, P. J., Rice, J. M., and Kakunaga, T. Dose-response studies on neoplastic transformation of BALB/3T3 clone A31-1-1 cells by aflatoxin B_1, benzidine, benzo[*a*]pyrene, 3-methylcholanthrene, and *N*-methyl-*N*-nitrosoguanidine. *Teratogen, Carcinogen, Mutagen., 3*:101, 1983.
37. Mirsalis, J. C., Tyson, C. K., and Butterworth, B. E. Detection of genotoxic carcinogens in the in vivo—in vitro hepatocyte DNA repair assay. *Environ. Mutagen., 4*:553, 1982.
38. Casto, B. C., Peiczynski, W. J., Janosko, N., and Dipaola, J. A. Significance of treatment interval and DNA repair in the enhancement of viral transformation by chemical carcinogens and mutagens. *Chem. Biol. Interact., 13*:105, 1976.
39. Sarasin, A. R., and Hanawalt, P. C. Carcinogens enhance survival of UV irradiated simian virus 40 in treated monkey kidney cells: induction of recovery pathway. *Proc. Natl. Acad. Sci. USA, 75*:346, 1978.
40. Hahon, N., Booth, J. A., and Stewart, J. D. Aflatoxin inhibition of viral interferon induction. *Agents Chemother., 16*:277, 1979.
41. Hamilton, J. W., and Bloom, S. E. Correlation between mixed-function oxidase enzyme induction and aflatoxin B_1-induced unscheduled DNA synthesis in the chick embryo, in vivo. *Environ. Mutagen., 6*:41, 1984.
42. Hsieh, D. P. H. Genotoxicity of mycotoxins. In *New Concepts and Developments in Toxicology* (P. L. Chambers, P. Gehring, and F. Sakai, Eds.), Elsevier, Amsterdam, 1986, p. 251.
43. Pullman, A., and Pullman, B. Electronic structure and carcinogenic activity of aromatic molecules. *Adv. Cancer Res., 3*:117, 1955.
44. Clayson, D. B. *Chemical Carcinogenesis*, Little, Brown, Boston, 1962 p. 1.
45. Heathcote, J. G., and Hibbert, J. R. Biological activity and electronic structure of the aflatoxins. *Br. J. Cancer, 29*:170, 1974.
46. Jerina, D. M., and Lehr, R. E. The bay region theory: a quantum mechanical approach to aromatic hydrocarbon-induced carcinogenicity. In *Microsomes and Drug Oxidations* (V. Ulrich, I. Roots, A. G. Hildebrant, R. W. Estabrook, and A. H. Cooney, Eds.), Pergamon, Oxford, 1977, p. 709.
47. Cooney, A. H., Levin, A. H., Wood, A. W., Yagi, H., Lehr, R. E., and Jerina, D. M. Biological activity of polycyclic hydrocarbon metabolites and the bay region theory. In *Toxicology*, Vol. 9 (Y. Cohen, Ed.), Pergamon, Oxford, 1979, p. 41.

48. Guengerich, F. P. Roles of cytochrome P-450 enzymes in chemical carcinogenesis and cancer chemotherapy. *Cancer Res.*, *48*:2946, 1988.
49. Caldwell, J., and Jackoby, W. B., Eds. *Biological Basis of Detoxification*, Academic, New York, 1983, p. 1.
50. White, R. E., and Coon, M. J. Oxygen activation by cytochrome P-450. *Ann. Rev. Biochem.*, *49*:315, 1980.
51. Caldwell, J., Conjugation mechanisms of xenobiotic metabolism: mammalian aspects. In *Xenobiotic Conjugation Chemistry* (G. D. Paulson, J. Caldwell, D. H. Huston, and J. J. Menn, Eds.), American Chemical Society, Washington, D.C., 1986, p. 2.
52. Karenlampi, S. O. Mechanism of cytotoxicity of aflatoxin B_1. *Biochem. Biophys. Res. Commun.*, *145*:854, 1987.
53. Lillehoj, E. B., and Ciegler, A. In *Mycotoxin Synergism in Microbiology—1975* (D. Schlesinger, Ed.), American Society of Microbiology, Washington, D.C., 1975, p. 344.
54. Lu, A. Y., and West, S. B. Multiplicity of mammalian microsomal cytochromes P-450. *Pharmacol. Rev.*, *31*:277, 1980.
55. Shimada, R., and Nakamura, S. Cytochrome P-450-mediated activation of procarcinogens and promutagens to DNA, damaging products by measuring expression of *umu* gene in *Salmonella typhimurium* TA1535/pSK1002. *Biochem. Pharmacol.*, *36*:1979, 1987.
56. Gurtoo, H. L., Dahms, R. P., and Paigen, B. Metabolic activation of aflatoxins related to their mutagenicity. *Biochem. Biophys. Res. Commun.*, *81*:965, 1978.
57. Guengerich, F. P. Separation and purification of multiple forms of microsomal cytochrome P-450. Activities of different forms of cytochrome P-450 towards several compounds of environmental interest. *J. Biol. Chem.*, *252*:3970, 1977.
58. Robertson, I. G. C., Zeiger, E., and Goldstein, J. A. Specificity of rat liver cytochrome P-450 isozymes in the mutagenic activation of benzo[*a*]pyrene, aromatic amines and aflatoxin B_1. *Carcinogenesis*, *4*:93, 1983.
59. Ishii, I., Maeda, K., Kamataki, T., and Kato, R. Mutagenic activation of aflatoxin B_1 by several forms of purified cytochrome P-450. *Mutat. Res.*, *174*:85, 1986.
60. Shimada, T., Nakamura, S.-I., Imaoka, S., and Funae, Y. Genotoxic and mutagenic activation of aflatoxin B_1 by several forms of constitutive forms of cytochrome P-450 in rat liver microsomes. *Toxicol. Appl. Pharmacol.*, *91*:13, 1987.
61. Essigmann, J. M., Croy, R. G., Bennett, R. A., and Wogan, G. N. Metabolic activation of aflatoxin B_1 and its metabolites: patterns of DNA adducts formation and removal in relation to biological effects. In *Chemical Carcinogenesis* (C. Nicolini, Ed.), Plenum, New York, 1982, p. 47.
62. Wogan, G. N., Croy, R. G., Essigmann, J. M., and Bennett, R. A. Aflatoxin—DNA interactions: qualitative, quantitative and kinetic features in relation to carcinogenesis. In *Quantum Chemistry and Biochemistry* (B. Pullman, P. O. P. Ts'o, and H. Gelboin, Eds.), Reidel, Dordrecht, Holland, 1980, p. 179.

63. Hertzog, P. J., Lindsay Smith, J. R., and Garner, R. C. A high pressure liquid chromatography study on removal of DNA-bound aflatoxin B_1 in rat liver and *in vitro*. *Carcinogenesis*, *1*:787, 1980.
64. Groopman, J. D., Croy, R. G., and Wogan, G. N. In vitro reactions of aflatoxin B_1 adducted DNA. *Proc. Natl. Acad. Sci. USA*, *78*:5445, 1981.
65. Hertzog, P. J., Lindsay Smith, J. R., and Garner, R. C. Characterization of the imidazole ring-opened forms of *trans*-8,9-dihydro-8-(7-guanyl)-9-hydroxy aflatoxin B_1. *Carcinogenesis*, *3*:723, 1982.
66. Stark, A. A., Essigmann, J. M., Demain, A. L., Skopek, T. R., and Wogan, G. N. Aflatoxin B_1 mutagenesis, DNA binding, and adduct formation in *Salmonella typhimurium*. *Proc. Natl. Acad. Sci. USA*, *76*:1343, 1979.
67. Muench, K. F., Misra, R. P., and Humayun, M. Z. Sequence specificity in aflatoxin B_1–DNA interaction. *Proc. Natl. Acad. Sci. USA*, *80*:6, 1983.
68. Misra, R. P., Muench, K. F., and Humayun, M. Z. Covalent and noncovalent interactions of aflatoxin with defined deoxyribonucleic acid sequences. *Biochemistry*, *22*:3351, 1983.
69. Leadon, S. A., Tyrrell, R. M., and Cerutti, P. A. Excision repair of aflatoxin B_1-DNA adducts in human fibroblasts. *Cancer Res.*, *41*:4125, 1981.
70. Swenson, D. H. Metabolic activation and detoxification of aflatoxins. In *Reviews in Biochemical Toxicology*, Vol. 3 (E. Hodgson, J. R. Bend, and R. M. Philpot, Eds.), Elsevier, Amsterdam, 1981, p. 155.
71. Craddock, V. M., and Henderson, A. R. De novo and repair replication of DNA in liver of carcinogen-treated animals. *Cancer Res.*, *38*:2135, 1978.
72. Bennett, R. A., Essigmann, J. M., and Wogan, G. N. Excretion of an aflatoxin-guanine adduct in the urine of aflatoxin B_1-treated rats. *Cancer Res.*, *41*:650, 1981.
73. Groopman, J. D., Roebuck, B. D., and Kensler, T. W. Application of monoclonal antibodies and dietary antioxidant-based animal models to define human exposure to aflatoxin B_1. *Prog. Exp. Tumor Res.*, *31*:52, 1987.
74. Hertzog, P. J., Shaw, A., Lindsay Smith, J. R., and Garner, R. C. Improved conditions for the production of monoclonal antibodies to carcinogen-modified DNA for use in enzyme-linked immunosorbent assays (ELISA). *J. Immunol. Methods*, *62*:49, 1983.
75. Helinski, D. R. Plasmid determined resistance to antibiotics: molecular properties of R factors. *Ann. Rev. Microbiol.*, *27*:437, 1973.
76. McCann, J., Spingarn, N. E., Kobori, J., and Ames, B. N. Detection of carcinogens as mutagens: bacterial tester strains with R factor plasmids. *Proc. Natl. Acad. Sci. USA*, *72*:979, 1975.
77. Strike, P., and Lodwick, D. Plasmid genes affecting DNA repair and mutation. *J. Cell Sci. Suppl.*, *6*:313, 1987.
78. McClintock, B. Mechanisms that rapidly reorganize the genome. *Stadler Genetic Symp.*, *10*:25, 1978.
79. Fedoroff, N. V. Controlling elements in maize. In *Mobile Genetic Elements* (J. A. Shapiro, Ed.), Academic, New York, 1983, p. 1.

80. Temin, H. M. Reverse transcription in the eukaryotic genome: retroviruses, pararetroviruses, retrotransposons, and retrotranscripts. *Mol. Biol. Evol.*, *2*:455, 1985.
81. Varmus, H. E. Retroviruses. In *Mobile Genetic Elements* (J. A. Shapiro, Ed.), Academic, New York, 1983, p. 411.
82. Rascati, R. J., and Tenant, R. W. Involvement of DNA damage in hydroxyurea-mediated induction of endogenous murine retrovirus. *Virology*, *94*:273, 1979.
83. Rascati, R. J., and McNeely, M. Induction of retrovirus gene expression by aflatoxin B_1 and 2-acetylaminofluorene. *Mutat. Res.*, *122*:235, 1983.
84. Baltimore, D. Retroviruses and retrotransposons: the role of reverse transcription in shaping the eukaryotic genome. *Cell*, *40*:481, 1985.
85. Blum, H. E., Haase, A. T., Harris, J. D., Walker, D., and Vyas, G. N. Asymmetric replication of hepatitis B virus DNA in human liver: demonstration of cytoplasmic minus-strand DNA by blot analyses and *in situ* hybridization. *Virology*, *139*:87, 1984.
86. Ganem, D., and Varmus, H. E. The molecular biology of hepatitis B viruses. *Ann. Rev. Biochem.*, *56*:651, 1987.
87. Shiba, T., and Saigo, K. Retrovirus-like particles containing DNA homologous to the transposable element *copia* in *Drosophila melanogaster*. *Nature*, *302*:119, 1983.
88. Flavell, A. J. Role of reverse transcription in the generation of extrachromosomal *copia* mobile genetic elements. *Nature*, *310*:514, 1984.
89. Fahmy, M. J., and Fahmy, O. G. Differential induction of altered gene expression by carcinogens at mutant alleles of a *Drosophila* locus with a transposable element. *Cancer Res.*, *43*:801, 1983.
90. Boeke, J. D., Garfinkel, C. A., and Fink, G. R. Ty elements transpose through an RNA intermediate. *Cell*, *40*:491, 1985.
91. Niggili, B., Hann, F. D., and Wurgler, F. E. Endogenous promutagen activation in the yeast *Saccharomyces cerevisiae* factors influencing aflatoxin B_1 mutagenicity. *Mutat. Res.*, *175*:223, 1986.
92. Tashiro, F., Morimura, S., Hayashi, K., Makino, R., Kawamura, H., Horikoshi, N., Nemoto, K., Ohtsubo, K., Sugimura, T., and Ueno, Y. Expression of the c-Ha-*ras* and cy-*myc* genes in aflatoxin B_1-induced hepatocellular carcinomas. *Biochem. Biophys. Res. Commun.*, *138*:858, 1986.
93. Barbacid, M. Oncogenes and human cancer: cause or consequence? *Carcinogenesis*, *7*:1037, 1986.
94. McMahon, G., Davies, E., and Wogan, G. N. Characterization of c-Ki-*ras* oncogene alleles by direct sequencing of enzymatically amplified DNA from carcinogen-induced tumors. *Proc. Natl. Acad. Sci. USA*, *84*:4974, 1987.
95. Yang, S. S., Taub, J. V., Modali, R. W., Yase, P., and Yang, G. C. Dose-dependency of aflatoxin B_1 binding on human high molecular weight DNA in the activation of a proto-oncogene. *Environ. Health Perspect.*, *62*:231, 1985.

96. Modali, R., and Yang, S. S. Specificity of aflatoxin B_1 binding on human proto-oncogene nucleotide sequence. In *Monitoring of Occupational Genotoxins* (M. Sorsa and H. Norppa, Eds.), A. R. Liss, New York, 1986, p. 147.
97. Foster, P. L., Eisenstadt, E., and Miller, J. H. Base substitution mutations induced by metabolically activated aflatoxin B_1. *Proc. Natl. Acad. Sci. USA, 80*:2695, 1983.
98. Adamson, R. H., Correa, P., and Dalgard, D. W. Occurrence of primary liver carcinoma in a rhesus monkey fed aflatoxin B_1. *J. Natl. Cancer Inst., 50*:549, 1973.
99. Adamson, R. H., Correa, P., Sieber, S. M., McIntire, R., and Dalgard, D. W. Carcinogenicity of aflatoxin B_1 in rhesus monkeys: two additional cases of primary liver cancer. *J. Natl. Cancer Inst., 57*:67, 1976.
100. Lin, J. J., Liu, C., and Svoboda, D. J. Long term effects of aflatoxn B_1 and viral hepatitis on marmoset liver. *Lab. Invest., 30*: 267, 1974.
101. Van Rensburg, S. J. Role of mycotoxins in endemic liver and oesophageal cancer. In *Mycotoxins and Phycotoxins* (P. S. Steyn and R. Vlegaarr, Eds.), Elsevier, Amsterdam, 1986, p. 483.
102. Adenan, H., Tsuboi, S., Kawamura, K., Cruz, M. L., Sulidai, H. W., and Soeharto, H. Determination of AFB_1 in serum samples in liver disease patients and normal subjects on Yugyakarta, Indonesia. *ICMR Ann., 4*:167, 1984.
103. Hendrickse, R. G., Coulter, J. B. S., Lamplugh, S. M., MacFarlane, S. B. J., Williams, T. E., Omer, M. I. A., and Suliman, G. I. Aflatoxins and kwashiorkor: a study in Sudanese children. *Br. Med. J., 285*:843, 1982.
104. Coulter, J. B. S., Hendrickse, R. G., Lamplugh, S. M., MacFarlane, S. B. J., Moody, J. B., Omer, M. I. A., Suliman, G. I., and Williams, T. E. Aflatoxins and kwashiorkor: clinical studies in Sudanese children. *Trans. Roy. Soc. Trop. Med. Hyg., 38*:945, 1982.
105. Apeagyei, F., Lamplugh, S. M., Hendrickse, R. G., Affram, K., and Lucas, S. Aflatoxins in the livers of children with kwashiorkor in Ghana. *Trop. Geogr. Med., 38*:273, 1986.
106. Enwonwu, C. O. The role of dietary aflatoxin in the genesis of hepatocellular cancer in developing countries. *Lancet, ii*:956, 1984.
107. Cole, R. J., Sanders, T. H., Hill, R. A., and Blankenship, P. D. Mean geocarposphere temperatures that induce preharvest aflatoxin contamination of peanuts under drought stress. *Mycopathology, 91*: 41, 1985.
108. World Health Organization. *Prevention of Liver Cancer,* WHO Tech. Ser. 691, Geneva, 1983, p. 1.
109. Deinhardt, F., and Gust, I. D. Viral hepatitis. *Bull. WHO, 60*: 661, 1982.
110. Martin, M. A., Bryan, T., McCutchan, T. F., and Chan, H. W. Detection and cloning of murine leukemia virus-related sequences from African green monkey liver DNA. *J. Virol., 39*:835, 1981.

111. Repaske, R., O'Neill, R. R., Steele, P. E., and Martin, M. A. Characterization and partial nucleotide sequence of endogenous type C retrovirus segments in human chromosomal DNA. *Proc. Natl. Acad. Sci. USA, 80*:678, 1983.
112. Beneviste, R. E., and Todaro, G. J. Evolution of type C viral genes: evidence for an Asian origin of man. *Nature, 261*:101, 1976.
113. Aaronson, S. A., Reddy, E. P., Robbins, K., Devare, S. G., Swan, D. C., Pierce, J. H., and Tronick, S. R. Retroviruses, *onc* genes, and human cancer. In *Human Carcinogenesis* (C. C. Harris and H. N. Autrup, Eds.), Academic, New York, 1983, p. 609.
114. Martin, C. N., Garner, R. C., Tursi, F., Garner, J. V., Whittle, H. C., Ryder, R. W., Sizaret, R., and Montesano, R. In *Monitoring Human Exposure to Carcinogenic and Mutagenic Agents* (A. Berlin, M. Draper, K. Hemminiski, and H. Vaino, Eds.), IARC Sci. Pub. 59, International Agency for Research on Cancer, Lyon, 1985, p. 313.
115. Groopman, J. D., Haugen, A., Goodrich, G. R., Wogan, G. N., and Harris, C. C. Quantitation of aflatoxin B_1-modified DNA using monoclonal antibody. *Cancer Res., 42*:3120, 1982.
116. Zhu, J. Q., Zhang, L. S., Hu, X., Xiao, Y., Chen, J. S., Xn, Y. C., Fremy, J., and Chu, F. S. Correlation of dietary aflatoxin B_1 levels with excretion of aflatoxin M_1 in human urine. *Cancer Res., 47*:1848, 1987.
117. Autrup, H., Bradley, K. A., Shamsuddin, A. K. M., Wakhisi, J., and Wasunna, A. Detection of putative adduct with fluorescence characteristics identical to 2,3-dihydro-2-(7'-guanyl)-3-hydroxy-aflatoxin B_1 in human urine collected in Muranag'a district, Kenya. *Carcinogenesis, 4*:1193, 1983.
118. Pier, A. C. Immunomodulation in aflatoxicosis in diagnosis of mycotoxicoses. In *Diagnosis of Mycotoxicoses* (J. L. Richard and J. R. Thurston, Eds.), Martinus Nijhoff, Boston, 1986, p. 143.
119. Pier, A. C. Aflatoxicoses and immunosuppression in mammalian animals. In *Aflatoxin in Maize: A Proceedings of the Workshop* (M. S. Zuber, E. B. Lillehoj, and B. L. Renfro, Eds.), CIMMYT, Mexico, D. F., 1987, p. 58.
120. Kaden, D. A., Call, K. M., Leong, P. M., Komives, E. A., and Thilly, W. G. Killing and mutation of human lymphoblast cells by aflatoxin B_1: evidence for an inducible repair response. *Cancer Res., 47*:1993, 1987.
121. Cheng-ya, W., Xin-sheng, Y., Jin-xin, H., Tao-zhen, L., and Yaos-qin, Y. HLA related genetic control of natural killer activity and aflatoxin B_1 suppression of lymphocytic blastogenesis. *Chin. Med. J., 100*:29, 1987.
122. Lutwick, L. I. Reaction between aflatoxin, hepatitis-B virus and hepatocellular carcinoma. *Lancet, i*:755, 1979.
123. Lederberg, J. Medical science, infectious disease, and the unity of mankind. *J. Am. Med. Assoc., 260*:684, 1988.
124. Dobzhansky, T. *Genetics of the Evolutionary Process.* Columbia Univ. Press, New York 1970, p. 1.

2

AFLATOXIN IN DEVELOPING MAIZE: INTERACTIONS AMONG INVOLVED BIOTA AND PERTINENT ECONICHE FACTORS

NEIL W. WIDSTROM *Agricultural Research Service, U.S. Department of Agriculture, Tifton, Georgia*

I. PROLOGUE TO RECOGNITION OF A FIELD PROBLEM

Contamination of maize by aflatoxins was considered to be a storage problem for at least ten years after the discovery and identification of these toxins in peanuts during the early 1960s [1]. Prior to the discovery of aflatoxins in stored maize [2], associated problems were generally referred to as "moldy corn poisoning" [3] or by some other descriptive term for the symptoms and effects of their presence in feed. The tremendous influence of environment on the *Aspergillus flavus* spp. molds associated with the feeding problems have always been recognized by researchers. Recognition of the influence of weather on fungus, insect, and maize plant was emphasized in early reports by Taubenhaus [4] and Koehler [5]. These reports, however, did not address the study of the complex interactions that exist among the three biological entities as they relate to environmental factors such as weather.

The difficulties encountered when moldy grain was ingested by farm animals [6] led to numerous studies investigating the effect of fungi on the quality of stored maize [7–9]. These studies also led to numerous reports of aflatoxin in stored or recently harvested maize [10–14]. Many of the reports suggested that differences in the incidence and levels of contamination were due to region of origin for the grain tested [15–17]. By the early 1970s, most researchers working with the aflatoxin problem were quite certain that contamination by aflatoxin was occurring along with the fungal infection phase, prior to harvest. Anderson et al. [18] were the first to conduct and report on a definitive study demonstrating the presence of aflatoxin in preharvest corn in South Georgia; their study also implicated insects as an exacerbating influence on both the infection and contamination processes. The stage was set, therefore, for numerous investigations into the cause-and-effect relationships of factors involved in the infection and contamination processes and the interrelationships or interaction among the associated environmental factors and biological entities.

II. PREHARVEST INFECTION BY *ASPERGILLUS* SPP. AND CONTAMINATION BY AFLATOXINS—A DIAGRAMMATIC MODEL

A basic knowledge of the chronology and epidemiology of the infection and contamination processes is necessary if the interactions among biological and environmental factors are to be understood. The plant is a good point of reference for discussion of this chronology, since it is the only stationary entity in the system and is also the ultimate recipient of aflatoxin contamination. A simplified diagrammatic representation of the infection and contamination process, using the maize plant developmental cycle as a time line, is depicted in Fig. 1. This frame of reference will help to bring other factors into focus in terms of their order and method of impacting the infection and contamination process.

The entire developmental period of the growing plant, including postmaturity, is illustrated in Fig. 1 because ecological factors exert an influence on the plant's total development. Ecological factors have an impact on all subsequent events during plant development. The obvious critical period occurs during ear and kernel development, but the processes being studied cannot be limited to that time frame.

Natural occurrences of infection of the maize plant by the *A. flavus* group prior to silking have not been reported and, if they exist, are not believed to significantly affect subsequent ear toxin concentrations. Superficial fungal colonization of leaves seems possible, however, provided environmental conditions involving wind, water, temperature, soil, and so on are favorable. The crucial stage of plant development occurs during an approximately 3-week period following full silk. During this period, the probability of ear infection is high when the ear is exposed to water- and insectborne spores. But colonization may occur at any time, including during the storage period, when environmental conditions permit.

Stress on the plant is believed to have a significant influence on the colonization process and subsequent levels of contamination, but evidence that stress on the fungus and plant induces accelerated aflatoxin production in the developing ear is circumstantial. It is probable that conditions that produce stress in the plant also favor fungal development and/or aflatoxin synthesis. Predisposing stresses on the developing plant from insect damage to foliage or stalks and ears after silking may occur at any time from planting to physiological maturity. Aflatoxin synthesis may occur at any time after infection and colonization, and it is a cumulative process until the substrate is exhausted, the fungus is killed, or permanent conditions unfavorable for synthesis are established.

Interactions among the three biological entities under discussion are most likely to occur at specific times during their respective life cycles because of the unique characteristics that associate them. Significant plant—insect interactions may occur throughout the plant and kernel life cycle; extensive leaf damage prior to silking may have a dramatic impact on kernal development. Plant—fungus interactions are probably most

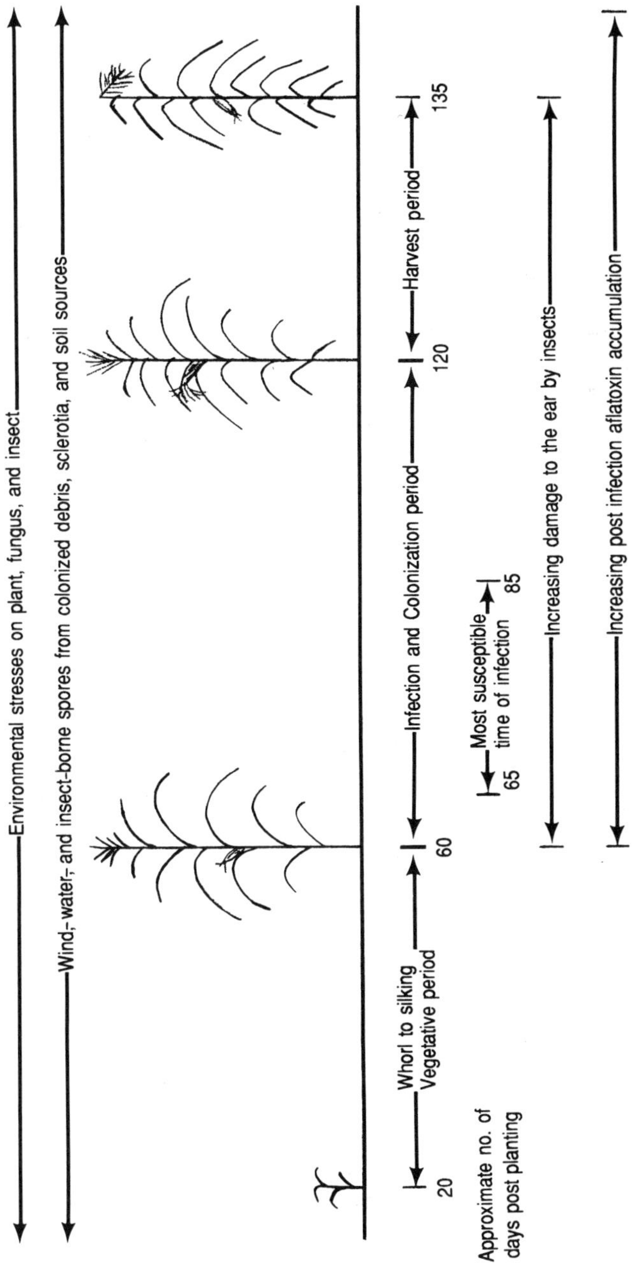

FIGURE 1. Preharvest kernel infection and aflatoxin contamination chronology.

important to aflatoxin contamination during the period from silking to physiological maturity. Critical fungus–insect interactions may occur throughout the insect's active life cycle, especially from host plant silking through physiological maturity. A second-order interaction of the plant–insect–fungus type can be a contributing factor to the contamination process from silking through drydown of the grain to less than 15% moisture, when pertinent fungi and insects are present.

The initial investigations into the preharvest maize infection and aflatoxin contamination processes were reviewed by Lillehoj and Zuber in 1975 [19]. Updates in our understanding of the *Aspergillus* spp.–plant–insect ecological system processes were made by Zuber and Lillehoj [20], Widstrom et al. [21], Lillehoj [22], and Wilson et al. [23]. Each of these reviews assessed the status and seriousness of the problem as well as the control of aflatoxin contamination in preharvest corn. The purpose here is to consolidate the information on first- and second-order interactions among the biological entities as they relate to the stresses imposed by nonbiological environmental factors. Much of the impetus for initial investigations was provided by the preharvest contamination findings of Anderson et al. [18], although several earlier surveys of postharvest and stored corn indicated possible preharvest contamination. Sampling of maize grown in 1964 and 1965 [10] and in 1967 [11] revealed aflatoxin contamination in more than 2% of the samples. Most of the contaminated samples tested contained less than 20 parts per billion (ng/g). However, Shotwell et al. [16, 24] reported that more than one-half of the contaminated samples from some areas had levels of aflatoxin exceeding 20 ng/g. Some surveys included corn from Corn Belt States; Illinois, Indiana, Iowa, and Missouri [2, 10, 25]. Yearly checks have been conducted in Georgia since 1977, with initial reports being made by McMillian et al. [26, 27] and Wilson et al. [28]. McMillian et al. [29] indicated that aflatoxin contamination of preharvest corn was chronic in Georgia and demonstrated that some samples exceeded 100 ng/g every year over a 6-year period. This pattern seems to be typical of corn grown in several southeastern states.

III. MAIZE PLANT GROWTH REQUIREMENTS AND MORPHOLOGICAL CHARACTERISTICS

A basic knowledge of plant characteristics and relevant interactions with the environment is necessary to study and understand plant involvement as a primary entity in the chronology of *A. flavus* infection and aflatoxin contamination processes. The maize plant is often referred to as a tropical plant, but modern, domesticated maize hybrids are really better adapted to temperate and subtropical conditions. Since maize hybrids are dependent upon humans for survival and maintenance, more is known of nutrient and growth requirements of maize than for most other species.

Modern maize hybrids are available for planting over a wide range of adaptation. Breeding stations have been located over the range of latitudes in the United States. The resulting developed hybrids have unique adaptation to specific areas. Adaptation might be defined as the plant's ability to fluorish in a specific area that has certain typical weather and other environmental characteristics that minimize the sum total of stresses imposed on that plant during its life cycle. The acceptable and desirable range of environmental conditions that delimit ideal adaptation are always exceeded at some time during the growth period, and the frequency and degree to which these conditions are exceeded determine the level of plant stress.

Some generalizations can be made about climate and other environmental factors that influence maize plant growth, even though certain hybrids are better adapted to specific extremes of these factors. Little maize is grown where average midsummer temperatures are less than 20°C; the greatest production occurs where the warmest month averages are between 21 and 27°C [30]. Little growth occurs at temperatures below 10°C, and growth approaches its maximum between 27 and 35°C. Growth slows and the optimal temperature recedes to the lower end of this range as water becomes more limited [31]. Growth entails the summation of accumulated dry matter due to photosynthesis during the day and loss due to respiration at night. A reduced dry matter accumulation rate results from periods of warm or hot nights, as respiration is temperature-dependent.

Maize is more water-efficient (372 weight units of water per weight unit of dry matter) than all other major U.S. crops except for sorghum [31]. It is estimated that maize requires more than 450 mm of water to produce a grain yield of 6.27 Mg/ha. Most environments where corn is grown do not provide adequate amounts of rainfall during the growing season, and occasional or extended periods of plant stress occur. Soils abundant in clay content may hold in their first meter of topsoil up to one-half the water needed to produce 6.27 Mg of corn per hectare, while many sandy soils in the southeastern United States are capable of holding less than one-half as much water as clay soils. High average midday humidities exceeding 60% in the southeastern states tend to reduce plant evapotranspiration and conserve some moisture in those sandy soils [31]. Low average midday humidities (less than 50%) across large areas of the U.S. Corn Belt reduce the temperature at which the plant begins to undergo moisture stress, that is, the temperature when plant evapotranspiration equals maximum water uptake by the roots at any given soil water availability level.

Nearly all conditions that are less than optimal for maize production also increase the time required to reach maturity and thus lengthen the time of exposure to the fungal infection and aflatoxin contamination processes. Optimal production conditions also require sufficient, but not excessive, amounts of major and minor nutrients. About 65 kg of N, 32 kg of P_2O_5, and 54 kg of K_2O are required to produce 6.27 Mg of maize grain per hectare [32]. Secondary and micronutrients are needed in lesser amounts, with molybdenum being used in quantities of <5 g per 6.27 Mg of grain. The major portion of secondary and micronutrients needed to produce the whole maize plant, with the exception of molybdenum, sulfur, and zinc, are returned to the soil when only grain is harvested [33]. Nutrients and micronutrients are supplied to the plant through fertilizers and soil amendments when they are not naturally available in the soil in sufficient quantities. Fertilizers and soil amendments influence the soil pH, which must be maintained between 5.6 and 7.0 to avoid stress on the plant [31].

The maize plant thrives in a wide range of soils maintained under numerous tillage practices in conjunction with highly varied crop rotations.

Little or no impact has been measured on *A. flavus* infection and aflatoxin contamination as related to cultural practices when the cultural procedures used are those that are also oriented toward optimal production of the crop. Individual cultural practices may on occasion affect other factors such as insects and disease that influence the aflatoxin contamination process.

The carbohydrate substrate of developing kernels on which the fungus develops and produces aflatoxin varies from simple sugars to a number of different starch types. The normal mature maize kernel contains an average of more than 70% carbohydrates [34]. The sugar content may vary from 3 to 36% and the starch content from 7 to 73% during the 4 weeks following pollination [35]; this stage corresponds to growth stages 5–7 as described by Hanway [36] and also to the period of greatest susceptibility to infection by *A. flavus* (Fig. 1). Several other plant characteristics have been implicated in the infection and aflatoxin contamination processes; however, the involvement has often been indirect because the plant traits were also those that influenced resistance to insects such as trichomes [37], maysin content [38], husk tightness [39, 40], and endosperm type [41, 42].

IV. *ASPERGILLUS FLAVUS* AND *A. PARASITICUS* OPTIMAL GROWTH CHARACTERISTICS AND OTHER PERTINENT PROPERTIES

Within the *A. flavus* group of fungi, only *A. flavus* and *A. parasiticus* isolates produce aflatoxin. Between 20 and 86% of the isolates produce aflatoxins [43–45]. The percentage of toxin-producing isolates is higher for *A. parasiticus* than for *A. flavus* [46]. *Aspergillus flavus* group isolates differ in their ability to produce aflatoxins at varying temperatures even when other factors are not limiting. The limiting range of temperatures for toxin production is 7.5–40°C [47], 11–40°C [48], or 12–41°C [49], depending on experimental conditions. Fungal growth, however, may continue at temperatures of 43–46°C. The limits of temperature for optimal aflatoxin production by the thermotolerant fungi are within the range of 25–35°C [50–52]. Information on temperature effects is of vital interest when consideration is given to a choice of control measures. A comparison of temperature limitations of the pertinent plant fungi and insects is given in Fig. 2.

Water is a limiting factor in aflatoxin production when availability falls below specified levels. The minimal level of water availability may be expressed as relative humidity of the environment, moisture percentage of the substrate, or water activity (a_w), the ratio of water vapor pressure of the substrate to the vapor pressure of pure water at equal temperature and pressure [53]. Relative humidity of less than 85% severely restricts or stops *A. flavus* and *A. parasiticus* growth and aflatoxin accumulation for nearly all isolates [54]; this limitation translates to values of approximately 16% for moisture content of maize and 0.85 for a_w [48, 55, 56]. Relative humidities above 92% are optimal for aflatoxin production [54, 57]. Complete saturation of the air or its equilibrium substrate environment does not significantly reduce fungal growth or toxin elaboration as long

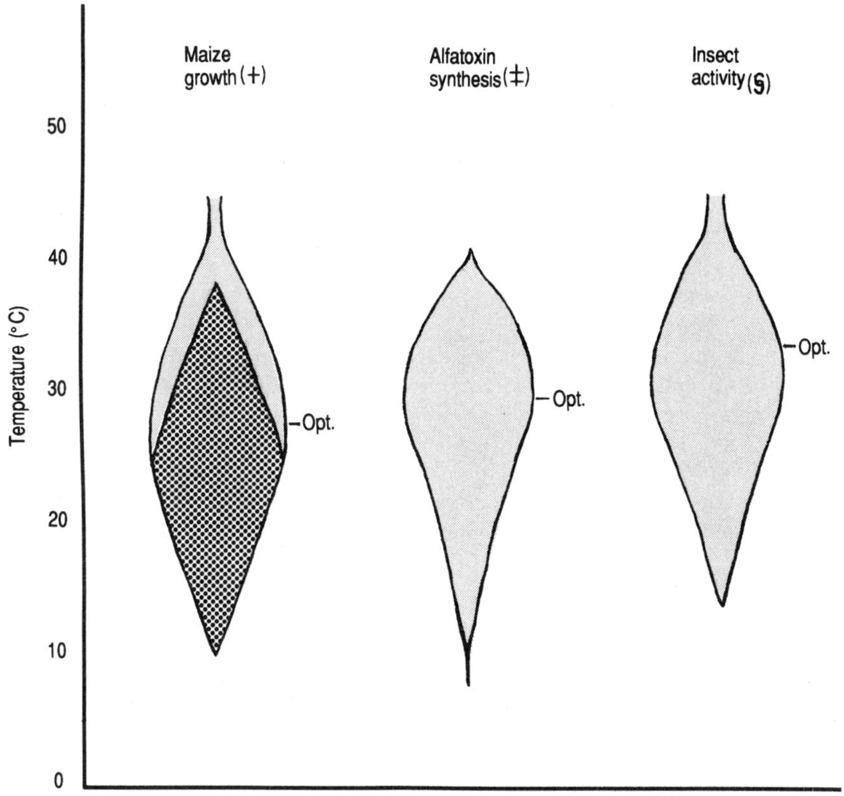

FIGURE 2. Limiting temperatures for maize growth, insect activity, and aflatoxin synthesis by *A. flavus* and *A. parasiticus*. (†) Plant growth will be more seriously expressed than indicated at temperatures greater than 27°C when water is limited due to soil and atmospheric factors (darkened area). (‡) *Aspergillus* spp. fungal growth continues at temperatures as high as 43–46°C. (§) Insect growth is often suppressed at temperatures greater than 30°C, but activity continues to increase as temperatures increase. Opt. = optimal temperature.

as there is periodic fluctuation below the maximum saturation point. Intermittent reductions in soil moisture between periods of saturation or inundation appear to be more important to germination of sclerotia [58]. This may explain why fungal propagules are so abundant and important in the sandy, well-drained soils of the Coastal Plain region in the southeastern United States.

Decreased amounts of aflatoxins have been observed when other fungi are grown in mixed culture with *A. flavus* and *A. parasiticus* [59]. Similar decreases have been observed when non-toxin-producing *Aspergillus* species were grown with *A. flavus* in mixed cultures [60]. Mixed inocula cultures have also produced different G_1/B_1 ratios of toxins, even though total amounts of B_1 remained constant [61].

Aflatoxin production on corn is severely restricted when storage atmospheres are modified to contain less than 0.5% O_2 [62]. Adequate aeration is needed to ensure fungal growth and aflatoxin production; these processes stop or are severely reduced when atmospheric gases are altered [54]. Similar aeration is apparently needed for germination of sclerotia [58] that are produced on harvest debris during the winter at southern locations [63].

An appropriate substrate is necessary for fungal growth and the elaboration of aflatoxin. This has been demonstrated by a near absence of toxin formation on soybean seed medium [54]. The reason for restricted contamination has not been determined, but it has been attributed to various sources, including inhibitors, that is, absence of trace elements such as zinc [64], enzymes, or a low molecular weight protein like that reported by Nagarajan and Bhat [65] in extracts of maize kernels. The quality or nutrient sufficiency of the substrate is obviously important because subtle differences such as those found in endosperm starch and sugar sources of similar genetic background influence the extent of aflatoxin contamination [42].

The *A. flavus* group of fungi are not considered to be aggressive plant pathogens but rather opportunistic in natural environments. Numerous chemical factors may influence aflatoxin elaboration, but levels at which they occur naturally do not influence the fungi to any extent when other factors are in their normal range. Relevant fungi grow best at a slightly acid pH (5.0), but will continue active growth and aflatoxin production over a range of pH from 3 to 7 [66]. Bracanto and Golding [67] reported an optimum pH of 5.5, with no detrimental effects on growth at a pH as high as 7.5.

V. OPTIMAL INSECT GROWTH AND ACTIVITY REQUIREMENTS

Insects probably have a greater range of tolerance to extremes of temperature and moisture in their environment than the maize plant or the *A. flavus* group of fungi. Although insect mobility enables them to modify their microenvironment when extremes are approached, emphasis will be placed on consideration of optimum conditions for growth and activity of insects implicated in the fungal infection and aflatoxin contamination processes in maize.

When insects and their damage were first shown to be associated with aflatoxin contamination [18], a host of maize insects immediately became suspect. The major insects of maize, as extensively listed by Petty and Apple [68] and Dicke [69], include numerous unlikely candidates for involvement in the aflatoxin problems, such as several soil and seedling insects; many of these same insects were listed by Barry [70] in 1987 as potentially being involved with aflatoxin contamination of maize grain. A limited list of maize insects known to have direct involvement with contamination, including the maize weevil (*Sitophilus zeamais* Motschulsky), corn earworm [*Helicoverpa zea* (Boddie)], fall armyworm [*Spodoptera frugiperda* (J. E. Smith)], and European corn borer (*Ostrinia nubilalis* Hübner), was reported by McMillian [71]. Insects of primary interest also include others that inflict injury upon the ear and/or feed upon the injured portions of the ear, such as pink scavenger caterpillar (*Sathrobrota rileyi*), southwestern corn borer (*Diatrea grandiosella Dgar*), and sap or picnic beetles (Nitidulidae) [69, 71].

Several insects attack the corn plant prior to ear formation, including the corn root aphid (*Aphis maidiradicus*), seed-corn maggot (*Hylemya platura*), and thrips (*Frankliniella* and *Aeolothrips* genera), while other insects that attack parts of the plant other than the ear such as the corn leaf aphid (*Rhopalosiphum maidis*) and corn flea beetle (*Chaetocnema pulicaria*) are probably not important vectors of the *A. flavus* group. A knowledge of optimal environmental conditions, especially those held in common by ear-damaging and -feeding insects, however, will be helpful in understanding the relationship between certain of these insects and aflatoxin contamination of the ear. In general, insect growth tends to be suppressed by conditions that deviate too far from the optimum. However, insect activity increases as temperatures rise well above those optimal for insect growth; this characteristic is important to consider when examining the interactions among pest, host, and fungal infection.

Optimal artificial rearing conditions have been determined for most of the major corn insects, especially those that feed on or damage the ear. Optimal rearing conditions include nutritional requirements [73], but these do not seem to be critical in the aflatoxin contamination process. Since the maize plant provides its insect pests with a complete nutritional diet, further consideration of insect development will be oriented toward environmental influences, especially temperature and moisture.

Most insects have greater tolerance to heat when in the egg, larval, and pupal stages than as adults. The optimal temperature for adults is almost always lower than for other stages, while the optimal relative humidity (RH) is usually higher for early stages than for adults. For example, ideal conditions for rearing corn earworm in the egg, larval, and pupal stages is about 29°C at 55–60% RH, whereas the adults perform best between 25 and 27°C at 75–80% RH [73]. Corn earworm eggs will tolerate temperatures as high as 43–45°C for short periods of less than 1 h without serious loss in percent hatch [74]. The earworm larvae grow well at 35°C, but growth is suppressed at 40°C [75]. Harrell et al. [76] showed that relative humidities of 50 and 70% influenced larval development more than temperatures ranging from 25 to 29°C. A relative humidity of 50% suppressed corn earworm pupal weight and emergence and delayed maturity compared to a 70% RH regime. The most rapid development of the corn earworm occurs at 30°C, but temperatures between 22 and 26°C optimize fertility and fecundity, respectively [77].

Patterns of growth and development in the corn earworm are somewhat similar to those for other lepidopterous species that infest maize ears. For example, a minimal development time of 18.4 days was observed for the fall armyworm when the temperature was maintained at 35°C. Optimum survival, however, was observed when temperatures were maintained between 24 and 32°C [78]. The activities of the corn earworm, fall armyworm, and European corn borer, when measured as development time, wing beat frequency, or flight speed, all increased significantly through the range of temperatures from 16 to 35°C [78, 79]; optimal relative humidity for these insects falls in the range of 50–90% [80].

The optimal temperatures for growth and development of the European corn borer are slightly lower than for the corn earworm and fall armyworm; this condition is probably related to the fact that the borer's primary area of adaptation is slightly cooler, especially in terms of nighttime temperatures. The borer thrives in temperatures ranging from 24 to 27°C at about 75% RH [81]. Rearing temperatures of 27–28°C and relative humidity of 75–80% are recommended for larvae [82]. The European corn borer survives well and reproduces in the 19–31°C range at relative humidity of 45–95%. Temperature cycling down to 18–20°C for 8 h each day is required for adequate mating and oviposition; this characteristic results in reduced adaptation in southern latitudes [82].

Moisture and temperature requirements for the maize weevil are surprisingly similar to those for the lepidopterous group of insects attacking corn. Harein and Soderstrom [83] recommended that this weevil be cultured and reared at 26.6°C and 70% RH. Numerous experiments have been carried out successfully at 27°C and 72% RH [84, 85]. Kirkpatrick [86] found that constant temperatures above 30°C increased mortality during oviposition and that temperatures exceeding 27.5°C reduced emergence in successive generations. Hwang et al. [87] reported that optimal conditions for weevil reproduction and growth are 28°C at 85% RH. They reported that temperature and relative humidity limits for development of maize weevil were 16–35°C and 45% to about 90% RH, and they observed fungal infection and growth on the substrate and insects at 93% RH.

Optimal temperature ranges for the maize plant, the *A. flavus* group of fungi, and the ear-feeding insect group are given in Fig. 2. The comparison helps to integrate information about the common temperature requirements of the three biological entities and to illustrate how stresses may occur at different times for each as temperatures deviate from the optimal range for each organism. Optimal temperatures are clearly higher for aflatoxin synthesis and insect activity than for plant growth, and differences widen when moisture is limiting on the plant. Relative humidities are omitted from the figure because macroenvironmental relative humidity is not comparable to microenvironmental relative humidity inside the ear where the plant is being invaded by both insects and fungi. However, the relative humidity does indeed contribute to the relationships among plant, fungus, and insect and must be considered when specific interactions are being studied.

VI. *ASPERGILLUS FLAVUS* GROUP–MAIZE PLANT INTERACTIONS

The existence of a relationship between the *A. flavus* group of fungi and ears of the maize plant in the field has been well documented [4]. The association received little attention until the discovery of aflatoxins as metabolites of the *A. flavus* group of fungi, and even then most reports suggested that the infection and aflatoxin contamination processes were primarily storage problems [88]. The reluctance of plant pathologists to clarify the role of the *A. flavus* group as pathogens is presumably due to these fungi being considered to be nonaggressive. Some may even consider these fungi to be near-saprophytic, or at worst noninvasive, fungi that are responsible for secondary infections of living tissue. In any event, many pathologists have even been reluctant to call these infections ear or kernel rots and have classified them separately as storage rots [89].

By the early 1970s many maize researchers were keenly aware of a field aflatoxin contamination problem [90, 91]. Nevertheless, the extent of the problem surprised everyone when the results from testing preharvest corn in 1977 became known [28, 92]. A field problem began to be reported from 1977–1979 surveys conducted at diverse locations including India [93]. Contamination of 1977 Corn Belt corn in the United States could no longer be considered a rare occurrence when approximately 7% of 493 samples taken from Indiana preharvest corn in 1983 were found to contain aflatoxin [25]. The most serious contamination was found in the southeastern states [13], a fact dramatically demonstrated by McMillian et al. [29] when they reported chronic contamination of the preharvest crop in Georgia over a 6-year period. Subsequent studies examined the relationship of the fungus to the maize ear in the field, beginning sometime prior to harvest of the grain.

The work that clearly established the existence of a preharvest problem in corn did not identify the routes by which infection and contamination occurred. The report by Anderson et al. [18] alluded to a prerequisite of kernel damage for field contamination, even though some samples obtained from surveyed corn had no obvious damage but yet were contaminated with aflatoxin. Jones et al. [94] demonstrated that developing kernels may indeed become infected and contaminated in an environment without insects to serve as either vectors or damaging agents; their silk inoculations occurred over a 4-week period after silk emergence, and temperatures of 38°C were near the upper limit of aflatoxin synthesis (Fig. 1) and the most conducive to ear infection. Inoculated silks produced infected kernels that were evenly distributed over the entire ear when the experiments were conducted in a controlled environment [95]. *Aspergillus flavus* growth may advance over the entire ear in as few as 4 days at temperatures exceeding 30°C in growth chambers, whereas the same process under field conditions may take as many as 13 days [96].

The elaboration of aflatoxin by the fungus may ensue rapidly after infection. Inoculated postharvest samples require about 3 days for toxin formation when incubated at room temperature [97]. Average levels of contamination exceeding 200 ng/g were detected just 2 days after inoculation of ears grown under controlled conditions in growth chambers [98]. Ears averaging more than 2000 ng/g were harvested only 9 days after inoculation when temperatures ranged from 26 to 34°C. Those field conditions that occur for corn inoculated in July are less than ideal, and approximately a week is required for aflatoxin to appear in the inoculated ears [99]. It may require a month or more under field conditions to obtain the same level of contamination that occurs in just 9 days under ideal conditions. In fact, levels as high as 2000 ng/g are not common in the field but do occur occasionally when conditions are favorable [29].

Wound-inoculated ears have more aflatoxin than those that are silk-inoculated [99]. This condition is probably due to more extensive colonization of wound-inoculated ears and a higher proportion of ears that escape infection when silks are inoculated. A third factor is the timing of inoculation. Marsh and Payne [96] found that silks inoculated prior to turning completely brown, that is, prior to complete drying, resulted in ears with a higher percentage of the kernels infected than those inoculated later. Lillehoj et al. [17] wound-inoculated ears in three places with a hypodermic needle at 20 days after silking. The efficacy of inoculating at 20 days after silking was later confirmed by Widstrom et al. [100]. Wound-inoculating with a spore-contaminated knife proved more effective than a needle injection in producing visibly infected ears with high levels of aflatoxin contamination. Wound-inoculating at 20 days after silking is now a generally accepted standard for study of the contamination process in the field. Tremendous variation for infection and contamination is encountered from kernel to kernel and from ear to ear [101], especially when natural inoculation is the source of infection. Differences in variation are important for assessment of different methods of inoculation and sampling field plots [102]. The inconsistency of interaction between plant and fungus makes precise studies of this relationship difficult.

A concerted effort has been made to identify the major factor(s) that influence the relationship between the *A. flavus* group of fungi and the maize plant and result in the production of large amounts of aflatoxin contamination. The first attempt to do this involved sampling of maize plots grown in a large number of different environments. Lillehoj and co-workers conducted a series of experiments [17, 103–105] at locations ranging in latitude from the Corn Belt to the deep south. Invariably, samples obtained from maize grown at southern locations had greater aflatoxin contamination than those grown at Corn Belt latitudes (39 to 46°). Numerous variables such as insect damage, hybrid adaptation, and endosperm type were considered in these studies, but the paramount condition of high toxin levels seemed clearly related to the interaction of plant and fungus with temperature and precipitation during corn development [104].

Cultural and management practices may have a profound effect on diseases, insects, and related factors that influence the vigor of the corn plant, such as nutrient and water availability. Any change in the environment that imposes stress on the plant should be suspected of placing the grain at risk for fungal infection and aflatoxin contamination [20, 21]. The well-drained soils of the Coastal Plain in many southeastern states are known to be subject to drought and low corn yields [106]. Years with greater than average drought stress in southern Georgia have reduced yields and have also resulted rather consistently in crops with higher than average amounts of aflatoxin. Typical examples of such years were 1977 and 1980 [29] and 1986 [107].

Attempts to determine the most important limiting factors that induce stress and result in high levels of aflatoxin contamination have met with only limited success. Stress conditions that limit water availablity to the plant and thus reduce yield seem to be those that most consistently place the crop at risk for contamination [108–111]. The limited availability [112] or excess [113] of nitrogen supplied to the plant has also been proposed as a significant contributor to increased toxin contamination. Water stress, however, is always associated with high temperatures in the field and is undoubtedly much more important than nutrition since aflatoxin synthesis is maximized at temperatures [28–32°C] that nearly always place the plant in a condition of stress (Fig. 2). Results of growth chamber studies [96, 114, 115] and field results interpreted over time [107] have shown that temperature has a critical effect on infection and contamination processes. Combined and pooled correlations of aflatoxin concentration with maximum or minimum temperatures during 20–60 days after full silk were significant at $P \leq 0.05$ in eight of eight instances. These values were obtained from a 5-year study and indicate that early plantings in the deep south are at greater risk for contamination because the critical grain filling period, beginning at 20 days after silking, falls during a time when maximum and minimum temperatures are highest. Interpolating the period from 20 days after silking to harvest for each planting date (Table 1), it can be seen that the March 18 planting is the only planting for which the period is entirely included between June 21 (longest yearly photoperiod date) and July 31 (July is the month with the warmest average daily temperatures at 27.3°C). Table 1 also shows a decline in aflatoxin concentrations on other planting dates when part or all of the critical 20-day postsilking to harvest period falls outside the June 21 through July 31 period.

The general loss of vigor manifested by the plant when attacked by disease is a well-known phenomenon, especially for diseases that attack the roots, stalks, and leaves. Such diseases presumably render the ear more susceptible to invasion by the *A. flavus* group of fungi. Several of the same genera that cause stalk rots also are the causative agents for ear rots—notably *Diplodia, Gibberella,* and *Fusarium* [89]; these ear invaders not only reduce the general vigor of the plant, predisposing the ear to other fungal invaders [116], but also compete directly with the *A. flavus* group for the host substrate. Concurrent colonization by several other fungi, along with *Aspergillus* species, is common [12, 117, 118], but limited information is available on interactions among them in the developing ear.

The *A. flavus* group of fungi is generally considered to include species that are poor competitors when they occur with other fungi because of their nonaggressiveness. Most reports of research in this area agree that the aflatoxin-producing capabilities of *A. flavus* are reduced when placed in competition with other fungi, although they disagree about the level of inhibition by various genera. Ehrlich et al. [59] found that *Penicillum oxalicum* was an effective inhibitor of aflatoxin production by *Aspergillus* species but reported that *Fusarium* species did not affect mycotoxin production. Wicklow et al. [119], however, found that *F. moniliforme* was an effective competitor; it inhibited *A. flavus* from infecting and contaminating uninjured kernels on the growing corn ear. The studies of Horn and Wicklow [120] demonstrated that *A. niger* could inhibit aflatoxin formation when placed in mixed culture on autoclaved corn. The mechanism of inhibition was attributed to the ability of *A. niger* to lower

TABLE 1. Planting Dates, Silking Dates, Harvest Dates, and Geometric Means for Aflatoxin Concentrations of Wound-Inoculated Corn Sampled at Tifton, Georgia, 1982–1987

Planting date	27 Feb.	18 Mar.	31 Mar.	15 Apr.	1 May	15 May	29 May	13 June	30 June	14 July	29 July
Silking date	24 May	1 June	9 June	18 June	29 June	13 July	24 July	8 Aug.	24 Aug.	9 Sept.	22 Sept.
Harvest date	19 July	28 July	8 Aug.	15 Aug.	24 Aug.	7 Sept.	18 Sept.	3 Oct.	18 Oct.	5 Nov.	17 Nov.
Aflatoxins (ng/g)[a]	1939	2018	1604	898	412	309	104	171	459	67	33

[a]The geometric means are the antilogarithms of the logarithmic means for concentration of aflatoxins.
Source: Adapted from Widstrom et al. [107].

pH of the substrate sufficiently to suppress aflatoxin production. Lesser levels of suppression were found in samples from preharvest ears that were wound-inoculated with equivalent numbers of *A. flavus* and *A. niger* propagules [121]. *Aspergillus flavus* was found to be an effective competitor when paired with *A. parasiticus*, suggesting at least one reason why *A. flavus* is the dominant invader of the two in preharvest corn [61].

The *A. flavus* group infection and aflatoxin synthesis processes have been demonstrated to be intricately involved in interactions with specific genetic plant types. The evidence for these interactions began to accumulate in the early investigations of preharvest contamination [122, 123]. Although differences were inconsistent, they were clearly evident among the popular hybrids grown in the southeastern states; these studies were quickly followed by others showing differences in aflatoxin accumulation among inbred lines when tested in hybrid combination [124, 125] or when tested as inbreds per se for percentage of infected kernels [126]. Similar differences were reported between population hybrids and open-pollinated varieties grown prior to 1940 in the southern states [127, 128]. In general, modern hybrids were less contaminated than the old open-pollinated varieties. The inheritance of resistance was also under simultaneous investigation in several of these early studies [124, 125, 129]. All of these reports point to additive genetic variation (the easiest type of variation to exploit during selection) as being of primary importance in conditioning resistance to aflatoxin accumulation. A 3-year study of preharvest aflatoxin accumulation by ears of a resistant and a susceptible hybrid revealed that the basis for differences in accumulation was the differing linear rates of accumulation during the period beginning at approximately 30 days post silking through harvest at physiological maturity [130] (Fig. 3); this finding confirms the wisdom of a common practice of sampling field plots at physiological maturity of the grain, a time when differences among test entries are maximized.

It is unknown at this time whether the principal source of resistance in maize to *A. flavus* infection and/or aflatoxin synthesis is chemical or physical in nature. The chemical basis for resistance has not been critically investigated beyond pH requirements and limitations on the fungus [66, 120] and the initial work on plant chemicals that influence infection and toxin synthesis. Studies by Wilson et al. [131] indicate that some metabolites such as β-ionone have a profound influence on *A. flavus* growth and toxin accumulation when tested at concentrations ≥ 50 μL/L; such concentrations far exceed those found naturally in corn and are not considered a primary source of resistance in the plant.

The basis for plant resistance has usually been sought among the physical characteristics of the plant. For example, good husk coverage has been demonstrated to be a significant source of resistance to both insects and aflatoxin contamination [39, 40]. When aflatoxin contamination of popcorn in Nebraska was reported by Riesselman and Doupnik [132], a number of investigators researched various kernel types to determine the influence of kernel characteristics on plant resistance. Popcorn lines showed significant differences for aflatoxin accumulation [133], as did sweet corn and dent corn types [125]. The dent types sustained significantly lower concentrations of aflatoxin than other types when tested under identical conditions. A thin-pericarp hybrid supported significantly

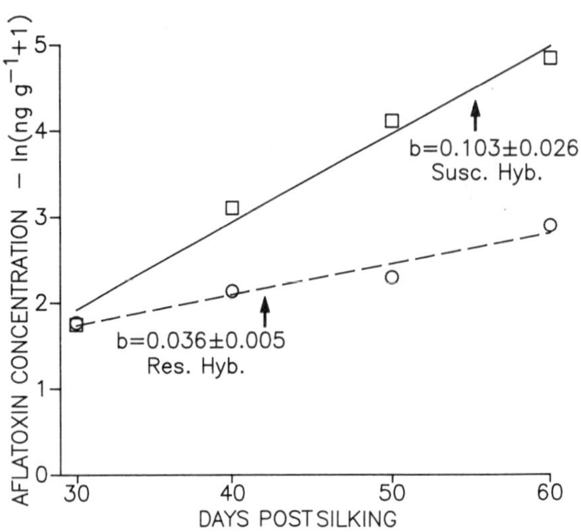

FIGURE 3. Regression of aflatoxin concentration on days postsilking, comparing two maize hybrids, one susceptible and one resistant, grown in the years 1982–1984.

greater aflatoxin production than one with a thick pericarp [61], but King and Scott [134] were unable to show a correlation between kernel hardness and kernel infection. A common thread through most of these studies, or with those showing inbred and hybrid differences, is inconsistency of findings from one location and year to the next. Tests involving hybrids with different endosperm characteristics either did not show significant differences in toxin accumulation due to endosperm type [17] or were in some way inconclusive [41]. A more recent report by Widstrom et al. [42] on kernel homogenates of endosperm types tested in a common genetic background clearly demonstrated greater susceptibility of the endosperm types with high sugar content. All genotypes with low starch content in their endosperm [those having the brittle (*bt*), shrunken (*sh*), or sugary (*su*) alleles] were more suceptible than the normal starchy genotypes.

Association between the *A. flavus* group of fungi and the corn plant in the field does not terminate when the grain crop is harvested and utilized. Shotwell et al. [135] found aflatoxin in kernels, cobs, husks, leaves, and stalks of postmature plants taken from a field heavily infested with *A. flavus*. Of 50 plants, 7 had no aflatoxin, 42 of the remaining 43 had kernel contamination, and 1 had only husk contamination. Each plant part, other than kernels, was contaminated in 19 or more of the 50 plants studied. No determination was made as to whether the plant parts were contaminated by translocation from one part to another or by the more likely in situ elaboration at infected parts. The corn plant is known to have a limited capacity to absorb aflatoxin from nutrient or soil solution through its roots and translocate the contamination throughout the plant [136]. Debris from the crop is not only liable to be contaminated but may also harbor the source of infection for the following season. The existence of *A. flavus* sclerotia on ears of corn left in a south Georgia corn field after harvest was demonstrated by Wicklow et al. [63], providing evidence of a probable source of propagules for overwintering in soil. Sclerotia were known to be produced by certain sclerotia-producing strains in the laboratory [137] and have been found in field debris (ears); this may increase the probability of infection for the crop following, especially if that crop is corn [138].

VII. *ASPERGILLUS* SPP.—INSECT INTERACTIONS

The historical association between insects and fungi that infect the corn ears is one of long standing but has been a poorly understood phenomenon. Reports in 1914 [139] and 1917 [140] emphasized the importance of this association, and Taubenhaus [4] specifically indicted *Aspergillus* spp. as being involved in the relationship. The discovery of aflatoxins as a secondary metabolite of these fungi and the fact that insects were associated with the contamination of corn by aflatoxins [18] kindled an intense scientific interest in the relationship between plant-damaging insects and mycotoxigenic fungi. Literature on the role of insects on aflatoxin contamination in corn was reviewed by Widstrom [141], and as more detailed studies began to be reported [44] the complexity of the problem became apparent. Among 205 *A. flavus* isolates identified visually from 871 insect and soil samples, only 62 tested positive on *Aspergillus* differential medium. Additionally, only about one-half of the confirmed isolates were toxin producers. Other studies by Lillehoj and his colleagues [142, 143] showed that insect-damaged and/or aflatoxin-contaminated kernels exhibited fluorescence under ultraviolet light and that the presence of *A. flavus* was associated with kernel damage. Shotwell et al. [24] found that 75% of the fluorescing kernels from samples harvested from the 1973 South Carolina crop also had insect damage. The damage was primarily attributed to weevils. Fractions of the total samples with insect damage comprised only 28% of the total weight but had 84% of the aflatoxin B_1 contamination. The 1977 Georgia crop contained a high level of contamination that was significantly associated with insect damage [26]. Similar conclusions were drawn from a 1977 study in South Carolina [123].

The basic questions persisted as to whether (1) insect damage was the primary reason for infection and contamination, (2) insects served only to exacerbate an already existing problem, or (3) insects not only provided infection courts for the fungus by damaging the kernels but also served as a primary vector of the fungus. Controlled experiments using mechanical means to simulate insect injury were begun in the early 1970s. Mechanical damaging of ears usually increased the incidence of aflatoxin-contaminated ears above that of the check [17] but did not always produce higher concentrations of aflatoxin [103, 144]. Researchers in South Carolina were very successful in obtaining higher concentrations of aflatoxin in ears for which insect damage was simulated by mechanical damage [110]. Little further effort was made to use mechanical damage to mimic insect injury because it would not help to answer questions about the toxicity of aflatoxin to insects or whether insects are an efficient vector of fungal spores for infection.

Insects are not immune to the toxic effects of the metabolites produced by the *A. flavus* group of fungi. Significant weight suppression occurred for second instar corn earworm larvae when they were fed diets containing as little as 0.25 μg of aflatoxin B_1 or G_1 per 10 mL of diet for a period of only 5 days [145]. Lesser amounts of suppression occurred for the fall armyworm and European corn borer at the same aflatoxin concentrations. In the same tests, mortalities exceeded 50% when 2.5 μg of aflatoxin per 10 mL of diet were fed (Table 2). Wicklow and Dowd [146] found that earworm larvae were less affected by maize kernel samples fermented for 8 days with *A. flavus* group fungi than were larvae of the fall armyworm; survival rates after 1 week were 57% (corn earworm) and 37% (fall armyworm). The maize weevil may be slightly less affected by the fungus and its metabolites than other insects because evolutionarily it is obligated to spend its entire life cycle on or in the kernel. Mortality rates did not exceed 25% when weevils were exposed for a period of 100 days to contaminated samples with end concentrations of aflatoxin as high as 2–13 μg/g [147].

McMillian [71] reviewed the insects known to be associated with the aflatoxin contamination problem in corn (maize weevils, corn earworms, fall armyworms, and European corn borers), and Barry [70] listed no fewer than 30 other insects that hold potential as candidates for involvement. Although most of these insects are probably not responsible for significant amounts of aflatoxin contamination of the corn crop, many are certainly worthy of testing.

TABLE 2. Percentage Mortality of Second Instar Larvae Fed on Diets Treated with Various Amounts of Aflatoxins B_1 and G_1

Aflatoxin (μg/10 mL diet)	Percent mortality[a]					
	Corn earworm		Fall armyworm		European cornborer	
	B_1	G_1	B_1	G_1	B_1	G_1
2.500	96a	78a	44a	58a	70a	88a
0.250	8b	0b	10b	2b	24b	12b
0.025	8b	0b	2b	0b	16b	18b
0.000	0b	0b	2b	2b	14b	14b

[a]Means in a column not followed by the same letter are significantly different at $P < 0.01$.
Source: Adapted from McMillian et al. [145].

Many of the early 1970s studies that associated insects with *A. flavus* or aflatoxin were conducted randomly on insects that were collected from harvested samples and examined for contamination. Fennell et al. [43] found *A. flavus* present in fewer than 5% of all insects collected from samples grown in a six-state experiment. Contamination was not species-specific in this study, but an early study of Missouri and Illinois corn [72] had yielded a significantly higher proportion of corn earworm larvae (37%) than corn borer larvae (14%) associated with the fungus. During a 12-year study by McMillian et al. [148], corn earworm moths were collected throughout the season from 1977 to 1988. The percentage of moths contaminated with *A. flavus* varied from 12% in 1982 to 87% in 1981. Lillehoj et al. [149] did not associate any species of insect with the presence of *A. flavus* but found that 3.3%, or 54, of 1621 insects had *A. flavus* contamination. Hesseltine et al. [14] reported that more than 50% of 375 insects collected from South Carolina corn were surface-contaminated and that 78 of 85 weevils were carrying *A. flavus* spores.

Experiments to determine which insects are most important to the aflatoxin contamination problem have been inconclusive. Fennell et al. [150] compared European corn borers and corn earworms from hand-infested ears of Georgia, Iowa, and Missouri corn but could not determine which produced the greater contamination, as both insects were effective at increasing the incidence and level of aflatoxin. In a similar test, Barry et al. [39] found no significant difference between the same two insects. Widstrom et al. [151] reported in a test among the earworms, armyworms, and corn borers that damage by the European corn borer had the highest concentrations of aflatoxin B_1, but again the differences were not significant. The first reports on the maize weevil as a vector of the fungus indicated that this insect was a nonvector or at best an efficient vector of *A. flavus* [152]. Subsequently, McMillian et al. [153] demonstrated that the maize weevil can contribute significantly to increased *A. flavus* infection by serving as a vector for spores and damaging corn kernels. Barry et al. [154] confirmed this finding in an experiment in which the maize weevil was compared with the wheat curl mite, *Eriophytes tulipae* (Kifer). Further confirmation of the maize weevil as an effective vector was reported by McMillian et al. [40] when they demonstrated in a 3-year study that ears hand-infested with weevils sustained significantly higher amounts of aflatoxin contamination (329 ng/g) than ears infested with earworms (60 ng/g), armyworms (80 ng/g), or corn borers (71 ng/g).

VIII. PLANT–INSECT INTERACTIONS

Plant–insect interactions may be the least important of the first-order interactions as far as their contribution to the aflatoxin contamination problem is concerned. However, the interactions are involved in the overall problem and cannot be summarily dismissed as unimportant.

The corn earworm might be considered somewhat typical as an ear-feeding insect with respect to its interactions with the corn plant.

Environmental factors, as influenced by planting date and inherited traits such as plant maturity, are known to be significantly related to the amount of damage sustained by the corn plant due to this insect [155]. Unlikely factors such as plant density [156] as well as the obvious ones involving husk characteristics [157] have some influence on ultimate damage to the ear. Taking account of the varied factors when studying the inheritance of resistance complicates interpretation considerably in identifying and locating major genes for resistance [158] or assessing germplasm for its amount and type of genetic variability [159, 160]. The plethora of relevant factors also complicates the selection process because of low heritabilities [161]. The breeder is forced to resort to such things as selection indices in order to make selection for resistance more effective [162].

The corn earworm is primarily an ear-feeding insect, making the silks an important aspect of resistance [163]. The silks are also important to the aflatoxin contamination process because they are a principal infection court for invasion by *A. flavus*. Chemical components of the silk, for example, maysin [38], which have an antibiotic effect on earworm larvae, may also have an effect on the invading fungus. Maysin is known to vary in concentration among hybrids; it has potential as another tool to control insects that serve as vectors for the fungus [164]. The search for chemical components of silks that restrict or inhibit *A. flavus* growth should certainly be a worthy effort with reasonable expectation of success.

The import of the influence of environmental factors on the fall armyworm and European corn borer is similar to that for the corn earworm, except that the corn borer is better adapted to cooler temperatures. The major difference between the earworm plant–insect interactions and those for the armyworm and European corn borer is that the latter two are primarily feeders on leaves and plant parts other than ears [165, 166]. Both of these species (fall armyworm and European corn borer) have been extensively evaluated for leaf-feeding resistance [167–169], a trait that would have only secondary effects on ear contamination by aflatoxin through reduction of plant vigor. The corn plant has a chemical defense against corn borer feeding in the seedling and whorl stage. The compound in leaves is commonly called dimboa and has an antibiotic effect on young larvae of the borer [170].

The pink scavenger caterpillar and maize weevil both damage corn kernels at about the same stage of development, that is, after the dough stage, but the types of damage produced by the two insects are somewhat different. The pink scavenger caterpillar tracks along the ear near the crown of the kernels or at the base of attachment, and a single larva inflicts damage to several kernels [171], while maize weevil larvae feed on the kernel internally in the field and in storage [172]. The measurement of maize weevil damage has been studied more extensively [84] than that for the pink scavenger caterpillar. Numerous sources of resistance to maize weevils have been identified [85], and inheritance of the resistance has been investigated [173]. Knowledge is limited on the interactions of the corn plant with maize weevils in the field because the majority of studies on the insect have been conducted in the laboratory under controlled conditions.

The pink scavenger caterpillar is less dependent on other insects (such as the earworm) than is the maize weevil for establishment [174].

All three insects commonly appear together in the ear and significantly reduce yields [175]. Husk coverage and tightness is important for protection against ear damage by the entire ear-feeding complex [176], with the possible exception of the corn borer, which often enters through the shank rather than through the husk or silk channel. Competition for food and cannibalism influence the survival of several ear-feeding insects, especially the Lepidoptera [177]. Probably due in part to a common evolutionary niche in the ear, maize resistance to some insect species such as the corn earworm and the European corn borer is sometimes found in the same corn lines [178].

IX. FUNGUS–PLANT–INSECT INTERACTIONS

Until the demonstration by Anderson et al. [18] that both fungus and insects are involved in the aflatoxin contamination process prior to maturity of the corn ear in the field, second-order interactions were not seriously considered as a functional piece of the aflatoxin puzzle. Prior studies were generally surveys attempting to assess the importance of insects in aflatoxin contamination [24, 143], while several of those following the report by Anderson and his colleagues emphasized the firm establishment of insect involvement on the fact that contamination was indeed occurring in the preharvest crop [142, 149, 150]. The 1977 aflatoxin contamination disaster for corn in Georgia [26], Alabama [179], and numerous other states clearly demonstrated that both insects and fungus were involved in interactions with the developing ear for some time before harvest. The discussion following will attempt to summarize what we now know about the nature of these second-order interactions.

In summarizing the results of the early experiments, Lillehoj and Hesseltine [180] proposed the identification of genetic resistance in the corn plant to provide a remedy for field infection. Zuber [181] recommended the use of genetic control through resistance to field infections and resistance to insects as an approach to control contamination in the same series of summary reports. The maximum protection is obviously needed for the all-important temperature window of 30–35°C (Fig. 2) when optimal insect activity and aflatoxin synthesis takes place and plant growth is less than optimal.

Attempts to control fungal infection by applying insecticide were helpful in limiting insect damage [182], but the control measures were unsuccessful in eliminating insect damage and preventing infection of corn ears by *A. flavus*. Iken [183] also suggested an insecticidal control program but pointed out how uneconomical such a program is for field corn. The results of a two-year experiment by Wilson et al. [184] point out the complexity of studying second-order interactions that influence aflatoxin concentrations in the corn crop. Plants were both inoculated with *A. flavus* spores and infested with corn earworm larvae, but neither affected insect damage, fungal recovery, kernal fluorescence, or aflatoxin contamination when the ears reached maturity in either of two years. Other overriding conditions apparently existed during the years 1975 and 1976 that prevented the influence of inoculation and infestation to be expressed. On the other hand, differences in toxin contamination attributable to insect damage are often detectable without artificial infestation [123, 185].

Various methods of eliminating insects or *A. flavus* from the ear have been attempted, such as bagging ears [150], but the procedures have been only partially successful. Bagging reduced insect damage but had no influence on the amount of aflatoxin production in the ear; this study illustrated a typical interaction between insect and fungus within the corn ear that produced the lowest aflatoxin concentration in kernels from earworm-infested plots, even though the highest incidence of aflatoxin-positive plots occurred for that insect rather than armyworms or corn borers (Table 3). The probable reason for high incidence and low concentrations is that corn earworm injury is usually confined to a smaller area than that of fall armyworm or European corn borer. Each insect, therefore, has its own unique interaction with the fungus, depending on the resistance of the corn hybrid, and there is a contrasting incidence and concentration of aflatoxins. Recent studies indicate that the natural defenses of the plant may indeed be the best source for reducing field contamination. Husk type and tightness [39, 40] are effective as characteristics that both limit insect damage and impede the progress of *A. flavus* toward the interior of the ear. McMillian et al. [186] have also demonstrated that resistance to both insects and aflatoxin production are present in certain corn hybrids. While this is encouraging, other environmental factors that cannot be controlled, such as mean temperature [29], and the heavy dews that are so prevalent in the southeast [186], are constantly exerting their influence on the fungus—plant—insect complex in preharvest corn.

The European corn borer has received special attention since 1980, partially because aflatoxin contamination of the crop in the Corn Belt (the insect's best area of adaptation) has occurred more frequently in recent years, but also because the wandering and tunneling type of ear damage inflicted by this insect usually results in higher concentrations of aflatoxin for infested ears (Table 3). In a series of experiments conducted

TABLE 3. Percentages of Plot Samples Testing Aflatoxin-Positive and Their Toxin Concentrations, as Associated with Each of Three Insect Species

Insect	Plots testing positive (%)	Aflatoxin B_1 concentration (ng/g)
Corn earworm	53	22
European corn borer	41	65
Fall armyworm	28	35

Source: Adapted from Widstrom et al. [151].

at locations in Iowa, Missouri, and Georgia [187, 188], contamination was found to be very location-dependent when a susceptible and a resistant hybrid were inoculated with *Aspergillus* spp. and infested with the European corn borer. The highest amounts of aflatoxin were found in Georgia corn, even though the highest percentages of aflatoxin-producing isolates obtained from corn borer larvae were found at the other locations. In another study by Lillehoj et al. [189], the aflatoxin concentration averages were three times as high in Missouri (419 ng/g) as in Georgia (140 ng/g). McMillian et al. [190] reported negligible contamination in Missouri samples in 1986, while Georgia corn that was similarly infested and inoculated was contaminated with 718 ng/g aflatoxin. The environmental component for the European corn borer seems to be larger than for other insects.

The most decisively indicted insect, as an efficient vector of *A. flavus*, has to be the maize weevil [153, 154]. This insect is decidedly more effective at southern latitudes than elsewhere, and these latitudes are specifically associated with chronic aflatoxin contamination [29]. Its interaction with the fungus inside the ear and its apparent capability to maintain activity over extended periods in the presence of high aflatoxin concentrations make the maize weevil an ideal exacerbating factor in the contamination equation [147]. Evolutionarily speaking, this insect must have developed under stressful conditions with the fungus and its metabolites, giving it a capacity to survive those conditions and thus effectively serve as a vector for the fungus. A 3-year study conducted at Tifton, Georgia, confirms the effectiveness of the weevil in southern latitudes [40]. The mean concentrations of aflatoxin for infested field plots was more than four times as high for weevils (329 ng/g) as for corn earworms (80 ng/g), European corn borers (71 ng/g), or fall armyyworms (60 ng/g).

The full cooperation of pathologists, geneticists, and entomologists will be required to resolve the intricate problem of aflatoxin contamination in the preharvest corn crop. The cooperative effort is critical because several environmental factors have a significant impact upon the multiplicity of fungus—plant—insect interactions. At this point, it appears that plant resistance, whether to *A. flavus* infection, aflatoxin synthesis, or insects, will be a major component in the control package for reducing aflatoxin contamination of corn.

REFERENCES

1. Detroy, R. W., Lillehoj, E. B., and Ciegler, A. Aflatoxin and related compounds, *Microbial Toxins*, Vol. 6 (A Ciegler, S. Kadis, and S. J. Ajl, Eds.), Academic, New York, 1971, p. 2.
2. Shotwell, O. L., Kwolek, W. F., Goulden, M. L., Jackson, L. K., and Hesseltine, C. W. Aflatoxin occurrence in some white corn under loan, 1971. I. Incidence and level. *Cereal Chem.*, *52*:373, 1975.
3. Schwarte, L. H. Moldy corn poisoning in horses, *J. Am. Vet. Med. Assoc.*, *92*:152, 1938.
4. Taubenhaus, J. J. A study of the black and the yellow molds of ear corn. *Texas Agric. Exp. Stn. Bull.*, *290*:3, 1920.

5. Koehler, B. Natural mode of entrance of fungi into corn ears and some symptoms that indicate infection. *J. Agric. Res., 64*:421, 1942.
6. Burnside, J. E., Sippel, W. L., Forgacs, J., Carll, W. T., Atwood, M. B., and Doll, E. R. A disease of swine and cattle caused by eating moldy corn. II. Experimental production with pure cultures of molds. *J. Am. Vet. Res., 18*:817, 1957.
7. Quasem, S. A., and Christensen, C. M. Influence of moisture content, temperature and time on the deterioration of stored corn by fungi. *Phytopathology, 48*:544, 1958.
8. Christensen, C. M., and Kaufmann, H. H. *Grain Storage: The Role of Fungi in Quality Loss*, Univ. Minnesota Press, Minneapolis, Minn., 1969.
9. Lillehoj, E. B., Fennell, D. I., and Hara, S. Fungi and aflatoxin in a bin of stored white maize. *J. Stor. Prod. Res., 11*:47, 1975.
10. Shotwell, O. L., Hesseltine, C. W., Burmeister, H. R., Kwolek, W. F., Shannon, G. M., and Hall, H. H. Survey of cereal grains and soybeans for the presence of aflatoxin. II. Corn and soybeans. *Cereal Chem., 46*:454, 1969.
11. Shotwell, O. L., Hesseltine, C. W., Goulden, M. L., and Vandegraft, E. E. Survey of corn for aflatoxin, zearalenone, and ochratoxin. *Cereal Chem., 47*:700, 1970.
12. Fennell, D. I., Bothast, R. J., Lillehoj, E. B., and Peterson, R. E. Bright greenish-yellow fluorescence and associated fungi in white corn naturally contaminated with aflatoxin. *Cereal Chem., 50*:404, 1973.
13. Lillehoj, E. B., Kwolek, W. F., Shannon, G. M., Shotwell, O. L., and Hesseltine, C. W. Aflatoxin occurrence in 1973 field corn. I. A limited survey in the southeastern U.S. *Cereal Chem., 52*:603, 1975.
14. Hesseltine, C. W., Shotwell, O. L., Kwolek, W. F., Lillehoj, E. B., Jackson, W. K. and Bothast, R. J. Aflatoxin occurrence in 1973 corn at harvest. II. Mycological studies. *Mycologia, 68*:341, 1976.
15. Shotwell, L. L., Hesseltine, C. W., Vandegraft, E. E., and Goulden, M. L. Survey of corn from different regions for aflatoxin, ochratoxin, and zearalenone. *Cereal Sci. Today, 16*:266, 1971.
16. Shotwell, O. L., Hesseltine, C. W., and Goulden, M. L. Incidence of aflatoxin in southern corn, 1969–1970. *Cereal Sci. Today, 18*:192, 1973.
17. Lillehoj, E. B., Kwolek, W. F., Vandegraft, E. E., Zuber, M. S., Calvert, O. H., Widstrom, N. W., Futrell, M. C., and Bockholt, A. J. Aflatoxin production in *Aspergillus flavus* inoculated ears of corn grown at diverse locations. *Crop Sci., 15*:267, 1975.
18. Anderson, H. W., Nehring, E. W., and Wichser, W. R. Aflatoxin contamination of corn in the field. *J. Agric. Food Chem., 23*:775, 1975.
19. Lillehoj, E. B., and Zuber, M. S. Aflatoxin problem in corn and possible solutions. *Proceedings*, 30th Annual Corn and Sorghum Research Conference, Chicago, Ill., 1975, pp. 230–250.

20. Zuber, M. S., and Lillehoj, E. B. Status of the aflatoxin problem in corn. *J. Environ. Qual.*, *8*:1, 1979.
21. Widstrom, N. W., McMillian, W. W., and Wilson, D. M. Contamination of preharvest corn by aflatoxin. *Proceedings*, 39th Annual Corn and Sorghum Research Conference, Chicago, Ill., 1984, pp. 68–83.
22. Lillehoj, E. B. The aflatoxin in maize problem: the historical perspective. In *Proceedings of the Workshop: Aflatoxin in Maize*, CIMMYT, Mexico, D.F., 1987, pp. 13–32.
23. Wilson, D. M., Widstrom, N. W., McMillian, W. W., and Beaver, R. W. Aflatoxins in corn. *Proceedings*, 44th Annual Corn and Sorghum Research Conference, Chicago, Ill., 1989, pp. 1–26.
24. Shotwell, O. L., Goulden, M. L., Lillehoj, E. B., Kwolek, W. F., and Hesseltine, C. W. Aflatoxin occurrence in 1973 corn at harvest. III. Aflatoxin distribution in contaminated, insect-damaged corn. *Cereal Chem.*, *54*:620, 1977.
25. Tuite, J., Sensmeier, R., Koh-Knox, C., and Noel, R. Preharvest aflatoxin contamination of dent corn in Indiana in 1983. *Plant Dis.*, *68*:893, 1984.
26. McMillian, W. W., Wilson, D. M., and Widstrom, N. W. Insect damage, *Aspergillus flavus* ear mold, and aflatoxin contamination in south Georgia corn fields in 1977. *J. Environ. Qual.*, *7*:564, 1978.
27. McMillian, W. W., Wilson, D. M., Widstrom, N. W., and Gueldner, R. C. Incidence and level of aflatoxin in preharvest corn in south Georgia in 1978. *Cereal Chem.*, *57*:83, 1980.
28. Wilson, D. M., McMillian, W. W., and Widstrom, N. W. Field aflatoxin contamination of corn in south Georgia. *J. Am. Oil Chem. Soc.*, *56*:798, 1979.
29. McMillian, W. W., Wilson, D. M., and Widstrom, N. W. Aflatoxin contamination of preharvest corn in Georgia: a six-year study of insect damage and visible *Aspergillus flavus*. *J. Environ. Qual.*, *14*:200, 1985.
30. Shaw, R. H. Climatic requirement. In *Corn and Corn Improvement* (G. F. Sprague, Ed.), Am. Soc. Agron., Madison, Wis., 1977, p. 591.
31. Aldrich, S. R., Scott, W. O., and Leng, E. R. *Modern Corn Production*, A & L Publications, Champaign, Ill., 1975.
32. Gurley, W. H. Growing Corn in Georgia. Georgia Coop. Ext. Serv. Bull. 547, 1965.
33. Olson, R. A., and Lucas, R. E. Fertility requirements: secondary and micronutrients. In *Advances in Corn Production: Principles and Practices* (W. H. Pierre, S. A. Aldrich, and W. P. Martin, Eds.). Iowa State Univ. Press, Ames, Iowa, 1966, p. 285.
34. Inglett, G. E. Kernel structure, composition, and quality. In *Corn: Culture, Processing, Products* (G. E. Inglett, Ed.), Avi, Westport, Conn., 1970, p. 123.
35. Alexander, D. E., and Creech, R. G. Breeding special industrial and nutritional types. In *Corn and Corn Improvement* (G. F. Sprague, Ed.), Am. Soc. Agron., Madison, Wis., 1977, p. 363.
36. Hanway, J. J. *How a Corn Plant Develops*, Iowa State Univ. Coop. Ext. Serv. Spec. Rep. No. 48, 1966.

37. Widstrom, N. W., McMillian, W. W., and Wiseman, B. R. Ovipositional preference of the corn earworm and the development of trichromes on two exotic corn selections. *Environ. Entomol.*, *8*:833, 1979.
38. Waiss, A. C., Jr., Chan, B. G., Elliger, C. A., Wiseman, B. R., McMillian, W. W., Widstrom, N. W., Zuber, M. S., and Keaster, A. J. Maysin, a flavone glycoside from corn silks with antibiotic activity toward corn earworm. *J. Econ. Entomol.*, *72*:256, 1979.
39. Barry, D., Lillehoj, E. B., Widstrom, N. W., McMillian, W. W., Zuber, M. S., Kwolek, W. F., and Guthrie, W. D. Effect of husk tightness and insect (Lepidoptera) infestation on aflatoxin contamination of preharvest maize. *Environ. Entomol.*, *15*:1116, 1986.
40. McMillian, W. W., Widstrom, N. W., and Wilson, D. M. Impact of husk type and species of infesting insects on aflatoxin contamination in preharvest corn at Tifton, Georgia. *J. Entomol. Sci.*, *22*:307, 1987.
41. Lillehoj, E. B., Zuber, M. S., Darrah, L. L., Kwolek, W. F., Findley, W. R., Horner, E. S., Scott, G. E., Manwiller, A., Sauer, D. B., Thompson, D., Warren, H., West, D. R., and Widstrom, N. W. Aflatoxin occurrence and levels in preharvest corn kernels with varied endosperm characteristics grown at diverse locations. *Crop Sci.*, *23*:1181, 1983.
42. Widstrom, N. W., McMillian, W. W., Wilson, D. M., Garwood, D. L., and Glover, D. V. Growth characteristics of *Aspergillus flavus* on agar infused with maize kernel homogenates and aflatoxin contamination of whole kernel samples. *Phytopathology*, *74*:887, 1984.
43. Fennell, D. I., Kwolek, W. F., Lillehoj, E. B., Adams, G. L., Bothast, R. J., Zuber, M. S., Calvert, O. H., Guthrie, W. D., Bockholt, A. M., Manwiller, A., and Jellum, M. D. *Aspergillus flavus* presence in silks and insects from developing and mature corn ears. *Cereal Chem.*, *54*:770, 1977.
44. Lillehoj, E. B., McMillian, W. W., Guthrie, W. D., and Barry, D. Aflatoxin-producing fungi in preharvest corn: inoculum source in insects and soil. *J. Environ. Qual.*, *9*:691, 1980.
45. Davis, N. D., and Diener, U. L. Some characteristics of toxigenic and nontoxigenic isolates of *Aspergillus flavus* and *Aspergillus parasiticus*. In *Aflatoxin and Aspergillus flavus in Corn* (U. L. Diener, R. L. Asquith, and J. W. Dickens, Eds.), Southern Coop. Ser. Bull. 279, Alabama Agric. Experimental Station, Auburn, Ala., 1983, p. 1.
46. Koehler, P. E., Hanlin, R. T., and Beraha, L. Production of aflatoxins B_1 and G_1 by *Aspergillus flavus* and *Aspergillus parasiticus* isolated from market pecans. *Appl. Microbiol.*, *30*:581, 1975.
47. Schindler, A. F. Temperature limits for production of aflatoxin by twenty-five isolates of *Aspergillus flavus* and *Aspergillus parasiticus*. *J. Food Prot.*, *40*:39, 1977.
48. Sauer, D. B., and Tuite, J. Conditions that affect growth of *Aspergillus flavus* and production of aflatoxin in stored maize. In *Proceedings of the Workshop: Aflatoxin in Maize*, CIMMYT, Mexico, D.F., 1987, pp. 41–50.

49. Diener, U. L., and Davis, N. D. Limiting temperature and relative humidity for aflatoxin production by *Aspergillus flavus* in stored peanuts. *J. Am. Oil Chem. Soc.*, *47*:347, 1970.
50. Sorenson, W. G., Hesseltine, C. W., and Shotwell, O. L. Effect of temperature on production of aflatoxin on rice by *Aspergillus flavus*. *Mycopathol. Mycol. Appl.*, *33*:49, 1967.
51. Diener, U. L., and Davis, N. D. Biology of *Aspergillus flavus* and *A. parasiticus*. In *Proceedings of the Workshop: Aflatoxin in Maize*, CIMMYT, Mexico, D.F., 1987, pp. 33–40.
52. Boller, R. A., and Schroeder, H. W. Influence of temperature on production of aflatoxin in rice by *Aspergillus parasiticus*. *Phytopathology*, *64*:283, 1974.
53. Northholt, M. D., Verhulsdonk, C. A. H., Soentoro, P. S. S., and Paulsch, W. E. Effect of water activity and temperature on aflatoxin production by *Aspergillus parasiticus*. *J. Milk Food Technol.*, *39*:170, 1976.
54. Lillehoj, E. B. Effect of environmental and cultural factors on aflatoxin contamination of developing corn kernels. In *Aflatoxin and Aspergillus flavus in Corn* (U. L. Diener, R. L. Asquith, and J. W. Dickens, Eds.), Southern Coop. Ser. Bull. 279, Alabama Agric. Experimental Station, Auburn, Ala., 1983, p. 27.
55. Trucksess, M. W., Stoloff, L., and Mislivec, P. B. Effect of temperature, water activity and other toxigenic mold species on growth of *Aspergillus flavus* and aflatoxin production on corn, pinto beans, and soybeans. *J. Food Prot.*, *51*:361, 1988.
56. Zuber, M. S., and Lillehoj, E. B. Aflatoxin contamination in maize and its biocontrol. In *Biocontrol of Plant Diseases*, Vol. 2 (K. G. Mukerji and K. L. Garg, Eds.), CRC Press, Boca Raton, Fla., 1987, p. 85.
57. Wieman, D. M., White, G. M., Taraba, J. L., Ross, I. J., Hicks, C. L., and Langlois, B. E. Production of aflatoxin on damaged corn under controlled environmental conditions. *Trans. Am. Soc. Agric. Eng.*, *29*:1150, 1986.
58. Wicklow, D. T. Taxonomic features and ecological significance of sclerotia. In *Aflatoxin and Aspergillus flavus in Corn* (U. L. Diener, R. L. Asquith, and J. W. Dickens, Eds.), Southern Coop. Ser. Bull. 279, Alabama Agric. Eexperimental Station, Auburn, Ala., 1983, p. 6.
59. Ehrlich, K., Ciegler, A., Klich, M., and Lee, L. Fungal competition and mycotoxin production on corn. *Experientia*, *41*:691, 1985.
60. Tsuruta, O., Sugimoto, T., and Minamisawa, M. Mycotoxin production on contaminated maize by several fungal species. *Rep. Natl. Food Res. Inst. (Tokyo)*, *32*:21, 1977.
61. Calvert, O. H., Lillehoj, E. B., Kwolek, W. F., and Zuber, M. S. Aflatoxin B_1 and G_1 production in developing *Zea mays* kernels from mixed inocula of *Aspergillus flavus* and *A. parasiticus*. *Phytopathology*, *61*:501, 1978.
62. Wilson, D. M., and Jay, E. Influence of modified atmosphere storage on aflatoxin production in high moisture corn. *Appl. Microbiol.*, *29*:224, 1975.

63. Wicklow, D. T., Horn, B. W., Burg, W. R., and Cole, R. J. Sclerotium dispersal of *Aspergillus flavus* and *Eupenicillium achrosalmoneum* from maize during harvest. *Trans. Br. Mycol. Soc., 83*:299, 1984.
64. Gupta, S. K., and Venkitasubramanian, T. A. Effect of zinc and phytic acid on incorporation of 1-^{14}C-acetate into aflatoxin by resting mycelia of *Aspergillus flavus*. *Z. Lebensm. Unters. Forsch., 159*:107, 1975.
65. Nagarajan, V., and Bhat, R. V. Factor responsible for varietal differences in aflatoxin production in maize. *J. Agric. Food Chem., 20*: 911, 1972.
66. Holmquist, G. U., Walker, H. W., and Stahr, H. M. Influence of temperature, pH, water activity and antifungal agents on growth of *Aspergillus flavus* and *A. parasiticus*. *J. Food Sci., 48*:778, 1983.
67. Bracanto, F. P., and Golding, N. W. The diameter of the mold colony as a reliable measure of growth. *Mycologia, 45*:848, 1953.
68. Petty, H. B., and Apple, J. W. Insects. In *Advances in Corn Production: Principles and Practices* (W. H. Pierre, S. A. Aldrich, and W. P. Martin, Eds.), Iowa State Univ. Press, Ames, Ia., 1966, p. 351.
69. Dicke, F. A. The most important corn insects. In *Corn and Corn Improvement* (G. F. Sprague, Ed.), Am. Soc. Agron., Madison, Wis., 1977, p. 50.
70. Barry, D. Insects of maize and their association with aflatoxin contamination. In *Proceedings of the Workshop: Aflatoxin in Maize*, CIMMYT, Mexico, D.F., 1987, pp. 201–211.
71. McMillian, W. W. Relation of insects to aflatoxin contamination in maize grown in the southeastern USA. In *Proceedings of the Workshop: Aflatoxin in Maize*, CIMMYT, Mexico, D.F., 1987, pp. 194–200.
72. Fennell, D. I., Lillehoj, E. B., and Kwolek, W. F. *Aspergillus flavus* and other fungi associated with insect-damaged field corn. *Cereal Chem., 52*:314, 1975.
73. Perkins, W. D., Jones, R. L., Sparks, A. N., Wiseman, B. R., Snow, J. W., and McMillian, W. W. Artificial diets for mass rearing the corn earworm (*Heliothis zea*). USDA Prod. Res. Rep. No. 154, 1976, pp. 1–7.
74. Guerra, A. A., and Ouye, M. T. Hatch, larval development, and longevity of four lepidopterous species after thermal treatment of eggs. *J. Econ. Entomol., 61*:14, 1968.
75. Hamm, J. J., Burton, R. L., Young, J. R., and Daniel, R. T. Elimination of *Nosema heliothidis* from a laboratory colony of the corn earworm. *Ann. Entomol. Soc. Am., 64*:624, 1971.
76. Harrell, E. A., Perkins, W. D., and Mullinix, B. G. Effects of temperature, relative humidity, and air velocities on development of *Heliothis zea*. *Ann. Entomol. Soc. Am., 72*:222, 1979.
77. Jones, R. L., Perkins, W. D., and Sparks, A. N. Corn earworm: effects of temperature variation on a colony. *Ann. Entomol. Soc. Am. 71*:393, 1978.
78. Barfield, C. S., Mitchell, E. R., and Poe, S. L. A temperature dependent model for fall armyworm development. *Ann. Entomol. Soc. Am., 71*:70, 1978.

79. Carpenter, J. E., Sparks, A. N., and Harrell, E. A. Effect of temperature on wingbeat frequency and sustained flight of certain lepidopteran insects. *J. Georgia Entomol. Soc.*, *16*:451, 1981.
80. Goodenough, J. L., and Parnell, C. B. Basic engineering design requirements for ventilation, heating, cooling, and humidification of insect rearing facilities. In *Handbook of Insect Rearing*, Vol. I (P. Singh and R. F. Moore, Eds.), Elsevier, New York, 1985, p. 137.
81. Raun, E. S. European corn borer. In *Insect Colonization and Mass Production* (C. N. Smith, Ed.), Academic, New York, 1966, p. 323.
82. Guthrie, W. D., Robbins, J. C., and Jarvis, J. L. *Ostrinia nubilalis*. In *Handbook of Insect Rearing*, Vol. II (P. Singh and R. F. Moore, Eds.), Elsevier, New York, 1985, p. 407.
83. Harein, P. K., and Soderstrom, E. L. Coleoptera infesting stored products. In *Insect Colonization and Mass Production* (C. N. Smith, Ed.), Academic, New York, 1966, p. 241.
84. Widstrom, N. W., Redlinger, L. M., and Wiser, W. J. Appraisal of methods for measuring corn kernel resistance to *Sitophilus zeamais*. *J. Econ. Entomol.*, *65*:31, 1972.
85. Widstrom, N. W., Redlinger, L. M., and Wiser, W. J. Dent corn inbred sources of resistance to the maize weevil (Coleoptera: Curculionidae). *J. Econ. Entomol.*, *76*:31, 1983.
86. Kirkpatrick, R. L. *Sitophilus oryzae*: survival and reproduction at 30–35°C. *J. Georgia Entomol. Soc.*, *13*:105, 1978.
87. Hwang, J. S., Hsieh, F. K., and Kung, K. S. Influences of temperature and relative humidity on the development and reproduction of the maize weevil *Sitophilus zeamais* Motschulsky. *Plant Prot. Bull. Taiwan, ROC*, *25*:41, 1983.
88. Shotwell, O. L. Aflatoxin in corn. *J. Am. Oil Chem. Soc.*, *54*:216A, 1977.
89. Shurtleff, M. C., Ed. *Compendium of Corn Diseases*, 2nd ed., The American Phytopathological Society, St. Paul, Minn., 1980.
90. Zuber, M. S., Calvert, O. H., Lillehoj, E. B., and Kwolek, W. F. Preharvest development of aflatoxin B_1 in corn in the United States, 1972–74. *Phytopathology*, *66*:1120, 1976.
91. Lillehoj, E. B. Toxin-producing fungi in corn. In *Proceedings*, 28th Annual Meeting Am. Assoc. Feed Microscopists, Chicago, Ill., 1980, pp. 29–31.
92. Hesseltine, C. W., Rogers, R. F., and Shotwell, O. L. Aflatoxin and mold flora in North Carolina in 1977 corn crop. *Mycologia*, *73*:216, 1981.
93. Bilgrami, K. S., Prasad, T., Misra, R. S., and Sinha, K. K. Aflatoxin contamination in maize under field conditions. *Indian Phytopathol.*, *34*:67, 1981.
94. Jones, R. K., Duncan, H. E., Payne, G. A., and Leonard, K. J. Factors influencing infection by silk-inoculated corn. *Plant Dis.*, *64*:859, 1980.
95. Thompson, D. L., Lillehoj, E. B., Leonard, K. J., Kwolek, W. F., and Zuber, M. S. Aflatoxin concentration in corn as influenced by kernel development stage and postinoculation temperature in controlled environments. *Crop Sci.*, *20*:609, 1980.

96. Marsh, S. F., and Payne, G. A. Preharvest infection of corn silks and kernels by *Aspergillus flavus*. *Phytopathology, 74*:1284, 1984.
97. Ilag, L. L., Madamba, L. S. P., Celino, L. P., and Malapitan, A. H. Comparative aflatoxin contents of pre-harvest and post-harvest corn. *Phil. J. Biol., 3*:83, 1976.
98. Thompson, D. L., Payne, G. A., Lillehoj, E. B., and Zuber, M. S. Early appearance of aflatoxin in developing corn kernels after inoculation with *Aspergillus flavus*. *Plant Dis., 67*:1321, 1983.
99. Payne, G. A., Hagler, W. M., Jr., and Atkins, C. R. Aflatoxin accumulation in inoculated ears of field-grown maize. *Plant Dis., 72*:422, 1988.
100. Widstrom, N. W., Wilson, D. M., and McMillian, W. W. Aflatoxin contamination of preharvest corn as influenced by timing and method of inoculation. *Appl. Environ. Microbiol., 42*:249, 1981.
101. Lee, L. S., Lillehoj, E. B., and Kwolek, W. F. Aflatoxin distribution in individual corn kernels from intact ears. *Cereal Chem., 57*:340, 1980.
102. Widstrom, N. W., Wilson, D. M., and McMillian, W. W. Evaluation of sampling methods for detecting aflatoxin contamination in small test plots of maize inoculated with *Aspergillus flavus*. *J. Environ. Qual., 11*:655, 1982.
103. Lillehoj, E. B., Kwolek, W. F., Zuber, M. S., Bockholt, A. J., Calvert, O. H., Findley, W. R., Guthrie, W. D., Horner, E. S., Josephson, L. M., King, S., Manwiller, A., Sauer, D. B., Thompson, D., Turner, M., and Widstrom, N. W. Aflatoxin in corn before harvest: interaction of hybrids and locations. *Crop Sci., 20*:731, 1980.
104. Lillehoj, E. B., Kwolek, W. F., Zuber, M. S., Calvert, O. H., Horner, E. S., Widstrom, N. W., Guthrie, W. D., Scott, G. E., Thompson, D. L., Findley, W. R., and Bockholt, A. J. Aflatoxin contamination of field corn: evaluation of regional test plots for early detection. *Cereal Chem., 55*:1007, 1978.
105. Lillehoj, E. B., Kwolek, W. F., Zuber, M. S., Horner, E. S., Widstrom, N. W., Guthrie, W. D., Turner, M., Sauer, D. B., Findley, W. R., Manwiller, A., and Josephson, L. M. Aflatoxin contamination caused by natural fungal infection of preharvest corn: variation between hybrids developed for specific regions but grown at diverse locations. *Plant Soil., 54*:469, 1980.
106. Sopher, C. D., McCracken, R. J., and Mason, D. D. Relationships between drouth and corn yields on selected South Atlantic Coastal Plain soils. *Agron. J., 65*:351, 1973.
107. Widstrom, N. W., McMillian, W. W., Beaver, R. W., and Wilson, D. M. Weather-associated changes in aflatoxin contamination of preharvest maize. *J. Prod. Agric., 3*:196, 1990.
108. Jones, R. K., Duncan, H. E., and Hamilton, P. B. Planting date, harvest date, and irrigation effects on infection and aflatoxin production by *Aspergillus flavus* in field corn. *Phytopathology, 71*:810, 1981.
109. Payne, G. A., Thompson, D. L., and Lillehoj, E. B. Effect of water stress on infection and aflatoxin production by *Aspergillus flavus* in corn. *Phytopathology, 72*:972, 1982.

110. Fortnum, B. A., and Manwiller, A. Effects of irrigation and kernel injury on aflatoxin B_1 production in selected maize hybrids. *Plant Dis.*, *69*:262, 1985.
111. Payne, G. A., Cassel, D. K., and Adkins, C. R. Reduction of aflatoxin contamination in corn by irrigation and tillage. *Phytopathology*, *76*:679, 1986.
112. Jones, R. K., and Duncan, H. E. Effect of nitrogen fertilizer, planting date, and harvest date on aflatoxin production in corn inoculated with *Aspergillus flavus*. *Plant Dis.*, *65*:741, 1981.
113. Wilson, D. M., Walker, M. E., and Gascho, G. J. Some effects of mineral nutrition on aflatoxin contamination of corn and peanuts. In *Soilborne Plant Pathogens: Management of Diseases with Macro- and Microelements* (A. W. Englehard, Ed.), APS Press, St. Paul, Minn., 1981, p. 137.
114. Payne, G. A., Thompson, D. L., and Lillehoj, E. B. Effect of silk age and incubation temperature on kernel infection by *Aspergillus flavus* in silk inoculated corn. *Phytopathology*, *71*:898, 1981.
115. Payne, G. A., Thompson, D. L., Lillehoj, E. B., Zuber, M. S., and Adkins, C. R. Effect of temperature on the preharvest infection of maize kernels by *Aspergillus flavus*. *Phytopathology*, *78*:1376, 1988.
116. Doupnik, B., Jr. Maize seed predisposed to fungal invasion and aflatoxin contamination by *Helminthosporium maydis* ear rot. *Phytopathology*, *62*:1367, 1972.
117. Wicklow, D. T., Hesseltine, C. W., Shotwell, O. L., and Adams, G. L. Interference competition and aflatoxin levels in corn. *Phytopathology*, *70*:761, 1980.
118. Wicklow, D. T. Patterns of fungal association within maize kernels harvested in North Carolina. *Plant Dis.*, *72*:113, 1988.
119. Wicklow, D. T., Horn, B. W., Shotwell, O. L., Hesseltine, C. W., and Caldwell, R. W. Fungal interference with *Aspergillus flavus* infection and aflatoxin contamination of maize grown in a controlled environment. *Phytopathology*, *78*:68, 1988.
120. Horn, B. W., and Wicklow, D. T. Factors influencing the inhibition of aflatoxin production in corn by *Aspergillus niger*. *Can. J. Microbiol.*, *29*:1087, 1983.
121. Wicklow, D. T., Horn, B. W., and Shotwell, O. L. Aflatoxin formation in preharvest maize ears inoculated with *Apsergillus flavus* and *Aspergillus niger*. *Mycologia*, *79*:679, 1987.
122. LaPrade, J. C., and Manwiller, A. Aflatoxin production and fungal growth on single cross corn hybrids inoculated with *Aspergillus flavus*. *Phytopathology*, *66*:675, 1976.
123. Manwiller, A., and Fortnum, B. *A Comparison of Aflatoxin Levels in Preharvest Corn from the South Carolina Coastal Plain for 1977–78*, Agronomy and Soils Res. Ser. No. 101, South Carolina Agric. Exp. Stn., Clemson, S.C., 1979.
124. Lillehoj, E. B., and Zuber, M. S. Variability in corn hybrid resistance to preharvest aflatoxin contamination. *J. Am. Oil Chem. Soc.*, *64*:970A, 1981.

125. Widstrom, N. W., Wilson, D. M., and McMillian, W. W. Ear resistance of maize inbreds to field aflatoxin contamination. *Crop Sci.*, *24*:1155, 1984.
126. Scott, G. E., and Zummo, N. Sources of resistance in maize to kernel infection by *Aspergillus flavus* in the field. *Crop Sci.*, *28*:504, 1988.
127. Stoloff, L., and Lillehoj, E. B. Effect of genotype (open-pollinated vs hybrid) and environment on preharvest aflatoxin contamination of maize grown in southeastern United States. *J. Am. Oil Chem. Soc.*, *58*:976A, 1981.
128. Zuber, M. S., Darrah, L. L., Lillehoj, E. B., Josephson, L. M., Manwiller, A., Scott, G. E., Gudauskas, R. T., Horner, E. S., Widstrom, N. W., Thompson, D. L., Bockholt, A. J., and Brewbaker, J. L. Comparison of open-pollinated maize varieties and hybrids for preharvest aflatoxin contamination in the southern United States. *Plant Dis.*, *67*:185, 1983.
129. Darrah, L. L., Lillehoj, E. B., Zuber, M. S., Scott, G. E., Thompson, D., West, D. R., Widstrom, N. W., and Fortnum, B. A. Inheritance of aflatoxin B_1 levels in maize kernels under modified natural inoculation with *Aspergillus flavus*. *Crop Sci.*, *27*:869, 1987.
130. Widstrom, N. W., Wilson, D. M., and McMillian, W. W. Differentiation of maize genotypes for aflatoxin concentration in developing kernels. *Crop Sci.*, *26*:935, 1986.
131. Wilson, D. M., Gueldner, R. C., and Hill, R. A. Effects of plant metabolites on the *Aspergillus flavus* group and aflatoxin production. In *Aflatoxin and Aspergillus flavus in Corn* (U. L. Diener, R. L. Asquith, and J. W. Dickens, Eds.), Southern Coop. Ser. Bull. 279, Alabama Agric. Experimental Station, Auburn, Ala., 1983, p. 13.
132. Riesselman, J., and Doupnik, B., Jr. Field occurrence of aflatoxins in strawberry popcorn. *Proc. Am. Phytopathol. Soc.*, *2*:38, 1975.
133. McMillian, W. W., Widstrom, N. W., and Wilson, D. M. Aflatoxin production on various popcorn genotypes. *Agron. J.*, *74*:156, 1982.
134. King, S. B., and Scott, G. E. Field inoculation techniques to evaluate maize for reaction to kernel infection by *Aspergillus flavus*. *Phytopathology*, *72*:782, 1982.
135. Shotwell, O. L., Goulden, M. L., Hesseltine, C. W., Dickens, J. W., and Kwolek, W. F. Aflatoxin: distribution in contaminated corn plants. *Cereal Chem.*, *57*:206, 1980.
136. Mertz, D., Lee, D., Zuber, M., and Lillehoj, E. Uptake and metabolism of aflatoxin by *Zea mays*. *J. Agric. Food Chem.*, *28*:963, 1980.
137. Wicklow, D. T., Horn, B. W., and Cole, R. J. Sclerotium production by *Aspergillus flavus* on corn kernels. *Mycologia*, *74*:398, 1982.
138. Wicklow, D. T., and Horn, B. W. *Aspergillus flavus* sclerotia form in wound-inoculated preharvest corn. *Mycologia*, *76*:503, 1984.
139. Garman, H., and Jewett, H. H. The life history and habits of the corn earworm (*Chloridea obsoleta*). *Kentucky Agric. Exp. Stn. Bull.*, *187*:513, 1914.

140. Bishopp, F. C. The boll worm or corn earworm. *USDA Farmers Bull.*, *872*:3, 1917.
141. Widstrom, N. W. The role of insects and other plant pests in aflatoxin contamination of corn, cotton, and peanuts — a review. *J. Environ. Qual.*, *8*:5, 1979.
142. Lillehoj, E. B., Fennell, D. I., and Kwolek, W. F. *Aspergillus flavus* and aflatoxin in Iowa corn before harvest. *Science*, *193*: 495, 1976.
143. Lillehoj, E. B., Kwolek, W. F., Peterson, R. E., Shotwell, O. L., and Hesseltine, C. W. Aflatoxin contamination, fluorescence, and insect damage in corn infected with *Aspergillus flavus* before harvest. *Cereal Chem.*, *53*:505, 1976.
144. Lillehoj, E. B., Kwolek, W. F., Manwiller, A., DuRant, J. A., LaPrade, J. C., Horner, E. S., Reid, J., and Zuber, M. S. Aflatoxin production in several corn hybrids grown in South Carolina and Florida. *Crop Sci.*, *16*:483, 1976.
145. McMillian, W. W., Wilson, D. M., Widstrom, N. W., and Perkins, W. D. Effects of aflatoxin B_1 and G_1 on three insect pests of maize. *J. Econ. Entomol.*, *73*:26, 1980.
146. Wicklow, D. T., and Dowd, P. F. Entomotoxigenic potential of wild and domesticated yellow-green aspergilli: toxicity to corn earworm and fall armyworm larvae. *Mycologia*, *81*:561, 1989.
147. McMillian, W. W., Widstrom, N. W., and Wilson, D. M. Rearing the maize weevil on maize genotypes when aflatoxin-producing *Aspergillus flavus* and *A. parasiticus* isolates were present. *Environ. Entomol.*, *10*:760, 1981.
148. McMillian, W. W., Widstrom, N. W., Wilson, D. M., and Evans, B. D. Annual contamination of *Heliothis zea* (Lepidoptera: Noctuidae) moths with *Aspergillus flavus* and incidence of aflatoxin contamination in preharvest corn in the Georgia coastal plain. *J. Entomol. Sci.*, *25*: 123, 1990.
149. Lillehoj, E. B., Fennell, D. I., Kwolek, W. F., Adams, G. L., Zuber, M. S., Horner, E. S., Widstrom, N. W., Warren, H., Guthrie, W. D., Sauer, D. B., Findley, W. R., Manwiller, A., Josephson, L. M., and Bockholt, A. J. Aflatoxin contamination of corn before harvest: *Aspergillus flavus* association with insects collected from developing ears. *Crop Sci.*, *18*:921, 1978.
150. Fennell, D. I., Lillehoj, E. B., Kwolek, W. F., Guthrie, W. D., Sheeley, R., Sparks, A. N., Widstrom, N. W., and Adams, G. L. Insect larval activity on developing corn ears and subsequent aflatoxin contamination of seed. *J. Econ. Entomol.*, *71*:624, 1978.
151. Widstrom, N. W., Sparks, A. N., Lillehoj, E. B., and Kwolek, W. F. Aflatoxin production and lepidopteran insect injury on corn in Georgia. *J. Econ. Entomol.*, *68*:855, 1975.
152. LaPrade, J. C., and Manwiller, A. Relation of insect damage, vector, and hybrid reaction to aflatoxin B_1 recovery from field corn. *Phytopathology*, *67*:544, 1977.
153. McMillian, W. W., Widstrom, N. W., Wilson, D. M., and Hill, R. A. Transmission by maize weevils of *Aspergillus flavus* and its survival on selected corn hybrids. *J. Econ. Entomol.*, *73*:793, 1980.

154. Barry, D., Zuber, M. S., Lillehoj, E. B., McMillian, W. W., Adams, N. J., Kwolek, W. F., and Widstrom, N. W. Evaluation of two arthropod vectors as inoculators of developing maize ears with *Aspergillus flavus*. *Environ. Entomol.*, *14*:634, 1985.
155. Widstrom, N. W., and Starks, K. J. Influence of environment on injury to corn by the corn earworm. *J. Econ. Entomol.*, *60*:181, 1967.
156. Widstrom, N. W. Corn earworm injury to maize as affected by plant density. *Agron. J.*, *61*:464, 1969.
157. Widstrom, N. W., McMillian, W. W., and Wiseman, B. R. Resistance in corn to the corn earworm and the fall armyworm. IV. Earworm injury to corn inbreds related to climatic conditions and plant characteristics. *J. Econ. Entomol.*, *63*:803, 1970.
158. Widstrom, N. W., and Wiseman, B. R. Locating major genes for resistance to the corn earworm in maize inbreds. *J. Hered.*, *64*:83, 1973.
159. Widstrom, N. W., and Davis, J. B. Analysis of two diallel sets of sweet corn inbreds for corn earworm injury. *Crop Sci.*, *7*:50, 1967.
160. Widstrom, N. W. Reciprocal differences and combining ability for corn earworm injury among maize single crosses. *Crop Sci.*, *12*: 245, 1972.
161. Widstrom, N. W., Wiser, W. J., and Bauman, L. F. Recurrent selection in maize for earworm resistance. *Crop Sci.*, *10*:674, 1970.
162. Widstrom, N. W., Wiseman, B. R., and McMillian, W. W. Responses to index selection in maize for resistance to ear damage by the corn earworm. *Crop Sci.*, *22*:843, 1982.
163. Widstrom, N. W., Wiseman, B. R., and McMillian, W. W. Responses of corn earworm larvae to maize silks. *Agron. J.*, *69*:815, 1977.
164. Widstrom, N. W., Wiseman, B. R., McMillian, W. W., Elliger, C. A., and Waiss, A. C., Jr. Genetic variability in maize for maysin content. *Crop Sci.*, *23*:120, 1983.
165. Morrill, W. L., and Greene, G. L. Distribution of fall armyworm larvae. I. Regions of field corn plants infested by larvae. *Environ. Entomol.*, *2*:195, 1973.
166. Guthrie, W. D., Russell, W. A., Neumann, F. L., Reed, G. L., and Grindeland, R. L. Yield losses in maize caused by different levels of infestation of second-brood European corn borers. *Iowa State J. Res.*, *50*:239, 1975.
167. Penny, L. H., and Dicke, F. F. European corn borer damage in resistant and susceptible dent corn hybrids. *Agron. J.*, *51*:323, 1959.
168. Widstrom, N. W., Wiseman, B. R., and McMillian, W. W. Resistance among some maize inbreds and single crosses to fall armyworm injury. *Crop Sci.*, *12*:290, 1972.
169. Williams, W. P., Davis, F. M., and Scott, G. E. Resistance of corn to leaf-feeding damage by the fall armyworm. *Crop Sci.*, *18*:861, 1978.
170. Klun, J. A., and Robinson, J. F. Concentration of two 1,4-benzoxazinones in dent corn at various stages of development of the

plant and its relation to resistance of the host plant to the European corn borer. *J. Econ. Entomol.*, *62*:214, 1969.
171. McMillian, W. W., Widstrom, N. W., and Wiseman, B. R. Pink scavenger caterpillar resistance among selected dent corn hybrids. *J. Georgia Entomol. Soc.*, *17*:93, 1982.
172. McMillian, W. W., Widstrom, N. W., and Starks, K. J. Rice weevil damage as affected by husk treatment within methods of artificially infesting field corn plots. *J. Econ. Entomol.*, *61*:918, 1968.
173. Widstrom, N. W., Hanson, W. D., and Redlinger, L. M. Inheritance of maize weevil resistance in maize. *Crop Sci.*, *15*:467, 1975.
174. Starks, K. J., Cox, H C, McMillian, W. W., and Burton, R. L. Damage to corn by the pink scavenger caterpillar and its relationship to corn earworm and rice weevil damage. *J. Econ. Entomol.*, *59*:931, 1966.
175. McMillian, W. W., Widstrom, N. W., and Wiseman, B. R. Yield losses in south Georgia field corn resulting from damage by several insects. *J. Georgia Entomol. Soc.*, *11*:208, 1976.
176. Wiseman, B. R., McMillian, W. W., and Widstrom, N. W. Husk and kernel resistance among maize hybrids to an insect complex. *J. Econ. Entomol.*, *63*:1260, 1970.
177. Wiseman, B. R., and McMillian, W. W. Competition and survival among the corn earworm, the tobacco budworm, and the fall armyworm. *J. Econ. Entomol.*, *62*:734, 1969.
178. Guthrie, W. D., and Walter, E. V. Corn earworm and European corn borer resistance in sweet corn inbred lines. *J. Econ. Entomol.*, *54*:1248, 1961.
179. Gray, F. A., Faw, W. F., and Boutwell, J. L. The 1977 corn-aflatoxin epiphytotic in Alabama. *Plant Dis.*, *66*:221, 1982.
180. Lillehoj, E. B., and Hesseltine, C. W. Aflatoxin control during plant growth and harvest of corn. In *Mycotoxins in Human and Animal Health* (J. V. Rodricks, C. W. Hesseltine, and M. A. Mehlman, Eds.), Pathotox Pub., Park Forest South, Ill., 1977, p. 107.
181. Zuber, M. S. Influence of plant genetics on toxic production in corn. In *Mycotoxins in Human and Animal Health* (J. V. Rodricks, C. W. Hesseltine, and M. A. Mehlman, Eds.), Pathotox Publ., Park Forest South, Ill, 1977, p. 173.
182. Widstrom, N. W., Lillehoj, E. B., Sparks, A. N., and Kwolek, W. F. Corn earworm damage and aflatoxin B_1 on corn ears protected with insecticide. *J. Econ. Entomol.*, *69*:177, 1976.
183. Iken, J. E. Interrelationships of genotype, insect damage and storage conditions in the development of aflatoxin in maize (*Zea mays* L.). Diss. Abstr. No. 8015262, Kansas State Univ., Manhattan, Kan., 1980.
184. Wilson, D. M., Widstrom, N. W., Marti, L. R., and Evans, B. D. *Aspergillus flavus* group, aflatoxin, and bright greenish yellow fluorescence in insect-damaged corn in Georgia. *Cereal Chem.*, *58*:40, 1981.
185. Lillehoj, E. B., Kwolek, W. F., Horner, E. S., Widstrom, N. W., Josephson, L. M., Franz, A. O., and Catalano, E. A. Aflatoxin

contamination of preharvest corn: role of *Aspergillus flavus* inoculum and insect damage. *Cereal Chem.*, *57*:255, 1980.

186. McMillian, W. W., Widstrom, N. W., and Wilson, D. M. Insect damage and aflatoxin contamination in preharvest corn: influence of genotype and ear wetting. *J. Entomol. Sci.*, *20*:66, 1985.

187. Guthrie, W. D., Lillehoj, E. B., McMillian, W. W., Barry, D., Kwolek, W. F., Franz, A. O., Catalano, E. A., Russell, W. A., and Widstrom, N. W. Effect of hybrids with different levels of susceptibility to second-generation European corn borers on aflatoxin contamination in corn. *J. Agric. Food Chem.*, *29*:1170, 1981.

188. Guthrie, W. D., Lillehoj, E. B., Barry, D., McMillian, W. W., Kwolek, W. P., Franz, A. O., Catalano, E. A., Russell, W. A., and Widstrom, N. W. Aflatoxin contamination of preharvest corn: interaction of European corn borer larvae and *Aspergillus flavus*-group isolates. *J. Econ. Entomol.*, *75*:265, 1982.

189. Lillehoj, E. B., Kwolek, W. F., Guthrie, W. D., Barry, D., McMillian, W. W., and Widstrom, N. W. Aflatoxin accumulation in preharvest maize kernels: interaction of three fungal species, European corn borers, and two hybrids. *Plant Soil*, *65*:95, 1982.

190. McMillian, W. W., Widstrom, N. W., Barry, D., and Lillehoj, E. B. Aflatoxin contamination in selected corn germplasm classified for resistance to European corn borer (Lepidoptera: Noctuidae). *J. Entomol. Sci.*, *23*:240, 1988.

3

AFLATOXINS: THEIR BIOLOGICAL EFFECTS AND ECOLOGICAL SIGNIFICANCE

KRISHNA S. BILGRAMI and KAUSHAL K. SINHA *Bhagalpur University, Bhagalpur, India*

I. INTRODUCTION

Although the existence of mycotoxins had been known for many years, their role in inducing diseases attracted worldwide attention only after the discovery of "turkey-X disease" in England in 1960, which was later attributed to aflatoxins. Because of their key role in realizing the significance of mycotoxins, the aflatoxins have been investigated by scientists devoted to various disciplines.

Aflatoxins constitute a group of acutely toxic and highly carcinogenic secondary metabolites produced mainly by *Aspergillus flavus* Link ex Fries and *A. parasiticus* Speare. These toxins, which have closely similar structures, constitute a unique group of highly oxygenated coumarin derivative heterocyclic compounds. Many individual aflatoxins induce both acute and chronic toxicity in mammals and other animals; these compounds act as carcinogens and teratogens as well as mutagens [1]. In this chapter we consider the biological and ecological effects of aflatoxins in animals, humans, and plants.

II. EFFECT OF AFLATOXINS ON ANIMAL SYSTEMS

The effects of aflatoxin exposure may be acute or chronic depending primarily on the test system, dosage, and frequency of exposure. Sensitivity toward aflatoxins varies with species, age, and sex of animals as well as the composition of diet and route of ingestion [2]. Aflatoxins have been found to be lethal to animals as well as to animal cells in culture when administered on a short-term basis. Long-term exposure for extended periods leads to induction of tumors in several animal species [3].

A. Acute and Chronic Toxicity

Hepatotoxicity

Of the first four aflatoxins to be recognized, aflatoxin B_1 is the most toxic, followed by G_1, B_2, and G_2 in order of decreasing potency [4]. Aflatoxins M_1 and Q_1 are also quite toxic, due to the presence of a 2,3-vinyl ether double bond in the compounds. The sublethal dose (LD_{50}) value of aflatoxin B_1 varies from 0.3 to 0.6 mg/kg body weight in ducklings to 9.0 mg/kg in mice [5]. Patterson and Allcroft [6] divided animal species into two groups: (1) susceptible—calves, chicks, ducklings, guinea pigs, and pigs; and (2) relatively resistant—goats, sheep, rats, and mice. Determination of LD_{50} values of different aflatoxins gives an indication of relative toxicity for a particular organism [7].

Aflatoxins induce a variety of symptoms in affected animals. Acute toxicity is expressed as the death of the animal within a limited time in which the liver generally becomes pale and discolored and increases in size [8]. If death does not occur within a week, there is a characteristic proliferation of undifferentiated cells in the portal spaces [9]. The kidney may show lesions like those of glomerular nephritis whereas the lung may have congestion. However, the changes are quite different in the case of chronic poisoning. Loss of appetite and reduction in growth of the affected animals are the first visible symptoms. The liver also is affected rapidly with congestion and hemorrhagic or necrotic zones caused by the toxin. Proliferation of epithelial cells of the bile ducts is also common in many experimental animals. The most characteristic indication of toxicity generally appears only a few days before death: animals exhibit melancholy, staggering, nervousness, and muscular spasms.

Initial manifestations of aflatoxicoses in turkeys were reported to be nonspecific in nature; birds developed anorexia, which terminated in death. Allcroft and Carnaghan [10] reported staggering gait as the earliest symptom of aflatoxicosis in turkeys. A similar picture was also observed in cattle, in which hair loss, eye twitching, and grinding of teeth were observed [11]. In a study by Carnaghan et al. [12], chickens fed a diet containing aflatoxin B_1 at a level of 1.5 mg/kg showed liver lesions after 4 weeks of treatment. Lesions included fatty change, biliary proliferation, and fibrosis in the affected animals. In cattle, the liver lesions (centrilobular degeneration, fibrosis, biliary proliferation) occurred in all animals after 4 months of aflatoxin feeding at a level of 2 mg/kg [13]. Hepatic lesions consisted mainly of cirrhosis with ascites accompanying visceral edema. Lynch et al. [14] recorded slight hyperplasia of the hepatic cells from the end of the first month of aflatoxin feeding in cattle; this phenomenon increased during the second and third months until centrolobular degeneration of the hepatic cells became apparent. However, during the fourth month central necrosis of the hepatic cells was marked by proliferation of the bile ducts and occlusion of the centrolobular veins.

Krogh et al. [15] recorded the development of centrolobular necrosis and fibrosis of the liver in pigs due to feeding of aflatoxin at a low level (300 μg/kg). Hemorrhagic enteritis and extensive midzonal necrosis in the liver were also recorded in a 35-kg Duroc pig that died after 3 days on a ration containing aflatoxins [16]. Acute aflatoxin poisoning causes hepatic necrosis, derangement of hepatic functions, coagulopathy, and extensive hemorrhagic lesions that frequently lead to the death of farm animals [17]. Subacute or chronic aflatoxicosis causes fatty changes in the liver, enlargement of the gall bladder, and periportal fibrosis with proliferative changes in bile duct epithelium. Liver lesions similar to those observed in farm animals have been experimentally induced by the administration of aflatoxin in a number of laboratory animals such as rats, guinea pigs, rabbits, and cats [18].

Hepatotoxicity has also been observed in rhesus monkeys. Madhavan et al. [19] recorded liver lesions having fatty infiltration, biliary proliferation, and portal fibrosis in two monkeys that were fed aflatoxins until their death on days 32 and 34. In another experiment, young (1.5–2.0-kg) rhesus monkeys were fed either 0.5 or 1.0 mg of aflatoxin per day for 18 days, then 1.0 mg/day thereafter. All test animals developed anorexia and died in 14–28 days after initiation of treatment. The principal histopathological finding included liver lesions of portal inflammation and fatty change with symptoms of biliary cirrhosis [20]. Various types of abnormalities of the liver have also been reported due to aflatoxin poisoning in various animals including horses [21, 22], dogs [23], trout [24], and rats [25].

Besides the liver, other important organs also are affected severely by high doses of aflatoxins. Butler [25] recorded bilateral adrenal hemorrhages and petechial hemorrhages in many organs, particularly in congested lungs. Occasionally, patchy necroses have been observed in the myocardium, kidney, and spleen during the first few days of aflatoxin B_1 feeding in male (7.2 mg/kg body weight) and female (17.9 mg/kg body weight) rats. Hemorrhages were more extensive in the lungs, kidneys, and adrenal glands after exposure to higher doses of aflatoxin. Butler [26] also recorded congested lungs with small petechial hemorrhages, hemorrhagic necrosis in the adrenal glands (localized in the inner zone of the reticularis), and patchy necroses in the kidneys, pancreas, and spleen of guinea pigs after 2–3 days of intraperitoneal injection of aflatoxin B_1 at 1.4 mg/kg body weight. With higher doses of aflatoxin, the hemorrhages were quite distinct. Many other animals exhibited marked ascites and edema of the omentum and subcutaneous tissue during the first week after injection [26].

Carcinogenicity

Aflatoxins are among the most potent hepatocarcinogens known [1]. Lancaster et al. [27] provided initial evidence for the carcinogenic properties of aflatoxin in contaminated, toxic peanut meals in rats. After feeding contaminated peanut meals for 6 months, 9 of 11 rats developed multiple liver tumors and two developed lung metastases. Although initial reports did not relate the concentration of aflatoxin present in the feed, later findings indicated that diets containing aflatoxins in the range of 0.07–4.0 ppm can induce liver tumors in rats [28, 29]. Wogan and Newberne [30] recorded liver tumor incidence in rats; occurrence was linearly correlated with dietary aflatoxin content over the range of 0.7–1.8 ppm. Higher levels of aflatoxin resulted in more than 90% tumor incidence when fed over a period of 370 days, whereas the lowest dose of aflatoxin (0.005 ppm) failed to induce liver tumors within a similar time period (384 days). Newberne and Wogan [31] recorded two types of changes in the liver of rats: (1) focal hyperplasia of basophilic staining parenchymal cells and (2) foci of large pale eosinophilic cells. In both types, initial changes were followed by nodule formation, hepatoma, and ultimately by liver carcinoma. Tumors affecting the hepatic parenchyma and the bile duct epithelium have also been described in ducklings that survived the acute effects of the aflatoxins [32]. Liver carcinoma has also been induced in nonhuman primates such as rhesus monkeys [33].

Comparative studies of aflatoxin severity were made by Platonow [34] and Wogan [35], who observed that mice were comparatively resistant to acute aflatoxin poisoning. However, Newberne [36] reported hepatoma formation among 6 out of 40 mice following ingestion of aflatoxin-contaminated meal for 16 months. Incidence of local tumors in 15 of 17 mice were observed by Dickens and Jones [37] after subcutaneous injection of 10 μg of aflatoxin biweekly for 76 weeks.

Orally administered aflatoxins, mainly B_1, have been hepatocarcinogenic in all species of test animals. A linear dose-response relationship was observed for the development of liver-cell carcinoma in male Fischer rats fed dietary concentrations of aflatoxin B_1 ranging from 1 to 100 μg/kg [38]. Tumor incidence was 10% at a level of 1 μg/kg, whereas 100% incidence was recorded at a dietary level of 100 μg/kg. A linear log (dose) response relationship has also been demonstrated in trout fed a dietary level of aflatoxin B_1 of 0.5–20.0 μg/kg for 20 months [39, 40].

The carcinogenic effects of different aflatoxins have been compared in a number of studies. Butler et al. [41] mixed aflatoxins B_1, B_2, or G_1 in drinking water and administered the liquid to 8–9-week-old male and female rats for 10 or 20 weeks. Renal tumors were recorded in both male and female rats receiving B_1 or G_1. None of the rats exposed to aflatoxin B_2 developed carcinomas. Wogan et al. [42] concluded that aflatoxin B_1 was more carcinogenic than aflatoxin G_1 and aflatoxin B_1 and G_1 together were much more active than aflatoxin B_2 alone. Sinnhuber et al. [43] reported carcinogenicity of aflatoxin M_1, although it was less potent than aflatoxin B_1.

Besides the liver, tumors have also been recorded in other parts of the animal body due to aflatoxin poisoning. Carcinoma of the colon has been reported by several workers [18, 44]. Newberne and Rogers [45] also recorded an increased incidence of tumors in the distal half of the colon in vitamin A-deficient rats fed aflatoxin B_1. Tumors have also been reported due to aflatoxin feeding in lacrimal glands [46, 47], squamous cells of the tongue [44], esophagus [41], and trachea [46].

In an experiment with white rats, Ranjan [48] observed tumors on the ventrolateral side in the lumbosacral region after 30 weeks of aflatoxin feeding; the tumors were attached just below the skin in the dermal adipose layer. The mesentery was also associated with the tumor. Histological studies of the tumors suggested it to be a case of secondary carcinoma.

Teratogenicity

Teratogenicity of aflatoxins is well established [5, 49, 50]. Elis and DiPaolo [51] reported that a single intraperitoneal injection of aflatoxin B_1

at the rate of 4 mg/kg body weight in female hamsters after 8 days of pregnancy resulted in a high proportion of malformed or dead reabsorbed fetuses. Aflatoxin B_1 at a dose of 2 mg/kg has no such effects. Teratogenicity produced in hamsters by aflatoxin B_1 poisoning was also reported by Di Paolo et al. [52]; however, no teratogenic effect was observed in C_3H mice at the same level of exposure.

Teratogenic effects of aflatoxin B_1 have also been observed in chick embryos including retarded growth, exencephalia, anophthalmia, microphthalmia, cleft palate, and malformation of the maxilla [53, 54]. Prenatal and early postnatal exposures of aflatoxin were also found to have teratogenic effects [55].

Mutagenicity

Aflatoxins are mutagenic in a wide variety of microorganisms and in cells in culture. Aflatoxin B_1 is mutagenic in histidine-requiring strains of *Salmonella typhimurium* [56, 57]. Mutagenic properties of aflatoxin B_1 have also been confirmed in tests with lysogenic bacteria [58].

Studies related to mutagenicity have demonstrated aflatoxin-induced chromosome aberrations in rat kangaroo cells [59]. Adzhigitov et al. [60] also recorded induced chromosomal aberrations and genome mutations due to aflatoxin B_1 in rhesus monkeys and baboons. Chromosomal aberrations were represented mainly by single and paired fragments. Comparison of the number of diploid and polyploid cells in the bone marrow of the control and experimental primates showed a toxin-induced increase in the percentage of polyploid cells.

Ueno et al. [61] examined the mutagenicity of aflatoxin B_1 in the presence of various microsomal proteins in *Salmonella typhimurium* (Ames tests). Bilgrami and Sinha [62] recorded chromosomal abnormalities such as clumping, fragmentation, meiotic polyploidy, and euploidy in mice fed with aflatoxin-contaminated diets; these results underscore the mutagenic potential of the toxin. The same authors also recorded dominant lethal mutations in mice. Significant decrease in the fertility of females mated to treated males was observed; this was caused by gross genetic damage (mutation) induced by the aflatoxin in the germ cells of treated males. Sinha [63] suggested a broad outline of methods for assay of mutagenicity of aflatoxins in mammalian and ancillary test systems.

Some attempts have been made to suppress the mutagenic activity of aflatoxin B_1 by modifying dietary regimes. Bhattacharya et al. [64] screened 19 vitamins for their effects on *Salmonella typhimurium* strain TA 100 that had been activated with a rat liver system. Several vitamins exhibited the ability to inhibit at least 50% of the mutagenic activity of aflatoxin B_1, including retinoids, riboflavin, folic acid, menadione, cyanocobalamin, ascorbic acid, and pyridoxine. Although inhibition by vitamins was apparent over a range of aflatoxin B_1 concentrations, their effect was more pronounced at the lower concentrations of toxin.

Effect on the Hematopoietic System

Acute aflatoxicoses are difficult to diagnose because the symptoms of intoxication in animals are almost identical to those that result from consumption of other hepatotoxins. Blood is the first and foremost body component to be tested for the verification of cases such as allergy, lyses, and toxicoses. Some work has also been done on the effect of aflatoxin on the hematological feature of affected animals. Panda et al. [65] studied the effect of aflatoxin B_1 on the blood features of albino rats and guinea pigs; their studies revealed a depression in the total red blood cell (RBC) and white blood cell (WBC) counts, including hemoglobin content of the blood. Toxin-fed animals exhibited anisocytosis and poikilocytosis, and the blood clotting time was prolonged. More or less similar results have been obtained by Ranjan [48] in aflatoxin-fed guinea pigs and white rats.

Decreases in the RBC count have been attributed to (1) retardation/disturbance in the erythropoiesis in bone marrow [65] and (2) a faster rate of destruction of peripheral red blood cells in the spleen [66]. Aflatoxins have also been found to produce clastogenic effects in maturing erythroblasts that were collected from bone marrow of femurs [67]; the results reflected the role of toxin in hampering RBC formation and was the reason for manifestation of an anemialike condition in most of the affected animals.

Clark et al. [68] studied the effect of aflatoxin on blood-clotting activity and platelets in white rabbits. Oral administration of aflatoxin B_1 at a dose of 0.05 mg of toxin per kilogram of body weight from day 0 to day 23 (subchronic study) or a single high dose of 0.4 mg/kg (acute study) produced marked reductions in the plasma activity of several blood coagulation factors and prolongation of PT (prothrombin time) and APTT (activated partial thromboplastin time). Platelet counts were significantly increased in subchronically exposed rabbits, and platelet size was decreased in animals given a single high toxin dose.

B. Biochemical Effects

Since aflatoxin B_1 is the most common and most potent toxin of the aflatoxin group, it has been studied extensively for biochemical effects on various experimental animals. Aflatoxin B_1 may be considered a biosynthetic inhibitor; large doses cause total inhibition, whereas lower doses progressively affect different systems [69]. Clifford and Rees [70] tried to tabulate the successive stages in biological activity of aflatoxin B_1 on the rat liver cell, each step being a consequence of the previous one. The stages included (1) interaction of aflatoxin B_1 with DNA and inhibition of the polymerases responsible for DNA and RNA syntheses, (2) suppression of DNA synthesis, (3) reduction of RNA synthesis and inhibition of messenger RNA, (4) alterations of nucleolar morphology, and (5) reduction in protein biosynthesis.

Effect on DNA Metabolism

It has been demonstrated in several experimental animals that aflatoxins can bind with DNA, causing a change in its characteristic spectrum [8]. Aflatoxin B_1 binds more effectively than other components of the aflatoxin family. Sporn et al. [71] recorded slight changes in the absorption shift maximum—from 363 nm to 368 nm—after binding of aflatoxin B_1 to calf thymus DNA; this shift was accompanied by marked hypochronism at 362 nm.

The criteria were used in demonstrating that aflatoxin B_1 bound to native (double-stranded, helical) calf thymus DNA and to a lesser extent to heat-denatured (single-stranded) calf thymus DNA and native *Escherichia coli* RNA. By the same criteria, aflatoxin B_1 did not bind to bovine serum albumin, calf thymus histone, or enzymatically hydrolyzed calf thymus DNA [35].

Black and Jirgensons [72] also used equilibrium dialysis techniques to determine the interaction of aflatoxin B_1 with calf thymus histones and DNA. They reported that aflatoxin B_1 bound to DNA and also to two highly purified lysine-rich histone preparations; the latter fractions were associated with increased viscosity of histone and DNA solutions. They concluded that aflatoxin binding resulted to gross conformational changes in the histone and DNA molecules.

Frayssinet et al. [73] made detailed studies on the effect of aflatoxin B_1 on DNA metabolism in rat liver. They reported that aflatoxin B_1 inhibited the compensatory hypertrophy following partial hepatectomy. Net synthesis of liver DNA was inhibited when the rats received 30 or 60 μg of toxin each prior to or immediately after partial hepatectomy. Similarly, aflatoxin B_1 at 100 μg/rat was shown to inhibit thymidine incorporation into liver DNA by 95% within 12 h [74]. It was demonstrated, however, that enzymes responsible for DNA synthesis were not affected by the toxin. The results indicated that aflatoxin affected DNA synthesis by impairing the ability of DNA to act as a template. Three milligrams of toxin was shown to be effective in suppressing thymidine labeling of liver nuclei [75]. At the enzymatic level, aflatoxin B_1 simulated mitomycin C activity in a DNA synthesis system that utilized DNA from *E. coli*; bacterial cells had been cultured in the presence of aflatoxin B_1 [76]. The priming ability of the toxin-exposed *E. coli* DNA was reduced significantly, and associated DNA synthesis was suppressed.

Effects on RNA Metabolism

Aflatoxin also affects RNA metabolism. Changes in hepatocyte nuclear fine structure after aflatoxin exposure in rat liver suggest that toxin affects RNA metabolism. At a dose of 100 μg/rat, aflatoxin B_1 was found to effect significant inhibition of [^{14}C]orotic acid incorporation into hepatocyte nuclear RNA within 30 min after dosing [77]. Subsequently, Clifford and Rees [70] demonstrated that an LD_{50} dose (7 mg/kg) of aflatoxin B_1 given to intact rats inhibited precursor incorporation into liver nuclear RNA by 80% within 3 h after dosing. In another study, Sporn et al. [71] reported that an LD_{50} dose of aflatoxin B_1 suppressed precursor incorporation to the extent of 92% within 70 min after dosing. Suppression of nucleolar RNA by varied doses of aflatoxin B_1 was demonstrated by Floyd et al. [78]. In vitro studies with liver-slice preparations also showed similar inhibitory actions of aflatoxin B_1 on RNA metabolism. Aflatoxins G_1 and G_2 inhibited RNA synthesis under similar conditions but with lower potency than aflatoxin B_1 [79].

Impaired RNA synthesis was attributed to the inhibition of RNA polymerase activity of aflatoxin B_1. Several workers demonstrated that a sublethal dose of aflatoxin B_1 can inhibit rat liver nuclear RNA polymerase activity [80, 81]; it was concluded that the suppression of RNA polymerase activity was temporally related to nucleolar fine structural alterations. Inhibition of RNA polymerase activity was related to dose level. Nucleolar disorganization was demonstrable at 0.2 mg/kg of aflatoxin B_1 but not observable at 0.1 mg/kg; the latter dose was capable of a 5% inhibition of both types of RNA polymerase activity [4].

Aflatoxin B_1 has also been shown to affect cytoplasmic RNA metabolism as well as nuclear RNA. A single sublethal dose of aflatoxin B_1 administered to rats resulted in a decrease of about 50% in total liver RNA content over 72 h following treatment [82]. Neal and Godoy [83] reported that aflatoxin is activated on the outer cell membranes to a form that inhibits RNA synthesis.

Effects on Protein Synthesis

Alterations in the mechanism of gene expression and changes in DNA-dependent RNA synthesis result in alterations in protein metabolism. Smith [84] studied the effects of aflatoxin B_1 on the rate of in vitro incorporation of [^{14}C]leucine into proteins by rat and duckling liver slices. Significant reduction in the rate of incorporation in 80-mg rat liver slices was caused by addition of 10 μg of aflatoxin B_1. Complete suppression was recorded for 200 μg of toxin. Duckling liver slices were more sensitive; they exhibited significant suppression in the rate of [^{14}C]leucine incorporation by only 3.5 μg of aflatoxin B_1. The results elucidated the role of aflatoxin B_1 in suppressing protein synthesis. In HeLa cell culture, aflatoxin B_1 at 40 μg/mL culture medium was shown to suppress leucine incorporation into protein by approximately 25% in 90 min [85]. Shank and Wogan [86] studied the in vivo incorporation of leucine into rat liver proteins after single sublethal doses of aflatoxin B_1. Inhibition was clearly evident within 30 min of dosing and was maximum at 6 h, by which time the rate of incorporation was reduced to nearly 50% of control values. In isolated perfused rat liver, 62.5 μg of aflatoxin B_1 added to the total perfusate did not affect net synthesis of albumin, fibrinogen, α_1-acid glycoprotein, or α_2-(acute phase) globulin, but doses of 125, 250, 500, or 1000 μg were found to bring about increasingly severe inhibition of protein synthesis [87]. Increasing doses of the toxin progressively suppressed lysine incorporation into liver protein.

The inhibitory action of aflatoxin B_1 on protein synthesis in vitro was not in conformity with its effect in vivo. Incorporation of [^{14}C]leucine into rat liver proteins was not affected by prior treatment of rats with LD_{50} doses of aflatoxin B_1 [79, 88]. However, aflatoxin B_1 inhibited the synthesis of certain liver enzymes under in vivo conditions. Characteristics of inhibition of protein synthesis appeared secondary to inhibition of RNA synthesis. The conclusions were based on studies of toxin effects on the inducibility of liver tryptophan pyrrolase and zoxazolamine hydroxylase [4]. Inhibitory action of aflatoxin B_1 on cortisone induction of tryptophan pyrrolase in rat liver was demonstrated by several workers [70, 79].

Aflatoxins have also been shown to have immunosuppressive properties probably related to their inhibitory effect on protein synthesis. A level of 0.25–0.5 mg/kg of aflatoxin in feed reduced the resistance of poultry birds toward infection by several bacteria, viruses, and pathogenic fungi [89–91].

III. EFFECT OF AFLATOXINS ON HUMANS

It is difficult to establish a direct correlation of aflatoxin ingestion with consequences in humans. However, circumstantial evidence identifies a role of the toxin in disease incidence. Direct ingestion of aflatoxins occurs through consumption of contaminated food of plant origin such as cereals, pulses, fruits, nuts, and spices, whereas indirect consumption takes place through contaminated milk and milk products, meat, and eggs. Direct exposure is more acute because of the predominant presence of elevated levels of aflatoxin B_1. In direct consumption, human beings are generally exposed to aflatoxin M_1 or other metabolic forms of aflatoxin B_1. In any case, it is rather difficult to observe the long-term effects of aflatoxins in human beings because of changes in the diet.

The possible acute and chronic effects of aflatoxin in human beings can only be estimated by observing suspected cases of aflatoxicosis and by analyzing pertinent diets. A positive correlation between aflatoxin ingestion and diseases such as primary liver cancer in humans has been found in populations with regular intake of aflatoxins; the first study of this kind was made in Uganda during 1964–1967, where 355 cases of primary liver cancer were registered in a general population of 5.8 million. Tumor incidence was highest in the age group of 35–45 years and was 3.3 times greater in males than in females. Food samples collected from home granaries and village markets exhibited the presence of aflatoxins. Levels of aflatoxins in most of the contaminated samples were above 100 ppb [92]. Another study of this type was carried out in Swaziland, where 90 cases of primary liver cancer were recorded from 1964 to 1968 [93]. Samples of peanuts, the main food, were collected from different parts of the country. Analysis of the samples showed that frequency of aflatoxin contamination was high in those areas in which there was a high frequency of primary liver tumors. The results suggested that members of the Shangaans tribe consumed larger amounts of peanuts more frequently than did the Swazis; increased peanut consumption was linked to the higher incidence of primary liver cancer among the Shangaans.

Shank et al. [94–96] carried out an extensive study for 3 years in Thailand to measure directly the impact of the extent of aflatoxin consumption on the incidence of hepatocellular carcinoma in selected populations. The studies began with a 23-month investigation of contamination of foods and foodstuffs by toxigenic molds and aflatoxins in order to establish a basis for epidemiological consideration. Subsequent investigations correlated the incidence of primary liver cancer with consumption of aflatoxins. Similar correlations were also established in Kenya [97], Mozambique [98], and Swaziland [99], where actual concentrations of aflatoxins consumed were related to the incidence of primary liver cancer.

Linsell and Peers [100] considered the possibility of infection by hepatitis B virus prior to the incidence of primary liver cancer in aflatoxin-exposed persons. The studies indicated that aflatoxin could be a major disease determinant in primary liver cancer but that hepatitis B virus served as a cofactor in the etiology [101]. Stoloff [102] reviewed the incidence of primary liver cancer in the United States caused by consumption of aflatoxin-contaminated diets. He also reported the presence of hepatitis B virus in the blood serum of affected individuals.

A study in the Philippines, where mean aflatoxin contamination levels in dietary items were established and individual levels of aflatoxins determined, demonstrated an increased, dose-related relative risk of developing primary liver cancer in persons with higher ingestion of aflatoxin [103]. Additional evidence for a causative association between aflatoxin exposure and human cancer was found in a retrospective cohort study of 71 workers in an oil seed plant in the Netherlands. In the aflatoxin-exposed group, mortalities over the entire 18-year study period were higher than in the general population. An increase in mortality from respiratory cancer in the exposed group was also evident.

Van Nieuwenhuize et al. [104] reported the results of their 11-year follow-up study of 67 men who had inhaled aflatoxin-contaminated particles while working in peanut crushing and other oil-seed mills. On first exposure to aflatoxins, 2 of 55 men aged above 39 years developed fatal liver disease and 11 developed cancers of other organs. The 13 afflicted men had inhaled doses of aflatoxins of between 160 and 395 $\mu g\ m^{-3}\ man^{-1}\ week^{-1}$.

A detailed in-depth investigation into an outbreak of acute hepatitis in tribal areas extending over 200 villages of Banswara and Panchmahal districts of western India was undertaken by the National Institute of Nutrition, Hyderabad, in 1974 [105–107]. In each of the affected villages, members of very poor families suffered from the disease, which was characterized by jaundice, rapidly developing ascites, and portal hypertension. About 500 persons, predominantly males (a sex ratio of 2:1), in the age groups 5–14 and over 30 years were affected, and 106 persons died of the disease. Maize, the staple food of the people in this area, was found to be visibly contaminated with *A. flavus*, with aflatoxin levels ranging from 6.5 to 15.5 ppm. Two of seven serum samples examined showed the presence of aflatoxin B_1. In an autopsied liver, bile duct proliferation and multinucleate giant cells were observed. On the basis of epidemiological, mycological, mycotoxic, and histopathological considerations, it was concluded that the disease was caused by the consumption of maize heavily contaminated with aflatoxins [108]. Reports from another group of scientists (All India Institute of Medical Sciences, New Delhi) supported these findings [109]; this was perhaps the first report directly implicating aflatoxin in food as a human health hazard.

Besides liver cancer, aflatoxin can also induce other abnormalities in human beings. As early as 1966, Becroft [110] suggested the role of aflatoxin ingestion in inducing Reye's syndrome in human beings. Becroft and Webster [111] subsequently reported the presence of aflatoxins B_1 and G_1 in the livers of two children who died of Reye's syndrome in New Zealand; this finding was also conformed by Dvorackova et al. [112] in Czechoslovakia and Chaves-Caballo et al. [113] in the United States, who detected aflatoxins in the livers of patients with Reye's syndrome. Aflatoxin B_1 was also reported in the blood serum of two patients with the disease in the United States [114]. Reports from Thailand [115, 116] showed the presence of aflatoxin B_1 in trace amounts in tissues, body fluids, gastrointestinal contents, stools, and livers.

Autopsy of a 15-year-old African child who died in Uganda in 1970 demonstrated pulmonary edema, a dilated heart, and diffuse centrolobular necrosis—all symptoms frequently associated with aflatoxicosis. Other members of the family had at the same time suffered violent abdominal pain. Analysis of cassava showed the presence of aflatoxins as high as 1.7 ppm [117].

Indian childhood cirrhosis (ICC), a serious disorder of liver in children, has also been correlated with ingestion of aflatoxin-contaminated foods [118]. According to Amla et al. [119, 120], there is sufficient evidence to suggest that children exposed to aflatoxins through mother's milk and foods such as parboiled rice and unrefined peanut oil may acquire ICC. These authors also reported that malnourished children who had consumed aflatoxin-contaminated peanut flour developed hepatic lesions similar to those of ICC [121]. Yadgiri et al. [122] analyzed the presence of aflatoxinlike compounds in urine and liver extracts of children suffering with ICC. Parpia et al. [123] confirmed the presence of aflatoxins in 7% urine samples of 332 children. However, later research by scientists of RNT Medical College, Udaipur, strongly contradicted such claims [124, 125]; the studies established that ICC was not produced by aflatoxins but was caused by the accumulation of excess copper in the liver.

The consumption of a high concentration of aflatoxins has been correlated with mental retardation among women and children. Caster et al. [126] surveyed a region of southern Georgia known for poor school performance. The consumption of large amounts of corn, rice, peanuts, and milk (food potentially high in aflatoxins) was significantly related to mental retardation of children in one county having high levels of aflatoxins in the food supply.

IV. EFFECT OF AFLATOXINS ON PLANT SYSTEMS

As compared to the studies on animals, comparatively little work has been done with regard to the effects of mycotoxins on plant systems [127–130]. However, available reports indicate that aflatoxins can induce a variety of physiological and biochemical disorders in various plant groups, including higher plants, ferns, algae, fungi, and bacteria.

A. Effect on Higher Plants

Aflatoxins generally inhibit plant growth by restricting seed germination and subsequent seedling formation. Schoental and White [131] recorded considerable inhibition in seed germination of *Lepidium sativum* due to treatment with crude aflatoxins. Reiss [132] observed that lower concentrations of aflatoxin (10 ppm) induced chlorophyll deficiency in the same plant, whereas higher concentrations completely inhibited seed germination and radicle development. Almost identical results were obtained with 30 varieties of lettuce (*Lactuca sativa*).

Seed germination of different species of crucifers was not affected by 100 ppm of aflatoxin B_1, although the formation of the hypocotyl and radicle was found to be severely affected by the toxin [133]. In cowpea, seed germination was inhibited both by aflatoxin B_1 and by crude aflatoxins [134]. Soaking of seeds of different crop plants in culture filtrates of toxigenic strains of *Aspergillus flavus* greatly reduced germination, and in *Phaseolus aureus* seeds, inhibition was up to 100%. Aflatoxin B_1 has been shown to inhibit seed germination of sorghum [135], maize [136–138], and mung [139]. A toxigenic strain of *A. flavus* can induce symptoms that are identical to responses caused by aflatoxins. Chohan and Gupta [140] reported a new disease of groundnut seedling, Aflaroot disease; the disorder is caused by *A. flavus* Link. Mehan and Chohan [141] subsequently confirmed that Aflaroot disease syndrome in groundnut seedlings was producing aflatoxins in cotyledons. Typical vein-clearing symptoms of the disease were induced by the application of crude aflatoxins and by pure aflatoxin B_1. Pathogenic features of *A. flavus* have also been confirmed by Bedi [142] on groundnut seedlings. Treatment of peanut seedling with pure aflatoxin B_1 as well as with the extracts of liquid cultures of *A. flavus* caused substantial reduction of plant growth [143, 144]. Application of toxin to the roots was more effective than treatment of cotyledons. Root elongation of *Kalanchoe daigremontiana* was inhibited at 100 μg of aflatoxin B_1 [145].

Besides seed germination and seedling growth, chlorophyll synthesis is adversely affected by crude and pure aflatoxins. Kang [146] observed marked inhibition in chlorophyll synthesis in the cotyledonary leaves of *Abelmoschus esculentum* due to aflatoxin B_1; the results were confirmed by Mehan [147], who suggested two bioassays for aflatoxins based on the principles of chlorosis and albinism in cotyledonary and primary leaves of *Raphanus sativus* and *Sorghum vulgare* seedlings. It has been proposed that aflatoxins inhibit chlorophyll synthesis, which produces virescence or albinism in affected plants. Electron microscopic studies demonstrated inhibition of grana formation in chloroplast of maize leaves treated with aflatoxins [148]. Jacquet et al. [149] observed similar disintegration of chloroplasts in *Vasturitium officinale*. Formation of chlorophyll in the expanding cotyledons of cucumber (*Cucumis sativus* var. *long green*) was also restricted by aflatoxins. Chohan [128] suggested that aflatoxin blocks synthesis of chlorophyll by inhibiting growth hormone–induced synthesis of RNA, DNA, and protein in the leaf.

Aflatoxin B_1 influences respiratory rates of maize seeds during germination and inhibits the gibberellic acid–stimulated synthesis of α-amylase and lipase in cottonseed [151]. Truelove et al. [152] studied the effect of aflatoxin B_1 on the uptake of [^{14}C]leucine and its incorporation into protein by cotyledons of cucumber (*Cucumis sativus*). Almost 50% reduction in uptake was recorded with preincubation of 5 μg toxin/mL for 6 h without affecting protein synthesis.

Aflatoxin B_1 suppressed the synthesis of RNA, DNA, and protein in germinating seeds of maize [136, 153] and mung [139]. Tripathi and Misra [136] also reported that at subtoxic concentrations in in vivo synthesis, aflatoxin inhibited chromatic-bound, DNA-dependent RNA polymerase activity. Synthesis of both polyadenylated and nonpolyadenylated RNA was inhibited by the toxin, but the effect on the former was more pronounced than on the latter. Dashek and Llewellyn [154] reported that aflatoxin B_1 interferes with protein synthesis in germinating pollen of *Lillium longiflorum*. Elongation of pollen tubes was suppressed by aflatoxin levels of more than 6 μg/mL [155].

Aflatoxins and aflatoxin-producing fungi have been reported to degrade the nutritive value of several plant and plant products. Significant changes in the levels of sugars, protein, phenols, ascorbic acid, and caloric value have been observed during aflatoxin elaboration in maize seeds [156, 157], fleshy fruits [158, 159], and dry fruits [160].

Lilly [161] observed aflatoxin-mediated cytological abnormalities such as an increase in the number of abnormal anaphase chromosomes and considerable inhibition of mitosis in the roots of *Vicia faba*. Various chromosomal aberrations, namely, fragmentation with bridges, were also frequent in his study with the roots of *Allium cepa*. Reiss [162] found chromosome bridges, C-mitosis, and reduction of mitotic index due to aflatoxin poisoning. Daradhiyar [163] studied cytological effects of aflatoxin B_1 in maize root tips and observed stimulatory effects of low doses of toxin on cell growth that was evidenced by enhancement of the mitotic index. Interestingly, elevated doses of toxin induced inhibitory effects on cell growth, and a reduction in mitotic index was also achieved. However, quite different results were obtained with *Allium cepa* root tips [164], where lower doses of aflatoxin B_1 were found to be mitoinhibitory and higher ones were mitopromotive.

Asahi et al. [165] reported aflatoxin-based reduction in the number of ribosomes and increase of dictyosomes as the prominent ultrastructural changes in meristematic root cells of *Allium sativum*. Crisan [166] also observed cytological abnormalities in aflatoxin-treated root cells of *Lepidium sativum*, including (1) occurrence of ring-shaped nucleoli with distinct nucleolar caps, (2) mitochondria with enlarged cisternae, (3) degranulation of the endoplasmic reticulum, and (4) enlarged and irregularly shaped nuclei. On the basis of these abnormalities, Crisan [166] concluded that aflatoxin B_1 acts as an antimetabolite that binds to DNA and interferes with RNA and protein biosynthesis.

B. Effect on Lower Plants

Cahill et al. [167] studied the effect of aflatoxin B_1 on germination and protonemal development of a fern, *Onclea sensibilis*. At 12.5 μM aflatoxin, 50% reduction in spore germination was noted. Increasing concentration of aflatoxin caused reduction in the total number of cells per protonema; the results were attributed to reduction in the rate of cell production rather than total inhibition of cell division.

Inhibition of the growth of *Chlorella pyrenoidosa* was recorded at 4 μg of aflatoxin B_1/mL in liquid culture and at 12.5 μg/disk in a disk assay method [168]. Sullivan and Ikawa [169] reported similar effects of aflatoxin B_1 on different strains of *Chlorella*.

Aflatoxins affect the metabolism of their producers. Lillehoj et al. [170] reported inhibition of the growth of *A. flavus*, *A. awamori*, *Penicillium chrysogenum*, and *P. duclauxi* by aflatoxin B_1. Even the sporulation of several fungi was prevented by aflatoxin [171]. Wildman [172] recorded abnormalities in the vegetative mycelium of *A. flavus* due to aflatoxins. Damaged mitochondria, restricted pinocytosis, and abnormal cell walls were observed in giant cells of *A. parasiticus* after aflatoxin treatment. Fermentation activity of the yeast *Saccharomyces cerevisiae* was also influenced by varied concentrations of aflatoxin [173].

A comprehensive study of the mutagenic activity of aflatoxin was made by Ong [174, 175] in *Neurospora crassa*. Aflatoxin B_1 was more mutagenic than aflatoxin G_1; the former toxin induced a 177-fold increase over spontaneous mutation frequency in the fungus.

After an exhaustive screening of 329 microorganisms for sensitivity to aflatoxins, Burmeister and Hesseltine [176] reported that growth of 14 was inhibited by the toxin. *Bacillus megatenium* and *B. brevis* were, however, found to be the most sensitive species to aflatoxins. Lillehoj et al. [177] studied the toxic effects caused by uptake of aflatoxin B_1 by *Flavobacterium aurantiacum*; a concentration of 5 ppm of aflatoxin B_1 inhibited the growth and caused the development of aberrant filamentous forms. The nature of the inhibition suggested that the toxin interfered with cell wall synthesis. Lillehoj et al. [178] also studied the effect of aflatoxin G_1 on the same bacteria; they found aflatoxin G_1 less toxic to the bacteria than B_1. The growth of *F. aurantiacum* was also inhibited by aflatoxin M_1; however, the compound was less toxic than aflatoxins B_1 and G_1 [179].

Aflatoxins also exert deleterious effects on the synthesis of nucleic acids and protein in bacteria. Lillehoj and Ciegler [180] reported that a concentration of 50 μg of aflatoxin B_1 per milliliter completely blocked DNA synthesis during a 4-h incubation, whereas RNA and protein syntheses were blocked by less than 15%. In *Bacillus cereus* and *B. licheniformis*, the formation of penicillinase and α-glucosidase was partially blocked by aflatoxin B_1 [181]. Wragg et al. [76] recorded aflatoxin-mediated decrease in the activity of DNA polymerase as well as in DNA synthesis in *E. coli*.

V. ECOLOGICAL SIGNIFICANCE OF AFLATOXINS

The ecological impact of aflatoxins has not been extensively studied, although their role in the ecosystem seems to be quite significant. It is not clear how aflatoxin production influences the physiology and ecology of the producing organism [182]. Batt [183] concluded that aflatoxin elaboration by *A. parasiticus* and asexual reproduction were independent, although the two factors are interlinked to some extent. Loss of toxigenicity and simultaneous reduction in asexual reproduction has been reported in *A. parasiticus*. Torres et al. [184] showed that 39 successive transfers of *A. flavus* on moist wheat resulted in loss of both activities. Mayne et al. [185] recorded instability of an aflatoxin-producing strain of *A. parasiticus* (M-3) after six transfers. A gradual decline in toxin production by four strains of *A. flavus* was noted after seven successive subcultures [186]. The decrease was almost rectilinear and was represented by lines of regression. Sclerotial development appears to be independent of aflatoxin production [187], although sclerotia frequently contain other secondary metabolites in addition to aflatoxins [188]. Aflatoxin production has been correlated with the ability of a specific strain to produce nitrate from ammonia or amino nitrogen [189].

Secondary metabolite production is controlled by the genetic makeup of the producing fungal strain. Fungal isolations from a wide variety of sources have shown that toxigenic and nontoxigenic strains of the same fungal species as well as some unrelated microorganisms coexist and share a common habitat. It is actually a "web of interactions" linked by biochemical factors under overall control of the environment [190]. Is it possible under such situations to ignore the role of other organisms? In highly specialized habitats like the phylloplane or rhizoplane, the toxin-producing organisms have to compete with toxin degraders, antibiotic producers, and a variety of other organisms with specialized functions. Are the toxins produced just to harm other organisms, or do they also provide a "survival benefit" to the producer against the predators? Inhibitory interactions have been suggested for aflatoxin production by *A. flavus* in the presence of other microorganisms. Boller and Schroeder [191, 192] observed marked inhibition in aflatoxin production by *A. parasiticus* in the presence of *A. candidus* and *A. chevalier*, although *A. parasiticus* grew well in the presence of both species. It has also been suggested that *A. niger* inhibits aflatoxin production. Ashworth et al. [193] reported that if *A. niger* colonizes groundnut first, no aflatoxin is produced by subsequent infection by *A. flavus*. Wicklow et al. [194] screened 13 fungal species against aflatoxin production on maize and reported marked inhibition by *A. niger* and *Trichoderma viride*. A combination of quantitative mycological examination and aflatoxin analysis demonstrated that when the ratio of *A. flavus* to *A. niger* exceeded 19:1, aflatoxin was present, but when it was less than 9:1 no aflatoxin could be detected [195].

In view of the substantial metabolic energy involved in the production of aflatoxins, fungal toxin production cannot be regarded as mere "biochemical effluvia." To quote Ernst Mayr [196], one of the grand masters of evolutionary biology, "Nature does not believe in extravagance." If secondary metabolites in green plants provide defense against herbivores, it can be assumed by making a simple extrapolation that toxins protect producers from their consumers. The antibiotic property of aflatoxin possibly provides the producer fungi with an advantage of inhibiting competitors in an econiche. It is also possible that aflatoxins aid survival by preventing the ingestion of infested seeds and other materials by animals.

Is there any interaction between toxigenic and nontoxigenic strains under natural conditions? The fact that the two types of fungi coexist in the same environmental niche suggests the possibility of regular or at least occasional interaction. The frequency of nontoxigenic conidia of *A. flavus* in air is invariably higher than that of the toxigenic ones. If results are considered in terms of studies made in population genetics, the nontoxigenic (abundantly distributed) forms may be regarded as the wild-type and those that are restricted in production and distribution (i.e., toxigenic ones) should be the mutant type. The mutants by and large have a lower selective advantage owing to increased genetic load (mutational load); therefore, their number is reduced compared to the nontoxigenic (wild) types. The abundantly trapped, nontoxigenic conidia of *A. flavus* do not seem to be mutants of the toxigenic ones; they are either unrelated to the toxigenic conidia or, if related, they may be the wild types because the ratio of toxigenic to nontoxigenic conidia differs with the habitat—air, soil, rhizosphere, organic substrates, and so on [197]. Do the results reflect a toxin—habitat relationship for aflatoxin producers? Possibly not; such variations appear to be due to differences in the methods of trapping or capturing conidia. The environment, biotic agencies, and the nutritional base appear to be important factors in determining the ratio.

On the basis of surveys of air mycoflora of susceptible crops, it is inferred [198] that there is ceaseless emergence of natural mutants from apparently saprophytic *Alternaria alternata* to differerent toxin-producing strains (also referred to as patho types). By analogy, can it be assumed that toxigenic strains of *A. flavus* are the mutants of nontoxigenic strains? These conjectures can be given a definite shape only when statistical data from different habitats are computed with the precise objective of determining the ratio of toxigenic to nontoxigenic conidia.

According to Lillehoj [199], aflatoxins may also function as chemical signals between species in an ecological niche, and they could perform a function in establishing the species or cultivar within a niche. He pointed out that insects collected from corn at various geographical locations show a general internal presence of *A. flavus*, the fungus often being present in pure culture. Lillehoj et al. [200] also demonstrated that *A. flavus* infections of developing corn kernels are linked to the development of the second-generation European corn borer (*Ostrinia nubilalis*) and that this developmental period coincides with the time span during which corn kernels are susceptible to *A. flavus* infection. Lillehoj also contended that current agricultural practices of intense monoculture actually constitute a system of ecological disequilibrium. This disequilibrium then provides the requisite milieu for the synthesis of toxic mutagenic interspecies chemical signals, that is, secondary metabolites.

REFERENCES

1. Anonymous. *IARC Monographs on the Evaluation of Carcinogenic Risk of Chemicals to Man*, Vol. 10, International Agency for Research on Cancer, Lyon, WHO Publication, 1976, p. 51.
2. Coker, R. D., Jones, B. D., and Nagler, M. J. *Mycotoxin Training Manual*, Tropical Development and Research Institute, London, 1984.
3. Wogan, G. N. Chemical nature and biological effects of the aflatoxins. *Bacteriol. Rev.*, *30*(2):460, 1966.
4. Wogan, G. N., and Pong, R. S. Aflatoxins. *Ann. N.Y. Acad. Sci.*, *174*:623, 1970.
5. Heathcote, J. G., and Hibbert, J. R. *Aflatoxins: Chemical and Biological Aspects*, Elsevier, Amsterdam, 1978, p. 212.
6. Patterson, D. S. P., and Allcroft, R. Metabolism of aflatoxin in susceptible and resistant animal species. *Food Cosmet. Toxicol.*, *8*:43, 1970.

7. Carnaghan, R. B. A., Hartley, R. D., and O'Kelly, J. Toxicity and fluorescence properties of aflatoxins. *Nature, 200*:1101, 1963.
8. Moreau, C. *Moulds, Toxins and Food.* Wiley, New York, 1979, p. 477.
9. Frayssinet, C., and Lafont, P. Les mycotoxines: un nouvel aspect de l'hygiene des aliments. *Cah. nutrit. diētēt., I*:21, 1966.
10. Allcroft, R., and Carnaghan, R. B. A. Groundnut toxicity: an examination for toxin in human food products from animals fed toxic groundnut meal. *Vet. Res., 75*:259–263, 1963.
11. Loosmore, R. M., and Markson, L. M. Poisoning of cattle by Brazilian groundnut meal. *Vet. Res., 73*:813, 1961.
12. Carnaghan, R. B. A., Lewis, G., Patterson, D. S. P., and Allcroft, R. Biochemical and pathological effects of groundnut poisoning in chicken. *Pathol. Vet. (Basel), 3*:601, 1966.
13. Allcroft, R., and Lewis, G. Groundnut toxicity in cattle: experimental poisoning of calves and a report on clinical effects in older cattle. *Vet. Rec., 75*:487, 1963.
14. Lynch, G. P., Todd, G. C., Shalkop, W. T., and Moore, L. A. Response of dairy calves to aflatoxin contaminated feed. *J. Dairy Sci., 53*:63, 1970.
15. Krogh, P., Hald, B., Hesselager, E., Madsen, A., Mortensen, H. P., Larsen, A. E., and Campbell, A. D. Aflatoxin residues in bacon pigs. *Pure Appl. Chem., 35*:275, 1973.
16. McGavin, M. D., and Knake, R. Hepatic midzonal necrosis in pig fed aflatoxin and a horse fed moldy hay. *Vet. Pathol., 14*:182, 1977.
17. Pier, A. C. Aflatoxicosis and immune suppression in mammalian animals. In *Aflatoxin in Maize: A Proceeding of the Workshop* (M. S. Zuber, E. B. Lillehoj, and B. L. Renfro, Eds.), CIMMYT, Mexico, D.F., 1987, p. 58.
18. Newberne, P. M., and Butler, W. H. Acute and chronic effects of aflatoxin on the liver of domestic and laboratory animals. A review. *Cancer Res., 29*:236, 1969.
19. Madhavan, T. V., Suryanarayana, K., and Tulpule, P. G. Effect of dietary protein level on susceptibility of monkeys to aflatoxin liver injury. *Indian J. Med. Res., 53*:985, 1965.
20. Tulpule, P. G., Madhavan, T. V., and Gopalan, C. Effect of feeding aflatoxin in young monkeys. *Lancet, i*:962, 1964.
21. Angsubhakorn, S., Poomvises, P., Romrnen, K., and Newberne, P. M. Aflatoxicosis in horses. *J. Am. Vet. Med. Assoc., 178*(3): 274, 1981.
22. Asquith, R. L. Biological effects of aflatoxin in horses. In *Aflatoxin and Aspergillus flavus in Corn* (U. L. Diener, R. L. Asquith, and J. W. Dickens, eds.), Auburn Univ., Auburn, Ala., 1983, p. 62.
23. Chaffee, V. W., Edds, G. T., Himes, J. A., and Neal, F. C. Aflatoxicosis in dogs. *Am. J. Vet. Res., 30*:1737, 1969.
24. Halver, J. E. Aflatoxicosis and trout hepatoma. In *Aflatoxin: Scientific Background, Control and Implications* (L. A. Goldblatt, Ed.), Academic, New York, 1969, p. 265.
25. Butler, W. H. Acute toxicity of aflatoxin B_1 in rats. *Br. J. Cancer, 18*:756, 1964.

26. Butler, W. H. Acute toxicity of aflatoxin B_1 in guinea pigs. *J. Pathol. Bacteriol.*, *91*:277, 1966.
27. Lancaster, M. C., Jenkins, F. P., and McL.Philp, J. Toxicity associated with certain samples of groundnuts. *Nature, 192*:1095, 1961.
28. Butler, W. H., and Barnes, J. M. Toxic effects of groundnut meal containing aflatoxin to rats and guinea pigs. *Br. J. Cancer, 17*:699, 1964.
29. Newberne, P. M., Harrington, D. H., and Wogan, G. N. Effects of cirrhosis and other liver insults on induction of liver tumours by aflatoxin in rats. *Lab. Invest.*, *15*:962, 1966.
30. Wogan, G. N., and Newberne, P. M. Dose-response characteristics of aflatoxin B_1 carcinogenesis in the rat. *Cancer Res.*, *27*:2370, 1967.
31. Newberne, P. M., and Wogan, G. N. Sequential morphologic changes in aflatoxin B_1 carcinogenesis in the rat. *Cancer Res.*, *28*:770, 1968.
32. Carnagham, R. B. A. Hepatic tumours in ducks fed a low level of toxic groundnut meal. *Nature, 208*:308, 1965.
33. Adamson, R. H., Correa, P., and Dalgard, D. W. Occurrence of a primary liver carcinoma in a rhesus monkey fed aflatoxin B_1. *J. Natl. Cancer Inst.*, *50*:549, 1973.
34. Platonow, N. Effect of prolonged feeding of toxic groundnut meal in mice. *Vet. Record, 76*:589, 1964.
35. Wogan, G. N. Metabolism and biochemical effects of aflatoxins. In *Aflatoxin* (L. A. Goldblatt, Ed.), Academic, New York, 1969, p. 151.
36. Newberne, P. M. Carcinogenicity of aflatoxin contaminated peanut meals. In *Mycotoxins in Foodstuffs* (G. N. Wogan, Ed.), MIT Press, Cambridge, Mass., 1965, p. 187.
37. Dickens, F., and Jones, H. E. H. Further studies on the carcinogenic action of certain lactones and related substances in the rat and mouse. *Br. J. Cancer, 19*:392, 1965.
38. Wogan, G. N. Paglialunga, S., and Newberne, P. M. Carcinogenic effects of low dietary levels of aflatoxin B_1 in rats. *Food Cosmet. Toxicol.*, *12*:681, 1974.
39. Ashley, L. M., Halver, J. E., and Wogan, G. N. Hepatoma and aflatoxicosis in trout. *Fed. Proc.*, *23*:105, 1964.
40. Ashley, L. M., Halver, J. E., Gardner, W. K., Jr., and Wogan, G. N. Crystalline aflatoxins cause trout hepatoma. *Fed. Proc.*, *24*:627, 1965.
41. Butler, W. H., Greenblatt, M., and Lijinsky, W. Carcinogenesis in rats by aflatoxins B_1, G_1, and B_2. *Cancer Res.*, *29*:2206, 1969.
42. Wogan, G. N., Edwards, G. S., and Newberne, P. M. Structure–activity relationships in toxicity and carcinogenicity of aflatoxins and analogs. *Cancer Res.*, *31*:1936, 1971.
43. Sinnhuber, R. O., Lee, D. J., Wales, J. H., Landers, M. K., and Keyl, A. C. Hepatic carcinogenesis of aflatoxin M_1 in rainbow trout (*Salmo gairdneri*) and its enhancement by cyclopropene fatty acids. *J. Natl. Cancer Inst.*, *53*:1285, 1974.
44. Ward, J. M., Soutag, J. M., Weisburger, E. K., and Brown, C. A. Effect of lifetime exposure to aflatoxin B_1 in rats. *J. Natl. Cancer Inst.*, *55*:107, 1975.

45. Newberne, P. M., and Rogers, A. E. Rat colon carcinomas associated with aflatoxin and marginal vitamin A. *J. Natl. Cancer Inst.*, *50*:439, 1973.
46. Dickens, F., Jones, H. E. H., and Waynforth, H. B. Oral, subcutaneous and intratracheal administration of carcinogenic lactones and related substances: the intratracheal administration of cigarette tar in the rat. *Br. J. Cancer*, *20*:134, 1966.
47. Goodal, C. M., and Butler, W. H. Aflatoxin carcinogenesis: inhibition of liver cancer induction in hypophysectomized rats. *Int. J. Cancer*, *4*:422, 1969.
48. Ranjan, K. S. Studies on effect of aflatoxin infested meal on some laboratory animals, Ph.D. thesis, Bhagalpur Univ., Bhagalpur, India, 1987.
49. Ong, T. Mutagenicity of mycotoxins. *Mutat. Res.* (*Edin.*), *32*:35 1975.
50. Chu, F. S. Mode of action of mycotoxins and related compounds. *Adv. Appl. Microbiol.*, *22*:83, 1977.
51. Elis, J., and DiPaolo, J. A. Aflatoxin B_1: induction of congenital abnormalities in hamsters. *Arch. Pathol.*, *83*:53, 1967.
52. DiPaolo, J. A., Elis, J., and Erwin, H. Teratogenic response by hamsters, rats and mice to aflatoxin B_1. *Nature*, *215*:638, 1967.
53. Verrett, M. J., Marliac, J. P., and McLaughlin, J., Jr. Use of the chicken enbryo in the assay of aflatoxin toxicity. *J. Assoc. Off. Anal. Chem.*, *47*:1003, 1964.
54. Vesely, D., Vesela, D., and Jelinek, R. Comparative assessment of the aflatoxin B_1, B_2, G_1, G_2 and M_1 embryotoxicity in the chick embryo. *Toxicol. Lett.*, *15*:297, 1983.
55. Grice, M. C., Moodie, C. A., and Smith, D. C. The carcinogenic potential of aflatoxin or its metabolites in rats from dams fed aflatoxin pre- and post-partum. *Cancer Res.*, *33*:262, 1973.
56. Wong, J. J., and Hsieh, D. P. H. Mutagenicity of aflatoxins related to their metabolism and carcinogenic potential. *Proc. Natl. Acad. Sci. USA*, *73*:2241, 1976.
57. Wong, J. J., Singh, R., and Hsieh, D. P. H. Mutagenicity of fungal metabolites related to aflatoxin biosynthesis. *Mutat. Res.*, *44*: 447, 1977.
58. Stoloff, L. *Carcinogens and Mutagens in the Environment*, Vol. 1, *Food Products* (H. F. Stitzh, Ed.), CRC Press, Boca Raton, Fla., 1982, p. 97.
59. Green, S., Legator, M., and Jacobson, C. Utilization of a cell line derived from rat kangaroo for cytogenetic studies. *Mamm. Chrom. Newslett.*, *8*:36, 1967.
60. Adzhigitov, F. I., Kosichenko, L. P., Popandopulo, P. G., Gubeladze, D. A., and Dzhemilev, Z. A. Frequency of chromosomal aberrations in bone marrow of primates and their offspring after the effect of aflatoxin B_1. *Tsitolog. Genet.*, *18*(3):173, 1984.
61. Ueno, Y., Tashiro, F., Haraikawa, K., and Yasuhiro, E. Metabolism of aflatoxins and their toxicity. In *Problems of Threshold in Chemical Mutagenesis* (Y. Tarina, S. Kondo, and Y. Kuroda, Eds.), The Environmental Mutagen Society of Japan, 1984, p. 61.

62. Bilgrami, K. S., and Sinha, S. P. Mutagenic effects of mycotoxins. Final Techn. Report, U.G.C. research project, Bhagalpur Univ., Bhagalpur, India, 1988, p. 48.
63. Sinha, S. P. Mycotoxins as mutagens—a need for confirmation using mammalian test-system. In *Mycotoxins in Food and Feed* (K. S. Bilgrami, T. Prasad, and K. K. Sinha, Eds.), Allied Press, Bhagalpur, India, 1983, p. 123.
64. Bhattacharya, R. K., Francis, A. R., and Shetty, T. K. Modifying role of dietary factors on the mutagenicity of aflatoxin B_1: *in vitro* effects of vitamins. *Mutat. Res.*, *188*:121, 1987.
65. Panda, P. C., Shankar Murti, A., Sreenavasa Murthy, V., and Murti, I. A. S. Effect of aflatoxin on the haematological picture of albino rats and guinea pigs. *Indian J. Exp. Biol.*, *13*(6):569, 1975.
66. Gupta, M., Bandyopadhyay, S., Sasmal, D., and Mazumder, S. K. Effect of citrinin on haematological parameters in mice. *IRCS Med. Sci.*, *7*:382, 1979.
67. Sahay, M. Histopathologic and cytogenetic effects of aflatoxins in a mammal. Ph.D. Thesis, Bhagalpur Univ., Bhagalpur, India, 1986.
68. Clark, J. D., Greene, C. E., Culpin, J. P., Roger, C. H., and Jan, A. V. Induced aflatoxicosis in rabbits: blood coagulation defects. *Toxicol. Appl. Pharmacol.*, *86*:353, 1986.
69. Frayssinet, C., and Lafarge, C. Action de L'aflatoxine sur la cellule hepatique du rat. *Cah. Nutr. Diet*, *5*:67, 1970.
70. Clifford, J. I., and Rees, K. R. Aflatoxin: a site of action in the rat liver cell. *Nature*, *209*:312, 1966.
71. Sporn, M. B., Dingman, C. W., Phelps, H. L., and Wogan, G. N. Aflatoxin B: binding to DNA *in vitro* and alteration of RNA metabolism *in vivo*. *Science*, *151*:1539, 1966.
72. Black, H. S., and Jirgensons, B. Interactions of aflatoxin with histones and DNA. *Plant Physiol.*, *42*:731, 1967.
73. Frayssinet, C., Lafarge, C., De Recondo, A. M., and Le Breton, E. Inhibition de l'hypertrophie compensatrice du foie chez le rat par les toxines d'*Aspergillus flavus*. *Compt. Rend. Acad. Sci.*, *259*:2143, 1964.
74. De Recondo, A. M., Frayssinet, C., Lafarge, C., and Le Breton, E. Action de l'aflatoxine sur le metabolisme du DNA au cours de l'hypertrophie compensatrice du foie apres hepatectomie partielle. *Biochim. Biophys. Acta*, *119*:322, 1966.
75. Rogers, A. E., and Newberne, P. M. Aflatoxin B_1 carcinogenesis in lipotrope-deficient rats. *Cancer Res.*, *29*:1965, 1969.
76. Wragg, J. B., Ross, V. C., and Legator, M. S. Effect of aflatoxin B_1 on the deoxyribonucleic acid polymerase of *Escherichia coli* (32274). *Proc. Soc. Exp. Biol. Med.*, *125*:1052, 1967.
77. Lafarge, C., Frayssinet, C., and De Recondo, A. M. Inhibition par l'aflatoxine de la synthese de RNA hepatique chez le rat. *Bull. Soc. Chim. Biol.* *47*:1724, 1965.
78. Floyd, L. R., Unuma, T., and Busch, H. Effects of aflatoxin B_1 and other carcinogens upon nucleolar RNA of various tissues in the rat. *Exp. Cell. Res.*, *57*:423, 1968.

79. Clifford, J. I., and Rees, K. R. The action of aflatoxin B_1 on the rat liver. *Biochem. J.*, *102*:65, 1967.
80. Gelboin, H. V., Wortham, J. S., and Wilson, R. G. Rapid and marked inhibition of rat-liver RNA-polymerase. *Science*, *154*:1205, 1966.
81. Friedman, M. A., and Wogan, G. N. Effects of aflatoxin B_1 on RNA polymerase activity and incorporation of cytidine into RNA of rat liver nuclei. *Fed. Proc.*, *26*:358, 1967.
82. Friedman, M. A., and Wogan, G. N. Effects of aflatoxin B_1 on enzyme induction and nuclear RNA metabolism in rat liver. *Fed. Proc.*, *25*:662, 1966.
83. Neal, G. E., and Godoy, H. M. The effect of pre-treatment with phenobarbitone on the activation of aflatoxin B_1 by rat liver. *Chem. Biol. Interactions*, *14*:279, 1976.
84. Smith, R. H. The influence of toxins of *Aspergillus flavus* on the incorporation of (C14) Leucine into proteins. *Biochem. J.*, *88*:50, 1963.
85. Harley, E. H., Rees, K. R., and Cohan, A. A comparative study of the effect of aflatoxin B_1 and actinomycin D on HeLa cells. *Biochem. J.*, *114*:289, 1969.
86. Shank, R. C., and Wogan, G. N. Effects of aflatoxin B_1 on some aspects of liver metabolism. *Fed. Proc.*, *23*:200, 1964.
87. John, D. W., and Miller, L. L. Effect of aflatoxin B_1 on net synthesis of albumin, fibrinogen and α_1-acid glycoprotein by the isolated perfused rat liver. *Biochem. Pharmacol.*, *18*:1135, 1969.
88. Shank, R. C., and Wogan, G. N. Acute effects of aflatoxin B_1 on liver composition and metabolism in the rat and duckling. *Toxicol. Appl. Pharmacol.*, *9*:468, 1966.
89. Pier, A. C., and Heddleston, K. L. The effect of aflatoxin on immunity in turkeys. I. Impairment of actively acquired resistance to bacterial challenge. *Avian. Dis.*, *14*:797, 1970.
90. Hamilton, P. B., and Harris, J. R. Interaction of aflatoxicosis with *Candida albicans* infections and other stresses in chickens. *Poult. Sci.*, *50*:906, 1971.
91. Edds, G. T., Nair, K. P. C., and Simpson, C. F. Effect of aflatoxin B_1 on resistance in poultry against cecal coccidiosis and Marek's disease. *Am. J. Vet. Res.*, *34*:819, 1973.
92. Alpert, M. E., Hutt, M. S. R., Wogan, G. N., and Davidson, C. S. Association between aflatoxin content of food and hepatoma frequency in Uganda. *Cancer*, *28*:253, 1971.
93. Keen, P., and Martin, P. Is aflatoxin carcinogenic in man? The evidence in Swaziland. *Trop. Geogr. Med.*, *23*:44, 1971.
94. Shank, R. C., Wogan, G. N., Gibson, J. B., and Nondasuta, A. Dietary aflatoxins in human liver cancer. II. Aflatoxins in market foods and foodstuffs of Thailand and Hong Kong. *Food Cosmet. Toxicol.*, *10*:61, 1972.
95. Shank, R. C., Gordon, J. E., Wogan, G. N., Nondasuta, A., and Subhamani, B. Dietary aflatoxins and human liver cancer. III. Field survey of rural Thai families for ingested aflatoxins. *Food Cosmet. Toxicol.*, *10*:71, 1972.

96. Shank, R. C., Bhamarpravati, N., Gordan, J. E., and Wogan, G. N. Dietary aflatoxins and human cancer. IV. Incidence of primary liver cancer in two municipal populations in Thailand. *Food Cosmet. Toxicol.*, *10*:171, 1972.
97. Peers, F. G., and Linsell, C. A. Dietary aflatoxins and liver cancer—a population based study in Kenya. *Br. J. Cancer*, *27*:473, 1973.
98. Van Rensburg, S. J., Vander Watt, J. J., Purchase, I. F. H., Pereira, C. L., and Markham, R. Primary liver cancer rate and aflatoxin intake in a high cancer area. *S. Afr. Med. J.*, *48*:2508a, 1974.
99. Peers, F. G., Gilman, G. A., and Linsell, C. A. Dietary aflatoxins and human liver cancer: a study in Swaziland. *Int. J. Cancer*, *17*: 167, 1976.
100. Linsell, C. A., and Peers, F. G. Field studies on liver cell cancer. In *Origins of Human Cancer* (H. H. Hiatt, J. D. Watson, and J. A. Wingsten, Eds.), Cold Spring Harbor, N.Y., 1977, p. 549.
101. Peers, F. G., and Linsell, C. A. Dietary aflatoxins and human primary liver cancer. *Ann. Nutr. Aliment.*, *31*:1005, 1977.
102. Stoloff, L. Aflatoxin as a cause of primary liver cell cancer in the United States: a probability study. *Nutr. Cancer*, *5*:165, 1983.
103. Bulatao-Jayme, J., Almero, E. M., Castro, M. C. A., Jardeleza, M. T. R., and Salamat, L. A. A case control dietary study of primary liver cancer risk from aflatoxin exposure. *Int. J. Epidemiol.*, *11*:112, 1982.
104. Van Nieuwenhuize, J. P., Herber, F. M., De Bruin, A., Meyer, P. M., and Duba, W. C. Cancers in men following a long term low-level exposure to aflatoxins in indoor air. *T. Soc. Geneeskd.*, *51*: 754, 1973.
105. Krishnamachari, K. A. V. R., Bhat, R. V., Nagarajan, V., and Tilak, T. B. G. Hepatitis due to aflatoxicoses: an outbreak in Western India. *Lancet*, *i*:1061, 1975.
106. Krishnamachari, K. A. V. R., Bhat, R. V., Nagarajan, V., and Tilak, T. B. G. Investigations into an outbreak of hepatitis in parts of Western India. *Indian J. Med. Res.*, *63*:1036, 1975.
107. Krishnamachari, K. A. V. R., Bhat, R. V., Nagarajan, V., and Tilak, T. B. G. Aflatoxicosis in humans. *Proc. Nutr. Soc. India*, *19*:18, 1975.
108. Krishnamachari, K. A. V. R., Bhat, R. V., Nagarajan, V., Tilak, T. B. G., and Tulpule, P. G. The problem of aflatoxic human disease in parts of India—epidemiological and ecological aspects. *Ann. Nutr. Alim.*, *31*:991, 1977.
109. Tandon, H. D., Tandon, B. N., and Ramalingaswami, V. Epidemic of toxic hepatitis in India of possible mycotoxic origin. *Arch. Pathol. Lab. Med.*, *102*:372, 1978.
110. Becroft, D. M. O. Syndrome of encephalopathy and fatty degeneration of viscera in New Zealand children. *Br. Med. J.*, *2*:135, 1966.
111. Becroft, D. M. O., and Webster, D. R. Aflatoxins and Reye's disease. *Br. Med. J.*, *14*:117, 1972.

112. Dvorackova, I., Brodsky, F., and German, J. Aflatoxin and encephalitic syndrome with fatty degeneration of viscera. *Nutr. Rep. Int.*, *10*:89, 1974.
113. Chaves-Caballo, E., Ellefson, R. D., and Gomez, M. R. An aflatoxin in the liver of a patient with Reye-Johnson syndrome. *Mayo Clin. Proc.*, *51*:48, 1976.
114. Hogan, G. R., Ryan, N. J., and Hayes, A. W. Aflatoxin B_1 and Reye's syndrome. *Lancet*, March 11:561, 1978.
115. Bourgeois, C. H., Shank, R. C., Grossman, R. A., Johnsen, D. O., Wooding, W. L., and Chandarimol, P. Acute aflatoxin B_1 toxicity in the macaque and its similarities in Reye's syndrome. *Lab. Invest.*, *24*:206, 1971.
116. Shank, R. C., Bourgeois, C. H., Keschamras, N., and Chandavimol, P. Aflatoxins in autopsy specimens from Thai children with an acute disease of unknown etiology. *Food Cosmet. Toxicol.*, *9*:501, 1971.
117. Serck-Hanssen, A. Aflatoxin-induced fatal hepatitis? A report from Uganda. *Arch. Environ. Health*, *20*:729, 1970.
118. Robinson, P. Infantile cirrhosis of the liver in India with special reference to probable aflatoxin etiology. *Clin. Pediatr.*, *6*:57, 1967.
119. Amla, I., Kumari, S., Sreenivasamurthy, V., Jayaraj, P., and Parpia, H. A. B. Role of aflatoxin in Indian childhood cirrhosis. *Indian Pediat.*, 7(5):262, 1970.
120. Amla, I., Sreenivasamurthy, V., Jayaraj, A. P., and Parpia, H. A. B. Aflatoxin in Indian childhood cirrhosis—a review. *J. Trop. Pediat. Environ. Child Health*, *20*:28, 1974.
121. Amla, I., Kamala, C. S., Gopalakrishna, G. S., Jayaraj, A. P., Sreenivasamurthy, V., and Parpia, H. A. B. Cirrhosis in children from peanut meal contaminated by aflatoxin. *Am. J. Clin. Nutr.*, *24*:609, 1971.
122. Yadgiri, B., Reddy, V., Tulpule, P. G., Srikantia, S. G., and Gopalan, C. Aflatoxin and Indian childhood cirrhosis. *Am. J. Clin. Nutr.*, *23*:94, 1970.
123. Parpia, H. A. B., Sreenivasamurthy, V., Shantha, T., Shankarmurthi, A., Srikantia, S., and Amla, I. Urinary aflatoxin B and effects of preservation on its detection. *Am. J. Clin. Nutr.*, *25*: 13, 1972.
124. Bhandari, B., and Sharda, B. Indian childhood cirrhosis: a search for cause and remedy. *Indian J. Pediat.*, *51*:135, 1984.
125. Bhandari, B., and Sharda, M. D. Indian childhood cirrhosis (ICC): recent observations. *Indian Pediat.*, *21*:569, 1984.
126. Caster, W. D., Burton, T. A., Irwin, T. R., and Tanner, M. A. Dietary aflatoxin, intelligence and school performance in Southern Georgia. *Int. J. Vit. Nutr.*, *56*:291, 1986.
127. Reiss, J. Effects of mycotoxins on higher plants, algae, fungi and bacteria. In *Mycotoxic fungi, Mycotoxins, Mycotoxicoses* (T. D. Wyllie and L. G. Morehouse, Eds.), Marcel Dekker, New York, 1978, p. 119.
128. Chohan, J. S. Effect of aflatoxins on plant systems. In *Mycotoxins in Food and Feed* (K. S. Bilgrami, T. Prasad, and K. K. Sinha, Eds.), Allied Press, Bhagalpur, India, 1983, p. 147.

129. Bilgrami, K. S., and Sinha, K. K. Mycotoxin induced abnormalities in plants. In *Frontiers in Applied Microbiology* (K. G. Mukerji, N. C. Pathak, and U. P. Singh, Eds.), Print House, Lucknow, India, 1985, p. 349.
130. Dashek, W. V., and Llewellyn, G. C. Mode of action of the hepatocarcinogens, aflatoxins in plant systems: a review. *Mycopathology, 81*:83, 1983.
131. Schoental, R., and White, A. F. Aflatoxins and albinism in plants. *Nature, 205*:57, 1965.
132. Reiss, J. Inhibition of germination of *Lepidium sativum* by aflatoxin B_1 and rubratoxin B. *Biochem. Physiol. Pflanz. 162*:363, 1971.
133. Crisan, E. V. Effects of aflatoxin on germination and growth of lettuce. *Appl. Microbiol., 25*:342, 1973.
134. Adekunle, A. A., and Bassir, O. The effects of aflatoxin B_1 and palmotoxins B_0 and G_0 on the germination and leaf colour of the cowpea (*Vigna sinensis*). *Mycopathol. Mycol. Appl., 51*(4):299, 1973.
135. Tripathi, R. K. Aflatoxins in sorghum grains infested with head moulds. *Indian J. Exp. Biol., 11*:361, 1973.
136. Tripathi, R. K., and Misra, R. S. Mechanism of inhibition of maize seed germination by aflatoxin B_1. In *Mycotoxins in Food and Feed* (K. S. Bilgrami, T. Prasad, and K. K. Sinha, Eds.), Allied Press, Bhagalpur, India, 1983, p. 129.
137. Jeswal, P. Effect of aflatoxin B_1 on germination and seedling growth of maize. *Natl. Acad. Sci. Lett., 10*(3):79, 1987.
138. Kumari, P. Studies on mycotoxin induced abnormalities in *Zea mays* Linn. and *Vigna radiata* (Linn.) Wilczek. Ph.D. Thesis, Bhagalpur Univ., Bhagalpur, India, 1988.
139. Sinha, K. K., and Kumari, P. Effect of aflatoxin B_1 on biochemical changes in mung seeds (variety Pusa-119). *Microbiol. Lett., 40*: 145, 1989.
140. Chohan, J. S., and Gupta, V. K. Afla root—a new disease of groundnut caused by *Aspergillus flavus* Link. *Indian J. Agric. Sci., 38*:568, 1968.
141. Mehan, V. K., and Chohan, J. S. Virulence of aflatoxigenic and non-toxigenic isolates of *Aspergillus flavus* Link ex. Fr. on groundnut seedlings. *Phytopathol. Medit., 13*:108, 1974.
142. Bedi, P. S. Ph.D. Thesis, Punjab Agricultural University, Ludhiana, 1978.
143. El-Khadem, M. Die Bedeutung von Aflatoxin fur die durch *Aspergillus flavus* verursachte keimlingskrankheit der Erdnuss. *Phytopathol. Z., 61*:218, 1965.
144. El-Khadem, M., Tewfik, M. S., and Hamdi, Y. A. Effect of aflatoxin B_1, aspergillic and kojic acid on peanut seedlings. *Zbl. Bakt. II, 130*:556, 1975.
145. Reiss, J. Effects of mycotoxins on the development of epiphyllous buds of *Kalanchoe diagremontiana*. *Z. Pflanzenphysiol., 82*:446, 1977.
146. Kang, M. S. Pathogenesis in groundnut (*Arachis hypogaea* L.) by *Aspergillus flavus* Link ex Fr. Ph.D. Thesis, Punjab Agricultural Univ., Ludhiana, India, 1970.

147. Mehan, V. K. Studies on growth requirements of toxigenic and nontoxigenic isolates of *Aspergillus flavus* Link ex Fries and their effects on different crop plants. M.Sc. Thesis (Plant Pathology), Punjab Agric. Univ., Ludhiana, India, 1971.
148. Slowatizky, I., Mayer, A. M., and Poljakoff-Mayber, A. The effect of aflatoxin on greening of etiolated leaves. *Israel J. Bot., 18*:31, 1969.
149. Jacquet, J., Boutibonnes, P., and Saint, S. Effects biologiques des flavacoumarines d'*Aspergillus parasiticus* ATCC 15517, *T. animaux. Rev. Immunol., 35*:159, 1971.
150. Misra, R. S., and Tripathi, R. K. Effect of aflatoxin B in germination, respiration and α-amylase in maize. *Plant Dis. Prot. 87*:155, 1980.
151. Black, H. S., and Altschul, A. M. Gibberellic acid induced lipase and α-amylase formation and their inhibition by aflatoxin. *Biochem. Biophys. Res. Commun., 19*(5):661, 1965.
152. Truelove, B., Davis, D. E., and Thompson, O. C. Effects of aflatoxin B_1 on protein synthesis by cucumber cotyledon discs. *Can. J. Bot., 48*:485, 1970.
153. Sinha, K. K., and Kumari, P. Effect of aflatoxin B_1 on some biochemical changes in germinating maize seeds. *Indian Phytopathol., 42*(4), 1989.
154. Dashek, W. V., and Llewellyn, G. C. The influence of the carcinogen aflatoxin B_1 on the metabolism of germinating lily pollen. In *Fertilization in Higher Plants* (H. F. Linskens, Ed.), North-Holland, Amsterdam, 1974, p. 351.
155. Llewellyn, G. C., and Dashek, W. V. The influence of aflatoxin B_1 on lily pollen germination and tube elongation. *Incompat. Newslett., 3*:18, 1973.
156. Bilgrami, K. S., Misra, R. S., and Sinha, K. K. Aflatoxin production and loss in caloric value of maize seeds due to *Aspergillus parasiticus. Curr. Sci., 48*:642, 1979.
157. Bilgrami, K. S., Prasad, T., Misra, R. S., and Sinha, K. K. Changes in protein and amino acid contents of maize seeds infested with aflatoxin producing strain of *Aspergillus parasiticus. Seed Res., 9*:52, 1981.
158. Singh, A., and Sinha, K. K. Biochemical changes in musambi fruits inoculated with species of aflatoxin producing *Aspergilli. Curr. Sci., 51*(17):841, 1982.
159. Sinha, K. K., and Singh, A. Biochemical changes in orange fruits during infestation with *Aspergillus flavus* and *Aspergillus parasiticus. Natl. Acad. Sci. Lett.,* 5(5):143, 1982.
160. Bilgrami, K. S., Sinha, K. K., and Singh, A. Chemical changes in dry fruits during aflatoxin elaboration by *Aspergillus flavus* Link ex Fries. *Curr. Sci., 52*:960, 1983.
161. Lilly, L. J. Induction of chromosome aberrations by aflatoxin. *Nature (Lond.), 207*:433, 1965.
162. Reiss, J. Mycotoxin poisoning of *Allium cepa* root tips. II. Reduction of mitotic index and formation of chromosomal aberrations

and cytological abnormalities by patulin, rubrotoxin B and diacetoxyscirpenol. *Cytologia, 40*:703, 1975.
163. Daradhiyar, S. K. Studies on aflatoxins associated with storage and standing crop of maize, poultry feed and cattle cakes in the tribal belts of Santhal Pargana District. Ph.D. Thesis, Bhagalpur Univ., Bhagalpur, India, 1980.
164. Bilgrami, K. S., Sinha, S. P., and Ranjan, K. S. Effect of aflatoxin B_1 on mitotic index. *Curr. Sci., 55*(21):1092, 1986.
165. Asahi, T., Mori, Z., Majima, R., and Uritani, I. The effects of aflatoxins on metabolic changes in plant tissue in response to injury. *J. Stored Prod. Res., 5*:219, 1969.
166. Crisan, E. V. Effects of aflatoxin on seedling growth and ultrastructure in plants. *Appl. Microbiol., 26*:991, 1973.
167. Cahill, S. G., Fisher, R. W., and Llewellyn, G. C. *Mycopathology, 64*:43, 1978.
168. Ikawa, M., Ma, D. S., Meeker, G. M., and Davis, R. P. Use of *Chlorella* in mycotoxin and phytotoxin research. *J. Agric. Food Chem., 17*:425, 1969.
169. Sullivan, J. D., Jr., and Ikawa, M. Variations in inhibition of growth of five *Chlorella* strains by mycotoxins and other toxic substances. *J. Agric. Food Chem., 20*:921, 1972.
170. Lillehoj, E. B., Cieglar, A., and Hall, H. H. Fungistatic action of aflatoxin B_1. *Experientia, 23*:187, 1967.
171. Reiss, J. Inhibition of fungal sporulation by aflatoxin. *Arch. Mikrobiol., 76*:219, 1971.
172. Wildman, J. D. Note on occurrence of giant cells in *Aspergillus flavus* Link. *J. Assoc. Off. Anal. Chem., 49*:562, 1966.
173. Reiss, J. Influence of the mycotoxins aflatoxin B_1, rubratoxin B, patulin and diacetoxyscirpenol on the formation activity of baker's yeast. *Mycopathol. Mycol. Appl., 51*:337, 1973.
174. Ong, T. Mutagenicity of aflatoxin in *Neurospora crassa. Mutat. Res., 9*:615, 1970.
175. Ong, T. Mutagenic activities of aflatoxin B_1 and G_1 in *Neurospora crassa. Mol. Gen. Genet., 111*:159, 1971.
176. Burmeister, H. R., and Hesseltine, C. W. Survey of sensitivity of micro-organisms to aflatoxin. *Appl. Microbiol., 14*:403, 1966.
177. Lillehoj, E. B., Ciegler, A., and Hall, H. H. Aflatoxin B_1 uptake by *Flavobacterium aurantiacum* and resulting toxic effects. *J. Bacteriol., 93*:464, 1967.
178. Lillehoj, E. B., Ciegler, A., and Hall, H. H. Aflatoxin G_1 uptake by cells of *Flavobacterium aurantiacum. Can. J. Microbiol., 13*:629, 1967.
179. Lillehoj, E. B., Stubblefield, R. D., Shannon, G. M., and Shotwell, O. L. Aflatoxin M_1 removal from aqueous solutions by *Flavobacterium aurantiacum. Mycopathol. Mycol. Appl., 45*:259, 1971.
180. Lillehoj, E. B., and Ciegler, A. Inhibition of deoxyribonucleic acid synthesis in *Flavobacterium aurantiacum* by aflatoxin B_1. *J. Bacteriol., 94*:787, 1967.
181. Lillehoj, E. B., and Ciegler, A. Aflatoxin B_1 induction of lysogenic bacteria. *Appl. Microbiol., 20*:782, 1970.

182. Moss, M. O., and Frank, J. M. The influence on mycotoxin production of interactions between fungi and their environment. In *Trichothecenes and Other Mycotoxins* (J. Lacey, Ed.), Wiley, New York, 1985, p. 257.
183. Batt, C. Relationship of asexual development to aflatoxin production in *Aspergillus parasiticus*. *J. Food Safety, 5*:31, 1983.
184. Torres, J., Guarro, J., Suarej, G., Sune, N., Calve, M. A., and Ramirez, C. Morphological changes in strains of *Aspergillus flavus* (Link) and *Aspergillus parasiticus* (Speare) related with aflatoxin production. *Mycopathology, 72*:171, 1980.
185. Mayne, R. Y., Bennett, J. W., and Tallant, J. Instability of an aflatoxin producing strain of *Aspergillus parasiticus*. *Mycologia, 63*:644, 1971.
186. Bilgrami, K. S., Sinha, S. P., and Jeswal, P. Loss of toxigenicity of *Aspergillus flavus* strains during subculturing—a genetic interpretation. *Curr. Sci., 57*(10):551, 1988.
187. Bennett, J. W., Horowitz, P. C., and Lee, L. S. Production of sclerotia by aflatoxigenic and nonaflatoxigenic strains of *Aspergillus flavus* and *A. parasiticus*. *Mycologia, 71*:415, 1979.
188. Wicklow, D. T., and Cole, R. J. Tremorgenic indole metabolites and aflatoxins in sclerotia of *Aspergillus flavus*: an evolutionary perspective. *Can. J. Bot., 60*:525, 1982.
189. White, J. P., and Johnson, G. T. Aflatoxin production correlated with nitrification in *Aspergillus flavus* group species. *Mycologia, 74*:718, 1982.
190. Bilgrami, K. S. Fungal toxins: their biological effects and ecological significance. Presidental address, Proceedings of 74th Indian Science Congress, Part II, Bangalore, India, 1987, p. 1.
191. Boller, R. A., and Schroeder, H. W. Influence of *Aspergillus chevalieri* on production of aflatoxin in rice by *Aspergillus parasiticus*. *Phytopathology, 63*:1507, 1973.
192. Boller, R. A., and Schroeder, H. W. Influence of *Aspergillus candidus* on production of aflatoxin in rice by *Aspergillus parasiticus*. *Phytopathology, 64*:121, 1974.
193. Ashworth, L. J., Schroeder, H. W., and Langley, B. C. Aflatoxins: environmental factors governing occurrences in Spanish peanuts. *Science, 148*:1228, 1965.
194. Wicklow, D. T., Hesseltine, C. W., Shotwell, O. C., and Adams, G. L. Interference competition and aflatoxin levels in corn. *Phytopathology, 70*:761, 1980.
195. Weckbach, L. S., and Marth, E. M. Aflatoxin production by *Aspergillus parasiticus* in a competitive environment. *Mycopathology, 62*:39, 1977.
196. Mayr, E. *Evolution and the Diversity of Life*, Harvard Univ. Press, Cambridge, Mass., 1978, p. 836.
197. Bilgrami, K. S., and Sinha, K. K. Mycotoxin contamination in food and its control. *Indian Rev. Life Sci., 4*:19, 1984.
198. Nishimura, S., and Kohmoto, K. Roles of toxins in pathogenesis. In *Toxins and Plant Pathogenesis* (J. M. Daly and B. J. Deverall, Eds.), Academic, New York, 1983, p. 137.

199. Lillehoj, E. B. Secondary metabolites as chemical signals between species in an ecological niche. *Proceedings*, Sixth International Fermentation Symposium, London, Ontario, 1980. In *Advances in Biotechnology*, Vol. 3 (M. Moo-Young, C. Vezina, and K. Singh, Eds.), Pergamon Press, Elmsford, N.Y., 1980, p. 397.
200. Lillehoj, E. B., Kowlek, W. F., Horner, E. S., Widstrom, N. W., Josephson, L. M., Franz, A. O., and Catalano, E. A. Aflatoxin contamination of preharvest corn: role of *Aspergillus flavus* inoculum and insect damage. *Cereal Chem.*, 57:255, 1980.

4

DEVELOPMENT AND USE OF IMMUNOASSAYS IN THE DETECTION OF ECOLOGICALLY IMPORTANT MYCOTOXINS

FUN S. CHU *University of Wisconsin—Madison, Madison, Wisconsin*

I. INTRODUCTION

Due to the widespread nature of fungi in the ecological system and the production of toxic metabolites by some fungal species, the presence of mycotoxins in foods and feeds is a great concern for human and animal health. Although the most effective way to control mycotoxin contamination is to prevent formation of the toxins in the field and during storage, technology does not yet exist for the complete prevention of toxin formation or the removal of mycotoxins from foods and feeds. Thus, measures for control of the mycotoxin problem rely on a rigorous monitoring program to detect the toxins in various ecological systems and prevent the consumption of contaminated food or feed by humans and animals. Consequently, a great deal of research has been carried out to search for sensitive, specific, simple methods for mycotoxin detection. The progress in this area can be seen from the vast number of research papers cited by the general referee on mycotoxins in the annual report on mycotoxins, which appears each year in either the February or March issue of the *Journal of the Association of Official Analytical Chemists* (*JOAC*) [1].

Among the many methods developed in recent years, immunochemical methods have been shown to be very promising because of their versatility, sensitivity, and specificity. An overall recognition of the potentially wide application of new powerful detection techniques has not only led to simple protocols that allow accurate detection of small amounts of mycotoxins in foods, feeds, and biological fluids but has also generated considerable interest in the preparation of commercial kits for mycotoxin analysis. Several reviews on this subject have appeared [2–18]. In this chapter, details on the development of new detection methods as well as the application of innovative techniques for the detection of mycotoxins in varied ecosystems are presented.

II. DEVELOPMENT AND CHARACTERIZATION OF ANTIBODIES AGAINST MYCOTOXINS

A. General Considerations

The development of immunoassays for mycotoxins generally involves four major phases: (1) conjugation of mycotoxins to protein carriers and preparation of marker ligands, (2) production and characterization of antibodies, (3) establishment of immunoassay protocols, and (4) establishment of cleanup protocols in complex matrices and validation of immunoassay protocols. Mycotoxins are low molecular weight secondary fungal metabolites that are not immunogenic. Thus, mycotoxins must first be conjugated to a protein or polypeptide carrier before subsequent use in immunization; with this approach, antibodies against a number of mycotoxins have been developed and characterized. In identifying pertinent antibodies, a mycotoxin marker should also be available. Most of the early research was focused on the first two phases [2–8]. With the availability of antibodies against mycotoxins, including both polyclonal and monoclonal antibodies, and mycotoxin markers, several types of immunoassays for mycotoxins have been developed. In the last few years, efforts in the first two phases have continued for selected mycotoxins, but work has also been directed to the third and fourth phases [9–13, 16, 18].

Introduction of a Reactive Group to Mycotoxins

Mycotoxins that have a reactive group can be directly conjugated to a protein carrier for antibody production. For example, conjugations of fusarochromanone [19] and ochratoxin A [20–22] to protein carriers were made through their reactive amino and carboxylic groups, respectively. However, for most mycotoxins a reactive group must first be introduced into the mycotoxin molecule before it can be coupled to a protein [2–12]. Additional reactive groups can be introduced into various analogs of a specific mycotoxin, so that the reactive group would be located at different positions in the related parent molecule. For mycotoxins such as aflatoxins (AFs), zearalenone (ZEARA), and group B trichothecenes (TCTCs), which contain a carbonyl group, a carboxymethyl oxime (CMO) can be readily prepared [12]. For mycotoxins containing a hydroxyl side chain such as some aflatoxin metabolites and mycotoxins in the group A TCTCs, acylation of the mycotoxin with bifunctional anhydrides such as succinic acid and glutaric acid anhydride to form the corresponding hemisuccinate (HS) and hemiglutarate (HG) are generally used. However, because some mycotoxin-HS derivatives are unstable and easily hydrolyzed [23, 24], the stability of the hemisuccinate and hemiglutarate of mycotoxins should be tested before conjugation to protein.

Since the ability of a hapten to elicit antibodies in an animal depends on both the amount of the hapten conjugated to the protein and the orientation of the hapten on the protein molecule, the presence of a spacer alkyl chain, such as glutarate [24], between the protein and the mycotoxin has an additional advantage in improving the quantity and quality of antibody production. Studies at the Food Research Institute, University of Wisconsin (FR1, UW), have also found that immunogenicity of mycotoxin-protein conjugates improved when additional spacer groups were introduced into bovine serum albumin (BSA) with ethylenediamine [25]. Other proteins modified with ethylenediamine have also been used as carriers [26].

An alternative conjugation method for mycotoxins containing a primary alcohol is oxidation of the OH group to a carbonyl group for subsequent derivatization [27]. Reactive chloride and bromide have been introduced into the aflatoxin B_1 (AFB_1) molecule to allow the halogenated AFB_1 to react with the protein directly [28]. Aflatoxin B_1 has also been activated with *m*-chloroperoxybenzoic acid and then reacted with protein [29, 30]. The varied approaches for the preparation of mycotoxin derivatives and methods of conjugation to protein carriers are summarized in Table 1.

TABLE 1. Approaches Used for the Conjugation of Mycotoxins to Protein

Derivatives prepared	Mycotoxins used	Methods of conjugation
CMO	AFB, AFM, FX, DAS, T-2, ZEARA	WSC, MA, NHS
HA	AFB_{2a}, AFG_{2a}, AFQ_{2a}, ST	RA
HG	AFB_{2a}, DAS, T-2	WSC, MA, NHS
HP	T-2	WSC, MA
HS	AFQ, AFB_3, DAS, DON, Ac-DON, DOVE, T-2, HT-2, acetyl-NIV, roridin, 3'-OH-T-2, T-2-tetraacetate	WSC, MA, NHS
AcCl	ST	Direct
None	PR toxin	RA
None	AFB_{2a}, KJ	TA, FA
None	OTA, rubratoxin B	WSC, MA
None	AFB-Cl, AFB-Br, SA	Direct
None	AFB *m*-Cl-PBA	Direct
None	TDP-1 (fusarochromanone)	WSC, GA
None	CPA	FA

Abbreviations: AcCl, acyl chloride; CMO, carboxymethyl oxime; DAS, diacetoxyscirpenol; DON, deoxynivalenol; DOVE, deoxyverrucarol; FA, formaldehyde; HA, hemiacetal; HG, hemiglutarate; HP, hemiphthalate; HS, hemisuccinate; GA, glutaraldehyde; MA, mixed anhydride, NHS, *n*-OH-succinimide; RA, reductive alkylation; TAB, tetrazobenzidine; WSC, water-soluble carbodiimide; AFB, AFM, AFB_{2a}, AFG_{2a}, AFG_{2a}, AFB_3, and AFQ are aflatoxins B_1, M_1, B_{2a}, G_{2a}, Q_{2a}, B_3, and Q_1, respectively; CPA, cyclopiazonic acid; FX, fusarenon-X; KJ, kojic acid; OTA, ochratoxin A; ST, sterigmatocystin; ZEARA, zearalenone; *m*-Cl-PBA, *m*-chloroperoxybenzoic acid; SA, secalonic acid.

Conjugation of Mycotoxins

Among available conjugation techniques, the water-soluble carbodiimide (WSC) method is most commonly used; in this procedure, a mycotoxin or mycotoxin derivative is generally reacted with a protein carrier in a molar ratio of 10:1 to 100:1 (mycotoxin/protein) at pH 5–8 in the presence of a coupling reagent such as 1-(3-dimethylaminopropyl)-3-ethylcarbodiimide (EDPC) [12, 21, 31, 32]. Alternatively, the mycotoxin or its derivative is first activated with *N*-OH-succinimide (NHS) in the presence of carbodiimide and then conjugated to protein directly [11, 12, 18]. The mixed anhydride method involves activation of mycotoxin derivatives containing a carboxylic acid to their corresponding anhydride by isobutyl chloroformate and then direct reaction with protein; this technique has been used for conjugation of some mycotoxins [24, 25]. For mycotoxins containing an active aldehyde group, reductive alkylation has been used; a Schiff base is generally formed between the amino group of the protein and the aldehyde group in the mycotoxin molecule. Stabilization of the bond is achieved by reduction with sodium borohydride or other reagents.

Since the furan ring of aflatoxins B_{2a} (AFB_{2a}) [33, 34], G_{2a} [35], Q_{2a} [36], diol [37], or hemiacetal sterigmatocystin (ST) [38] opens up and makes two aldehyde groups available at pH above 7.0, the reductive alkylation method has been used for conjugation of aflatoxin and sterigmatocystin to protein. Cross-linking agents such as glutaraldehyde have been used for mycotoxins containing an amino group, such as fusarochromanone [19]. The Mannisch reaction, for which formaldehyde acts as a cross-linking reagent, has been successfully used in the preparation of immunogen eliciting antibody against cyclopiazonic acid (Chu and Huang, in preparation) and for condensation of the phenolic group of AFB_{2a} with the amino group of certain amino acid derivatives [39].

After pertinent reactions, unreacted mycotoxins are routinely removed by dialysis or Sephadex gel filtration, and the quantity of conjugated mycotoxin is determined [3, 8]. Conjugates with a ratio of 10–20 moles of mycotoxin per mole of protein have generally been found to be good immunogens. Although BSA has been commonly used as the carrier protein for the preparation of immunogens, other proteins such as γ-globulin, human serum albumin, ovalbumin, lysozyme, polylysine, and keyhole limpet hemocyanin have also been used successfully [12].

Production of Antibodies

Two approaches have been followed for the production of antibodies against mycotoxins. The classic approach is to immunize laboratory animals and subsequently recover polyclonal antibodies from their sera; antibodies against many mycotoxins have been successfully produced by this procedure. Immunogens are mixed with Freund's complete adjuvant and injected into the back of rabbits at multiple intradermal sites or subcutaneously [3, 12, 16, 18]. Goats [40] and pigs [41] have also been used as the immunization animals. If the conjugate is highly immunogenic, antibody titer with binding of a constant amount of mycotoxin-marker ligand usually starts to increase at 5–7 weeks. Subsequent booster injections of immunogen are generally given once each month. Since polyclonal antibodies are heterogeneous, a specific mycotoxin marker must be used in the titer determination to identify the presence of relevant antibodies. Inadequate mycotoxin markers fail to provide unequivocal information on whether the immunogen and immunization protocols are effective.

Although production of polyclonal antibodies is relatively simple and the affinity of polyclonal antibodies is generally high, the antibodies are heterogeneous and the production is limited. The demand for large amounts of antibodies with uniform properties for immunoassays and recent advances in hybridoma technology have led to the production of specific monoclonal antibodies for several major mycotoxins. In monoclonal antibody production, spleen cells are obtained from mice that have been immunized with mycotoxin-protein conjugates and are subsequently fused with myeloma cells; the hybrid cells are propagated, and selected clones are characterized in terms of elaboration of relevant antibodies. Production of antibodies can be carried out either in tissue culture or in the ascites fluids of mice. Thus, when a hybridoma cell line is obtained there is a potential for an unlimited supply of homogeneous antibodies with unique properties. Similar to the production of polyclonal antibodies, the key to success in obtaining a useful hybridoma cell line also lies in the effectiveness of the immunogens used and the mycotoxin marker used in the selection of positive clones. Both polyclonal and monoclonal antibodies against most mycotoxins have been obtained [12, 16, 18]. Antibodies that have been prepared against various mycotoxins and mycotoxin metabolites are listed in Table 2.

Characterization of Antibodies

Antibodies should be well characterized before they are used in the development of immunoassays. In general, both the type of immunoglobulin and the specificity of the antibody to various mycotoxin analogs should be determined. Information regarding the affinity of the antibody to various mycotoxins is also important and can be obtained from data on specificity; cross-reactivity of an antibody is determined primarily by the type of mycotoxin or its metabolites that have been used in the antibody production as well as by the conjugation site in the mycotoxin molecule. The degree of cross-reactivity of antibodies with their respective structural analogs varies considerably [8, 12].

The accuracy of immunoassay for mycotoxins in naturally contaminated samples generally is affected by both the specificity of the antibody and the presence of structurally related analogs of the mycotoxin that may react with the antibody [98]. Experimentally, the cross-reactivity of an antibody is determined by an immunoassay in which various structurally related analogs of a specific mycotoxin are used at a wide range of concentrations to compete with the binding of the marker ligand with the relevant antibody. A typical example of competitive radioimmunoassay (RIA) and enzyme-linked immunosorbent assay (ELISA) for aflatoxin B_1 is shown in Fig. 1. The specificity of antibodies against various mycotoxins has been documented in a number of reviews [8, 12, 14, 57].

TABLE 2. Antibodies Against Mycotoxins

Mycotoxin	Type	Ref.
Aflatoxins		
B_1, B_2, M_1	M, P	12, 16, 18, 26, 42–55
G_1, B_3 (B_1 and G_1)	M, P	12, 35, 53, 54
Aflatoxin metabolites		
B_{2a}, Q_1, M_1, R_0	M, P	12, 50, 52
DNA adducts	M, P	56–63
Cyclopiazonic acid	P	a
Fusarochromanone	P	19
Kojic acid	P	64
Ochratoxin A	M, P	20, 65–68
Patulin	P	69
PR toxin	P	70
Rubratoxin B	P	71
Secalonic acid	P	72
Sterigmatocystin	P	38, 73
Trichothecene		
DAS, DON, DOVE, FX	M, P	12, 74–78
AcDON, NIV, roridin A	M, P	79–84
T-2 toxin	M, P	12, 85–90
T-2 toxin metabolites		
HT-2, T-2-tetraacetate	M, P	12, 91–92
3'-OH-T-2, de-ep-T-2	M, P	93–96
Zearalenone	M, P	41, 97

Abbreviations: DAS, diacetoxyscirpenol; DON, deoxynivalenol; DOVE, deoxyverrucarol; NIV, nivalenol; de-ep-T-2, deepoxide T-2 toxin; M, monoclonal antibody; P, polyclonal antibodies.
[a]Huang and Chu (in preparation).

B. Development of Antibodies Against Aflatoxins and Their Metabolites

Antibodies Against Aflatoxins

Conjugation of aflatoxin to protein carrier can be achieved either through the cyclopentanone ring by conversion of the carbonyl group of aflatoxins such as B_1 [31, 32], M_1 [42], and B_2 [26] to carboxymethyl oxime (CMO) derivatives before conjugation to protein [8] or through the dihydrofuran ring portion of the molecule by conversion of AFB_1 or AFG_1 to

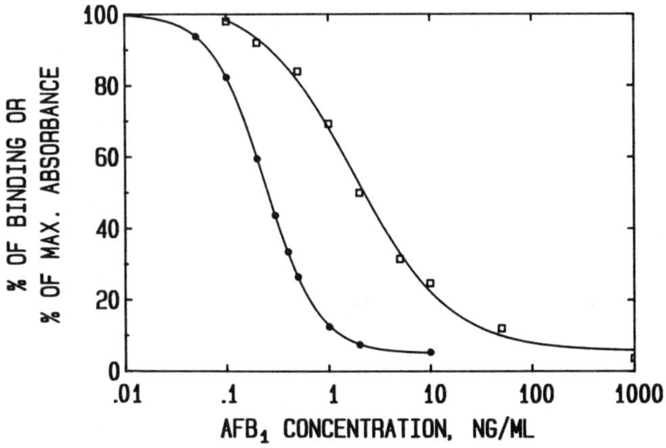

FIGURE 1. Standard curve for (□) RIA and (●) ELISA of aflatoxin B_1.

AFB_{2a} [34] or AFG_{2a} [35], which is then conjugated to protein by the reductive alkylation method (Fig. 2). Other approaches through the dihydrofuran ring portion of the molecule include conversion of AFB_1 to its chloride or bromide and activation of AFB_1 with *m*-chloroperoxybenzoic acid [28–30]. Conjugation through other portions of aflatoxin molecules could be made by chemical conversion to AFQ_1 followed by acylation to form the hemisuccinate or hemiglutarate and linkage to protein carriers.

The specificity of various polyclonal and monoclonal antibodies is summarized in Tables 3 and 4, respectively. Immunization of aflatoxin-protein conjugates in rabbits led to antibodies with different specificities (Table 3). When rabbits were immunized with conjugates prepared by linking the CMO of AFB_1 and AFM_1, the antibodies generally recognized the dihydrofuran ring of the molecule as well as structural variations in the dihydrofuran portion of the molecules. Thus, antibodies highly specific against AFB_1 and AFM_1 were produced; these two types of antibodies are currently used for most of the immunoassays for both aflatoxins. Antibodies for AFB_1 cannot recognize AFM_1, but antibodies for AFM_1 bind AFB_1 effectively; this property has been used as an approach for the immunoassay of AFM_1 so that radioactive AFB_1 or enzyme-linked AFB_1 can be used as the marker in the assays [8, 12, 14, 42, 107]. It is apparent that the side-chain hydroxyl group in AFM_1 plays an important role in eliciting antibodies that can recognize AFM_1. It is also apparent that there is a slight variation in antibody specificity when the same immunogen is used (Table 3). The variations are produced by the use of different markers in the analysis; the data reinforce the importance of the selection of the marker ligand in immunoassays.

When conjugates prepared through the dihydrofuran portion of the aflatoxin molecule were used in immunization, the antibodies exhibited specificity toward the cyclopentanone ring. Antibodies with similar specificity

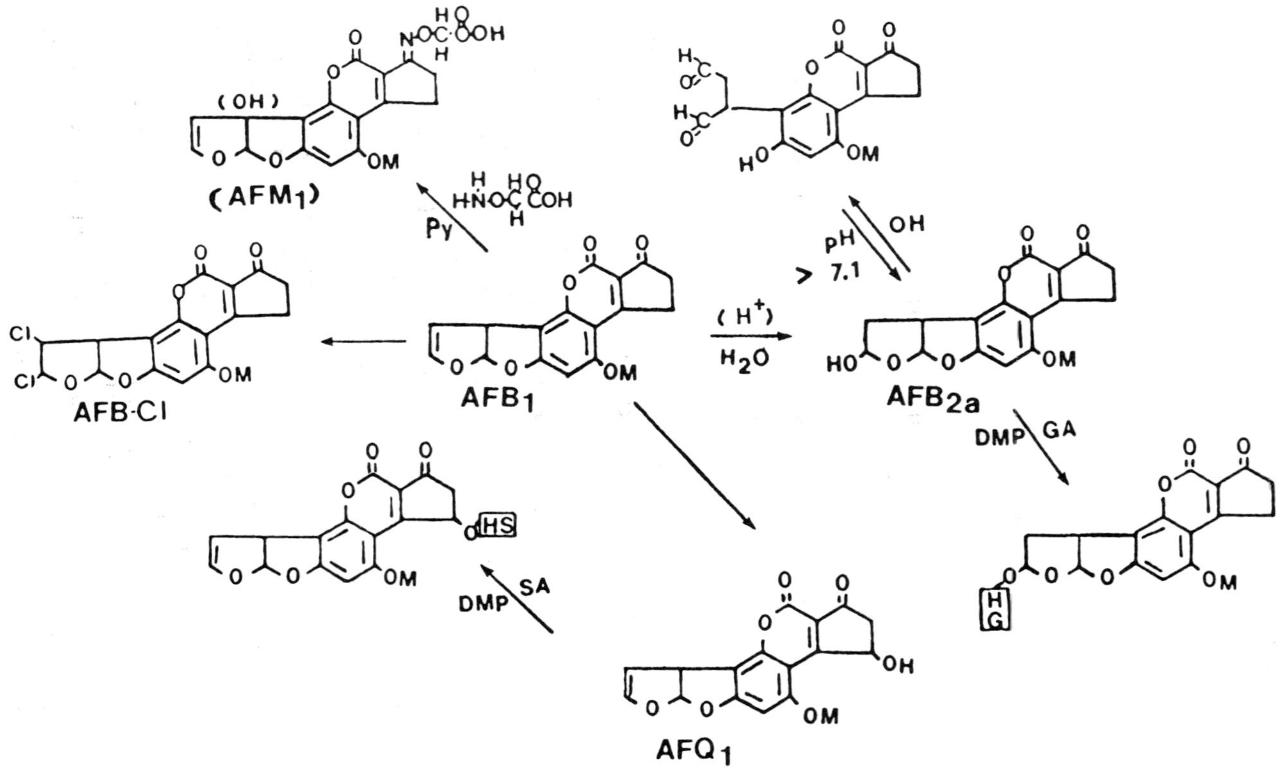

FIGURE 2. Approaches used for the preparation of various aflatoxin B_1 derivatives. Abbreviations are given in Tables 1 and 2. Py: pyridine; DMP: 4-dimethylaminopyridine. [After Chu (12).]

TABLE 3. Specificity of Antibodies Obtained from Rabbits After Immunization with Aflatoxin Conjugates[a]

Conjugate	Aflatoxin							Ref.
	B_1	B_2	G_1	G_2	M_1	B_{2a}	R_o	
BSA-CMO-B_1	100	11	9	1	<1	—	3	31
	100[b]	125[b]	31[b]	1[b]	0.1[b]		167[b]	99
	100	28	100	22	0			100
	100	0.2	1	0.1	0.1			14
	100	28	3.4	0.2	0.5			101
	100	10	37	6	—			102
	100	24	44	10	1			103
	100	<10	<3	<0.2	<0.1			14
EDA-BSA-CMO-B_1	100[b]	6[b]	1[b]	<1[b]	<0.1[b]			104
PLL-CMO-B_1	100	43	15	5	—	1		32
BSA-B_{2a}	100	13	3	<1	6	8	<1	34
	100[b]	25[b]	<1[b]	<1[b]	1[b]	125[b]	1[b]	34
	100[b]		7.5[b]					52
	56[b]	20[b]	0.1[b]	0.1[b]	1[b]	100[b]		105
BSA-B_1-Cl[c]	100	238	3	2	3	—	—	101
BSA-B_{2a}-HG	100	10	6	<1	2	11	<1	24
BSA-G_{2a}	2[b]	2[b]	8[b]	15[b]	2[b]	—	—	35
BSA-CMO-M_1	56	<1	<1	<1	100	<1	<1	42
	1[b]	<1[b]	<1[b]	<1[b]	100[b]	<1[b]	<1[b]	106
	100	33	21	4	65			14
BSA-Q_1-HS	100	1	<1	1	—	—	33	36
BSA-Q_{2a}	<1[b]	—	—	—	—	1[b]	—	36
MA-B_1	100[b]				75[b]			39
BSA-B_3	100	8	78	6	—	—	—	54
SA-M_1	12[b]	0.2[b]	<0.1[b]	<0.1[b]	100[b]			107

Abbreviations: See Tables 1 and 2.

[a]Expressed as percent of reactivity relative to aflatoxin B_1 except for BSA-G_{2a}, BSA-CMO-M_1, and BSA-Q_{2a} antibodies, which are relative to aflatoxins G_{2a} (100%), M_1, and Q_{2a} (100%), respectively. Data obtained from RIA except as noted.

[b]Data obtained from ELISA.

[c]BSA-B_1-Cl: aflatoxin dichloride conjugated to BSA directly.

TABLE 4. Specificity of Monoclonal Antibodies Against Aflatoxins[a]

Conjugate	Aflatoxin							Ref.
	B_1	B_2	G_1	G_2	M_1	B_{2a}	R_o	
BSA-CMO-B_1								
DX1	100	20	34	17	16	n.d.	—	49
KV	100	20	22		13			47
KW1	100	48	39	37	6	<1	1	50
KW3	100	2	6	4	8	<1	33, 112	50
KW4	100	2	8	10	10	2	98, 200	50
KW5	100	<19	<19	<19	<19	<19	46, 39	50
BSA-B_{2a}	100	18	67	50	6	100	<1	108
KLH-CMO-B_1								
CD1	100	13	14	1	7	—	—	45
EDA-BSA-CMO-B_2								
HA1	19	100	<1	<1	<1	<1		61
BSA-CMO-M_1								
WY1	10	<2	10	<2	100			44
DX2	160	13	3	7	100			49
BGG-AFB_1								
GP	100[b]	100[b]	5[b]	4[b]	100[b]			29
BSA-AFB_3								
He1	100[b]	5[b]	153[b]	6[b]	—			53
He2	100[b]	40[b]	154[b]	40[b]	—			53
He3	100[b]	40[b]	152[b]	23[b]				53

Abbreviations: BSA, bovine serum albumin; CMO, carboxymethyl oxime; KLH, keyhole limpet hemocyanin; BGG, bovine γ-globulin; n.d., not determined.

[a]Expressed as percent of reactivity relative to aflatoxin B_1 except BSA-B_{2a} and BSA-CMO-M_1, which are relative to AFB_{2a} (100%) and AFM_1 (100%), respectively. Data obtained from ELISA except as noted.

[b]Data from RIA.

[c]BSA-B_1-Cl: aflatoxin dichloride conjugated to BSA directly.

and a wide range of cross-reactivities were obtained from rabbits when protein conjugates were prepared by coupling AFB_{2a} (designed as B_{2a} antibody [8, 34, 57]), AFG_{2a} [35], AFB_1 diol (designated as AFB-diol antibody [37]), AFB_1 activated with *m*-chloroperoxybenzoic acid [29, 30], and AFB_1 chloride (designated AFB-Cl antibody [28, 101]). The antibodies also cross-react with AFB_1-DNA adducts and other related compounds [108, 109]. Most of the antibodies against aflatoxin are highly specific to AFB_1 and only partially cross-reactive with AFG_1, another important naturally occurring aflatoxin in some commodities such as peanuts. Therefore, attempts to produce antibodies that have good cross-reactivity to both AFB_1 and AFG_1 were carried out at the Food Research Institute, University of Wisconsin. To achieve the objective, AFB_3-HS was conjugated to protein for immunization. Because the structure of AFB_1 and AFG_1 is similar to that of AFB_3 (parasiticol), the antibodies elicited showed good reaction with both AFB_1 and AFG_1 and were very useful for immunoassays for both aflatoxins [53, 54].

A number of monoclonal antibodies against aflatoxins have been obtained in the last few years [26, 29, 43–47, 49, 50, 53, 55, 57]. The specificity of different monoclonal antibodies for aflatoxins also depends on the type of conjugate used in the immunization (Table 4). When the same immunogens are used for the preparation of either polyclonal or monoclonal antibody, most of the high-affinity monoclonal antibodies have almost the same specificity as polyclonal antibodies. Although clones that elicit antibodies with different cross-reactivities have been found, the antibodies generally have reduced affinities. For example, Kawamura et al. [50] found that some of their monoclonal antibodies for AFB_1 had high cross-reactivity with aflatoxicol (AFR_0), one of the aflatoxin metabolites. Nevertheless, affinity of the antibodies to aflatoxicol was lower than the affinity of the best monoclonal antibody to AFB_1.

Antibodies Against Aflatoxin-Macromolecule Adducts

Two types of antibodies against AFB_1-DNA adducts have been prepared [29, 56–59, 62]. One type of antibody was obtained from animals after immunization with AFB_1-DNA adducts complexed with methylated BAS (MBSA) [60, 61] or methylated hemocyanin (MH) [62, 63]. Another type of antibody involved immunization of aflatoxin-protein conjugates that were coupled through the 8,9-dihydrofuran (or 2,3-dihydrofuran) portion of the molecule [33–37]. Monoclonal and polyclonal antibodies of both types have been obtained. With AFB_1-DNA immunogens, the monoclonal antibodies prepared against AFB_1-guanine-modified DNA and AFB_1-ring-opened modified DNA [62, 63] showed virtually no cross-reactivity with aflatoxins B_1, B_{2a}, diol, and AFB_1-guanine; the antibodies exhibited a degree of recognition for epitopes in AFB_1-DNA adducts regardless of whether they were AFB_1-DNA or AFG_1-DNA [57]. However, the aflatoxin-protein immunogen yielded antibodies that cross-reacted with a number of aflatoxins in addition to aflatoxin-DNA adducts and the hydrolyzed products of aflatoxin-macromolecule adducts including hydrolyzed products of both aflatoxin-protein and aflatoxin-DNA adducts. The cross-reactivity of various antibodies to aflatoxin-macromolecule adducts is summarized in Table 5.

C. Development of Antibodies Against Trichothecene Mycotoxins and Their Metabolites

Antibodies Against Group A Trichothecenes

Production of antibodies against group A TCTCs can be illustrated by the preparation of various derivatives of T-2 toxin and its metabolites (Fig. 3). The functional hydroxyl group at the C-3 position has been used exclusively

TABLE 5. Antibodies Cross-Reactive with Aflatoxin-Macromolecule Adducts

	Aflatoxin-macromolecule adduct							
Conjugate[a]	B_1	B_{2a}	Diol	B-DNA	B-Gua	B-FRY-DNA	B-FaYr	Ref.
BSA-B_{2a} (P)	80	100	80	50	18	40	22	109
BSA-B_1-diol (P)	200	100	100	32	0.6	—	—	37
BSA-B_1-Cl (P)	100	—	—	—	2	—	4	57
BGG-AFB (M)	100	—	—	—	3.0	—	33	29
MBSA-B_1-DNA (M)	<0.1	<0.1	<1	100	<1		<1	60
MHB_1-FA-Pyr-DNA (M)	—	—	<0.1	53	<0.1	100	<0.1	62
				10	<0.1	100		63

[a]P = polyclonal antibody; M = monoclonal antibody.

as the site for the introduction of a reactive group in the molecule in most studies. A carboxylic group was introduced at this position either by converting the hydroxyl group to hemisuccinate (routes 1 and 8), hemiglutarate (route 2) [9, 61], or hemiphthalate [110] or by converting it to the ketone [27] and subsequently to the CMO (route 3). In a comparative study, Ohtani et al. [110] found that hemiglutarate-protein conjugate was most effective for antibody production. For the deacetylated T-2 toxin metabolites, T-2 toxin was first converted to the CMO-T-2 toxin, which was then hydrolyzed to the respective CMO derivatives (route 4—5) [91]. With a reactive group in the mycotoxin, the compound can be conjugated to the protein or enzymes via the mixed-anhydride method or carbodiimide techniques [8]; these approaches have also been used for the preparation of diacetoxyscirpenol-protein conjugates [23, 27]. A generic antibody against group A TCTCs was produced by utilization of an immunogen that was prepared by conjugation of a 3-acetyl-neosolaniol hemisuccinate (3-Ac-NEOS-HS) to protein at the C-8 position instead of by using conjugates that had been coupled to the protein through the C-3 position of TCTCs (routes 9 and 10) [92, 95].

The relative cross-reactivity of different types of polyclonal antibodies obtained from rabbits after immunization with various TCTC mycotoxin-BSA conjugates with various group A TCTCs is summarized in Table 6. It is apparent that the pertinent antibodies are very specific for the toxins that were originally used in the conjugation. Since T-2 antibodies cross-reacted with HT-2 [12, 91, 111], it is possible to use the antiserum for monitoring HT-2; however, it is useless for the analysis of other important metabolites of T-2 toxin. Likewise, it is also possible to use the HT-2 and 3'-OH-T-2

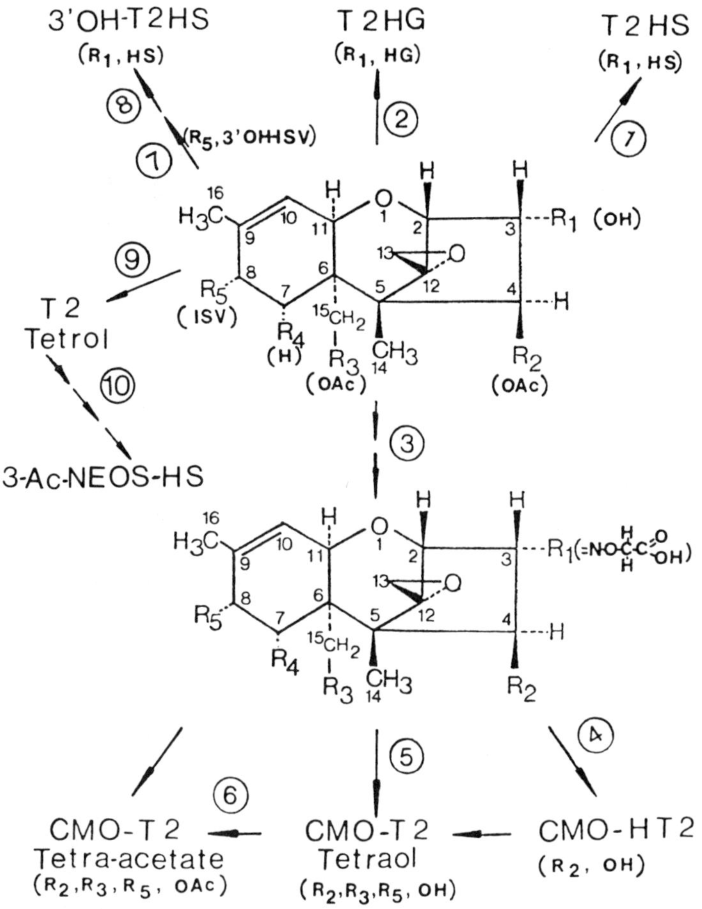

FIGURE 3. Approaches used for the preparation of various T-2 toxin derivatives. Abbreviations are given in Tables 1 and 2. [After Chu (12).]

antibodies to measure T-2 toxin [61, 91]. Antibodies for diacetoxyscirpenol (DAS) [23, 27, 75–77] are highly specific for DAS itself and cannot be used for monitoring DAS metabolites including either 4-monoacetyl-DAS (4-MAS) or 15-monoacetyl-DAS (15-MAS) as well as scirpentriol. Antibodies against T-2 tetraol tetraacetate [91] were also obtained, with specificity to the tetraacetate of T-2 toxin but not to the tetraol. It was found that the side chain of TCTC group plays an important role in elicitation of antibody against this group of mycotoxins. Consequently, 3-Ac-NEOS-HS conjugated to BSA was used as an immunogen for antibody production [92, 95]. Antibodies, both polyclonal and monoclonal, obtained from animals after immunization with this immunogen showed cross-reactivities with most of the group A TCTCs. The affinity of the antibodies for the acetylated group A TCTCs was almost 10 times higher than for the nonacetylated group A TCTCs; this

TABLE 6. Specificity of Polyclonal Antibodies Against Major Group A Trichothecenes[a]

	Trichothecene										
Conjugate	Ac-T2	T2	IsoT2	HT2	T2-3ol	3'OH-T2	T2-4Ac	DAS	MAS	TAS	Ref.
T-2-HS-BSA	100	100	50	17	2	n.d.					85
	1000[b]	100[b]		13[b]	1[b]			0.1[b]			86
		100		10	0.1						111
T-2-HG-BSA	4	100	—	20							6
O-CMO-T-2-BSA	25	100	—	20	1						27
3-Ac-NEOS-HS-BSA	100	100	1	—	—	50	100	14			95
	0.3[b]	0.2[b]					100[b]			395[b]	112
O-CMO-HT-2-BSA	3	25	10	100							91
O-CMO-T-2-rAc-BSA		—					100				91
3'-OH-T-2-HS-BSA	34	25	3	2		100					94
DAS-HG		0.1						100	1		23
DAS-*O*-CMO		1						100			27
FX-*O*-CMO	6[b]	367[b]		4[b]				275[b]			76

Abbreviations: Same as Tables 1 and 2. MAS and TAS represent monoacetoxyscirpenol and triacetoxyscirpenol, respectively.

[a]Expressed as percent of reactivity relative to T-2 toxin of DAS except FX-*O*-CMO, which is relative to FX (100%). Data obtained from RIA except as noted.

[b]Data from ELISA.

[c]100% cross-reactive with FX.

property has become very useful for the development of an ELISA method for the detection of less than 10 ppb of group A TCTC in urine and serum [112]. The epoxide group in the TCTC does not appear to be an important epitope for the production of antibody because two types of antibodies raised against T-2 toxin (anti-T-2-HS-BSA and anti-3-Ac-NEOS-HS-BSA) showed cross-reactivity with deepoxy T-2 toxin [96].

The specificity of monoclonal antibodies against several group A TCTCs is summarized in Table 7. Several monoclonal antibodies were found to have specificities that differ from those of the polyclonal antibodies obtained with the same type of immunogen. For example, immunization with T-2-HS-BSA conjugates resulted in monoclonal antibodies from different laboratories with varied specificities. Some monoclonal antibodies have high specificity against HT-2 toxin [87], whereas others have specificity against 3'-OH-T-2 toxin [93]. Since TCTCs have different side chains, the diversity of specificity of these monoclonal antibodies might be due to the hydrolysis of either the immunogen in the immunization process or the hydrolysis of the T-2 toxin conjugate in the microtiter plate during selection of specific clones.

Antibodies Against Group B Trichothecenes

Difficulties have been encountered in the production of antibodies against type B TCTCs such as deoxynivalenol (DON), in which a carbonyl group is present at the C-8 position. The main problem was the presence of many hydroxyl groups in the molecules; such groups not only create problems in conjugation to proteins but also appear to decrease the immunogenicity. For example, no antibody against DON was obtained in rabbits when the CMO-DON-BSA conjugate was used in the immunization [80]. Conjugation of DON to protein through the hydroxyl group at the C-3 position was achieved by protecting other hydroxyl groups, but the affinity of the antibody for DON was low [74]. Antibody against 3-acetyl-DON was obtained after immunization of rabbits with an immunogen that was prepared by conjugation of 3-acetyl-DON at the C-15 position [79]; this antibody also had low affinity for DON. Nevertheless, antibody with high affinity to triacetyl-DON was obtained after immunization with the hemisuccinate of 7,8-dihydroxycalonectrin (DHC) conjugated to BSA [80]. Similar observations were found with nivalenol (NIV); thus, antibodies against tetraacetyl-NIV was used for the immunoassay of nivalenol [81, 82, Wang and Chu, 1991, in press]. Since antibody against triacetyl-DON [80], 3-acetyl-DON [79], and tetraacetyl-NIV [81, 82] can be used only for measuring triacetyl-DON, 3-acetyl-DON, and tetraacetyl-NIV, DON and NIV and related acetyl derivatives must be first converted to the triacetyl-DON and tetraacetyl-NIV before analysis of DON and NIV can be made [81, 82, 113, 114]. Nevertheless, the simultaneous use of the two antibodies in immunoassays provides for determination of most of the group B TCTCs. Both polyclonal and monoclonal antibodies against fusarenon-X (FX) were obtained by Martlbauer et al. [76] when fusarenon-X oxime conjugated to HSA was used as an immunogen. Since conjugation of the ligand to HSA was made through the C-8 position, both antibodies showed good cross-reactivities with several group A TCTCs including T-2 toxin, DAS, NEOS, and fusarenon-X. Thus, the fusarenon-X antibodies could be used for monitoring only group A TCTCs and fusarenon-X.

TABLE 7. Specificity of Monoclonal Antibodies Against Major Group A Trichothecenes[a]

	Trichothecenes										
Conjugate	Ac-T2	T2	IsoT2	HT2	T2-3ol	3'OH-T2	T2-4Ac	DAS	MAS	TAS	Ref.
T-2-HS-BSA											
Hu1		100[b]		200[b]	n.d.						87
HU2		100[b]		26[b]							87
Ha	75	100	35	35	22						90
Ge	100	100		2		200					93
T-2-HS-KLH											
Gol	—	100	—	25							89
3Ac-NEOS-HS-BSA											
Fa		100[b]					200[b]	20[b]			92
DAS-HG-BSA											
Ha		6						100		2224	75
Pa		—						100	533/13		77

Abbreviations: Same as Tables 1 and 2.
[a]Expressed as percent of reactivity relative to T-2 toxin (100%) or DAS (100%). The data were obtained from ELISA except as noted.
[b]Data from RIA.

Antibodies Against Macrocyclic and Other Types of Trichothecenes

Both polyclonal and monoclonal antibodies against a macrocyclic TCTC, roridin A, have been recently obtained [83, 84] by immunization of rabbits and mice with an immunogen prepared by conjugation of hemisuccinate-roridin A to human serum albumin. The antibodies exhibited the highest cross-reaction with roridin family TCTCs. For the polyclonal antibodies, the relative cross-reactivities with roridin A (RA), roridin J (RJ), verrucarin A (VA), satratoxin H (SH), and satratoxin G (SG) were 100, 41, 15, 15, and 7, respectively. Among 16 nonmacrocyclic TCTCs, only DAS and verrucarol had less than 0.2% cross-reactivity. With the same immunogen, two monoclonal antibodies were obtained [83, 84]. One clone produced antibodies with cross-reactivities to different macrocyclic TCTCs similar to those of the polyclonal antibodies. The other antibody had the highest binding with roridin A, but its cross-reactivity to verrucarin A was 64% that of roridin A, and with roridin J, satratoxin H, and satratoxin G, it was less than 10% that of roridin A [84]. In an earlier effort to produce generic-type antibodies against TCTCs, deoxyverrucarol (DOVE), a TCTC that contains only one OH at the C-15 position, was selected as a starting compound for the preparation of immunogen by making a hemisuccinate of the derivative with subsequent conjugation to BSA [78]. However, the antibody elicited by the immunogen was very specific for DOVE. Since the cross-reactivity of the antibody with verrucarol was about 10% of that with DOVE, it is possible to use the antibody for the analysis of verrucarin A after hydrolysis of verrucarin A to verrucarol.

D. Development of Antibodies Against Other Mycotoxins

Ochratoxin A

Both monoclonal and polyclonal antibodies against ochratoxin A (OTA) were obtained after immunization of animals with OTA-BSA or OTA-human γ-globulin conjugate [20, 21, 67]. The polyclonal antibodies for OTA were most specific for OTA (100%) and ochratoxin C (OTC) (37.5%) and were not specific for ochratoxin B (0.8% for RIA and 14% for ELISA), OT-α, or other coumarins (less 0.1%) [115]. Thus, the antibodies cannot be used for monitoring OT-α, which has been reported to be a major ochratoxin metabolite in animals. Two different monoclonal antibodies against OTA have been obtained; one monoclonal antibody that had a lower affinity to OTA ($\sim 2 \times 10^6$) was shown to have good cross-reactivity with OTA (100%), OTC (150%), and OT-α (39%) [65]. However, the cross-reactivity of a monoclonal antibody with high affinity to OTA ($\sim 1 \times 10^8$) was similar to that of polyclonal antibodies [66]. Results obtained from both monoclonal and polyclonal studies indicate that the phenylalanine and chloride in OTA play an important role in eliciting OTA antibody.

Sterigmatocystin (ST)

Conjugation of sterigmatocystin (ST) to protein was made either through the phenolic hydroxyl group in the anthraquinone portion of the molecule by reacting sterigmatocystin with sebacoyl dichloride in the presence of pyridine followed by conjugation to protein [73] or through the dihydrofuran ring moiety after it was converted to the hemiacetal of sterigmatocystin similar to the reaction with AFB_{2a} [38]. Immunization of rabbits with conjugate prepared through the phenolic group of sterigmatocystin led to the production of polyclonal antibodies that are highly specific for ST-hemiacetal with less cross-reactivity with sterigmatocystin (less than 0.1%), *O*-methyl-ST, 5-methoxy-ST, and aflatoxins [73]. The antibody obtained from rabbits immunized with sterigmatocystin hemiacetal, that is, coupling through the dihydrofuran portion of the sterigmatocystin molecule, also shows the best cross-reactivity with sterigmatocystin hemiacetal. Some cross-reactivity of the antibody with dehydrosterigmatocystin was demonstrated (16 times less than with sterigmatocystin hemiacetal), but it did not cross-react with aflatoxins or 6-*O*-methyl-ST [38]. Similar to the antibodies against AFB_{2a}, antibodies against sterigmatocystin also effectively cross-react with sterigmatocystin-DNA adducts (Olson and Chu, unpublished observations).

Zearalenone

The 6-*O*-CMO derivative of zearalenone conjugated to BSA was used as an immunogen. Both rabbits and pigs were used in eliciting polyclonal antibodies against zearalenone [41, 116]. The polyclonal antibodies obtained from pigs were most specific for zearalenone (100%) and α-zearalenol (100%), with good cross-reactivity with α-zearalenol (53%) and β-zearalanol (44%) [41, 116]. Polyclonal antibodies obtained from rabbits also showed good cross-reactivity with α-zearalenol (50% of zearalenol) but less cross-reaction with β-zearalenol (12%) and zearalanols (6% for α and 3% for β) in the indirect ELISA. In the direct competitive ELISA in which an *N*-OH-succinimide method was used in the preparation of the enzyme conjugate, the antiserum showed enhanced affinity with α-zearalenol, and cross-reactivity with β-zearalenol was reduced [117]. However, monoclonal antibodies against zearalenone show good cross-reaction with most pertinent metabolites, and the relative cross-reactivity (zearalenone = 100) for α-zearalenol, β-zearalenol, α-zearalanol, and β-zearalanol was 107, 29, 35, and 25, respectively. Thus, the antibodies could be used both for the determination of zearalenone in agricultural commodities and for monitoring zearalenone metabolites in serum and urine [41, 117].

Others

Antibodies against several other mycotoxins including kojic acid [64], patulin [69], fusarochromanone [19], PR toxin [70], rubratoxin B [71], secalonic acid [72], and cyclopiazonic acid (Huang and Chu, unpublished observations) have also been produced. Most of the mycotoxins were conjugated to protein directly without any derivatization, except for PR toxin, which was conjugated to protein by a reductive akylation method [70].

III. DEVELOPMENT OF IMMUNOASSAY PROTOCOLS FOR MYCOTOXINS

Although a number of immunochemical methods have been used for the analysis of low molecular weight biological substances, only radioimmunoassay

(RIA), enzyme-linked immunosorbent assay (ELISA), and immunoaffinity assay (IAA) have been developed for mycotoxin assays. Recent developments have led to several quick screening tests, and more than 10 types of commercial kits have become available in the last few years [11, 13, 18, 118]. With the exception of IAA, almost all the immunoassays for mycotoxin detection involve competition between the binding of the target mycotoxin to the antibody and the binding of mycotoxin marker to the antibody. An efficient method for the separation of the free and bound mycotoxin or mycotoxin marker is important in the techniques.

Samples after extraction from the solid matrix and dilution in buffer can be used directly in most immunoassays for mycotoxins. However, the sensitivity of detection increases considerably if a sample cleanup is included. Consequently, decisions should be made regarding cleanup in order to reach desired detection levels. A study should always be done to determine the maximum sample level that may interfere with the immunoassay for each mycotoxin.

A. Radioimmunoassay for Mycotoxins

Radioimmunoassay involves the use of a radioactive mycotoxin marker in the assay to allow competition of the mycotoxin in the sample and the marker for the binding of the antibody [8]. Thus, the RIA procedure generally includes the incubation of specific antibody with a solution of unknown sample or known standard and a constant amount of labeled toxin. After separation of free and bound toxin, radioactivity in the fractions is determined. The toxin concentration of the unknown sample is determined by comparing test results with a standard curve; the latter is established by plotting the ratio of radioactivities in the bound fraction and the free fraction versus the logarithm of the concentration of unlabeled standard toxin. Several methods, including ammonium sulfate precipitation [31], double antibody technique [8, 32], a solid-phase RIA method in which the immunoglobulin G (IgG) is conjugated to CNBr-activated Sepharose gel [119], a dextran-coated charcoal column [120], albumin-coated charcoal [111], and polyethylene glycol 6000 [120, 121], have been used for the separation of free and bound toxins in the RIA. With the introduction of controlled-size fine-particle magnetic gels and linked specific antibody, RIA is now simpler to perform (Wei and Chu, unpublished observations).

Since a radioactive mycotoxin marker is used in the RIA, the specific activity of the radioactive ligand plays an essential role in the sensitivity of RIA; both tritiated mycotoxins and iodinated mycotoxin derivatives have been used. High-specific-activity tritiated AFB_1 and ochratoxin are commercially available and have been used in the RIA of AFB_1 [31] and OTA [21]. Methods for the preparation of high-specific-activity tritiated TCTCs have also been established by oxidation of the hydroxyl group and subsequent reduction with tritiated $NaBH_4$ [23]. Tritiated mycotoxins are generally used in the RIA with a sensitivity of 0.20–0.5 ng/assay (2–5 ppb), but recent developments have led to the use of more sensitive iodinated markers. For example, 3'-[^{125}I]tyramine-AFB_1-1-*O*-carboxymethyl oxime (2300 Ci/nmol) was used in the RIA of both AFB_1 and AFM_1 [102, 121–126]. As little as 10 pg of AFB_1 could be measured in each assay in the system. Both ^{125}I-OTA [127, 128] and [^{14}C]OTA [68] have also been used for OTA analysis.

In general, RIA can detect as little as 0.05–0.5 ng of purified mycotoxin in a standard preparation [8]. The lower limits for mycotoxin detection in food or feed samples by RIA is about 1–5 ppb. Whereas RIA generally does not require an extensive cleanup, the sensitivity can be improved if a simple cleanup is done. For example, after a Sep-Pak treatment, the minimum detection level for T-2 toxin in corn and wheat by RIA decreased to 1 and 2.5 ppb from 2.5 and 5 ppb, respectively [129, 130]. As low as 10 ppt of AFB_1 in liver tissue could be detected after Sep-Pak treatment of the sample extract [131, 132]. Using antibodies against acetyl-DON and T-2 tetraol tetraacetate, RIA for DON [113] in corn and wheat and for T-2 tetraol in urine [91] have a sensitivity of 50 ppb.

Although RIA is a simple and sensitive method for mycotoxin analysis, it is limited by the need of a marker toxin with high specific radioactivity, instruments for measuring radioisotopes, and licenses for the use and disposal of radioactive materials. Thus, not all laboratories are equipped to do RIA. Consequently, ELISA has become more popular in the last few years. Nevertheless, it should be pointed out that because the radioactive marker ligand has the same structural features as the compound to be analyzed [8, 13], RIA is an effective method in the initial phase for screening of antibodies and could be used by laboratories equipped to perform such assays.

B. Competitive Enzyme-Linked Immunosorbent Assays (ELISA)

Two types of ELISA have been used for analysis of mycotoxins; both types are heterogeneous competitive assays that involve the separation of free (unbound) toxin in one phase (generally in liquid) from the bound toxin in another (solid phase). One type, direct ELISA, involves the use of a mycotoxin-enzyme conjugate, and the other system, indirect ELISA, involves the use of a protein-mycotoxin conjugate and a secondary antibody to which an enzyme has been conjugated. Although horseradish peroxidase (HRP) is most commonly used as the enzyme for conjugation, other enzymes such as alkaline phosphatase and β-galactosidase are available [12, 18, 51].

Direct Competitive ELISA

Specific antibodies are initially coated on a solid phase such as the wells of a microtiter plate [8, 11, 12]; other solid phases such as nylon beads, Terasaki plates [133], and polystyrene tubes [105] have also been used. Either a sample or standard toxin solution is generally incubated simultaneously with enzyme conjugate or incubated separately in two steps. After washing, the amount of enzyme bound to the plate is determined by incubation with a substrate solution [12]. The resulting color [12] or fluorescence [81], which is inversely proportional to the mycotoxin concentration present in the sample, is measured instrumentally or by visual comparison with standards. In general, direct ELISA is approximately 10–100 times as sensitive as RIA with purified mycotoxin; as little as 2.5–5 pg of pure mycotoxin can be measured. However, the sample matrix sometimes causes interference with the assay; this problem is generally overcome by dilution of the sample to a range that does not affect the assay [134] or by using sample extract in the preparation of the standard curve [8, 12, 16, 18]. Since a cleanup step is generally not necessary, many samples can be analyzed within a relatively short period. The sensitivity of direct ELISA for the analysis of mycotoxins in foods and feeds is generally in the range of 0.05–50 ppb [3–12, 16, 18]. Like RIA, the sensitivity of ELISA can be improved when a cleanup treatment is included [8–12]. Due to the use of better antibody and toxin-enzyme conjugates, the time required to run the ELISA has improved considerably. Thus, the entire ELISA procedure can be completed within an hour. By shortening the incubation time and also by using removable microtiter wells, the method can be used as a screening test for mycotoxins [11, 12, 18, 134–136].

Indirect Competitive ELISA (or Double Antibody ELISA)

In the indirect ELISA, a mycotoxin-protein (or polypeptide) conjugate—an immunogen but generally with a different carrier protein or peptide rather than the original immunogen conjugate—is coated to the microtiter plate [8–12, 16, 18]. The plate is incubated with specific antibody against the mycotoxin in the presence or absence of the homologous mycotoxin. The amount of antibody bound to the plate coated with mycotoxin-protein conjugate is then determined by reaction with a second antibody–enzyme complex such as goat anti-rabbit IgG–enzyme complex or goat anti-mouse IgG (IgM)–enzyme complex that is commercially available. Toxin in the samples and toxin in the solid phase compete for the same binding sites with the specific antibody in solution. The sensitivity of the indirect ELISA is comparable to or slightly better than that of the direct ELISA, with the advantage that much less antibody (1% as much) is required. In addition, it is not necessary to prepare a toxin-enzyme conjugate. The disadvantage of the method is that an additional incubation step is necessary for the assay, and thus it requires a longer analytical time (2–3 h).

Two modifications on the indirect ELISA have been made for the analysis of several mycotoxins. One approach involves conjugation of antibody, especially monoclonal antibody, to an enzyme and direct use in the assay [137, 138]; this modification converts the indirect ELISA format to a direct assay, and some studies have called this "direct ELISA" instead of indirect ELISA [45, 89]. In the second approach the antibody is premixed with the second antibody-enzyme conjugate [138]. Both modifications have shortened the assay time without sacrificing sensitivity. Because it requires only small amounts of antibody, the indirect ELISA has also been used extensively for monitoring the antibody titer of hybridoma culture fluids for the screening of monoclonal antibody–producing cells [8–12, 16, 18].

C. Immunoaffinity Assay

The immunoaffinity assay, for which an antibody column is used to trap mycotoxins, has been used for AFB_1, AFM_1, and OTA [11, 12, 58, 139–141]. After the sample is loaded and the column is washed, the toxin can be eluted for subsequent analysis or adsorbed to a solid phase for fluorescence determination. The affinity column serves as a specific cleanup and concentration tool for analysis. Recent improvements in the instrumentation for fluorescence detection and postcolumn derivitization have led to wider application of the method for screening various commodities for aflatoxins. An AOAC collaborative study with immunoaffinity assay has shown promising results [142]; aflatoxin extracted from the sample is diluted with buffer at pH 7.0 and then placed in a disposable affinity column containing anti-aflatoxin antibody-Sepharose gel. After washing, aflatoxin is removed from the column with methanol and subjected to treatment with iodine or bromine solution, and fluorescence is determined. The method cannot be used for mycotoxins, such as TCTCs, that do not exhibit high fluorescence or possess a chromophore.

D. New Developments

Several new approaches to the analysis of mycotoxins have been made in recent years. One method, called the "hit-and-run" assay [143], was developed for T-2 toxin. In this assay, an affinity column is prepared by conjugating T-2 toxin to Sepharose gel and is then equilibrated with fluorescein isothiocyanate (FITC)-labeled Fab fractions of IgG (anti-T-2 toxin). Samples containing T-2 toxin are injected into the column. The FITC-Fab that elutes together with the samples containing T-2 toxin is then determined in a standard flow-through fluorometer. The detection limit for T-2 toxin was found to be between 25 and 50 ng/mL. Although the assay is rapid and the column can be used many times, it is very selective and requires specific monoclonal antibodies (Chu, unpublished observations). In addition, the sensitivity is not very high.

Another approach that may lead to the development of a biosensor is a homogenous immunoassay for T-2 toxin that involves the use of liposomes and complement [144]. Unfortunately, the system is not very sensitive. Antibody against T-2 toxin has been conjugated to fiber optics for further development of a biosensor [145]. Recently, a time-resolved fluoroimmunoassay was developed for aflatoxin analysis [146]; this method involves the use of europium-labeled antibodies and has a sensitivity similar to most ELISA methods with a I-50 (50% inhibition of binding) concentration of 0.2 ng AFB_1/mL.

E. Quick Screening Tests

By shortening the incubation time and adjusting the antibody and enzyme concentrations in the direct or modified indirect competitive microtiter plate ELISA assay systems, it is possible to obtain a positive or negative test at certain toxin levels (such as 20 ppb) as a quick screening test for aflatoxin. The approach has been adapted by the AOAC as first action for screening aflatoxin at 15—30 ppb in several commodities [11, 12, 18, 135, 136, 147, 148].

Several other types of immunoscreening tests, whose sensitivity is similar to that of ELISA, were developed for mycotoxin analysis in recent years [1, 11, 12, 149]. Rather than coating the antibody onto the microtiter ELISA plate, other approaches immobilize the antibody on a paper disk or membrane that is mounted either in a plastic card ("card screen test" [136, 148, 151]) or a plastic cup ("cup test" [148, 149]) or on the top of a plastic tube (Cite-Probe [152]). The reaction can be carried out on the wetted membrane disk; the principle of the reaction in this assay format is similar to that of direct ELISA. In practice, a few drops of sample extract can be applied to the test spot on a plastic card or in a cup followed by reaction with aflatoxin-enzyme conjugate, washing, and finally reaction with the substrate. In the Cite-Probe test, the probe, which has antibody coated on it, is dipped into the wells containing sample, toxin-enzyme, and substrate in that order [152]. Absence of color at the spot on the card, cup, or probe compared to a negative control spot indicates that the sample contains mycotoxins; the reaction is generally very rapid, taking less than 10 min to complete. Thus, the rapid screening test kits permit monitoring of mycotoxins semiquantitatively in the field.

Good correlation between the quick-screening methods with conventional methods such as TLC and HPLC has been found [136, 142, 149, 151, 153]. Collaborative studies on several new screening tests are currently being carried out [1, 135, 136, 142, 149, 151].

F. Complementarity of Chemical Analyses and Immunochemical Methods

Immunoaffinity Chromatography As a Cleanup Tool

The application of affinity column chromatography for the immunoassay of mycotoxins began when Sun and Chu [119] immobilized antibody on Sepharose gel in an RIA. Subsequent studies led to use of the affinity column as a tool for preconcentration and cleanup of samples of mycotoxins before RIA or HPLC analysis [51, 58, 140, 154–157]. With the availability of different monoclonal antibodies against AFB_1 and AFM_1, affinity chromatography has gained wide application as a cleanup tool for aflatoxin in recent years. Because the first step in affinity chromatography involves the binding of toxin with the column at neutral pH, the method is ideal for biological fluids; it is used frequently for the analysis of aflatoxin metabolites, especially AFM_1, in biological fluids including urine and milk [11, 12, 18]. A robotics system that uses an affinity column and HPLC was recently developed for routine analysis of AFM_1 in milk [158].

Immunochromatography (IC)

An approach that makes use of immunochemical methods for detection of mycotoxin after separation of the toxin by either high-performance liquid chromatography or thin-layer chromatography was developed recently. The term *immunochromatography* expresses the idea that a chromatogram can be generated by TLC or liquid chromatography when any immunochemical method is used to determine the immunoreactive compounds in the relevant fractions after chromatographic separation. Accordingly, the terms HPL-IC and TL-IC are used for immunochromatography after separation by high-performance liquid chromatography and thin-layer chromatography, respectively. Various group A TCTCs can be separated on a C-18 reversed-phase column; individual fractions eluted from the column can be analyzed by ELISA using generic antibodies against group A TCTCs. The approach cannot only identify each individual group A TCTC but can also determine concentrations quantitatively. Two nanograms of T-2 toxin, related TCTCs, and their metabolites can be monitored by the method. Combination of HPLC and ELISA technology is an efficient, sensitive, and specific method for the analysis of TCTCs and other mycotoxins [159]. Crude samples of various TCTCs have been separated by TLC; each fraction from the thin-layer chromatogram was eluted, acetylated, and then subjected to ELISA for mycotoxin quantitation.

IV. USE OF IMMUNOCHEMICAL METHODS FOR THE DETECTION OF ECOLOGICALLY IMPORTANT MYCOTOXINS

A. Application of Immunochemical Methods for Monitoring Mycotoxins in Foods and Feeds

Immunoassays for Aflatoxins

With the availability of various immunochemical reagents and the development of efficient immunoassay protocols, immunoassays have gained wide application for analyzing various commodities for mycotoxins, especially aflatoxins [10]. Chu [11] and Lee [118] have described more than 10 types of commercially available immunoassay kits for mycotoxin analysis. Most of the kits were designed for analysis or screening of aflatoxins. The efficacy of some commercially prepared kits has been studied [153, 160, 161]. Criteria for selection of appropriate kits for mycotoxin analysis have been suggested by Chu [10, 11] and Pestka [18].

Immunoassay protocols have been established for analysis of aflatoxins in barley [162], corn [98, 104, 134, 136, 149, 163–166], cottonseed [135, 149, 164, 166], peanuts [50, 98, 139, 149, 164, 167], peanut butter [50, 98, 104, 136, 149, 163, 164, 167–170], eggs [121], figs [171], mixed feeds [149], and various dairy products [51]. In most of the analyses, aflatoxins were extracted from the commodities with a solvent containing a mixture of water, methanol, and/or acetonitrile, diluted in buffer, and subsequently subjected to immunoassay. Interference problems were generally overcome by diluting the sample either in buffer or in buffer containing sample matrix. Several protocols also include a very simple cleanup procedure. The sensitivities of various immunoassays and screening methods for aflatoxins in diverse agricultural commodities are summarized in Table 8. Although RIA is a less popular method than ELISA, with high-specific-activity iodinated AFB_1, it is an efficient and accurate procedure. Commercial RIA kits for AFB, AFM, and OTA analysis are also available [123, 175]. Both direct and indirect ELISAs detect AFB_1 accurately at about 10 ppb (ng/kg). For AFB_1 levels less than 10 ppb, protocols should be included for concentration of the toxin and sample cleanup.

For practical application of immunoassay, the sample matrix interference problem should never be overlooked; this problem varies in different matrices as well as with relevant immunochemical reagents. It can be overcome by sample dilution in assay systems that use either high-affinity antibody or specific marker ligand, but the approach may not be applicable for other systems, and sample cleanup may be necessary [103, 162].

Thus, rigorous preliminary tests should be carried out [11, 12]. Rauch et al. [176, 177] have pointed out that caffeic acid and chlorgenic acid, two common natural plant constituents, interfere with the RIA of aflatoxin.

Various immunochemical methods for the analysis of AFM_1 in dairy products were recently reviewed by Fremy and Chu [51]. A simple, rapid, and specific direct ELISA with a detection limit in the range of 0.05–0.1 ppb in milk has gained wide application in recent years [39, 47, 49, 106, 107, 178, 179]. Because milk samples can be used directly in the assay, the entire ELISA protocol can be completed in less than an hour. The sensitivity of ELISA for AFM_1 in milk increased to 10–25 ppt when milk samples were subjected to a cleanup treatment with a C-18 reversed-phase Sep-Pak cartridge [172] or through an affinity column [8, 51, 47, 154]. Analysis of AFM_1 in cheese involves extraction of the toxin from cheese followed by a requisite cleanup treatment before the ELISA [173]. A radioimmunoassay, which permitted detection of 0.1 ng AFM_1/mL, was developed for analysis of milk and eggs at the 1–13 ppb level [121]. Commercial kits for quick screening of AFM_1 in milk are also available [118]. Most of the kits have detection limits of 0.25–0.5 ppb of AFM_1 in milk, and screening is generally completed in less than 10 min.

Immunoassays for Other Mycotoxins

Immunoassays have also been used for the analysis of a number of other mycotoxins with varied sensitivities (Table 9). With the exception of ochratoxin A, however, immunoassays for mycotoxins other than AFB_1 and AFM_1 have not been widely used. Efficient RIAs and ELISAs have been established for analysis of ochratoxin A in barley [66, 67, 162, 141], wheat [180, 188], cereals and feed [128], and animal tissues [68, 126, 185, 188] with a sensitivity range of 10–50 ppb with no sample cleanup and 2–5 ppb after cleanup. In the TCTC group, both RIA and ELISA have been used for the analysis of T-2 toxin, diacetoxyscirpenol, deoxynivalenol, 3-acetyl-DON, and nivalenol in rice [79], corn [74, 129, 182, 183], fungal cultures [189], wheat [113, 114, 129, 183, 184], barley [82, 162], and animal organs [190], with a sensitivity range of 50–1000 ppb [12, 191].

Most of the antibodies against TCTC have a higher affinity to their acetylated derivatives, and the sensitivity of immunoassays of TCTCs improves considerably when acetylated derivatives are used in the assay. Thus, several protocols include acetylation of the samples before assay [82, 113, 114]. The use of antibodies specifically directed against the acetylated TCTCs also permits detection of total TCTC in the sample. Nonetheless, if high sensitivity is not needed, TCTCs can be analyzed without acetylation and cleanup. For example, a monoclonal antibody–based ELISA for DON with a sensitivity in the range of 10–250 ng per assay (0.2–5.0 μg/mL) could be used for detection of high levels of DON (>1 ppm) in the sample [74]. Effective ELISAs for T-2 and diacetoxyscirpenol are available that can detect the toxins at levels above 50 ppb without acetylation [129, 182, 183]. Zearalenone in corn, mixed feed, and wheat can be extracted with a solvent containing methanol and buffer and then directly tested by ELISA. The detection limit is about 1 ng/mL in the absence of sample matrix; in the presence of sample matrix, the estimated limit for detection of zearalenone is around 150 ppb [117].

TABLE 8. Sensitivity of Various Immunoassays for Aflatoxins in Agricultural Commodities

Aflatoxin	Commodity	Extraction solvents	Cleanup	Standard range (pg/assay)	Detection limits (μg/kg)	Ref.
RIA						
B_1-H	C, Wh	Chl	N	250–5000	5	163, 119
	PB	Hx, 55% MeOH	N	250–5000	5	163
B_1-I	E	Chl	N	4	0.1	121
M_1-H	M	—	N	5000–50,000	5	106
	M		Y		0.003	131
M_1-I	M, E	Chl	N	0.1–1 ng/mL	1	121
D-ELISA						
AFB_1/Afs	C, Wh	Chl (CB)	N	5–1000	5	163
	PB	Hx, 55% MeOH	N	5–1000	3–5	163
	P, PB	Hx, 55% MeOH	N	12.5–300	5–10	166
	PB	Chl/W	Y	10–1000	1	170
	C, PB	70% MeOH	N	5–50	5–10	134
	P, PB	60% MeOH	N	2.5–1000	1	50
M_1	M	—	N	25–1000	0.10	47, 106
	M	—	Y	25–1000	0.01	172
	M	—	Y	5–1000	0.10	39
	Ch	MeCl	Y	1.25–12.5	0.01	173
	F	Chl	Y	1.25–12.5	5	39
ID-ELISA						
AFs	C, P	Hx, 55% MeOH	N	20–1000	5	104
	PB	50% MeCN	N	0.2–10	0.25	100

Md-ID-ELISA						
B_1	P, PB, C	Hx, 55% MeOH	N	10–1500	10	167
B_1	P, C	55% MeOH	N	50–1000	10	174
B_1	B	$MeCN/KCl/H_2SO_4$	N		10	162
Screening tests						
Affc-AFs	C, P, PB	70% MeOH	N		>10	142
Affc-M_1	M	—	N		0.1	47, 58
Card	C, P	MeOH	N		20	118, 151
	P	80% MeOH	N		10	153
Cite-M_1	M	—	N		0.25	152
Cite-Probe	C, PB	80% MeOH	N		20	152
Cup-AFs	PB	Hx, 55% MeOH	N		>20	149
	CS	80% MeOH	N		>20	149
	C, P	80% MeOH	N		>30	149
	P	80% MeOH	N		>10	153
MTP-AFs	C, P, F	70% MeOH	N		1–3	134
	C, P, PB	MeOH	N		>20	136
	CS, MF	Hx/MeOH/W	N		>15	135
MTP-M_1	M		N		0.2	118
	M		Y		0.01	118

Abbreviations: Affc, affinity column; B, barley; C, corn; Ch, cheese; Chl, chloroform; CS, cottonseed; E, eggs; F, feed; M, milk; MF, mixed feed; P, raw peanuts; PB, peanut butter; Hx, hexane; MeOH, methanol; MeCl, methylene chloride; MeCN, acetonitrile; MTP, microtiter plate; W, water; Wh, wheat; H, tritiated; I, iodinated.

TABLE 9. Sensitivity of Various Immunoassays for Mycotoxins Other Than Aflatoxins in Agricultural Commodities

Mycotoxin	Commodity	Extraction solvent	Cleanup	Standard range (pg/assay)	Detection limits (μg/kg)	Ref.
RIA						
DON	C, W	84% MeCN	Y	50–5000	30	113
NIV	B	70% MeOH	Y	50–5000	60	82
OTA	K	Chl/PO_4	Y	100–200	0.2	68
PR toxin	Ch	EtOAc	N	1000–10,000	100	70
T-2	C, W		Y	200–10,000	1	129
			N		2.5	129
	C	Acetone	Y	100–5000	10	95
	S, U	—	Y	250–10,000	5.0	130
ZEARA	S					
D-ELISA						
DON	C, W	84% MeCN	Y	12–1250	10	114
	C	70% MeOH	N	10–250 ng/A	1000	9
OTA	W	MeOH	N	25–500	30	180
	W	MeOH	Y	25–500	1–2	180
T-2	C, W	MeOH/Hx	Y	2.5–200	1	181
	C, W	MeOH/Hx	N	2.5–200	2.5–5	181
	C	MeOH/W	N	—	50	182

ZEARA	C	Me/PBS	N	25–2500	100	117
ID-ELISA						
DAS	C, W	Acetone	Y	5–5000	1	183
	W	MeOH	N	5–250	300	184
DON	C, W	84% MeOH	Y	12–1250	10	114
	C	70% MeOH	N	10–150/A	1000	9
3 Ac-DON	R	60% MeOH	N	5–1000	1	79
NIV	B	70% MeCN	Y	50–5000	30	81
OTA	B	Chl/W	N	250–2500	5	66
	B	Chl/PO_4	N	20–1000	0.06	67
	K	Chl/PO_4	Y	50–10,000	1–10	185
	B	MeCN/KCl/H_2SO_4	N	5–5000	50	162
T-2	M, S, U	—	Y	2–200	0.2–1	186
	S	—	N	2–200	5	186
HT-2	U	—	Y	5–500	0.5	187
ST	B	CHl/HCl	Y	10–1000	—	73
ZEARA	C, W, MF	60% MeOH	N	50–2500	1–60	116
MD-ID-ELISA						
T-2	B	MeCN/KCl/H_2SO_4	Y	50–5000	50	162

Abbreviations: B, barley; C, corn; CH, cheese; CHl, chloroform; M, milk; MF, mixed feed; Hx, hexane; K, kidney; MeOH, methanol; MeCN, acetonitrile; R, rice; S, serum; U, urine; W, water; ZEARA, zearalenone; ST, sterigmatocystin; A, assay; EtOAc, ethylacetate.

B. Application of Immunochemical Methods for Monitoring Exposure of Humans and Animals to Mycotoxins

In addition to monitoring mycotoxins in foods and feeds, detection of mycotoxins in human and animal body fluids, tissues, and organs would provide direct evidence for exposure to the toxins. Two immunochemical approaches have been used for this purpose. One approach involves the analysis of residual mycotoxin and/or specific metabolites in body fluids and tissues by RIA or ELISA; the procedure has been used in both humans and experimental animals exposed to mycotoxins. The other approach involves immunohistochemical analysis of mycotoxins and their metabolites in target organs [5, 8, 11].

Monitoring AFB_1 and Its Major Metabolites and Macromolecule Adducts in Body Fluids

Use of immunoassay as a diagnostic tool requires prior knowledge on the metabolism of specific toxins as well as the kinetics of excretion and accumulation of the pertinent substance in the body [5, 11, 192, 193]. Although immunoassays of several mycotoxins in body fluids have been available for some time, they have not been widely used until recently. Epidemiological application of immunoassay has been limited to aflatoxins.

Monitoring Aflatoxin Metabolites in Human Urine and Milk: Various immunoassay protocols for detection of aflatoxins in human urine, milk, and serum are summarized in Table 10. Several types of antibodies have been used for monitoring aflatoxins and their metabolites in body fluids. Aflatoxin M_1 has been considered to be an important AFB_1 metabolite, and it has been found in both milk and urine samples after human and animal

TABLE 10. Immunoassay of Mycotoxins in Human Body Fluids

Body fluid	Mycotoxin	Method used	Detect. limits[a]
Serum	AFB_1	RIA	0.03
	AFB_1	ID-ELISA	0.02
	AFB_{2a}	DC-ELISA	0.02
	OTA	RIA	0.10
	OA	ELISA	0.03
Urine	AFM_1	Affc, ELISA	0.01
	AFB_{2a}	Affc, ELISA	0.01
	AFB-DNA	Affc, ELISA	—
Milk	AFM_1	DC-ELISA	0.01
	OA	ELISA	0.03

[a]Expressed as ng/mL.

ingestion of AFB_1. The ELISA method developed for the analysis of AFM_1 in cow's milk [51] has been used successfully in an epidemiological study in China [194, 195]. A direct correlation was found between total dietary AFB_1 intake and total AFM_1 excretion in urine collected from people in regions of high liver cancer incidence in China [194]. Between 1.23 and 2.18% of dietary AFB_1 was found to be present as AFM_1 in human urine. A total of 0.04–4.84 μg of AFM_1 (0.03–3.2 ng/mL) was excreted by each person per day in the endemic regions compared to less than 0.04 μg (0.03 ng/mL) in a control region. Results from the study support previous observations by Wu et al. [155]; their studies used an immunoaffinity column and HPLC method or ELISA method to monitor AFM_1 in human urine (detection limit of 3.0 pg/mL) obtained from a different district of high cancer incidence. The results suggested that analysis of AFM_1 in urine by ELISA could be used as an index for human exposure to AFB_1 in an extensive epidemiological study. Aflatoxin M_1 was also found in human breast milk obtained from women in Zimbabwe and Sudan at levels of 10–50 ppt by ELISA [196].

Rather than analyze specifically for AFM_1 in human urine, antibodies such as anti-AFB_{2a} or anti-AFB_1 dichloride, which have a wider spectrum of cross-reactivities with different aflatoxins, have been used in ELISA [57, 197, 198] to monitor aflatoxin exposure in humans. Although the lower detection limit in estimating the "overall" aflatoxin metabolite level in urine was approximately 10 ppt, undiluted human urine from control subjects sometimes gave false positive results. For example, urine samples obtained from people with hepatocellular carcinoma and their immediate contacts in Gambia had a mean "total aflatoxin" concentration of 312 ng/mL whereas the control group had a level of 81.5 ng/mL. Nonspecific inhibition could be prevented either by dilution or by a cleanup procedure using C-18 reverse-phase treatment [47]. In another study, ELISA analysis of 29 urine samples from the Philippines showed AFB_1 levels in the range of 0–4.25 ng/mL, with a mean of 0.88 ng/mL compared with a mean of 0.066 ng/mL in samples obtained from France [198]. Thus, the detection limit for aflatoxin in human urine in the denoted ELISA methods was between 50 and 100 ppt when no sample cleanup treatment step was included.

Monitoring Aflatoxin-DNA Adducts in Urine and Tissues: Immunochemical analysis of aflatoxin metabolites such as AFM_1 in human urine and milk is an effective approach for monitoring human exposure to aflatoxin, but pharmacokinetically the method suffers from the disadvantage of the short half-life of aflatoxin metabolites in urine. Unless there is continuous exposure, short-term exposure is difficult to determine. In contrast, the macromolecule adducts of aflatoxin generally have a longer half-life but occur at a lower level in body fluids [193]. Based on the differential effects of ethoxyquin on the kinetics of aflatoxin-DNA adduct and γ-glutamyl transpeptidase positive foci formation, Groopman and his colleagues [58, 59] found that measurement of the major, rapidly excised AFB-7-guanine (Gua) adduct in tissues and fluids is an appropriate dosimeter for estimating exposure status and risk in individuals consuming aflatoxin. Monoclonal antibodies against AFB_1-DNA as well as a major urinary product, formamidopyrimidine (AF-FAPYr) [2,3-dihydro-2(*N*-formyl)-2',5',6'-triamino-4'-oxy-*N*-pyrimidyl-3-hydroxy-AFB_1] have been used either as immunoaffinity agents or in immunoassays [29, 58–61, 140, 197]. Immunoaffinity chromatography in combination with other analytical methods [58, 59, 197] has been used to detect aflatoxin metabolites in urine samples obtained from regions of high liver cancer incidence [140] and in liver samples from hepatocellular carcinoma patients [63, 197].

Monitoring Aflatoxin in Human Serum: In addition to monitoring aflatoxins in urine and milk, immunoassays have been used to detect AFB_1 in human serum [170, 199, 200]. Tsuboi et al. [200] found that 29 of 80 samples obtained from subjects 2–3 h after eating had an AFB_1 level by RIA of 218.1 ± 268.3 ppt whereas 5 of 20 samples in a fasting group had 33.4 ± 14.6 ppt. The presence of AFB_1 in the samples was confirmed by HPLC as well as by mass spectral analysis in the study [200]. Radioimmunoassay was also used by Czechoslovakian scientists for analysis of AFB_1 in human serum [127]. In a recent survey, 1.8% of 227 human sera collected from Czechoslovakia had AFB_1 at levels between 30 and 100 ppt (ng/L). A double antibody indirect ELISA was used by a U.K. research team [199, 201–203]. They found that serum samples obtained from Nigeria had a mean AFB_1 level of 665 pg/mL with a standard error of 95 pg/mL (range from less than 20 pg/mL to 3.1 ng/mL) in contrast with an AFB_1 level of less than 20 pg/mL serum in 78 men from the United Kingdom. Thus, a background of 20–40 ppt, either due to interference—that is, false positives—or true exposure, was observed in both studies. These investigators also demonstrated that serum AFB_1 (ELISA) of neonates immediately after birth was significantly higher than that of their mothers in some Thai subjects [199].

Since aflatoxin-serum albumin (AF-SA) adducts have a longer half-life in animals [193] and possibly in humans than low molecular weight aflatoxin derivatives, Gan et al. [204] analyzed the aflatoxin-albumin adducts in the blood of 42 residents of Quangxi Province, People's Republic of China [58, 204]. Blood specimens were obtained during the same period in which urine was collected to study the correlation of intake of AFB_1 and excretion of AFM_1 in urine [194]. Analysis of AF-SA adducts was achieved by affinity purification of the albumin, enzymatic proteolysis, and immunoaffinity isolation of the digested adducts followed by RIA. A highly significant correlation of adduct level with AFM_1 excretion was observed. About 1.4–2.3% of ingested AFB_1 was found to be covalently bound to serum albumin; it is interesting to note that this value is similar (0.98–2.15%) to observations by Wild et al. [205] of rats fed AFB_1. Thus, a large portion of the aflatoxins found in human serum probably exist in adduct form. Formation of aflatoxin-macromolecule adducts in humans and animals raises an intriguing question of antibody formation against aflatoxin after animal or human exposure. Autrup and Wakhisi [206] suggested that such antibody response is highly likely.

In a recent study, Wild et al. [207] compared three immunochemical approaches for quantitation of AFB-albumin adducts. In the first two approaches ELISA was used to determine the intact albumin directly (D-ELISA) and the hydrolysate of albumin (H-ELISA), and the third approach used HPLC and fluorometric analysis of the AFB-lysine adduct after hydrolysis of albumin and affinity purification. The latter approach was the most sensitive (0.5–5.0 pg AFB/mg human albumin compared with 5.0 pg for H-ELISA and 100 pg for D-ELISA). Analysis of samples obtained from people in Thailand, Gambia, and Kenya showed levels of 7–338 AFB/mg albumin compared to nondetectable levels in human albumin obtained from France. Wild et al. [207] suggested use of the H-ELISA for large-scale screening and the third approach as a confirmatory test.

Monitoring Other Mycotoxins in Human and Animal Body Fluids and Tissues

Monitoring Ochratoxin A in Human Serum and Milk: Sensitive immunochemical methods have been developed for the analysis of OTA in serum and kidney at levels less than 1 ppb [67, 188]. Due to the high affinity of OTA with serum albumin and its associated extended half-life in animals and humans, analysis of OTA in serum has long been considered an effective monitoring system for human exposure to the mycotoxin [208, 209]. Detectable levels of OTA were found in human serum in some European countries. Fukal and Reisnerova [127] found that 13.3% and 10.5% of 143 human serum samples collected in Czechoslovakia contained OTA at levels of 100–500 and 500–1000 ppt (ng/L), respectively; about 0.7% had levels greater than 1000 ppb. A sensitive ELISA method (0.03 ng OTA/mL at 50% inhibition) has been used for the analysis of human serum and milk samples. Ochratoxin A was also demonstrated in human milk by ELISA [210]; these data suggest that dietary OTA was transmitted into human milk.

Monitoring Trichothecenes in Biological Fluids: Whereas immunoassays have been applied to monitoring aflatoxin and OTA in human serum and urine, the approach has not been used for the TCTC mycotoxins. Nevertheless, studies have led to development of both RIA and direct and indirect ELISA with sensitivity in the range of 0.5–2 ppb for the analysis of diacetoxyscirpenol, T-2 toxin, and some of its metabolites such as 3'-OH-T-2, HT-2, and T-2 tetraol in urine, serum, and milk [94, 130, 186, 187]. Methods for analysis of total deoxynivalenol have also become available in recent years [113, 114]. Recently, a direct competitive ELISA and RIA for the analysis of total group A TCTCs and total T-2 toxin metabolites in urine were developed [112]; the assays involve the use of a generic antibody against T-2 toxin with a wide spectrum of cross-reactivity [95]. All group A TCTCs and their metabolites were converted to T-2 tetraol tetraacetate (T-2-4ol-4Ac) before ELISA. The detection limits for T-2-4ol-4Ac in the presence and absence of urine samples were 10–20 ppb and 0.02 ppb (0.4 pg/assay), respectively; this method has been successfully applied in a study for the metabolism of T-2 toxin in rats and monkeys [211]. To test the adequacy of the method in monitoring TCTCs in human urine, a collaborative study between Food Research Institute, University of Wisconsin and Dr. C. P. Wild of the International Agency for Research on Cancer has been carried out. Human urine was obtained from three population groups in Kenya; the results (Chu, Lee, and Wild, unpublished observations) showed that (1) all 16 samples in the control volunteers in the Chogoria regions were negative for group A TCTC mycotoxins, (2) 8 of 16 samples in volunteers in the Kaloeni regions had low levels of toxin in the urine (0.23–2.31 ppb), and (3) 7 of 21 samples obtained from patients showing gastric pain in the Chogoria region had toxin levels in the range of 0.44–5.10 ppb. Although the significance of the present data is not known, the results are interesting, especially the elevated toxin levels in the urine of the third group; the results also proved that ELISA is adequate for future epidemiological study for TCTC exposure.

Monitoring Zearalenone in Serum and Urine: As early as 1983, an RIA was developed that allowed detection of 5 ppb zearalenone and its metabolites in serum [41]. Subsequent investigation by Dixon et al. [97] led to the development of a monoclonal antibody–based ELISA for zearalenone in urine. Improvement of the direct ELISA led to a protocol that detected 1 ppb or less of zearalenone and its metabolites in urine. In the assay, urine samples were hydrolyzed with β-glucuronidase, extracted with ethyl acetate, concentrated, and redissolved in 40% methanol before ELISA [212]. This protocol has been used for monitoring the excretion of zearalenone and its metabolites in urine after a gilt was administered zearalenone.

Immunohistochemical Studies

Two immunohistochemical methods, indirect immunostaining [213] and peroxidase–antiperoxidase (PAP) [214, 215], have been tested for monitoring mycotoxins in organs and tissues of animals that received the toxins. In the indirect system, organs or tissues are fixed, sectioned, and subsequently incubated with specific antibody against the mycotoxin and are then reacted with respective secondary antibody conjugated to an enzyme such as peroxidase. After washing, the bound peroxidase is localized by incubation with a chromogenic substrate such as H_2O_2 and diaminobenzidine to yield a dense, brown, water-insoluble product that can be visualized under a light microscope. This method has been used for localization of AFB_1 [213] and OTA [214] in animal organs and tissue. Aflatoxin B_{2a} antibodies [34] yield positive stains, particularly in the nuclear region, of hepatocytes in AFB_1-treated rats [213]. In contrast, positive stain for OTA was found in the esophagus, stomach, small intestine (most intense), kidney (intermediate, primarily in the distal tubules), and liver (least) of mice treated with OTA [214]. Instead of an enzyme conjugate, an indirect immunofluorescence technique was used by Elling [216], who demonstrated an accumulation of OTA in the kidney proximal tubular cells of pigs and rats that had ingested OTA. In both experiments, accumulation of mycotoxins in specific organs was consistent with pathological lesions that have been documented in other studies. An enzyme-linked immunocytochemical method was used by Lawellin et al. [217] to visualize the deposition of AFB_1 in the hyphae of *Aspergillus parasiticus*.

An additional step of incubation with a peroxidase–antiperoxidase (PAP) complex before staining with the substrate was needed in a PAP method for staining OTA [214] and T-2 toxin [215] in animal tissues. In the OTA study, results were similar to those obtained by an indirect stain. However, the PAP method was more sensitive and needed less antiserum to produce an optimal stain density [214]. In the study with

T-2 toxin, accumulation of T-2 toxin in the esophagus, stomach, duodenum, and jejunum (weak stain) occurred, but OTA was not found in the ileum. In contrast to the observation for OTA, the kidney medulla contained most of the T-2 toxin stain. T-2 toxin was not found in any of the hepatic tissue examined [5, 215].

V. CONCLUDING REMARKS

The development of immunoassays for detection of mycotoxins in ecological systems has progressed rapidly in the last few years. Various antibodies, including both monoclonal and polyclonal antibodies, with diverse specificity against mycotoxins have been developed. With the availability of such antibodies, good and reliable immunoassay protocols for mycotoxins in foods, feeds, and biological samples have been established. The methods are simple, sensitive, and specific. Several immunoassay kits for aflatoxin and other mycotoxins are currently available. It is imperative for those who perform the immunoassay to know the specificity of the antibody used in each assay. Depending on the purpose of research or routine analysis, the selection of appropriate antibody plays an important role in determining the success of a typical assay. Antibodies highly specific for certain mycotoxins and antibodies that cross-react with a group of mycotoxins (generic type) are both important, and both should be considered for immunoassays. Since analysis of mycotoxins in agricultural commodities is complicated by the uneven distribution of toxins in the samples, a detailed sampling program [218] should be considered. Inadequate sampling often yields meaningless data regardless of whether a good immunochemical method is used. The effect of sample matrix on the immunoassays should also be considered. Thus, whether or not cleanup of the sample should be included in the assay protocol will have to be determined.

Immunochemical methods, including affinity chromatography as a cleanup tool and immunoassay as a monitoring system for HPLC and TLC, have been used in conjunction with other analytical methodology. Immunohistochemical methods have also been used to localize mycotoxins in animal tissues [5]. Such innovative approaches have led to a new dimension in the immunochemical method for mycotoxins; this type of research should be continued and extended. In addition, new antibodies must be developed for other important mycotoxins and mycotoxin metabolites, especially generic-type antibodies, and existing immunoassay protocols must be improved.

ACKNOWLEDGMENTS

This work was supported by the College of Agricultural and Life Sciences, the University of Wisconson—Madison, Madison, Wisconsin. The work described in this contribution was supported in part by Public Health Service grant CA 15064 from the National Cancer Institute, contract DAMD17-86-C-6173 from the U.S. Army Medical Resarch and Development Command of the Department of Defense, and U.S. Department of Agriculture North

Central Regional Project NC-129. I thank Ms. Susan Hefle for her help in preparing the manuscript.

REFERENCES

1. Scott, P. M. Mycotoxins—general referee reports. *J. Assoc. Off. Anal. Chem.*, *73*:98, 1990.
2. Chu, F. S. Immunochemical methods for mycotoxin analysis. In *Proceedings of an International Symposium on Mycotoxins*, Cairo, Egypt (K. Naguib, M. M. Naguib, D. L. Park, and A. E. Pohland, Eds.), National Information Documentation Center, Dokki, Cairo, Egypt, 1983, pp. 177–194.
3. Chu, F. S. Immunoassays for analysis of mycotoxins. *J. Food Prot.*, *47*:562, 1984.
4. Chu, F. S. Immunochemical studies on mycotoxins. In *Toxigenic Fungi—Their Toxins and Health Hazard* (H. Kurata and Y. Ueno, Eds.), Kodansha, Tokyo, and Elsevier, New York, 1984, p. 234.
5. Chu, F. S. Immunochemical methods for diagnosis of mycotoxicoses. In *Diagnosis of Mycotoxicoses* (J. L. Richard and J. R. Thurston, Eds.), Martinus Nijhoff, Dordrecht, 1986, pp. 163–176.
6. Chu, F. S. Recent studies on immunochemical analysis of mycotoxins. In *Mycotoxins and Phytotoxins* (P. S. Steyn and R. Vleggaar, Eds.), Elsevier, Amsterdam, 1986.
7. Chu, F. S. *Aspergillus flavus* toxins: aflatoxin B_1 and M_1. In *Methods of Enzymatic Analysis*, Vol. XI, *Antigens and Antibodies 2* (H. U. Bergmeyer, Ed.), 3rd ed., VCH, Weinheim, Federal Republic of Germany, 1986, pp. 245–256.
8. Chu, F. S. Immunoassay for mycotoxins. In *Modern Methods in the Analysis and Structural Elucidation of Mycotoxins* (R. J. Cole, Ed.), Academic, New York, 1986, pp. 207–237.
9. Chu, F. S. Current techniques in mycotoxin analysis. In *Proceedings*, 23rd Annual Nutrition Conf. for Feed Manufactures, Canadian Feed Industries Association, Ottawa, Canada, 1987, pp. 16–30.
10. Chu, F. S. Current immunochemical methods for analysis of aflatoxin in groundnut and groundnut products. In *Aflatoxin Contamination of Groundnuts* (D. McDonald, V. K. Mehan, and S. D. Hall, Eds.), Proceedings, International Workshop in Aflatoxin Contamination of Groundnuts, 6–9 Oct., 1987, ICRISAT Center, Hyderabad, India, 1989, pp. 161–172.
11. Chu, F. S. Immunoassay for mycotoxins, current state of the art, commercial and epidemiological applications. *Vet. Human Toxicol.*, 32 (Suppl.) 42, 1990.
12. Chu, F. S. Current immunochemical methods for mycotoxin analysis. In *Immunoassays for Trace Chemical Analysis* (M. Vanderlaan, L. H. Stankers, B. E. Watkins, and D. W. Roberts, Eds.), American Chemical Society, Washington, D.C., 1991, pp. 140–157.
13. Fukal, L., and Kas, J. The advantages of immunoassay in food analysis. *Trends Anal. Chem.*, *8*:112, 1989.
14. Fukal, L., Rauch, P., and Kas, J. Utilization of immunochemical methods in food production analytical chemistry. I. Determination

of nonimmunogenic-low molecular compounds. *Chem. Listy*, *82*:959, 1988.

15. Martlbauer, E., and Terplan, G. Development of enzyme immunoassays for analysis of mycotoxins. In *Proceedings, Second World Congress of Foodborne Infections and Intoxications*, Vol. II, 1986, pp. 920–923.
16. Morgan, M. R. A. Newer techniques in food analysis–immunoassays and their application to small molecules. *J. Assoc. Pub. Anal.*, *23*: 59, 1985.
17. Morgan, M. R. A., and Lee, H. A. Mycotoxins and natural food toxicants. In *Development and Application of Immunoassay for Food Analysis* (J. H. Rittenburg, Ed.), Elsevier, New York, 1990, pp. 143–170.
18. Pestka, J. J. Enhanced surveillance of foodborne mycotoxins by immunochemical assay. *J. Assoc. Off. Anal. Chem.*, *71*:1075, 1988.
19. Yu, J., Wei, R. D., Mirocha, C. J., and Chu, F. S. Production and characterization of antibody against fusarochromanone. *Food Agric. Immunol.*, *2*:55, 1990.
20. Aalund, O., Brundeldt, K., Hald, B., Krogh, P., and Poulson, K. A radioimmunoassay for ochratoxin A: a preliminary investigation. *Acta Pathol. Microbiol. Scand. (Sect. C)*, *83*:390, 1975.
21. Chu, F. S,. Chang, F. C. C., and Hinsdill, R. D. Production of antibody against ochratoxin A. *Appl. Environ. Microbiol.*, *31*:831, 1976.
22. Worsaae, H. Production of an ochratoxin A antigen with high hapten/carrier molar ratio. *Acta Pathol. Microbiol. Scand. (Sect. C)*, *86*:203, 1978.
23. Chu, F. S., Liang, M. Y., and Zhang, G. S. Production and characterization of antibody against diacetoxyscirpenol. *Appl. Environ. Microbiol.*, *48*:777, 1984.
24. Lau, H. P., Gaur, P. K., and Chu, F. S. Preparation and characterization of aflatoxin B_{2a}-hemiglutarate and its use for the production of antibody against aflatoxin B_1. *J. Food Safety*, *3*:1, 1981.
25. Chu, F. S., Lau, H. P., Fan, T. S., and Zhang, G. S. Ethylenediamine modified bovine serum albumin as protein carrier in the production of antibody against mycotoxins. *J. Immunol. Methods*, *55*:73, 1982.
26. Hastings, K. L., Tulis, J. J., and Dean, J. H. Production and characterization of a monoclonal antibody to aflatoxin B_2. *J. Agric. Food Chem.*, *36*:404, 1988.
27. Zhang, G. S., Schubring, S. L., and Chu, F. S. An improved method for the production of antibody against T-2 toxin and diacetoxyscirpenol in rabbits. *Appl. Environ. Microbiol.*, *51*:132, 1986.
28. Sizaret, P., and Malaveille, C. Preparation of aflatoxin B_1-BSA conjugate with high hapten/carrier molar ratio. *J. Immunol. Methods*, *63*:159, 1983.
29. Groopman, J. D., Trudel, L. J., Donahue, P. R., Marshak-Rothstein, A., and Wogan, G. N. High-affinity monoclonal antibodies for aflatoxins and their application to solid-phase immunoassays. *Proc. Natl. Acad. Sci. USA*, *81*:7728, 1984.

30. Martin, C. N., Garner, R. C., Tursi, F., Garner, J. V., Whittle, H. C., Ryder, R. W., Sizaret, P., and Montesano, R. An ELISA procedure for assaying aflatoxin B_1. In *Monitoring Human Exposure to Carcinogenic and Mutagenic Agents* (A. Berlin, M. Draper, K. Hemminki, and H. Vainio, Eds.), IARC Scientific Publication No. 59, Lyon, France, 1984, p. 313.
31. Chu, F. S., and Ueno, I. Production of antibody against aflatoxin B_1. *Appl. Environ. Microbiol.*, *33*:1125, 1977.
32. Langone, J. J., and van Vunakis, H. Aflatoxin B_1: specific antibodies and their use in radioimmunoassay. *J. Natl. Cancer Inst.*, *56*: 591, 1976.
33. Ashoor, S. H., and Chu, F. S. Reduction of aflatoxin B_{2a} with sodium borohydride. *Agric. Food Chem.*, *23*:445, 1975.
34. Gaur, P. K., Lau, H. P., Pestka, J. J., and Chu, F. S. Production and characterization of aflatoxin B_{2a} antiserum. *Appl. Environ. Microbiol.*, *41*:478, 1981.
35. Chu, F. S., Steinert, B. W., and Gaur, P. K. Production and characterization of antibody against aflatoxin G_1. *J. Food Safety*, *7*:161, 1985.
36. Fan, T. S., Zhang, G. S., and Chu, F. S. Production and characterization of antibody against aflatoxin Q_1. *Appl. Environ. Microbiol.*, *47*:526, 1984.
37. Pestka, J. J., and Chu, F. S. Aflatoxin B_1 dihydrodiol antibody: production and specificity. *Appl. Environ. Microbiol.*, *47*:472, 1984.
38. Li, Y. K., and Chu, F. S. Production and characterization of antibody against sterigmatocystin. *J. Food Safety*, *6*:119, 1984.
39. Jackman, R. Determination of aflatoxins by ELISA with special reference to aflatoxin M_1 in raw milk. *J. Sci. Food Agric.*, *36*:685, 1985.
40. Gaur, P. K., El-Nakib, O., and Chu, F. S. Comparison of antibody production against aflatoxin B_1 in goats and in rabbits. *Appl. Environ. Microbiol.*, *40*:678, 1980.
41. Thouvent, D., and Morfin, R. F. A radioimmunoassay for zearalenone and zearalanol in human serum: production, properties and use of porcine antibodies. *Appl. Environ. Microbiol.*, *45*:16, 1983.
42. Harder, W. O., and Chu, F. S. Production and characterization of antibody against aflatoxin M_1. *Experientia*, *35*:104, 1979.
43. Sun, T., Wu, Y., and Wu, S. Monoclonal antibody against aflatoxin B_1 and its potential applications. *Chin. J. Oncol.*, *5*:401, 1983.
44. Woychik, N. A., Hinsdill, R. D., and Chu, F. S. Production and characterization of monoclonal antibody against aflatoxin M_1. *Appl. Environ. Microbiol.*, *48*:1096, 1984.
45. Candlish, A. A. G., Stimson, W. H., and Smith, J. E. A monoclonal antibody to aflatoxin B_1: detection of the mycotoxin by enzyme immunoassay. *Lett. Appl. Microbiol.*, *1*:57, 1985.
46. Blankford, M. B., and Doerr, J. A. The development and characterization of monoclonal antibodies against aflatoxin. *Hybridoma*, *5*: 57, 1986.
47. Kaveri, S. V., Fremy, J. M., Lapeyre, C., and Strosberg, A. D. Immunodetection and immunopurification of aflatoxins using a high

affinity monoclonal antibody to AFB_1. *Lett. Appl. Microbiol.*, *4*:71, 1987.
48. Karveri, S. V., Lapeyre, C., Reboul, S., and Fremy, J. M. Aflatoxin determination in feed and food samples by indirect ELISA using monoclonal antibody. Abstracts of the Hybridoma Conference, Paris, France, 25–27 Nov. 1987, Innobio Association, Paris, France, 1987.
49. Dixon-Holland, D. E., Pestka, J. J., Bridigare, B. A., Casale, W. L., Warner, R. L., Ram, B. P., and Hart, J. P. Production and sensitive monoclonal antibodies to aflatoxin B_1 and aflatoxin M_1 and their application to ELISA of naturally contaminated foods. *J. Food Prot.*, *51*:201, 1988.
50. Kawamura, O., Nagayama, S., Sato, S., Ohtani, K., Ueno, I., and Ueno, Y. A monoclonal antibody-based enzyme-linked immunosorbent assay for aflatoxin B_1 in peanut products. *Mycotoxin Res. 1988, 4*: 75, 1988.
51. Fremy, J. M., and Chu, F. S. Immunochemical methods of analysis for aflatoxin M_1. In *Mycotoxin in Dairy Products* (H. P. van Egmond, Ed.), Elsevier, London, 1989, p. 97.
52. Lawellen, D. W., Gant, D. W., and Joyce, B. K. Enzyme-linked immunosorbent analysis of aflatoxin B_1. *Appl. Environ. Microbiol.*, *34*: 94, 1977.
53. Hefle, S. L., and Chu, F. S. Production and characterization of monoclonal antibodies cross-reactive with major aflatoxins. *Food Agric. Immunol.*, *2*:181–188, 1990.
54. Zhang, G. S., and Chu, F. S. Production and characterization of antibodies cross-reactive with major aflatoxins. *Experientia*, *45*:182, 1989.
55. Lubet, M. T., Olson, D. F., Yang, G., Ting, R., and Steuer, A. Use of a monoclonal antibody to detect aflatoxin B_1 and M_1 in enzyme immunoassay. *Abstracts*, 97th AOAC Annual Meeting, Oct. 3–6, 1983, p. 71.
56. Garner, R. C., Dvorackova, I., and Tursi, F. Immunoassay procedures to detect exposure to aflatoxin B_1 and benzo[*a*]pyrene in animals and man at the DNA level. *Int. Arch. Occup. Environ. Health*, *60*:145, 1988.
57. Garner, C., Ryder, R., and Montesano, R. WHO-IARC meeting report: monitoring of aflatoxins in human body fluids and application of field studies. *Cancer Res.*, *45*:922, 1985.
58. Groopman, J. D., and Donahue, K. F. Aflatoxin, a human carcinogen: determination in foods and biological samples by monoclonal antibody affinity chromatography. *J. Assoc. Off. Anal. Chem.*, *71*: 861, 1988.
59. Groopman, J. D., and Kemsler, T. W. The use of monoclonal antibody affinity column for assessing DNA damage and repair following exposure to aflatoxins. *Pharmacol. Ther.*, *34*:321, 1987.
60. Groopman, J. D., Haugen, A., Goodrich, G. R., Wogan, G. N., and Harris, C. C. Quantitation of aflatoxin B_1-modified DNA using monoclonal antibodies. *Cancer Res.*, *42*:3120, 1982.

61. Haugen, A., Groopman, J. D., Hsu, I. C., Goodrich, G. R., Wogan, G. N., and Harris, C. C. Monoclonal antibody to aflatoxin B_1-modified DNA detected by enzyme immunoassay. *Proc. Natl. Acad. Sci. USA*, *78*:4124, 1981.
62. Hertzog, P. J., Lindsay Smith, J. R., and Garner, R. C. Production of monoclonal antibodies to guanine imidazole ring opened aflatoxin B_1-DNA, the persistent DNA adduct in vivo. *Carcinogenesis*, *38*:825, 1982.
63. Hsieh, L. L., Hsu, S. W., Chen, D. S., and Santella, R. M. Immunological detection of aflatoxin B_1-DNA adducts formed in vivo. *Cancer Res.*, *48*:6328, 1988.
64. Abdalla, A. E., and Grant, D. W. An immunoanalysis of kojic acid. *Sabouraudia*, *18*:191, 1981.
65. Candlish, A. A. G., Stimson, W. H., and Smith, J. E. A monoclonal antibody to ochratoxin A. *Lett. Appl. Microbiol.*, *3*:9, 1986.
66. Candlish, A. A., Stimson, W. H., and Smith, J. E. Determination of ochratoxin A by monoclonal antibody-based enzyme immunoassay. *J. Assoc. Off. Anal. Chem.*, *71*:961, 1988.
67. Morgan, M. R. A., McNerney, R., and Chan, H. W. S. Enzyme-linked immunosorbent assay of ochratoxin A in barley. *J. Assoc. Off. Anal. Chem.*, *66*:472, 1983.
68. Rousseau, D. M., Candlish, A. G., Slegers, G. A., Van Peteghem, C. H., Stimson, W. H., and Smith, J. E. Detection of ochratoxin A in porcine kidneys by a monoclonal antibody-based radioimmunoassay. *Appl. Environ. Microbiol.*, *53*:514, 1987.
69. Mehl, M., Strke, R., Jacobi, H. D., Schleinitz, K. D., and Wasicki, P. Immunologic detection of the mycotoxin patulin. *Pharmazie*, *41*:147, 1986.
70. Wei, R. D., and Chu, F. S. Production and characterization of antibody against PR toxin. *J. Food Prot.*, *51*:463, 1988.
71. Davis, R. M., and Stone, S. S. Production of anti-rubratoxin antibody and its use in a radioimmunoassay for rubratoxin B. *Mycopathologia*, *67*:27, 1979.
72. Neucere, J. N. Enzyme-linked immunosorbent assay for secalonic acid. *Mycopathologia*, *93*:39, 1986.
73. Morgan, M. R. A., Kang, A. S., and Chan, H. W. S. Production of antisera against sterigmatocystin hemiacetal and its potential for use in an enzyme-linked immunosorbent assay for sterigmatocystin in barley. *J. Sci. Food Agric.*, *37*:837, 1986.
74. Casale, W. L., Pestka, J. J., and Hart, P. Enzyme-linked immunosorbent assay employing monoclonal antibody specific for deoxynivalenol (vomitoxin) and several analogues. *J. Agric. Food Chem.*, *36*:663, 1988.
75. Hack, R., Klaffer, U., and Terplan, G. A monoclonal antibody to the trichothecene mycotoxin diacetoxyscirpenol. *Lett. Appl. Microbiol.*, *8*:71, 1989.
76. Martlbauer, E., Hack, R., and Terplan, G. Preparation and characterization of polyclonal and monoclonal antibodies against fusarenon-X. *Food Agric. Immunol.*, *1*:137, 1989.

77. Pauly, J. U., Bitter-Suermann, D., and Dose, K. Production and characterization of a monoclonal antibody to the trichothecene mycotoxin diacetoxyscirpenol. *Biol. Chem. Hoppe-Seyler*, *369*:487, 1988.
78. Chu, F. S., Zhang, G. S., Williams, M. D., and Jarvis, B. B. Production and characterization of antibody against deoxyverrucarol. *Appl. Environ. Microbiol.*, *48*:781, 1984.
79. Kemp, H. A., Mills, E. N. C., and Morgan, M. R. A. Enzyme-linked immunosorbent assay of 3-acetyldeoxynivalenol applied to rice. *J. Sci. Food Agric.*, *37*:888, 1986.
80. Zhang, G. S., Li, S. W., and Chu, F. S. Production and characterization of antibody against deoxynivalenol triacetate. *J. Food Prot.*, *49*:336, 1986.
81. Ikebuchi, H., Teshima, R., Hirai, K., Sato, M., Ichinoe, M., and Terao, T. Production and characterization of monoclonal antibodies to nivalenol tetraacetate and their application to enzyme-linked immunoassay of nivalenol. *Biol. Chem. Hoppe-Seyler*, *371*:31, 1990.
82. Teshima, R., Hirai, K., Sato, M., Ikebuchi, H., Ichinoe, M., and Terao, T. Radioimmunoassay of nivalenol in barley. *Appl. Environ. Microbiol.*, *56*:764, 1990.
83. Martlbauer, E., Gareis, M., and Terplan, G. Enzyme immunoassay for the macrocyclic trichothecene roridin A: production, properties, and use of rabbit antibodies. *Appl. Environ. Microbiol.*, *54*:225, 1988.
84. Hack, R., Martlbauer, E., and Terplan, G. Production and characterization of monoclonal antibodies to the macrocyclic trichothecene roridin A. *Appl. Environ. Microbiol.*, *54*:2328, 1988.
85. Chu, F. S., Grossman, S., Wei, R.-D., and Mirocha, C. J. Production of antibody against T-2 toxin. *Appl. Environ. Microbiol.*, *37*:104, 1979.
86. Peters, H., Dierich, M. P., and Dose, K. ELISA for detection of T-2 toxin. *Z. Physiol. Chem.*, *363*:1437, 1982.
87. Hunter, K. W., Jr., Brimfield, A. A., Miller, M., Finkelman, F. D., and Chu, F. S. Preparation and characterization of monoclonal antibodies to trichothecene mycotoxin T-2. *Appl. Environ. Microbiol.*, *49*:168, 1985.
88. Chiba, J., Kawamura, O., Kajii, H., Ohtani, K., Nagayama, S., and Ueno, Y. A sensitive enzyme-linked immunosorbent assay for detection of T-2 toxin with monoclonal antibodies. *Food Addit. Contam.*, *5*:629, 1988.
89. Goodbrand, I. A., Stimson, W. H., and Smith, J. E. A monoclonal antibody to T-2 toxin. *Lett. Appl. Microbiol.*, *5*:97, 1987.
90. Hack, R., Martlbauer, E., and Terplan, G. A monoclonal antibody to the trichothecene T-2 toxin: screening for antibody by a direct enzyme immunoassay. *J. Vet. Med. B*, *34*:538, 1987.
91. Fan, T. S. L., Zhang, G. S., and Chu, F. S. Production and characterization of antibodies against HT-2 toxin and T-2 tetraol tetraacetate. *Appl. Environ. Microbiol.*, *53*:17, 1987.
92. Fan, T. S. L., Schubring, S. L., Wei, R. D., and Chu, F. S. Production and characterization of a monoclonal antibody cross-reactive

with most group A trichothecenes. *Appl. Environ. Microbiol.*, *54*: 2959, 1988.
93. Gendloff, E. H., Pestka, J. J., Dixon, D. E., and Hart, L. P. Production of a monoclonal antibody to T-2 toxin with strong cross-reactivity to T-2 metabolites. *Phytopathology*, *77*:57, 1987.
94. Wei, R. D., Bischoff, W. B., and Chu, F. S. Production and characterization of antibodies against 3'-OH-T-2 toxin. *J. Food Prot.*, *49*:267, 1986.
95. Wei, R. D., and Chu, F. S. Production and characterization of a generic antibody against group A trichothecenes. *Anal. Biochem.*, *160*:399, 1987.
96. Wei, R.-D., Swanson, S., and Chu, F. S. Cross-reactivity of antibodies against T-2 with deepoxide T-2 toxin. *Mycotoxin Res.*, *4*:15, 1988.
97. Dixon, D. E., Warner, R. L., Ram, B. P., Hart, L. P., and Pestka, J. J. Hybridoma cell line production of a specific monoclonal antibody to the mycotoxins zearalenone and α-zearalenol. *J. Agric. Food Chem.*, *35*:122, 1987.
98. Chu, F. S., Lee, R. C., Trucksess, M. W., and Park, D. L. Evaluation by enzyme-linked immunosorbent assay of cleanup for thin-layer chromatography of aflatoxin B_1 in corn, peanuts and peanut butter. *J. Assoc. Off. Anal. Chem.*, *71*:126, 1988.
99. Pestka, J. J., Gaur, P. K., and Chu, F. S. Quantitation of aflatoxin B_1 and aflatoxin B_1 antibody by an enzyme-linked immunosorbent microassay. *Appl. Environ. Microbiol.*, *40*:1027, 1980.
100. Morgan, M. R. A., Kang, A. S., and Chan, H. W. S. Aflatoxin determination in peanut-butter by enzyme-linked immunosorbent assay. *J. Sci. Food Agric.*, *37*:908, 1986.
101. Sizaret, P., Malaveille, C., Montesano, R., and Frayssinet, C. Detection of aflatoxins and related metabolites by R.I.A. *J. Natl. Cancer Inst.*, *69*:1375, 1980.
102. Fukal, L., Prosek, J., Rauch, P., Sova, Z., and Kas, J. Selection of the separation step in the radioimmunoassay for aflatoxin B_1 using ^{125}I as marker. *J. Radioanal. Nucl. Chem., Articles*, *109*: 383, 1987.
103. Fukal, L., Sova, Z., and Reisnerova, H. Some interferences in radioimmunoassay of aflatoxin B_1. *J. Radioanal. Nucl. Chem., Lett.*, *108*:297, 1986.
104. Fan, T. S., and Chu, F. S. An indirect ELISA for detection of aflatoxin B_1 in corn and peanut butter. *J. Food Prot.*, *47*:263, 1984.
105. Biermann, V. A., and Terplan, G. Nachweis von aflatoxin B_1 mittels ELISA. *Arch. Lebensmitelhyg.*, *31*:51, 1980.
106. Pestka, J. J., Lee, Y. K., Harder, W. O., and Chu, F. S. Comparison of a radioimmunoassay and an enzyme-linked immunosorbent assay for the analysis of aflatoxin M_1 in milk. *J. Assoc. Off. Anal. Chem.*, *64*:294, 1981.
107. Martlbauer, E., and Terplan, G. Ein hochempfindlicher heterologer enzymimmunologischer nachweis von aflatoxin M_1 in milch under milchpulver. *Arch. Lebensmittelhyg.*, *36*:53, 1985.

108. Dragsted, L. O., Bull, I., and Autrup, H. Substances with affinity to a monoclonal aflatoxin B_1 antibody in Danish urine samples. *Food Chem. Toxicol.*, *26*:233, 1988.
109. Pestka, J. J., Li, Y. K., and Chu, F. S. Reactivity of aflatoxin B_{2a} antibody with aflatoxin B_1-modified DNA and related metabolites. *Appl. Environ. Microbiol.*, *44*:1159, 1982.
110. Ohtani, K., Kawamura, O., and Ueno, Y. Improved preparation of T-2 toxin-protein conjugates. *Toxicon*, *26*:1107, 1988.
111. Fontel, P. A., Beheler, J., Bunner, D. L., and Chu, F. S. Detection of T-2 toxin by an improved radioimmunoassay. *Appl. Environ. Microbiol.*, *45*:640, 1983.
112. Lee, R. C., Wei, R.-D., and Chu, F. S. Enzyme-linked immunosorbent assay for T-2 toxin metabolites in urine. *J. Assoc. Off. Anal. Chem.*, *72*:345, 1989.
113. Xu, Y. C., Zhang, G. S., and Chu, F. S. Radioimmunoassay of deoxynivalenol in wheat and corn. *J. Assoc. Off. Anal. Chem.*, *69*:967, 1986.
114. Xu, Y. C., Zhang, G. S., and Chu, F. S. Enzyme-linked immunosorbent assay for deoxynivalenol in corn and wheat. *J. Assoc. Off. Anal. Chem.*, *71*:945, 1988.
115. Pestka, J. J. Steinert, B. W., and Chu, F. S. An enzyme-linked immunosorbent assay for detection of ochratoxin A. *Appl. Environ. Microbiol.*, *41*:1472, 1980.
116. Liu, M. T., Ram, B. P., Hart, P., and Pestka, J. J. Indirect enzyme-linked immunosorbent assay for the mycotoxin zearalenone. *Appl. Environ. Microbiol.*, *50*:332, 1985.
117. Warner, R., Ram, B. P., Hart, H. P., and Pestka, J. J. Screening for zearalenone in corn by competitive direct enzyme-linked immunosorbent assay. *J. Agric. Food Chem.*, *34*:714, 1986.
118. Lee, L. S. Aflatoxin. *J. Am. Oil Chem. Soc.*, *66*:1398, 1989.
119. Sun, P., and Chu, F. S. A simple solid-phase radioimmunoassay for aflatoxin B_1. *J. Food Safety*, *1*:67, 1977.
120. Fukal, L., Prosek, J., Rauch, P., Sova, Z., and Kas, J. Choice of procedure conditions for radioimmunoassay of aflatoxin B_1 using ^{125}I as marker and dextran-coated charcoal as a separation matrix. *J. Radioanal. Nucl. Chem., Lett.*, *108*:259, 1986.
121. Rauch, P., Fukal, L., Prosek, J., Bresina, P., and Kas, J. Radioimmunoassay of aflatoxin M_1. *J. Radioanal. Nucl. Chem., Lett.*, *117*:163, 1987.
122. Bludovsky, R. Determination of aflatoxin B_1 in foods by RIA method. *Radioizotopy*, *27*:339, 1986.
123. Fukal, L. Utilization of commercial RIA-set for milk contamination with M_1 aflatoxin. *Melkarske Listy*, *14*:196.36, 1988.
124. Fukal, L., Reisnerova, H., and Rauch, P. Application of radioimmunoassay with ^{125}I for determination of aflatoxin B_1 in foods. *Sci. Aliments*, *8*:397—403, 1988.
125. Rauch, P., Brezina, P., Lohnisky, J., et al. Radioimmunochemical determination of aflatoxin B_1 in feeds and foods. *Potravin. Vedy.*, *5*:273, 1987.

126. Ruprich, J., Veres, K., and Schmiedova, D. Radioimmunologic determination of the mycotoxin ochratoxin A—a new diagnostic method in Czechoslovakia. *Vet. Med. Praha, 33*:101, 1988.
127. Fukal, L., and Reisnerova, H. Monitoring of aflatoxins and ochratoxin-A in Czechoslovak human sera by immunoassay. *Bull. Environ. Contam. Toxicol., 44*:345, 1990.
128. Ruprich, J., Piskac, A., and Mala, J. Using the radioimmunological assays of the ochratoxin A mycotoxin for the screening control of cereal-type foods and feeds of plant origin. *Vet. Med. Praha, 33*: 165, 1988.
129. Lee, S., and Chu, F. S. Radioimmunoassay of T-2 toxin in corn and wheat. *J. Assoc. Off. Anal. Chem., 64*:156, 1981.
130. Lee, S., and Chu, F. S. Radioimmunoassay of T-2 toxin in biological fluids. *J. Assoc. Off. Anal. Chem., 64*:684, 1981.
131. Qian, G. S., Yasei, P., and Yang, G. C. Rapid extraction and detection of aflatoxin M_1 in cow's milk by high-performance liquid chromatography and radioimmunoassay. *Anal. Chem., 56*:2079, 1984.
132. Qian, G. S., and Yang, G. C. Rapid extraction and detection of aflatoxin B_1 and M_1 in beef liver. *Anal. Chem., 56*:1071, 1984.
133. Pestka, J. J., and Chu, F. S. Enzyme-linked immunosorbent assay of mycotoxins using nylon-bead Terasaki plate solid phases. *J. Food Prot., 47*:305, 1984.
134. Chu, F. S., Fan, T. S. L., Zhang, G. S., Xu, Y. C., Faust, S., and McMahon, P. L. Improved enzyme-linked immunoassay for aflatoxin B_1 in agricultural commodities. *J. Assoc. Off. Anal. Chem., 70*:854, 1987.
135. Park, D. L., Miller, B. M., Hart, P., Yang, G., McVey, J., Page, S. W., Pestka, J., and Brown, L. H. Enzyme-linked immunosorbent assay for screening aflatoxin in cottonseed products and mixed feed: collaborative study. *J. Assoc. Off. Anal. Chem., 72*:326, 1989.
136. Park, D. L., Miller, B. M., Nesheim, S., Trucksess, M. W., Vekich, A., Bidigare, B., McVey, J. L., and Brown, L. H. Visual and semiquantitative spectrophotometric screening method for aflatoxin B in corn and peanut products: follow-up study. *J. Assoc. Off. Anal. Chem., 72*:638, 1989.
137. Itoh, Y., Nishimura, M., Hifumi E., Uda, T., Ohtani, K., Kawamura, O., Nagayama, S., and Ueno, Y. Detection of ochratoxin A and T-2 toxin by one step ELISA. *Proc. Jpn. Assoc. Mycotoxicol., 24*:25, 1986.
138. Sidwell, W. J., Chan, H. S. W., and Morgan, M. R. A. Reduction in assay time for a double antibody, indirect enzyme-linked immunosorbent assay applied to ochratoxin A. *Food Agric. Immunol., 1*: 111, 1989.
139. Candlish, A. A., Haynes, C. A., and Stimson, W. H. Detection and determination of aflatoxins using affinity chromatography. *Int. J. Food Sci. Technol., 23*:479, 1988.
140. Groopman, J. D., Donahue, P. R., Zhu, J. Q., Chen, J. S., and Wogan, G. N. Aflatoxin metabolism in human: detection of metabolites and nucleic acid adduct in urine by affinity chromatography. *Proc. Natl. Acad. Sci. USA, 82*:6492, 1986.

141. Rousseau, D. M., Slegers, G. A., and van Peteghem, C. H. Radioimmunoassay of ochratoxin A in barley. *Appl. Environ. Microbiol., 50*:529, 1985.

142. Trucksess, M. W., Stack, M. E., Nesheim, S., Page, S. W., Albert, R. H., Hansen, T. J., and Donahue, K. F. Immunoaffinity column coupled with solution fluorometry or LC postcolumn derivatization for aflatoxins in corn, peanuts, and peanut butter: collaborative study. *Abstracts*, 103rd AOAC Annual International Meeting, St. Louis, Mo., Sept. 25–28, 1989.

143. Warden, B. A., Allam, K., Sentissi, A., Cecchini, D. J., and Giese, R. W. Repetitive hit-and-run fluoroimmunoassay for T-2 toxin. *Anal. Biochem., 162*:363, 1987.

144. Ligler, F. S., Bredehorst, R., Talebian, A., Shriver, L. C., Hammer, C. F., Sheridan, J. P., Vogel, C., and Gaber, B. P. A homongeneous immunoassay for the mycotoxin T-2 utilizing liposomes, monoclonal antibodies, and complement. *Anal. Biochem., 163*:369, 1987.

145. Williamson, M. L., Atha, D. H., Reeder, C. L., and Sundaram, P. V. Anti-T2 monoclonal antibody immobilization on quartz fibers: kinetics of inactivation and recognition of T-2 mycotoxin. *Anal. Lett., 22*: 803, 1989.

146. Degan, P., Montagnoli, G., and Wild, C. P. Time-resolved fluoroimmunoassay of aflatoxins. *Clin. Chem., 35*:2308, 1989.

147. Neogen Bulletins. Agri-Screen diagnostics tests for aflatoxin, zearalenone, T-2, and vomitoxin (DON). Neogen Co., Lansing, Mich., 1989.

148. International Diagnostics System Bulletins. Afla-10 cup tests and related products. International Diagnostics System Co., St. Joseph, Mo., 1989.

149. Trucksess, M. W., Stack, M. E., Nesheim, S., Park, D. L., and Pohland, A. E. Enzyme-linked immunosorbent assay of aflatoxin-B_1, aflatoxin-B_2, and aflatoxin G_1 in corn, cottonseed, peanuts, peanut butter, and poultry feed—collaborative study. *J. Assoc. Off. Anal. Chem., 72*:957, 1989.

150. Environmental Diagnostics Bulletins. EZ-Screen Quick card. Environmental Diagnostics Inc., Burlington, N.C., 1989.

151. Park, D. L., Nesheim, S., Trucksess, M. W., Stack, M., Morris, D., Lewis, E., and Romer, T. ELISA "quick card" screening method for aflatoxins B_1, B_2, G_1, and G_2: collaborative study. *Abstracts*, 101st AOAC Annual International Meeting, San Francisco, Calif., Sept. 14–17, 1987, p. 64.

152. Idexx Inc. Bulletins. Cite aflatoxin M_1 and Cite-Probe aflatoxin B_1. Idexx Corp., Portland, ME., 1989.

153. Dorner, J. W., and Cole, R. J. Comparison of two ELISA screening tests with liquid chromatography for determination of aflatoxins in raw peanuts. *J. Assoc. Off. Anal. Chem., 72*:962, 1989.

154. Mortimer, D. N., Gilbert, J., and Sheperd, M. J. Rapid and highly sensitive analysis of aflatoxin M_1 in liquid and powdered milks using an affinity column cleanup. *J. Chromatogr., 407*:393, 1987.

155. Wu, S., Yang, G., and Sun, T. Studies on immuno-concentration and immunoassay of aflatoxins. *Chin. J. Oncol., 5*:81, 1983.

156. Oxoid Co. Bulletins. Aflatoxin test kits and affinity columns. Oxoid USA, Inc., Columbia, MD., 1989.
157. Vicam Co. Bulletins. Aflatest and affinity columns. Vicam Co., Somerville, MA., 1989.
158. Carman, A. S., Kuan, S. S., Ware, G. M., Francis, O. J., and Miller, K. V. Automated assay of aflatoxin M_1 in fluid milk using robotics. *Abstracts*, 103rd AOAC Annual International Meeting, St. Louis, Mo., Sept. 25–28, 1989.
159. Chu, F. S., and Lee, R. C. Immunochromatography of group A trichothecene mycotoxins. *Food Agric. Immunol., 1*:127, 1989.
160. Goto, T., and Manabe, M. Application of ELISA to aflatoxin analysis: results of preliminary tests of analysis kits on the market. *Rep. Natl. Food Res. Inst., 52*:53, 1988.
161. Patey, A. L., Sharman, M., Wood, R., and Gilbert, J. Determination of aflatoxin concentrations in peanut butter by enzyme-linked immunosorbent assay (ELISA)—study of 3 commercial ELISA kits. *J. Assoc. Off. Anal. Chem., 72*:965, 1990.
162. Ramakrishna, N., Lacey, J., Candlish, A. A. G., Smith, J. E., and Goodbrand, I. A. Monoclonal antibody-based enzyme linked immunosorbent assay of aflatoxin-B_1, T-2-toxin, and ochratoxin-A in barley. *J. Assoc. Off. Anal. Chem., 73*:71, 1990.
163. El-Nakib, O., Pestka, J. J., and Chu, F. S. Determination of aflatoxin B_1 in corn, wheat, and peanut butter by enzyme-linked immunosorbent assay and solid phase radioimmunoassay. *J. Assoc. Off. Anal. Chem., 64*:1077, 1981.
164. Hart, L. P., Park, D. L., Miller, B. M., Yang, G., McVey, J., and Page, S. W. Enzyme-linked immunosorbent assay for the determination of aflatoxin B_1 in peanut, corn, and cottonseed products and mixed feed: collaborative study. *Abstracts*, 100th AOAC Annual International Meeting, Scottsdale, Ariz., Sept. 15–17, 1986.
165. Li, S. F., Liu, S. J., Hu, X., Liu, M. Y., Li, Y. W., Zhang, L. S., Liang, G. K., and Chu, F. S. Determination of aflatoxin B_1 in corn by enzyme-linked immunoassay. *J. Hyg. Res., 18*:29, 1987.
166. Ram, B. P., Hart, L. P., Shotwell, O. L., and Pestka, J. J. Enzyme-linked immunosorbent assay of aflatoxin B_1 in naturally contaminated corn and cottonseed. *J. Assoc. Off. Anal. Chem., 69*:904, 1986.
167. Candlish, A. A. G., Stimson, W. H., and Smith, J. E. The detection of aflatoxin B_1 in peanut kernels, peanut butter and maize using a monoclonal antibody based enzyme immunoassay. *Food Microbiol., 4*:147, 1987.
168. Mortimer, D. N., Shepherd, M. J., and Gilbert, J. Enzyme-linked immunosorbent (ELISA) determination of aflatoxin B_1 in peanut butter: collaborative trial. *Food Addit. Contam., 5*:601, 1988.
169. Mortimer, D. N., Shepherd, M. J., and Gilbert, J. A survey of the occurrence of aflatoxin B_1 in peanut butters by enzyme-linked immunosorbent assay. *Food Addit. Contam., 5*:127, 1988.
170. Ueno, I. A simple and improved enzyme-linked immunsorbent assay method for microquantitation of aflatoxin B_1 in peanuts and blood plasma. *Proc. Jpn. Assoc. Mycotoxicol., 1986*:24, 1986.

171. Reichert, N., Steinmeyer, S., and Weber, R. Bestimmung von aflatoxin B_1 in trockenfeigen durch visuelles screening, dunnschicht chromatographie und ELISA. *Z. Lebensm. Unters. Forsch., 186*: 505, 1988.
172. Hu, W. J., Woychik, N., and Chu, F. S. ELISA of picogram quantites of aflatoxin M_1 in urine and milk. *J. Food Prot., 47*:126, 1983.
173. Fremy, J. M., and Chu, F. S. A direct ELISA for determing aflatoxin M_1 at ppt levels in various dairy products. *J. Assoc. Off. Anal. Chem., 67*:1098, 1984.
174. Itoh, Y., Hifumi, E., Sudoh, K., Uda, T., Ohtano, K., Kawamura, O., Nagayama, S., Satoh, S., and Ueno, Y. Determination of aflatoxin B_1 by direct ELISA. *Proc. Jpn. Assoc. Mycotoxicol., 26*:31, 1988.
175. Penicillin Assays Inc. Bulletin. Charm aflatoxin test. Penicillin Assays Inc., Malden, MA., 1989.
176. Rauch, P., Fukal, L., Brezina, P., and Kas, J. Interferences in radioimmunoassay of aflatoxins in foods and fodder samples of plant origin. *J. Assoc. Off. Anal. Chem., 71*:491, 1988.
177. Rauch, P., Viden, I., Davidek, T., Velisek, J., and Fukal, L. Caffeine as main interfering compound in radioimmunoassay of aflatoxin B_1 in coffee samples. *J. Assoc. Off. Anal. Chem., 72*:1015, 1989.
178. Kornfeld, S., Hurpet, D., and Bonneau, M. Determination of aflatoxins in foodstuffs and milk by ELISA. *Abstracts*, Hybridoma Conference, Paris, France, 25–27 Nov. 1987, Innobio Association, Paris, France, 1987, p. 29.
179. Li, Y. K., and Chu, F. S. Kinetics of transformation of aflatoxin M_1 from aflatoxin B_1 in lactating mouse: an ELISA analysis. *Experientia, 38*:842, 1982.
180. Lee, S., and Chu, F. S. Enzyme-linked immunosorbent assay of ochratoxin A in wheat. *J. Assoc. Off. Anal. Chem., 67*:45, 1981.
181. Pestka, J. J., Lee, S. S., Lau, H. P., and Chu, F. S. Enzyme-linked immunosorbent assay for T-2 toxin. *J. Am. Oil Chem. Soc., 58*:940A, 1981.
182. Gendloff, E. H., Pestka, J. J., Swanson, S. P., and Hart, L. P. Detection of T-2 in *Fusarium sporotrichioides*-infested corn by enzyme-linked immunosorbent assay. *Appl. Environ. Microbiol., 47*: 1161, 1984.
183. Schubring, S. L., and Chu, F. S. An indirect enzyme-linked immunosorbent assay for the detection of diacetoxyscirpenol in wheat and corn. *Mycotoxin Res., 3*:97, 1987.
184. Mills, E. N. C., Johnston, J. M., Kemp, H. A., and Morgan, M. R. A. An enzyme-linked immunosorbent assay for diacetoxyscirpenol applied to the analysis of wheat. *J. Sci. Food Agric., 42*:225, 1988.
185. Morgan, M. R. A., McNerney, R., Chan, H. W. S., and Anderson, P. H. Ochratoxin A in pig kidney determination by enzyme-linked immunosorbent assay (ELISA). *J. Sci. Food Agric., 37*:475, 1986.

186. Fan, T. S., Zhang, G. S., and Chu, F. S. An indirect ELISA for analysis of T-2 toxin in biological fluids. *J. Food Prot.*, *47*:964, 1984.
187. Fan, T. S. L., Xu, Y. C., and Chu, F. S. Simultaneous analysis of T-2 toxin and HT-2 toxin in urine by an indirect ELISA. *J. Assoc. Off. Anal. Chem.*, *70*:657, 1987.
188. Sato, S., Nagayama, S., Kawamura, O., and Ueno, Y. Detection of ochratoxin A in meat, wheat and plasma by an enzyme-linked immunosorbent assay. *Maikotosishin (Tokyo)*, *26*:47, 1987.
189. Nagayama, S., Kawamura, O., Ohtani, K., Ryu, J., Latus, D., Sudheim, L., and Ueno, Y. Application of an enzyme-linked immunosorbent assay for screening of T-2 toxin-producing *Fusarium* spp. *Appl. Environ. Microbiol.*, *54*:1302, 1988.
190. Hewetson, I. F., Pace, J. G., and Beheler, J. E. Detection and quantitation of T-2 toxin in rat organs by radioimmunoassay. *J. Assoc. Off. Anal. Chem.*, *70*:654, 1987.
191. Isohata, E., Toyoda, M., and Saito, Y. Studies on chemical analysis of mycotoxin (XIX). Enzyme immunoassay of T-2 toxin in foods. *Eisei shikenjo Hokoku*, *106*:117, 1988.
192. Wild, C. P., Chapot, B., Scherer, E., Den-Engelse, L., and Montesano, R. Application of antibody methods to the detection of aflatoxin in human body fluids. *IARC-Sci-Publ.*, *89*:67, 1988.
193. Henderson, R. F., Bechtold, W. E., Bond, J. A., and Sun, J. D. The use of biological markers in toxicology. *CRC Crit. Rev. Toxicol.*, *20*:65, 1989.
194. Zhu, J. Q., Zhang, L. S., Hu, X., Xiao, Y., Chen, J. S., Xu, Y. C., Fremy, J., and Chu, F. S. Correlation of dietary aflatoxin B_1 levels with excretion of aflatoxin M_1 in human urine. *Cancer Res.*, *47*:1848, 1987.
195. Yu, S. Z., Cheng, Z. Q., Liu, X. K., Zhang, L. S., and Zhao, Y. F. The aflatoxins and water contaminated in the etiological study of primary liver cancer. In *Mycotoxins and Phycotoxins '88* (S. Natori, K. Hashimoto, and Y. Ueno, Eds.), Elsevier, Amsterdam, 1989, pp. 37–44.
196. Wild, C. P., Pionneau, F. A., Montesano, R., Mutiro, C. F., and Chetsanga, C. J. Aflatoxin detected in human breast milk by immunoassay. *Int. J. Cancer*, *40*:328, 1987.
197. Garner, R. C. Monitoring aflatoxin exposure at a macromolecular level in human with immunological methods. In *Mycotoxins and Phycotoxins '88* (S. Natori, K. Hashimoto, and Y. Ueno, Eds.), Elsevier, Amsterdam, 1989, pp. 29–35.
198. Wild, C. P., Umbenhauer, D., Chapot, B., and Montesano, R. Monitoring of individual human exposure to aflatoxins and *N*-nitrosamines by immunoassay. *J. Cell. Biochem.*, *30*:171, 1986.
199. Morgan, M. R. A., Wilkinson, A. P., and Denning, D. W. Human exposure to mycotoxins monitored by immunoassay. In *Mycotoxins and Phycotoxins '88* (S. Natori, K. Hashimoto, and Y. Ueno, Eds.), Elsevier, Amsterdam, 1989, pp. 45–50.
200. Tsuboi, S., Nakagawa, T., Tomita, M., Seo, T., Ono, H., Kawamura, K., and Iwamura, N. Detection of aflatoxin B_1 in serum

samples of male Japanese subjects by radioimmunoassay and high performance liquid chromatography. *Cancer Res.*, *44*:1231, 1984.
201. Denning, D. W., Onwubalili, J. K., Wilkinson, A. P., and Morgan, M. R. A. Measurement of aflatoxin in Nigerian sera by enzyme-linked immunosorbent assay. *Trans. Roy. Soc. Trop. Med. Hyg.*, *82*:169, 1988.
202. Wilkinson, A. P., Denning, D. W., and Morgan, M. R. A. Analysis of UK sera for aflatoxin by enzyme-linked immunosorbent assay. *Human Toxicol.*, *7*:353, 1988.
203. Wilkinson, A. P., Denning, D. W., and Morgan, M. R. A. An ELISA method for the rapid and simple determination of aflatoxin in human serum. *Food Addit. Contam.*, *5*:609, 1988.
204. Gan, L. S., Skipper, P. L., Peng, X. C., Groopman, J. D., Chen, J. S., Wogan, G. N., and Tannenbaum, S. R. Serum albumin adducts in the molecular epidemiology of aflatoxin carcinogenesis, correlation with aflatoxin B_1 intake and urinary excretion of aflatoxin M_1. *Carcinogenesis (Lond.)*, *9*:1323, 1988.
205. Wild, C. P., Garner, R. C., Montesano, R., and Tursi, F. Aflatoxin B_1 binding to plasma albumin and liver DNA upon chronic administration to rats. *Carcinogenesis (Lond.)*, *7*:853, 1986.
206. Autrup, H., and Wakhisi, J. Evidence for human antibodies that recognize an aflatoxin epitope in groups with high and low exposure to aflatoxins. *Arch. Environ. Health*, *45*:31, 1988.
207. Wild, C. P., Jiang, Y. Z., Sabbioni, G., Chapot, B., and Montesano, R. Evaluation of methods for quantitation of aflatoxin-albumin adducts and their application to human exposure assessment. *Cancer Res.*, *50*:245, 1990.
208. Chu, F. S. Mode of action of mycotoxins and related compounds. *Adv. Appl. Microbiol.*, *22*:83–143, 1977.
209. Hult, K., and Fuchs, R. Analysis and dynamics of ochratoxin A in biological systems. In *Mycotoxins and Phycotoxins* (P. S. Steyn and R. Vleggaar, Eds.), Elsevier, Amsterdam, 1986, pp. 365–376.
210. Gareis, M., Martlbauer, E., Bauer, J., and Gedek, B. Determination of ochratoxin A in human milk. *Abstracts*, 7th Int. Symp. Mycotoxins and Phycotoxins (Aug. 16–19, 1988, Tokyo, Japan), 1988, p. 82.
211. Lee, R. C., Bunner, D. L., Wannemacher, R. W., Jr., and Chu, F. S. Immunochemical studies on the excretion of T-2 metabolites in rat and *Cynomolgus* monkey urine. *J. Agric. Chem.*, *38*:444, 1990.
212. MacDougald, O. A., Thulin, A. J., and Pestka, J. J. Determination of zearalenone and related metabolites in porcine urine by modified enzyme-linked immunosorbent assay. *J. Assoc. Off. Anal. Chem.*, *76*:65, 1990.
213. Pestka, J. J., Beery, J. T., and Chu, F. S. Indirect immunoperoxidase localization of aflatoxin B_1 in rat liver. *Food Chem. Toxicol.*, *21*:41, 1983.
214. Lee, S. C., Beery, J. T., and Chu, F. S. Immunohistochemical fate of ochratoxin A. *Toxicol. Appl. Pharmacol.*, *72*:218, 1984.

215. Lee, S. C., Beery, J. T., and Chu, F. S. Immunoperoxidase localization of T-2 toxin. *Toxicol. Appl. Pharmacol.*, *72*:228, 1984.
216. Elling, F. Demonstration of ochratoxin A in kidney of pigs and rats by immunofluorescence microscopy. *Acta Pathol. Microbiol. Scand.* (*Sect. A*), *85*:151, 1977.
217. Lawellin, D. W., Grant, D. W., and Joyce, B. K. Aflatoxin localization by enzyme-linked immunocytochemical techniques. *Appl. Environ. Microbiol.*, *34*:88, 1977.
218. Park, D. L., and Pohland, A. E. Sampling and sample preparation for determination and quantitation of natural toxicants in food and feed. *J. Assoc. Off. Anal. Chem.*, *72*:399, 1989.

5

INSECT INTERACTIONS WITH MYCOTOXIN-PRODUCING FUNGI AND THEIR HOSTS

PATRICK F. DOWD *Agricultural Research Service, National Center for Agricultural Utilization Research, U.S. Department of Agriculture, Peoria, Illinois*

I. INTRODUCTION

Both insects [1] and fungi [2] arose at roughly the same time, around 400 million years ago. Thus, their interactions have spanned a time some 100 million years longer than that of fungus—mammal interactions [2] and some 275 million years longer than insect—flowering plant interactions [3]. Although insect—plant interactions have been extensively studied, comparatively little is known about insect—fungus interactions. Other chapters in this volume explore the potential role of fungus-derived mycotoxins as defensive agents. I will concentrate on the role of mycotoxins as defensive agents against insects (primarily caterpillars) and examine cases where insects have developed resistance to mycotoxins, as well as other adaptations that permit some insects to feed on and act as vectors for mycotoxin-producing fungi.

II. INSECT-DRIVEN FUNGAL DEFENSES

Microorganisms (including fungi) are generally more nutritious than plant tissue, especially leaves, because they have higher levels of protein and sterols [4]. Thus, fungi and materials contaminated by fungi are potentially desirable sources of food for insects. Indeed, many insects have made appropriate adjustments for feeding on fungi, a strategy that is likely to have existed before the advent of modern vascular plants, due to the longer coexistence of insects and fungi. Therefore, it is not surprising that fungi would develop secondary metabolites that play a defensive role against insects. This concept of a defensive role for fungal

---

Note: The mention of firm names or trade products does not imply that they are endorsed or recommended by the U.S. Department of Agriculture over other firms or similar products not mentioned.

secondary metabolites has developed relatively slowly [5] compared to the now widely accepted defensive role attributed to the secondary metabolites of vascular plants [6]. The reader is directed to prior speculative discussions [7–11] for more information on a general theory of defensive roles of fungal secondary metabolites as well as to other chapters in this volume.

Part of the problem in considering fungal metabolites as defensive metabolites as opposed to storage metabolites or waste products is that much of the discussion is canted toward the all-or-nothing rule that a fungal metabolite cannot be a storage product if it has a defensive role. There is no reason to make this assumption that each secondary metabolite has only one specific role. Rather, it seems more likely that a secondary metabolite can have multiple roles that have been developed through evolutionary requirements involving factors other than predation. The activity of mycotoxins and other fungal metabolites across this gradient of potential roles will be considered in light of what is known about their biosynthesis and toxicity to insects in the following discussion in order to gain a greater appreciation of their effects. Similarly, the defensive role of a secondary metabolite is likely to be mediated by the spectrum of predators or competitors to which the fungus is exposed. Fungi predominantly exposed to insect predators would be expected to contain relatively greater numbers or amounts of compounds with activity against insects than fungi that have microorganisms as a chief source of competition. Again, a gradient of activity of a particular secondary metabolite is likely to exist. In terms of predators, the longer period of association of fungi and insects, compared to that of fungi and vertebrates, suggests that fungi ancestral to those that produce mycotoxins are likely to have developed defenses against insects initially. In cases where these defenses also worked against vertebrates, the defenses would be strongly selected for in fungi exposed to both sets of predators. In the following discussion I will point out the remarkable similarity in mycotoxin activity profile for vertebrates and insects.

It should be kept in mind that of the tremendous variety of secondary metabolites produced by fungi [12, 13], a relatively small proportion are toxic to vertebrates at naturally occurring concentrations. Another small proportion have antimicrobial activity. If one is to accept the concept of a defensive role for fungal secondary metabolites, there still seem to be a large number for which no role is known. It should be kept in mind that the structure of many of these has been determined during a search for these two types of activity (antimicrobial and vertebrate toxins), suggesting an overrepresentation in these areas. This consideration again brings us to the concept of a role for secondary fungal metabolites in insect control. For example, nearly all compounds considered to be mycotoxins and tested on insects have had activity against insects, although the relative toxicity of an individual mycotoxin for different insects varies considerably (see Ref. 14 for review). More about the reasons for this will be presented later. An increasing number of fungal metabolites that have activity against insects have been discovered in recent years, including compounds that apparently have no vertebrate toxicity (and thus no role in defense against vertebrate predation), such as the *Aspergillus*-derived dihydroxyflavinine [15, 16] and analogs [17, 18], and the *Penicillium* metabolites cyclopenol and brevanamide A [19]. The relative toxicity and potential modes of action of mycotoxins in insects will now be explored.

III. MYCOTOXINS AS A DEFENSE AGAINST INSECTS

The defensive role of mycotoxins against insects can be readily understood when considered by analogy with the defensive role of the secondary metabolites of vascular plants, and this understanding is valuable in identifying areas in need of further research. From early speculative discussions on the defensive role of these plant secondary metabolites (e.g., Ref. 6) through today's current research directions, the insect–plant secondary metabolite interactions have been found increasingly to be a complex phenomenon. Insects have apparently adapted to feed on plants containing a truly remarkable spectrum of compounds by means of detoxification involving excretion, sequestration, and/or enzymatic conversion (see Ref. 20 for review). Furthermore, insects and plants appear to undergo a "coevolutionary conflict" whereby insects adapt to plant toxins, plants counteradapt by producing new compounds, and so on [21]. Other plant secondary metabolites that are not particularly toxic in themselves can synergize the toxicity of the co-occurring compounds, a phenomenon that has been little demonstrated but is thought to be widespread [22]. Given the long history of insect–fungus interactions, one would expect that insects and fungi should interact in a similar manner. Again, although little documented, evidence continues to be uncovered that these interactions occur. The interactions between mycotoxins and insects serve as appropriate examples and will be considered next.

A. Mycotoxin Toxicity to Insects

If mycotoxins do indeed play a defensive role against insects, they should be toxic at naturally occurring concentrations. Here, it is appropriate to correctly consider the conditions under which these interactions may occur. In situations where insects are frequently exposed to (or adapted to feed upon) mycotoxin-producing fungi, mycotoxins would be expected to be relatively less toxic, due to the development of resistance in these insects. A good example of this toxicity spectrum is the range of toxicity of aflatoxin B_1 and T-2 toxin, probably the mycotoxins that have been tested most widely on insects, to different insects (see Ref. 14). While the methodology of administration is quite variable, it is still possible to get some idea of relative toxicity of these compounds by considering thresholds for toxicity of orally administered doses. Cockroaches, which are among the most ancient of insects (and potentially have been interacting with fungi for a longer period of time), are relatively resistant to aflatoxin B_1 [23]. Stored-product insects, which may frequently encounter toxin-producing fungi and would be limited in their ability to find food uncontaminated once the fungi invade, are relatively resistant to T-2 [24, 25]. The fruit flies, such as *Drosophila melanogaster*, which naturally feed on material potentially contaminated by a variety of microorganisms, are nevertheless more susceptible to aflatoxin B_1 [26]. Caterpillars, such as the silkworm (*Bombyx mori*) [27], the corn earworm (*Heliothis zea*), and the fall armyworm (*Spodoptera frugiperda*) [28–30], are quite susceptible to aflatoxin B_1 and T-2 [30]. These caterpillars would not be expected to encounter microorganisms on a frequent basis and so would not be expected to have much resistance to their toxins (although mechanisms present to deal with plant secondary metabolites may be applicable, depending on structural similarity).

Insects that feed on or serve as vectors for mycotoxin-producing fungi should be especially resistant to aflatoxin B_1. Indeed, the adult of the driedfruit beetle (*Carpophilus hemipterus*), which has been implicated as a vector for a variety of mycotoxin-producing fungi (see below), is unaffected by 25 ppm of aflatoxin B_1, and the larvae are only slightly affected (Dowd, unpublished data). Comparatively speaking, other mycotoxins may also be most effective as a defense against insects that do not frequently encounter them, such as caterpillars, especially those that could feed on material that may become contaminated by fungi, such as corn. In examining a spectrum of mycotoxins against the corn earworm and fall armyworm (Table 1), it is clear that most of them are toxic (albeit sublethally) at doses of 25 ppm or less.

The question remains, however: Is the dose required to cause toxic effects in insects a naturally relevant dose? Although quantitation of mycotoxins in naturally contaminated material is predominantly based on bulk samples comprising many uninfected kernels and a few infected kernels, it should be kept in mind that the individual kernels often have much higher levels of toxins [e.g., 31, 32]. Since for insect–fungus interactions it is the individual kernels that can be a make-or-break situation for the fungus, concentrations in individual kernels, although rarely reported in the literature, are most relevant in considering concentrations of mycotoxins useful in deterring insects. As such, a 25 ppm level may be conservatively considered as a relevant naturally occurring concentration for individual kernels, yet this concentration has been reported for several mycotoxins in bulk samples [e.g., 33–36], and in some cases even higher levels may be present [e.g., 37]. As indicated above, a 25 ppm level of many mycotoxins is toxic to caterpillars, suggesting that they play a natural role in defending against these insects.

The utility of mycotoxins in defending against predation by stored-grain insects is less clear. Those that have been tested are generally toxic to representative stored-product caterpillars [38], but most stored-product beetles appear relatively resistant to mycotoxins. For example, the confused flour beetle (*Tribolium confusum*) and the cigarette beetle (*Lasioderma serricorne*) show little or no ill effects from ochratoxin A at 1000 ppm [25]. As discussed previously, it is possible that these insects have a longer history of exposure to mycotoxins and thus are relatively resistant to their effects. It is also possible that other fungal secondary metabolites that are not recognized as vertebrate toxins may be relatively more important in defending against these insects.

B. Mycotoxin Modes of Action

The mode of action of mycotoxins on insects is also worth considering. Since very little information is available on their mode of action in insects,

TABLE 1. Effects of Mycotoxins on Corn Earworms and Fall Armyworms

Mycotoxin	Corn earworm PM	Corn earworm PGR	Fall armyworm PM	Fall armyworm PGR
	Aspergillus			
Aflatoxin B_1	—	94.1	45.9	16.1
Ochratoxin A	38.5	78.6	92.3	66.7
Sterigmatocystin	10.0	94.7	23.5	95.5
	Penicillium			
Citrinin	22.5	48.6	79.7	62.2
Cyclopiazonic acid	—	47.6	—	13.9
Penicillic acid	—	26.4	12.8	31.9
	Fusarium			
Deoxynivalenol	—	52.0	—	85.3
Diacetoxyscirpenol	—	91.5	12.8	85.1
Dihydrodeoxynivalenol	—	39.9	—	38.1
Dihydroxycalonectrin	—	47.0	—	50.5
Hydroxycalonectrin	10.5	52.3	—	71.2
Moniliformin	—	—	—	22.1
Sambucinol	—	27.1	—	16.7
T-2 toxin	12.8	84.3	—	67.4
Zearalenone	—	—	—	34.9
	Tremorgens and related compounds			
Chaetoglobosin C	18.0	18.1	—	10.1
Cytochalasin H	—	57.6	—	13.0
Paspaline	—	20.3	—	26.2
Paxilline	10.0	28.9	—	35.8
Penitrem A	28.2	96.7	—	76.6
Roseotoxin B	38.7	99.7	100.0	—
Verruculogen	—	79.8	—	79.9

All compounds tested at 25 ppm except for aflatoxin B_1 (2.5 ppm), moniliformin and zearalenone (both at 250 ppm). PM = percent mortality, PGR = percent growth reduction. Values less than 10% are not reported.
Source: Compiled from Refs. 15, 29, 30, 44, and 57 and Dowd, unpublished data.

mycotoxins will be assumed to function in the same manner as in vertebrate or cellular models. Insects are generally considered to be free of cancer-like problems, although a few instances of tumors in insects have been reported [39, 40]. Thus, the carcinogenic activity of mycotoxins is probably not relevant in their defensive role against insects. Carcinogenic activity would possibly seem most relevant in long-term, stable environments (e.g., stored grain) rather than in unstable environments (e.g., field crops). Thus, those mycotoxins that cause acute toxicity to insects (indeed, very few mycotoxins are carcinogenic and not otherwise toxic) may be assumed to be acutely toxic and not carcinogenic by design.

It is probably most advantageous for the fungus to make toxins that affect as wide a variety of organisms as possible. Since potential predators may have a varied spectrum of target sites, there is an obvious advantage to producing a toxin that affects target sites in common to the widest variety of organisms. This strategy is effectively demonstrated by those mycotoxins that interfere with processes governing cellular multiplication, such as the aflatoxins and trichothecenes [41]. More specialized target site activity may be advantageous for other reasons. For example, a rapidly growing fungus may have less need to defend against potentially competing microorganisms and thus a proportionately greater need to defend against predatory multicellular organisms. The specialized abilities of differentiated cells in higher organisms produce corresponding "specialized" target sites. This may explain the somewhat selective toxicity of mycotoxins such as ochratoxin A and citrinin to cells involved in nitrogen ion regulation such as the vertebrate kidney [42, 43] and insect Malpighian tubules [44].

With increasing complexity of multicellular organisms, a centralized controlling network is necessary—the nervous system. This area is particularly critical to the functioning of higher organisms and thus represents a good target site for defensive toxins. The widespread distribution of neuroactive mycotoxins, such as the tremorgens, thus can be explained readily. Although the actual target site for many of these tremorgens has not been established conclusively, they appear to interact with a diverse number of neurotransmitters/receptors, such as glycine (penitrem A [45]), γ-aminobutyric acid (GABA) (penitrem A and verruculogen [46], paspalanine [47], aflatrem [48]), and glutamate (verruculogen [46]). Although glycine has not been established conclusively as a transmitter in the insect nervous system, the rest of the aforementioned neurotransmitters are present in the insect nervous system [49]. More similarities between insect and vertebrate nervous systems continue to be discovered as investigations progress. Thus, neuroactive mycotoxins also have a logical role in defending against insects, and indeed many of the aforementioned compounds are quite active against insects [15].

One mycotoxin that does not appear to have an obvious parallel target site in insects is zearalenone. An estrogen-related target site is not known to exist in insects. However, insects do have a steroid-based hormone (ecdysone) as well as carefully regulated steroid metabolism, so the few instances of zearalenone activity in insects [24] may be due to its interference with other functions that are steroid-based.

Mycotoxins may also have insect-antifeedant properties (again a spectrum of antifeedant activity and toxicity is likely for any particular mycotoxin) that are also important in considering a defensive role. Certainly it is advantageous for the fungus to be detected by the insect before it is consumed in significant quantities. A "tasteless" toxin that requires significant consumption of the fungus before it has any effect appears less desirable. A very potent toxin that is less readily detected may be equally acceptable as a defense if a limited amount of fungus is consumed or if the toxin is secreted away from the fungus. Again, a gradient of responses to a particular mycotoxin is likely to be involved, and responses should vary depending on the target organism.

Thus, other than carcinogenic effects, the acutely toxic effects of mycotoxins (as defined by effects on vertebrates) seem to be parallelled in insects. Returning to the earlier theme on the greater likelihood of insect-targeted defenses of fungi due to longer periods of coevolution, it appears that vertebrates may be victims of defenses ancestrally derived to deter insects. The ability to produce insecticidal compounds is retained by domesticated forms of *Aspergillus flavus* and *A. parasiticus*, although the ability to produce compounds toxic to vertebrates appears to have been lost [50]. An extremely high percentage of mycotoxins are also toxic to the insects that have not selectively adapted to them, such as caterpillars. If further investigations indicate that there are a large number of fungal secondary metabolites that are toxic to insects than there are toxic to vertebrates, this would further support the concept that fungal toxins are ancestrally derived to defend against insects versus vertebrates.

C. Interactions Between Mycotoxins and Other Co-occurring Chemicals

Earlier, I mentioned that higher plants are thought to produce relatively nontoxic secondary metabolites that synergize the toxicity of co-occurring toxins to insects (and presumably other organisms). Such a scenario would be expected to occur with mycotoxins as well. Before considering synergistic interactions, it is worth considering other types of interactions.

Nutrition can be a powerful modifier of the toxicity of many toxins, including mycotoxins. For example, the toxicity of the trichothecenes, which inhibit protein synthesis, can be partly counteracted with high-protein diets [51]. Insects fed on nutritionally inferior material also appear to be more susceptible to mycotoxins [24, 52]. In cases where multiple toxins co-occur, the resulting toxicity of the combination should be additive. This relationship is seen when the mycotoxins deoxynivalenol and aflatoxin B_1 are coadministered [53].

However, when two (or more) toxins are combined, the resulting toxicity may be greater than anticipated, or synergized (sometimes referred to as *potentiated* when two toxins are involved). Both additivity and synergistic activity of mycotoxins are cause for concern, since materials can be contaminated with single species of fungi that produce multiple toxins, such as citrinin and ochratoxin A [34, 54], penicillic acid and citrinin [54], or ochratoxin A and penicillic acid [33]. Similarly, more than one species of mycotoxin-producing fungi may co-occur, so multiple combinations of toxins would again be expected, such as aflatoxin B_1 and deoxynivalenol [55], or *Aspergillus flavus* and *A. niger* toxins [11]. All possible combinations of ochratoxin A, citrinin, and penicillic acid caused synergized toxicity in vertebrates [56]. Insects are similarly affected by these compounds. For example, ochratoxin A and citrinin showed synergized toxicity to the fall armyworm [44]. Interestingly, this combination had the greatest toxicity to both the fall armyworm and corn earworm, and it is a commonly occurring combination as well, suggesting adaptive selection. Combining deoxynivalenol with sublethal amounts of other *Fusarium graminearum* trichothecenes also resulted in synergized toxicity to these two insect species [57]. Once again, it is important to consider that a particular fungus has to deal with a spectrum of predators/competitors, so combinations that are effective on one predator may not be effective on another. The final complex the fungus arrives at is likely to give the best overall defense against all organisms involved. However, as described above, the large number of mycotoxin target sites that insects and vertebrates have in common suggest that many synergistic interactions will also occur in common in these two groups of organisms.

Mycotoxins often co-occur with a number of additional fungal secondary metabolites for which no role has been determined. By analogy with the situation in higher plants, it is also possible that these co-occurring fungal metabolites can synergize (or antagonize) the toxicity of co-occurring mycotoxins. These types of interactions may be an explanation for levels of toxicity seen in fungus-contaminated materials that cannot be duplicated by spiking uncontaminated materials with relevant levels of mycotoxins detected in the contaminated materials. For example, the *F. graminearum* metabolite culmorin synergized the toxicity of deoxynivalenol to the corn earworm [57]. The relatively nontoxic kojic acid, which is produced by many species of mycotoxin-producing *Aspergillus* and *Penicillium*, will synergize the toxicity of aflatoxin B_1 to the corn earworm and fall armyworm such that toxicity is approximately equal to a tenfold higher concentration of aflatoxin B_1 alone [29, 30]. The synergistic activity appears to be due to the ability of kojic acid to inhibit unspecific monooxygenases (mixed-function oxidases, polysubstrate monooxygenases) responsible for detoxification [30]. The exact route of aflatoxin B_1 detoxification in insects is presently unknown. The aflatoxin may be directly detoxified by oxidative enzymes or by glutathione transferases that act on the activated epoxide form that is produced by oxidative enzymes, as occurs in mammals [58], since insects also contain these enzymes [59]. Since kojic acid is widely produced by mycotoxin-producing fungi, kojic acid may act as a synergist in other situations as well. Other fungal secondary metabolites that contain functional groups known to interact with unspecific monooxygenases may also act as synergists. Other types of detoxifying enzymes (e.g., hydrolytic, conjugating) may also be inhibited by co-occurring fungal metabolites. These metabolites could also act as synergists for various mycotoxins. Although these synergistic phenomena may be more oriented toward insects, the similarities of detoxifying enzymes in vertebrates and insects suggest that fungal secondary metabolites may synergize the toxicity of mycotoxins in vertebrates as well as other organisms.

However, since insects and mycotoxin-producing fungi often interact within the framework of a complex of the secondary metabolites produced by a host plant, potential interactions between these chemicals and mycotoxins cannot be ignored. The nutritional influence of the host material has already been discussed. However, host plants have their own spectrum of secondary metabolites designed to defend against insects and other predators or pathogens, as well as other secondary metabolites for which no function has been determined. In cases where mycotoxin-producing fungi invade higher plants, an additional complex of chemical interactions must be considered. Many classes of plant secondary metabolites can defend against both insects and fungi. Classes of these dual-role metabolites include phenolics, stilbenes, naphthoquinones, terpenes, lignins, flavonoids, *N*-aromatics, sesquiterpene lactones, and phytoalexins (compiled from Refs. 60–62). In some cases the same compound is known to defend against insects and microorganisms; examples of such compounds are gossypol, hydroxamic acids, and phenolic acids. Again we see that single compounds can defend against a complex of potential predators. Whether mycotoxins and other fungal secondary metabolites can interact with plant secondary metabolites will depend on the timing of the co-occurrence, which may be influenced by the detoxification of plant toxins that are toxic to the invading fungus (and potentially to insects the fungus may wish to defend against), or the relative presence or absence in the plant, which may depend on plant health [63]. However, interactions between fungal and plant secondary metabolites may occur and may be important. Plant phenolics can alter aflatoxin B_1 toxicity to vertebrates [64]. Fusaric acid, which is widely produced by many species of plant pathogenic *Fusarium*, was found to synergize the toxicity to the corn earworm of some insecticidal plant defense compounds with which it could co-occur [65]. Thus, a spectrum of additivity, synergism, or antagonism may occur in insect–fungus–plant interactions involving mycotoxins, which may explain incongruities of toxicity of material infested by mycotoxin-producing fungi that cannot be explained on the basis of the concentration of known mycotoxins alone.

There is an adaptive reason for the presence of secondary metabolites that synergize mycotoxins. In cases where preexisting plant-based insect defenses are present, the fungus may not need to exert as much "effort" in defending itself against insects. Whether this is something that can be sensed by the fungus and serves to regulate mycotoxin biosynthesis remains to be tested. The synergism of mycotoxins, which are frequently biosynthetically complex (and thus costly) to produce (e.g., aflatoxins [66] and trichothecenes [67]) by biosynthetically simpler co-occurring fungal metabolites (such as kojic [68] and fusaric [69] acids), would appear to result in a net savings in effort for the fungus involved [39, 65]. Similarly, when the fungus can make use of preexisting plant defenses to synergize mycotoxins, or produce biosynthetically simple compounds such as fusaric acid to synergize plant toxins, the fungus could again conserve its resources [65]. However, existing plant defenses may also decrease the toxicity of mycotoxins [70]. Further investigation of this area could be fruitful in explaining some mycotoxin problems.

IV. INSECT ADAPTATIONS TO MYCOTOXINS

Once again we will make use of the analogy between fungal and plant secondary metabolites as defensive compounds. Insects have developed resistance to plant defensive metabolites and synthetic insecticidal compounds, so there is no reason to assume that they do not have the same capability to adapt to mycotoxins. Examples of insects that appear to have adapted to mycotoxins were briefly examined when the relative toxicity of mycotoxins to insects was discussed. Presumably, where there is a selective advantage for insects to adapt to mycotoxins, whether this is because they are frequently encountered in the insect's environment or because the insects are targeting material contaminated with these fungi as a food source, insects will develop resistance. I will again use the apparently unadapted caterpillars and well-adapted nitidulids for comparisons.

The infrequent and unpredictable encounters of caterpillars with mycotoxin-producing fungi do not provide sufficient selective drive to promote resistance in these insects. This does not mean that existing defenses against other toxins (plant-derived or synthetic) will not coincidentally work on mycotoxins as well. If the functional groups targeted by detoxifying mechanisms developed for plant secondary metabolites also occur in an enzymatically acceptable orientation in mycotoxins, the mycotoxins should also be detoxified. For example, underivatized conjugated ring systems should be attacked by unspecific monooxygenases regardless of the synthetic source (fungus, plant, or human), based on the substrate specificity of these enzymes [71]. However, of several mycotoxins tested for NADPH-associated oxidation by midgut homogenates of the apparently unadapted fall armyworm and corn earworm, only zearalenone and sterigmatocystin were acted on (Dowd, unpublished data), and these examples of metabolism could involve activation reactions as well. The model trichothecene 4-monoacetoxyscirpenol is metabolized (primarily hydrolyzed) by the midguts of the fall armyworm and corn earworm [72]. However, in order for the detoxification mechanisms to be effective, they must be active enough to prevent substantial toxicity. Rates of detoxification may also be modified by the degree of enzyme induction that occurs upon exposure to a particular toxin or co-occurring compound. Although activity of an unspecific monooxygenase and an esterase were induced by 250 to 25 ppm of deoxynivalenol, T-2 toxin, and diacetoxyscirpenol incorporated into caterpillar diets, the compounds were still significantly toxic at these levels [73]. Thus, the relatively high toxicity of mycotoxins to these caterpillars appears to be exacerbated by relatively poorly or inappropriately adapted detoxifying systems.

As indicated earlier, some insects appear to be relatively resistant to mycotoxins, which may be the result of exposure-driven adaptation or actual adaptation driven by pursuit of the fungi as food. Those insects that may have developed resistance due to simple exposure include many of the stored-product beetles. However, these beetles do not feed very successfully on mycotoxin-producing fungi such as *A. flavus* or *P. citrinum* [14, 74]. Even the foreign grain beetle (*Cynaeus angustus*), which feeds on many stored-product fungi [75, 76], is unable to feed succesfully on *A. flavus* [76] and, as would be suspected, is still susceptible to aflatoxin B_1 at 2.5 ppm (Dowd, unpublished data). However, adults of a representative nitidulid, *Carpophilus hemipterus*, that has been implicated as a vector for *A. flavus* (Lussenhop and Wicklow, unpublished data) are apparently unaffected, and larvae are affected by only a few mycotoxins at 25 ppm (Table 2). The adaptations of nitidulids that promote the carrying of mycotoxin-producing fungi to corn (*Zea mays*) will be considered next.

Although many insects have been implicated as vectors of mycotoxin-producing fungi [77], nitidulids have received relatively little consideration. Yet these insects have been recognized as vectors of many yeasts and fungi for several decades [78]. Their importance as potential vectors for mycotoxin-producing fungi has been recognized only recently [79–81]. These insects have a number of traits that promote their role as vectors of mycotoxin-producing fungi as well as yeast and other fungi in general. Their resistance to mycotoxins has already been mentioned. This resistance may be associated with greater enzymatic detoxifying activity, since the model trichothecene 4-monoacetoxyscirpenol was metabolized (primarily hydrolyzed) by the guts of driedfruit beetle larvae at about 10 times the rate at which it is metabolized by the guts from corn earworms and fall armyworms [72]. Many nitidulids are relatively hairy, which appears to promote the transmission of both sticky and dry spores [82, 83]. The gut is adapted (or the fungal spores counteradapted) so that a significant percentage of fungal spores, including conidia of *A. flavus* (Wicklow, personal communication), will germinate after passing through the gut of the

TABLE 2. Effects of Mycotoxins on Driedfruit Beetle Larvae

Mycotoxin	Growth reduction (%)
Aspergillus	
Aflatoxin B_1	24.1
Ochratoxin A	77.8
Sterigmatocystin	—
Penicillium	
Citrinin	—
Cyclopiazonic acid	—
Penicillic acid	—
Fusarium	
Diacetoxyscirpenol	19.2
Deoxynivalenol	—
T-2 toxin	—

All compounds tested at 25 ppm. Values less than 10% are not reported.

insect [84, 85]. These insects are attracted to general fermentative volatiles such as ethanol [86], vinegar [87, 88], and small-molecule esters that may be produced in fermentation [86, 89–91]. These insects also promote the spread of these fungi due to their tendency to aggregate, which would allow many insects to become contaminated when only one individual brings in inoculum. This aggregation is in part controlled by an aggregation pheromone [91, 92] that can be synergized by fungal volatiles ([82]; Dowd and Bartelt, unpublished data). Long-chain alcohols produced by corn [93] and many species of *Aspergillus* and *Penicillium* [94, 95] will also attract or synergize pheromonal attraction of these insects (Dowd and Bartelt, unpublished data).

The high incidence of the most common nitidulid pest of corn in temperate areas of North America, the dusky sap beetle (*C. lugubris*), as a pest of corn has been attributed to the use of loose-husked corn and mechanical harvesting, which promote more readily available access [78]. However, mechanical damage of corn will more readily promote entry of these insects to the ears [78, 96]. In fact, nitidulids are more attracted to wounds caused by the corn earworm than to simple mechanically caused wounds [97]. However, there appears to be something of a conflict between nitidulids and caterpillars. Nitidulids such as spotted picnic beetles (*Grischrochilus* spp.) are reported to prey on European corn borer (*Ostrinia nubilalis*) larvae in corn [98], whereas larvae of other nitidulids are reportedly preyed upon by larvae of the corn earworm [99]. Since these two types of insects are competing for the same resource (corn), serving as a vector of mycotoxin-producing fungi may prove advantageous for the nitidulids because the fungi may produce mycotoxins that are toxic to the caterpillars but not the nitidulids. However, since the nitidulids also eat the fungi, there appears to be a trade-off similar to that of other fungi that are spread by insects (e.g., ambrosia fungi [100]) in that the insects consume some of the fungi but enough remains to be carried by the insects. A mutualistic association has been proposed for their role as vectors of oak wilt [87]. The presence of chemical defenses in the long-term survival structures of some fungi, the sclerotia, that strongly deter nitidulids [16–18] also supports the concept that nitidulids and mycotoxin-producing fungi have a somewhat mutually adaptive relationship. These fungi can tolerate feeding on their short-term and very numerous conidia by nitidulids, but their long-term survival sclerotia needs to be protected from even these insects. Although it appears that nitidulids and mycotoxin-producing fungi have undergone coadaptations, the relationship does not at present appear to be one of obligate dependence.

V. SUMMARY

Mycotoxins can be likened to plant allelochemicals in that they can play a defensive role against insects. Since insects and fungi have been interacting for possibly 100 million years longer than vertebrates and fungi, many mycotoxins may have their origins as defensive substances against insects. Owing to the number of target sites insects and vertebrates have in common, vertebrates have also been affected. However, the defensive orientation of mycotoxins is likely to involve a wide spectrum of potential predators/competitors, and so their relative proportion and activity in any given situation are likely to be mediated by the relative importance of the different predators/competitors. Unadapted insects, such as caterpillars, are highly susceptible to most mycotoxins. Mycotoxins can interact with co-occurring fungal metabolites or plant metabolites such that toxicity to insects (and other organisms) can be altered and potentially synergized.

Certain insects appear to have adapted to mycotoxins, presumably through natural selection driven by various degrees of exposure. These adaptations may involve enhanced detoxification through sequestration, excretion, resistant target sites, and/or metabolic conversion, although few mechanisms have been determined. From the standpoint of mycotoxin control, these resistant insects represent potential vectors for mycotoxin-producing fungi. For example, nitidulids appear to be well adapted or coadapted for carrying mycotoxin-producing fungi to corn. They are resistant to naturally occurring levels of many mycotoxins, are attracted to volatiles produced by fungi and their hosts, are well synchronized with the U.S. corn crop in many areas, have aggregating tendencies that could promote the spread of mycotoxin-producing fungi by enhancing contamination, and have external and internal adaptations that promote transmission of these fungi. Serving as vectors for mycotoxin-producing fungi also appears to be adaptively desirable for these insects in that mycotoxin contamination of corn is likely to inhibit potential competitors/predators such as caterpillars.

REFERENCES

1. Borror, D. J., DeLong, D. M., and Triplehorn, C. A. *An Introduction to the Study of Insects*, Holt, Rinehart, and Winston, New York, 1976, p. 139.
2. Brock, T. D. *Biology of Microorganisms*, Prentice-Hall, Englewood Cliffs, N.J., 1974, p. 599.
3. Smart, J., and Hughes, N. F. The insect and plant: progressive palaeoecological integration. In *Insect/Plant Relationships* (H. F. van Emden, Ed.), Wiley, New York, 1973, p. 143.
4. Southwood, T. R. E. The insect/plant relationship—an evolutionary perspective. In *Insect/Plant Relationships* (H. F. van Emden, Ed.), Wiley, New York, 1973, p. 3.
5. Miller, J. D. Toxic metabolites of epiphytic and endophytic fungi of conifer needles. In *Microbiology of the Phyllosphere* (N. J. Fokkema and J. van den Heuvel, Eds.), Cambridge Univ. Press, New York, 1985, p. 223.
6. Whittaker, R. H., and Feeny, P. P. Allelochemicals: chemical interactions between species. *Science, 171*:757, 1971.
7. Janzen, D. H. Why fruits rot, seeds mold, and meat spoils. *Am. Nat., 111*:691, 1977.
8. Wicklow, D. T., and Cole, R. J. Tremorgenic indole metabolites and aflatoxins in sclerotia of *Aspergillus flavus*: an evolutionary perspective. *Can. J. Bot., 60*:525, 1982.

9. Wicklow, D. T. Ecological approaches to the study of mycotoxigenic fungi. In *Toxigenic Fungi — Their Toxins and Health Hazard* (H. Kurata and Y. Ueno, Eds.), Elsevier, Amsterdam, The Netherlands, 1984, p. 33.
10. Wicklow, D. T. Adaptation in wild and domesticated yellow-green aspergilli. In *Toxigenic Fungi — Their Toxins and Health Hazard* (H. Kurata and Y. Ueno, Eds.), Elsevier, Amsterdam, The Netherlands, 1984, p. 78.
11. Wicklow, D. T. Metabolites in the coevolution of fungal chemical defence systems. In *Coevolution of Fungi with Plants and Animals* (K. A. Pirozynski and D. L. Hawksworth, Eds.), Academic, New York, 1988, p. 173.
12. Turner, W. B. *Fungal Metabolites*, Academic, New York, 1971, p. 1.
13. Turner, W. B., and Aldridge, W. B. *Fungal Metabolites*, Vol. 2, Academic, New York, 1983, p. 1.
14. Wright, V. F., Vesonder, R. F., and Ciegler, A. Mycotoxins and other fungal metabolites as insecticides. In *Microbial and Viral Pesticides* (E. Kurstak, Ed.), Marcel Dekker, New York, 1982, p. 559.
15. Dowd, P. F., Cole, R. J., and Vesonder, R. F. Toxicity of selected tremorgenic mycotoxins and related compounds to *Spodoptera frugiperda* and *Heliothis zea*. *J. Antibiot., 41*:1868, 1988.
16. Wicklow, D. T., Dowd, P. F., Tepaske, M. R., and Gloer, J. B. Sclerotial metabolites of *Aspergillus flavus* toxic to a detritivorous maize insect (*Carpophilus hemipterus*, Nitidulidae). *Trans. Br. Mycol. Soc., 91*:433, 1988.
17. Gloer, J. B., Rinderknecht, B. L., Wicklow, D. T., and Dowd, P. F. Nominine: a new insecticidal indole diterpene from the sclerotia of *Aspergillus nomius*. *J. Org. Chem., 54*:2530, 1989.
18. Gloer, J. B., TePaske, M. R., Sima, J. S., Wicklow, D. T., and Dowd, P. F. Antiinsectan aflavinine derivatives from the sclerotia of *Aspergillus flavus*. *J. Org. Chem., 53*:5457, 1988.
19. Patterson, R. R. M., Simmonds, M. S. J., and Blaney, W. M. Mycopesticidal effects of characterized extracts of *Penicillium* isolates and purified secondary metabolites (including mycotoxins) on *Drosophila melanogaster* and *Spodoptera littoralis*. *J. Invert. Pathol., 50*:124, 1987.
20. Dowd, P. F. Detoxification of plant substances by insects. In *Handbook of Natural Pesticides*, Vol. 5 (E. D. Morgan and L. D. Mandava, Eds.), CRC Press, Boca Raton, Fla., 1990, p. 181.
21. Feeny, P. Coevolution of plants and insects. In *Natural Products for Innovative Pest Management* (D. L. Whitehead and W. S. Bowers, Eds.), Pergamon, New York, 1983, p. 167.
22. Berenbaum, M. R. Brementown revisited: allelochemical interactions in plants. *Recent Adv. Phytochem., 19*:139, 1985.
23. Sheretz, P. C., Mills, R. R., and Llewellyn, G. C. Comparative dietary patterns and body weight changes in adult male American cockroaches fed pure aflatoxin B_1. *Bull. Environ. Contam. Toxicol., 20*:687, 1978.
24. Wright, V. F., De las Casas, E., and Harein, P. K. The response of *Tribolium confusum* to the mycotoxins zearalenone (F-2) and T-2 toxin. *Environ. Entomol.*, 5:371, 1976.

25. Wright, V. F., De las Casas, E., and Harein, P. K. Evaluation of *Penicillium* mycotoxins for activity in stored-product Coleoptera. *Environ. Entomol., 9*:217, 1980.
26. Lalor, J. H., Chinnici, J. P., and Llewellyn, G. C. Mutagenicity of aflatoxins. *Devl. Ind. Microbiol., 17*:443, 1976.
27. Murakoshi, S., Ohtomo, T., and Kurata, H. Toxic effects of various mycotoxins to silkworm (*Bombyx mori* L.) larvae in ad libitum feeding tests. *J. Food Hyg. Soc. Jpn., 14*:65, 1973.
28. McMillian, W. W., Wilson, D. M., Windstrom, N. W., and Perkins, W. D. Effects of aflatoxin B_1 and G_1 on three insect pests of maize. *J. Econ. Entomol., 73*:26, 1980.
29. Dowd, P. F. Synergism of aflatoxin B_1 toxicity with the co-occurring fungal metabolite kojic acid to two caterpillars. *Entomol. Exp. Appl., 47*:69, 1988.
30. Dowd, P. F. Toxicological and biochemical interactions of the fungal metabolites fusaric acid and kojic acid with xenobiotics in *Heliothis zea* (F.) and *Spodoptera frugiperda* (J. E. Smith). *Pestic. Biochem. Physiol., 32*:123, 1988.
31. Wicklow, D. T., Horn, B. W., and Shotwell, O. L. Aflatoxin formation in preharvest maize ears coinoculated with *Aspergillus flavus* and *Aspergillus niger*. *Mycologia, 79*:679, 1987.
32. Bennett, G. A., Wicklow, D. T., Caldwell, R. W., and Smalley, E. B. Distribution of trichothecenes and zearalenone in *Fusarium graminearum*: rotted corn ears grown in a controlled environment. *J. Agric. Food Chem., 36*:639, 1988.
33. Ciegler, A. Bioproduction of ochratoxin A and penicillic acid by members of *Aspergillus ochraceus* group. *Can. J. Microbiol., 17*:631, 1972.
34. Scott, P. M., van Walbeek, W., Kennedy, B., and Anueti, D. Mycotoxins (ochratoxin A, citrinin, and sterigmatocystin) and toxigenic fungi in grains and other agricultural products. *J. Agric. Food Chem., 20*:1103, 1972.
35. Puls, R., and Greenway, J. A. Fusariotoxicosis from barley in British Columbia. II. Analysis and toxicity of suspected barley. *Can. J. Comp. Med., 40*:16, 1976.
36. Vesonder, R. F., Ciegler, A., Jensen, A. H., Rohwedder, W. K., and Weisleder, D. Co-identity of the refusal and emetic principle from *Fusarium*-infected corn. *Appl. Environ. Microbiol., 31*:280, 1976.
37. Miller, J. D., Young, J. C., and Trenholm, H. L. *Fusarium* toxins in field corn. I. Time course of fungal growth and production of deoxynivalenol and other mycotoxins. *Can. J. Bot., 61*:3080, 1983.
38. Wright, V. F., and Harein, P. K. Effects of some mycotoxins on the Mediterranean flour moth. *Environ. Entomol., 11*:1043, 1982.
39. Stanton, M. F. *Neoplasms and Related Disorders of Invertebrate and Lower Vertebrate Animals*, National Cancer Institute, Bethesda, Md., 1969, p. 200.
40. Gateff, E. Malignant and benign neoplasms of *Drosophila melanogaster*. In *The Genetics and Biology of Drosophila*, Vol. 2b (M. Ashburner and T. R. F. Wright, Eds.), Academic, New York, 1978, p. 123.

41. Betina, V. Biological effects of mycotoxins. In *Mycotoxins – Production, Isolation, Separation and Purification* (V. Betina, Ed.), Elsevier, Amsterdam, 1984, p. 25.
42. Purchase, I. F. H., and Nel, W. Recent advances in research on ochratoxin. Part 1. Toxicological aspects. In *Biochemistry of Some Foodborne Microbial Toxins* (R. I. Mateles and G. N. Wogan, Eds.), MIT Press, Cambridge, Mass., 1967, p. 153.
43. Krogh, P., Hasselager, E., and Friis, P. Studies on fungal nephrotoxicity: II. Isolation of two nephrotoxic compounds from *Penicillium virdicatum*, Westling: citrinin and oxalic acid. *Acta Pathol. Microbiol. Scand.*, *B78*:401, 1970.
44. Dowd, P. F. Toxicity of naturally occurring levels of the *Penicillium* mycotoxins citrinin, ochratoxin A, and penicillic acid to the corn earworm, *Heliothis zea*, and the fall armyworm, *Spodoptera frugiperda* (Lepidoptera: Noctuidae). *Environ. Entomol.*, *18*:24, 1989.
45. Catovic, S., Filipovic, N., and Stern, P. The effect of penitrem-A upon the level of glycine in the CNS. *Bull. Sci.*, *A20*:284, 1975.
46. Norris, P. J., Smith, C. C. T., De Belleroche, J., Bradford, H. F., Mantle, P. G., Thomas, A. J., and Penny, R. H. C. Actions of tremorgenic fungal toxins on neurotransmitter release. *J. Neurochem.*, *34*:33, 1980.
47. Gant, D. B., Cole, R. J., Valdes, J. J., Eldefrawi, M. E., and Eldefrawi, A. T. Action of tremorgenic mycotoxins on $GABA_a$ receptor. *Life Sci.*, *41*:2207, 1987.
48. Yao, Y., Peter, A. B., Baur, R., and Sigel, E. The tremorigen aflatrem is a positive allosteric modulator of the *gamma*-aminobutyric acid receptor channel expressed in *Xenopus* oocytes. *Mol. Pharmacol.*, *35*:319, 1989.
49. Leake, L. D., and Walker, R. J. *Invertebrate Neuropharmacology*, Wiley, New York, 1980, p. 195.
50. Wicklow, D. T., and Dowd, P. F. Entomotoxigenic potential of wild and domesticated yellow-green aspergilli: toxicity to corn earworm and fall armyworm larvae. *Mycologia*, *81*:561, 1989.
51. Hayes, M. A., and Schiefer, H. Subacute toxicity of dietary T-2 toxin in mice: influence of protein nutrition. *Can. J. Comp. Med.*, *44*:219, 1980.
52. Davis, G. R. F., and Schiefer, H. B. Effects of dietary T-2 toxin concentrations fed to larvae of the yellow mealworm at three dietary protein levels. *Comp. Biochem. Physiol.*, *73C*:13, 1982.
53. Harvey, R. B., Kubena, L. F., Huff, W. E., Corrier, D. E., Clark, D. E., and Phillips, T. D. Effects of aflatoxin, deoxynivalenol, and their combinations in the diets of growing pigs. *Am. J. Vet. Res.*, *50*:602, 1989.
54. Frisvad, J. C., and Filtenborg, O. Classification of terverticillate penicillia based on profiles of mycotoxins and other secondary metabolites. *Appl. Environ. Microbiol.*, *46*:1301, 1983.
55. Hagler, W. M., Tyczkowska, K., and Hamilton, P. B. Simultaneous occurrence of deoxynivalenol, zearalenone, and aflatoxin in 1982 scabby wheat from the midwestern United States. *Appl. Environ. Microbiol.*, *47*:151, 1984.

56. Sansing, G. A., Lillehoj, E. B., Detroy, R. W., and Miller, M. A. Synergistic toxic effects of citrinin, ochratoxin A, and penicillic acid in mice. *Toxicon, 14*:213, 1976.
57. Dowd, P. F., Miller, J. D., and Greenhalgh, R. Toxicity and interactions of some *Fusarium graminearum* metabolites to caterpillars. *Mycologia, 81*:646, 1989.
58. Degen, G. H., and Neumann, H.-G. The major metabolite of aflatoxin B_1 in the rat is a glutathione conjugate. *Chem-Biol. Interactions, 22*: 239, 1978.
59. Ahmad, S., Brattsten, L. B., Mullin, C. A., and Yu, S. J. Enzymes involved in the metabolism of plant allelochemicals. In *Molecular Aspects of Insect–Plant Associations* (L. B. Brattsten and S. Ahmad, Eds.), Plenum, New York, 1986, p. 73.
60. Rosenthal, G. A., and Janzen, D. H. *Herbivores: Their Interactions with Plant Secondary Metabolites*, Academic, New York, 1979, p. 1.
61. Stoessel, A. Secondary plant metabolites in preinfectional and postinfectional resistance. In *The Dynamics of Host Defense* (J. A. Bailey and B. J. Deverall, Eds.), Academic, New York, 1983, p. 71.
62. Wink, M. Plant breeding: importance of plant secondary metabolites for protection against pathogens and herbivores. *Theor. Appl. Genet., 75*:225, 1988.
63 Gershenzon, J. Changes in the levels of plant secondary metabolites under water and nutrient stress. In *Phytochemical Adaptations to Stress* (B. N. Timmermann, C. Steelink, and F. A. Loewus, Eds.), Plenum, New York, 1983, p. 273.
64. Francis, A. R., Shetty, T. K., and Bhattacharya, R. K. Modification of the mutagenicity of aflatoxin B_1 and *N*-methyl-*N'*-nitro-*N*-nitrosoguanidine by certain phenolic compounds. *Cancer Lett., 45*:177, 1989.
65. Dowd, P. F. Fusaric acid, a secondary fungal metabolite that synergizes the toxicity of cooccurring host aellelochemicals to the corn earworm, *Heliothis zea* (Lepidoptera). *J. Chem. Ecol., 15*:249, 1989.
66. Betina, V. Mycotoxins as secondary metabolites. In *Mycotoxins — Production, Isolation, Separation and Purification* (V. Betina, Ed.), Elsevier, Amsterdam, 1984, p. 13.
67. Tamm, C., and Breitenstein, W. The biosynthesis of trichothecene mycotoxins. In *The Biosynthesis of Mycotoxins: A Study in Secondary Metabolism* (P. S. Stein, Ed.), Academic, New York, 1980, p. 69.
68. Wilson, B. J. Miscellaneous *Aspergillus* toxins. In *Microbial Toxins*, Vol. 6 (A Ciegler, S. Kadis, and S. J. Ajl, Eds.), Academic, New York, 1971, p. 207.
69. Hill, R. D., Unrau, A. M., and Canvin, D. T. The biosynthesis of fusaric acid from ^{14}C-labelled acetate in *Gibberella fujikuroi*. *Can. J. Chem., 44*:2077–2082, 1966.
70. Santhanam, K., and Lotlikar, P. D. Effect of β-naphthoflavone on the metabolism of aflatoxin B_1 in hamsters. *Cancer Lett., 45*:129, 1989.
71. International Union of Biochemistry, Nomenclature Committee. *Enzyme Nomenclature*, Academic, New York, 1984, p. 128.

72. Dowd, P. F., and Van Middlesworth, F. L. In vitro metabolism of the trichothecene 4-monoacetoxyscirpenol by fungus- and non-fungus-feeding insects. *Experientia, 45*:393, 1989.
73. Dowd, P. F. Responses of representative midgut detoxifying enzymes from *Heliothis zea* and *Spodoptera frugiperda* to trichothecenes. *Insect Biochem., 20*:349, 1990.
74. Wright, V. F., De las Casas, E., and Harein, P. K. The nutritional value and toxicity of *Penicillium* isolates for *Tribolium confusum. Environ. Entomol., 9*:204, 1980.
75. Sinha, R. N. Development and mortality of *Tribolium castaneum* and *T. confusum* (Coleoptera: Tenebrionidae) on seed-borne fungi. *Ann. Entomol. Soc. Am., 59*:192, 1966.
76. Sinha, R. N. Fungus as food for some stored-product insects. *J. Econ. Entomol., 64*:3, 1971.
77. Widstrom, N. W. The role of insects and other plant pests in aflatoxin contamination of corn, cotton, and peanuts — a review. *J. Environ. Qual., 8*:5, 1979.
78. Connell, W. A. Nitidulidae of Delaware. *Univ. Del. Agric. Exper. Sta. Bull., 318*:1, 1956.
79. Windels, C. E., Windels, M. B., and Kommedhal, T. Association of *Fusarium* species with picnic beetles on corn ears. *Phytopathology, 66*:328, 1976.
80. Attwater, W. A., and Busch, L. V. Role of sap beetle *Grischrochilus quadrisignatus* in the epidemiology of giberella corn ear rot. *Can. J. Plant Pathol., 5*:158, 1983.
81. Trenholm, H. L., Prelusky, D. B., Young, J. C., and Miller, J. D. A practical guide to the prevention of *Fusarium* mycotoxins in grain and animal feedstuffs. *Arch. Environ. Contam. Toxicol., 18*:443, 1989.
82. Yount, W. L., Jeffery, A. R., and Thompson, H. E. Spores of *Endoconidiophora fagacearum* on the external surfaces of the body of nitidulids. *Plant Dis. Rep., 39*:54, 1955.
83. Juzwik, J., and French, D. W. *Ceratocystis fagacearum* and *C. piceae* on the surfaces of free-flying and fungus-mat inhabiting nitidulids. *Phytophatology, 73*:1164, 1983.
84. Jewell, F. F. Insect transmission of oak wilt. *Phytopathology, 46*: 244, 1956.
85. Moller, W. J., and DeVay, J. E. Insect transmission of *Ceratocystis fimbriata* in deciduous fruit orchards. *Phytopathology, 58*:1499, 1968.
86. Smilanick, J. M., Ehler, L. E., and Birch, M. C. Attraction of *Carpophilus* spp. to volatile compounds of figs. *J. Chem. Ecol., 4*:701, 1978.
87. Dorsey, C. K., and Leach, J. G. The bionomics of certain insects associated with oak wilt, with particular reference to the Nitidulidae. *Plant Dis. Rep., 37*:419, 1956.
88. Neel, W. W., Glick, B. D., May, L. L., and True, R. P. Attractiveness to Nitidulidae (Coleoptera) of natural attractants of tree and fungus origin supplemented with vinegar and water in Appalachian hardwood forest. *J. Econ. Entomol., 60*:1104, 1967.

89. Alm, S. R., Hall, F. R., Ladd, T. L., Jr., and Williams, R. N. A chemical attractant for *Grischrochilus quadrisignatus* (Coleoptera: Nitidulidae). *J. Econ. Entomol., 79*:1276, 1985.
90. Alm, S. R., Hall, F. R., McGovern, T. P., and Williams, R. N. Attraction of *Grischrochilus quadrisignatus* (Coleoptera: Nitidulidae) to semiochemicals: butyl acetate and propyl propionate. *J. Econ. Entomol., 79*:654, 1986.
91. Bartelt, R. J., Dowd, P. F., Plattner, R. D., and Weisleder, D. Aggregation pheromone of the driedfruit beetle, *Carpophilus hemipterus*: wind tunnel bioassay and identification of two novel tetraene hydrocarbons. *J. Chem. Ecol., 16*:1015, 1990.
92. Bartelt, R. J., Dowd, P. F., and Plattner, R. D. Aggregation pheromone of *Carpophilus lugubris*: New pest management tools for the nitidulid beetles. In *Naturally Occurring Pest Bioregulators* (P. E. Hedin, Ed.), American Chemical Society, Washington, D.C., 1991, p. 27.
93. Buttery, R. G., Ling, L. C., and Teranishi, R. Volatiles of corn tassels: possible corn earworm attractants. *J. Agric. Food Chem., 28*:771, 1980.
94. Kaminski, E., Libbey, L. M., Stawicki, S., and Wasowicz, E. Identification of the predominant volatile compounds produced by *Aspergillus flavus*. *Appl. Microbiol., 24*:721, 1972.
95. Kaminski, E., Stawicki, S., and Wasowicz, E. Volatile flavor compounds produced by molds of *Aspergillus, Penicillium,* and fungi imperfecti. *Appl. Microbiol., 27*:1001, 1974.
96. Tamaki, G., Fox, L., and Featherston, P. Laboratory biology of the dusky sap beetle and field interaction with the corn earworm in ears of sweet corn. *J. Entomol. Soc. Br. Columbia, 79*:3, 1982.
97. Sanford, J. W., and Luckman, W. H. Observations on the biology and control of the dusky sap beetle in Illinois. *Proc. N. Cent. Br. Entomol. Soc. Am., 18*:39, 1963.
98. McCoy, C. E., and Brindley, T. A. Biology of the four spotted fungus beetle, *Grischrochilus quadrisignatus* and its effect on European corn borer populations. *J. Econ. Entomol., 54*:713, 1961.
99. Daugherty, D. M., and Brett, C. H. Nitidulidae associated with sweet corn in North Carolina and influences affecting their damage to this crop. *N. Carolina Tech. Bull., 171*:1, 1966.
100. Leach, J. G. *Insect Transmission of Plant Diseases*, McGraw-Hill, New York, 1940, p. 213.

6

SPECIES- AND DIET-RELATED DIFFERENCES IN AFLATOXIN BIOTRANSFORMATION

DAVID L. EATON and HOWARD S. RAMSDELL* *University of Washington, Seattle, Washington*

I. INTRODUCTION

The aflatoxins encompass a series of structurally related compounds produced by *Aspergillus flavus* and *A. parasiticus*. The principal compounds found in plant food products are aflatoxins B_1, B_2, G_1, and G_2 (Fig. 1). Aflatoxin B_1 (AFB_1) has been the subject of the most extensive study, as its acute and chronic toxicity is substantially greater than that of the other aflatoxins in most biological systems. This chapter focuses on the biotransformation of AFB_1 because of the crucial role that metabolic products play in both activation and detoxification of AFB_1. Differences in biotransformation appear to be important determinants of the variation in response to AFB_1 exposure among different species; these processes are also subject to modulation by exogenous factors, including dietary components. Dietary treatments have been used both to probe the biochemical mechanisms of AFB_1 toxicity and to identify effective chemopreventive regimens that could decrease the susceptibility of human populations to AFB_1 carcinogenicity.

II. TOXIC ACTIONS OF AFLATOXINS

A. Acute Toxicity of Aflatoxins

The sensitivity of poultry to the acute effects of aflatoxins was responsible for the initiation of research on this class of mycotoxins in the process of determining the etiologic factor of turkey X disease in the early 1960s [1, 2]. Subsequent investigations demonstrated the toxicity of aflatoxins in other species. Ducklings and trout are very sensitive to the hepatotoxic effects of AFB_1 with LD_{50} values of 0.4 mg/kg (p.o.) and 0.8 mg/kg (i.p.), respectively [3, 4], whereas the rat and mouse are relatively less sensitive to the acute effects of AFB_1. Rat LD_{50} values,

*Present affiliation: Colorado State University, Fort Collins, Colorado.

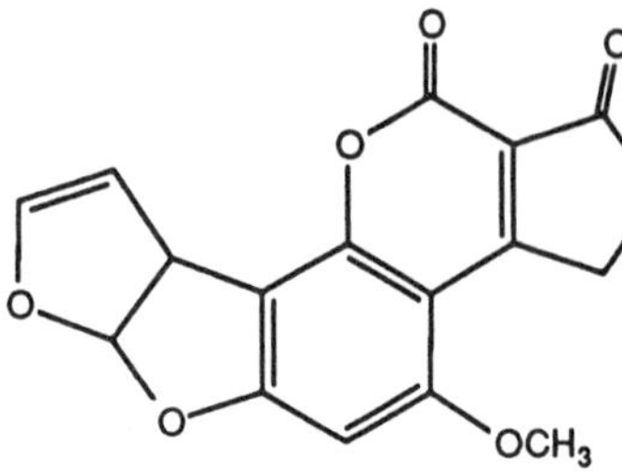

AFB$_1$

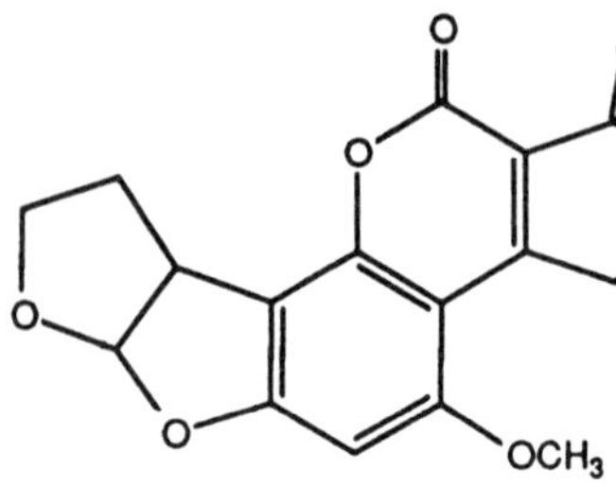

AFB$_2$

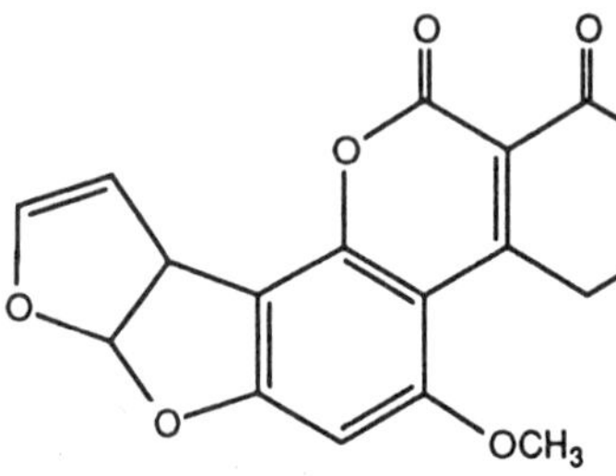

AFG$_1$

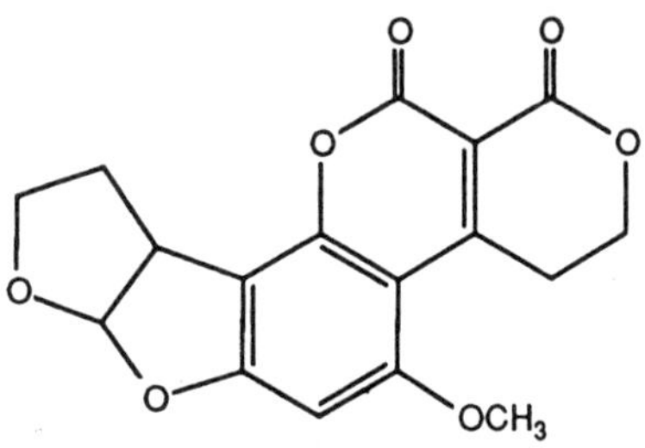

AFG$_2$

FIGURE 1. Aflatoxin structures.

which are influenced by age, sex, strain, and route of administration, vary from 1 to 18 mg/kg [5–7]. Mice appear to be more resistant than rats; in one study in male C57BL/6 mice given 60 mg/kg AFB$_1$ (i.p.), no mortality was observed [8]. In another study in CFW Swiss mice, LD_{50} values from 1 to above 150 mg/kg were obtained, with 1-day-old animals being the most sensitive [7]. Evidence for the toxicity of aflatoxins to humans is provided by a report of fatal acute hepatitis that was linked to consumption of corn containing 12 ppm AFB$_1$ [9]. The liver is the principal target organ in acute aflatoxicosis; characteristic pathological changes include hemorrhagic necrosis. Parenchymal cell necrosis and bile duct proliferation, as well as fatty infiltration in some species, are common microscopic findings in toxin-damaged liver [10].

Early biochemical effects of acute AFB$_1$ exposure include inhibition of both DNA and RNA synthesis [11, 12]. Subsequently, a prolonged decrease in protein synthesis is observed, probably as a result of polysomal disaggregation [13]. Inhibition of RNA synthesis may be due, in part, to the inhibition of tRNA synthesis [14, 15]. It has been suggested that interference with ATP production may also play a role in hepatotoxicity of AFB$_1$. Administration of AFB$_1$ (4 mg/kg, i.p.) results in a modest transient decrease in hepatic glutathione levels in rats about 12 h after treatment; this is accompanied by an increase in lipid peroxide levels (thiobarbituric acid—reactive material) at 24 h, particularly in hepatic mitochondria, followed by a rise in intracellular Ca^{2+} levels at 48 h after dosing [16].

Disruption of protein synthesis (directly or via inhibition of RNA synthesis) may be related to the acute toxicity of AFB_1. In isolated hepatocytes, the dose dependence of cellular toxicity, as measured by lactate dehydrogenase release, was paralleled by decreases in both protein and RNA synthesis [17]. Rats are more sensitive than mice to the inhibition of hepatic RNA synthesis by AFB_1 in liver slices in vitro, consistent with the observed species differences in in vivo aflatoxin sensitivity [18]. There is a substantial body of information demonstrating that biotransformation of AFB_1 plays a critical role in acute toxicity. For example, an inverse relationship between LD_{50} and the extent of total hepatic microsomal oxidation of AFB was observed for a number of species [19]. Other studies have shown that modulators of hepatic drug-metabolizing enzymes alter the acute toxicity of AFB_1 [20—22]. Inducers of cytochrome P-450 have also been shown to affect the cytotoxicity of AFB_1 in hepatic cell cultures [23, 24].

B. Carcinogenicity of Aflatoxins

Aflatoxin B_1 is a potent carcinogen in many species, most notably the rat and trout, whereas mice (and perhaps other species) are relatively resistant to the carcinogenic effects of AFB. Liver is the most common target of aflatoxin carcinogenesis [7], although renal tumors have been observed in rats given AFB_1 by long-term oral administration [25, 26]. Aflatoxin G_1 exposure was even more effective at producing kidney tumors than AFB_1 [25].

Trout are very sensitive to hepatoma induction by AFB_1; exposure to 8 ppb AFB_1 in the diet for 12 months produced a 70% incidence of hepatic tumors [27]. Trout fed a diet containing 20 ppb AFB_1 for as few as 5 days had a 10% incidence of hepatomas 12 months later, with the incidence rising to 40% in fish fed the diet for 20 days [27]. The extreme sensitivity of trout to AFB_1-induced hepatoma is also demonstrated by the development of tumors in 66% of 12-month-old fish exposed to a 0.5 ppm solution of AFB_1 for just 30 min as 21-day-old embryos [28].

Rats are also relatively sensitive to the hepatocarcinogenic effects of AFB_1. Continuous feeding of as little as 15 ppb AFB_1 for 52—80 weeks in a purified diet induced hepatic carcinomas in a significant number of male Fischer rats [29, 30]. Liver tumors have been reported in rats given a single oral dose of 5—7 mg/kg of AFB_1 [30, 31]. Mice are relatively resistant to carcinogenesis by AFB_1; a 1 ppm level of toxin in the diet for 80 weeks following weaning did not elicit liver tumors [32]. However, infant mice given 2 mg/kg (i.p.) at 4, 7, and 10 days of age developed a significant incidence of hepatomas after 82 weeks [33]. Hamsters are also relatively resistant to AFB_1-induced hepatocellular carcinomas [34]; 2 mg/kg of AFB_1 given 5 days/week for 6 weeks resulted in a low incidence of hepatocellular carcinomas (<5%), although most of the animals developed cholangiocellular tumors [35].

Epidemiological studies have suggested that aflatoxins are carcinogenic in humans [36, 37]. Primary liver cancer is relatively common among populations consuming food highly contaminated with AFB_1 [38]. There is some controversy regarding the role of AFB_1 in human hepatocarcinogenesis due to confounding effects of hepatitis B infection in the populations studied [39]. Two independent studies have shown, however, that AFB_1 exposure is an independent risk factor for human liver cancer [40, 41].

It appears that the biochemical mechanisms of acute aflatoxin toxicity are distinct from aflatoxin-mediated carcinogenesis. Rats are far more susceptible to AFB_1-induced hepatocellular carcinoma than are hamsters, yet the acute LD_{50} values for these species are similar (1–18 and 10 mg/kg, respectively) [42].

III. BIOTRANSFORMATION AND AFLATOXIN TOXICITY

Quantitative, and in some instances qualitative, differences in biotransformation of aflatoxins appear to be responsible for the variations in toxic responses that have been observed in various species. Modulation of aflatoxin biotransformation has been utilized in experimental studies to investigate mechanisms of toxicity. Enhancement of detoxification pathways or inhibition of metabolic activation of AFB_1 have been studied to develop useful approaches for the protection of high-risk populations against aflatoxin carcinogenicity [43].

A. Pathways of Aflatoxin Biotransformation

Bioactivation reactions catalyzed by both microsomal and cytosolic enzymes of liver are important in the disposition of AFB_1 (Fig. 2).

Microsomal Oxidation

Formation of the highly reactive AFB_1-8,9-epoxide by cytochrome P-450-mediated oxidation is regarded as the major bioactivation pathway for AFB_1 [44], although co-oxidation of AFB_1 by prostaglandin H synthase has also been demonstrated [45]. The oxidative metabolism of AFB_1 to the epoxide is the initial critical step leading to covalent binding to cellular macromolecules, including DNA [46]. The cytochrome P-450 mixed-function oxidase system is also responsible for formation of the hydroxylated metabolites AFM_1, AFP_1, and AFQ_1 [47, 48]. Further microsomal oxidation of AFM_1 or AFP_1 results in the formation of 4,9a-dihydroxyaflatoxin B_1 [49]. Aflatoxin Q_1, a major oxidative metabolite of AFB in human microsomal incubations [50, 51], is a relatively poor substrate for secondary oxidation to the epoxide by human liver microsomes (Ramsdell and Eaton, unpublished observations). Acute toxicities of the hydroxylated metabolites are less than those of the parent compounds [52–55]; the derivatives are not likely to contribute significantly to the acute effects of AFB_1 because they occur in relatively small amounts in vivo. Mutagenic potencies of hydroxylated metabolites of AFB_1 in the Ames *Salmonella typhimurium* system are all much lower than that of AFB_1 [56, 57]. Dietary AFM_1 was approximately 30% as carcinogenic as AFB_1 in trout [58] and about 10% as carcinogenic in rats [59], whereas AFQ_1 was about 1% as carcinogenic to trout as AFB_1 [60].

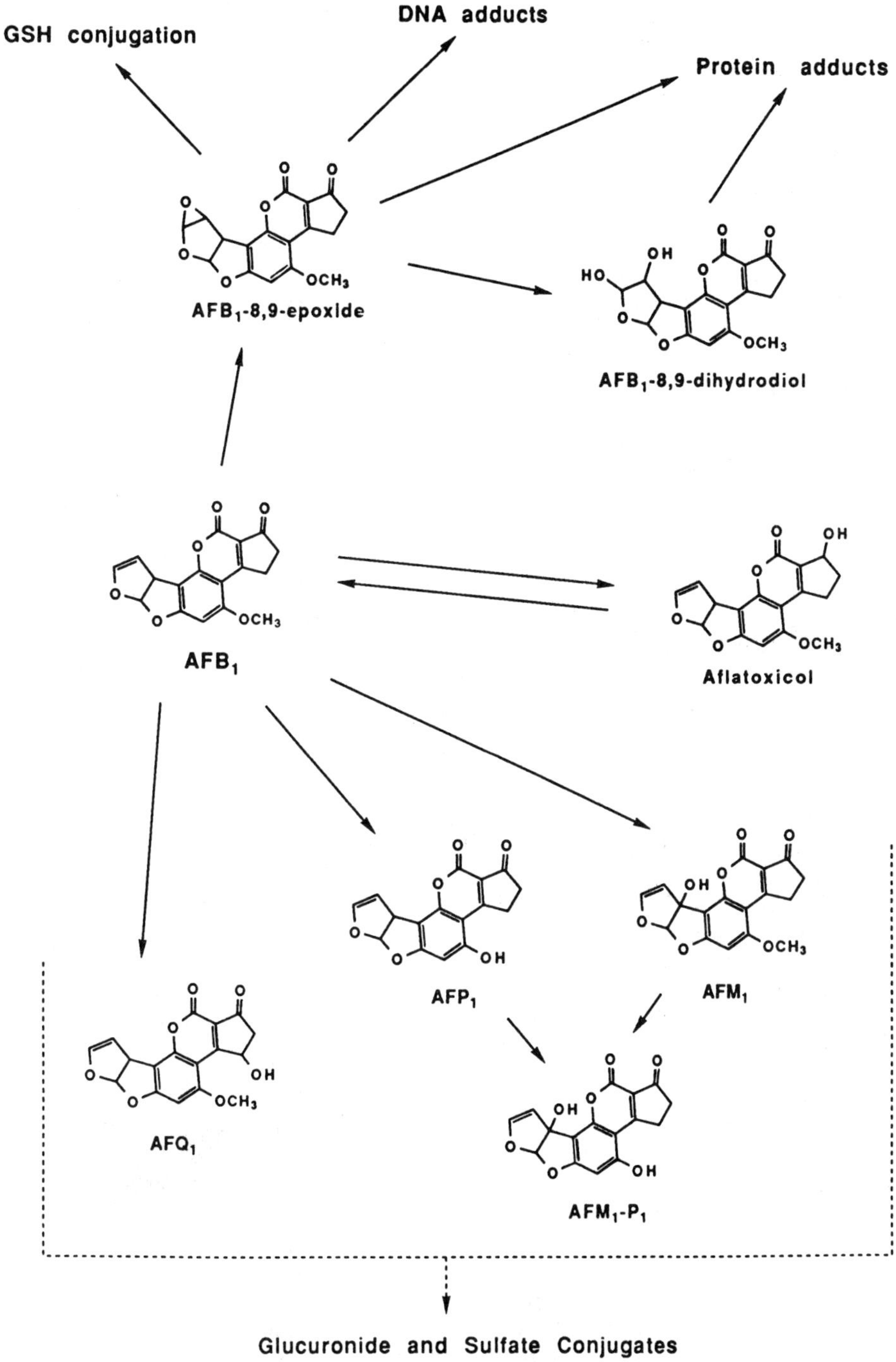

FIGURE 2. Aflatoxin biotransformation pathways.

Aflatoxicol is an important reductive metabolic product of AFB_1 in trout [61, 62] but is not a major metabolite of AFB in most mammalian species [63]. A cytosolic NADPH-dependent reductase is responsible for the conversion of AFB_1 to aflatoxicol [64]. Aflatoxicol elicits lower acute toxicity than the parent compound [65, 66] but is relatively mutagenic in bacterial revertant assays that utilize metabolic activation systems (23% and 66% as potent as AFB_1 using S-9 fractions from rat and trout liver, respectively) [56, 57]. Aflatoxicol is carcinogenic in trout and has a potency of approximately half that of AFB_1 [67]. Formation of aflatoxicol is not an effective detoxification reaction because the compound is readily reoxidized to AFB_1 by microsomal dehydrogenation [68–70]. It is likely that the carcinogenic effects of aflatoxicol are due to its conversion to AFB_1. Aflatoxicol may represent a metabolic reservoir for AFB_1 in vivo, prolonging its elimination half-life [70].

Recent studies have demonstrated the importance of consideration of the enzyme kinetics of AFB_1 biotransformation reactions when assessing the relative fractional disposition of AFB between biologically active and inactive metabolites [50]. Concentrations of AFB_1 commonly used in in vitro biotransformation experiments ($\approx 100\ \mu M$) [71–73] greatly exceed the levels that would be attained in the liver of a human exposed to dietary AFB_1. Patterns of biotransformation observed at a single substrate concentration are unlikely to be valid over a wide range when a system of multiple isoenzymes with varying substrate affinities (such as cytochromes P-450) is involved. The effect of substrate concentration on the oxidation of AFB_1 by liver microsomes from rats, mice, monkeys, and humans has been examined (Fig. 3). The data are expressed as a rate fraction of the total oxidative biotransformation rate to allow comparison between species with appreciably different overall biotransformation rates and to permit convenient assessment of enzymatic processes that do not follow simple Michaelis–Menten kinetics. With rat and human microsomes, epoxidation represents an increasing proportion of AFB_1 biotransformation with decreasing substrate concentrations. The fraction of AFB_1 oxidized to AFQ_1 by human microsomes decreases substantially at low concentrations, suggesting that AFQ_1 is formed by cytochrome P-450 enzyme(s) with relatively low affinity for AFB_1. Previous studies had suggested that humans are relatively resistant to AFB_1-induced hepatocarcinogenesis because of a high rate of AFQ_1 formation [48, 51, 55, 60, 74, 75], but this does not appear to be the case at low levels of AFB_1 (Fig. 3). These data suggest that human liver may activate a relatively large proportion of an ingested dose of AFB_1 to the epoxide. However, extrapolation of such in vitro data to low-dose dietary exposures of interest to human populations must be done with great caution, as many other biological processes operative in vivo are important in the ultimate expression of toxicity and/or carcinogenicity.

The results presented in Fig. 3 are consistent with the involvement of multiple cytochrome P-450 enzymes in the epoxidation of AFB_1 by human liver microsomes; this conclusion is supported by a recent report of the ability of at least five different human P-450 enzymes to activate AFB_1 to genotoxic metabolites [76]. Recent studies indicate that the contribution to AFB_1-8,9-epoxide formation by one of these enzymes, cytochrome P-450IIIA4, decreases significantly at low substrate concentrations [165].

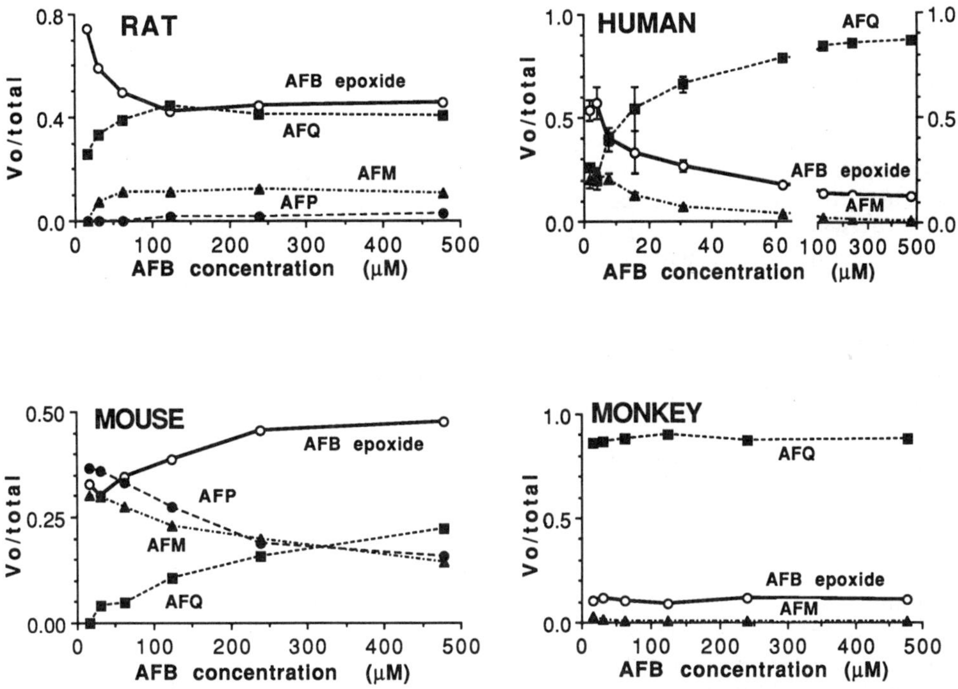

FIGURE 3. Effect of aflatoxin B_1 concentration on biotransformation by hepatic microsomes. Aflatoxin B_1 metabolites were quantified in incubations with pooled liver microsomes by an HPLC assay [68]. Initial rates of formation were determined and values are expressed in terms of a rate fraction (ν_0/total): the rate of formation of a metabolite divided by the sum of the rates for the metabolites detected for a given species.

Further research is needed to identify which other human P-450 enzyme(s) are involved in the activation of AFB_1 at the dose levels that result from human dietary exposure.

The significance of enzymatic hydrolysis of AFB_1-8,9-epoxide to the dihydrodiol by an epoxide hydrolase has not been firmly established. Studies in isolated hepatocytes from rats [17] and mice [77] suggest that epoxide hydrolase is important in the deactivation of AFB_1-8,9-epoxide, whereas other studies with subcellular fractions suggest that epoxide hydrolase is not important [73, 78, 79]. In species with significant cytosolic glutathione *S*-transferase (GST) activity toward AFB_1-8,9-epoxide, it appears unlikely that epoxide hydrolase plays an important role in the biotransformation of AFB_1.

Conjugation Reactions

Enzyme-catalyzed conjugation of AFB_1-8,9-epoxide with reduced glutathione represents the most important detoxification pathway for AFB_1 in mammalian systems. Diversion of the active epoxide metabolite appears to be a crucial determinant of species differences in susceptibility to AFB_1 hepatocarcinogenesis [19, 72, 79]. The glutathione (GSH) conjugate of AFB_1-8,9-epoxide is the major biliary metabolite of AFB_1 in rats [80–82].

Human liver cytosol appears to have relatively low GST activity toward AFB_1-8,9-epoxide [51]. Comparative GST activity data have been obtained for rat, mouse, monkey, and human cytosol (Table 1). Mouse liver cytosol has a total specific activity of GST toward AFB_1-8,9-epoxide about two orders of magnitude greater than rat or monkey liver. The activity of human liver cytosol is approximately an order of magnitude lower than that of rat and monkey liver cytosol. The relatively low activity of human liver for conjugation of the reactive AFB_1-8,9-epoxide combined with the relatively efficient conversion of AFB_1 to the exoxide at low concentrations suggest that humans could be relatively susceptible to AFB_1-induced hepatocarcinogenesis.

Conjugation of hydroxylated metabolites of AFB_1 appears to play a role in their excretion. The glucuronide of AFP_1 has been identified in the bile of rats treated with AFB_1 [49, 82], and 4,9a-dihydroxyaflatoxin B_1 is conjugated with glucuronic acid by rat liver and also excreted in the bile [49]. There is evidence for the urinary excretion of glucuronide or sulfate conjugates of AFM_1 and AFP_1 by the monkey and rat [63]; conjugation of these less carcinogenic AFB_1 metabolites is unlikely to significantly alter the toxicokinetics of AFB_1 in species that have appreciable levels of glutathione *S*-transferase activity toward AFB_1-8,9-epoxide.

TABLE 1. Cytosolic Glutathione *S*-Transferase Activity

	AFB_1-8,9-epoxide [pmol/(mg/min)]	1-Chloro-2,4-dinitrobenzene [μmol/(mg/min)]	Ratio $\frac{AFBO \times 10^6}{CDNB}$
Rat[a]	30	1.9	16
Mouse[b]	2400	5.0	480
Monkey[c]	26 ± 10	1.20 ± 0.16	21
Human[d]	<2[e]	0.77 ± 0.18	11

[a]Values represent average of numerous experiments in male Sprague-Dawley rats. SE less than 10% of the mean.
[b]Values represent average of numerous experiments in male Swiss-Webster mice. SE less than 10% of the mean.
[c]Mean ± SE of two males and one female adult *Macacca nemestrina*.
[d]Mean ± SE of nine samples.
[e]Below the limits of detection of about 2 pmol/(mg/min) in all human liver cytosol assays.

B. Molecular Targets of Aflatoxin Binding

AFB_1-8,9-epoxide reacts readily with nucleophiles, including the N^7 group of guanine [83], whereas the phenolate ion resonance form of the hydrolyzed epoxide, 8,9-dihydro-8,9-dihydroxy AFB_1, can form Schiff bases with protein amines [84, 85]. Hydrolysis of DNA incubated with AFB_1 and liver microsomes yielded as the major product an adduct identified as 8,9-dihydro-8-(N^7-guanyl)-9-hydroxyaflatoxin B_1 [86, 87]; the compound was also identified in hydrolysates of liver DNA from rats treated with AFB_1 [87, 88]. Determination of the AFB_1-DNA adducts in rat liver after in vivo treatment with AFB_1 showed that the N^7-guanine adduct is removed from the DNA relatively rapidly, with a half-life of approximately 7.5 h [89]. An imidazole ring-opened derivative, 8,9-dihydro-8-(N^5-formyl)-2',5',6'-triamino-4'-oxo-N^5-pyrimidyl)-9-hydroxyaflatoxin B_1 (AFB_1-formamidopyrimidine adduct) was found as a minor product of DNA—aflatoxin reactions in vitro [87]. The AFB_1-formamidopyrimidine adduct was also a minor product in vivo within a few hours after administration of AFB_1 to rats and was present at about the same levels after 3 days. The adduct became the predominant adduct as the N^7-guanine product was removed by excision repair and/or converted to the stable formamidopyrimidine product [89]. The persistence of the AFB_1-formamidopyrimidine adduct may be attributed to the increased stability of its bond to the deoxyribose of the DNA chain [90].

Noncovalent interactions of aflatoxins with DNA appear to have a significant effect on the reaction of the epoxides with guanine. There is evidence to suggest that the planar aflatoxin molecule intercalates with DNA and that the transition state for N^7-guanine adduct formation involves an intercalated complex of the aflatoxin epoxide [91, 92]. The less planar nature of AFG_1 and AFG_2 results in less efficient noncovalent binding to DNA than binding with AFB_1 and AFB_2; the variation in binding has been suggested as an explanation for the lower genotoxicity of AFG_1 relative to that of AFB_1 [92].

As would be expected from the highly organotropic effects of AFB_1, the greatest toxin binding is found in liver, with appreciable levels also observed in kidney [46, 93]. The greatest binding of AFB_1 in liver of rats, hamsters, and guinea pigs is to RNA and DNA, with relatively low amounts bound to protein [46, 94]. In mouse liver, however, overall binding of AFB_1 is much lower than in liver of other animals, and roughly equal amounts are found in the three macromolecular fractions [94]. Differences in the binding of AFB_1 to DNA appear to reflect interspecies variations in susceptibility to the toxin's carcinogenic effects [46, 73, 94—96].

Binding of radiolabeled AFB_1 to protein, RNA, and DNA in rat liver has been detected at in vivo doses as low as 10 ng/kg (i.p.) [97]. These results confirm that bioactivation of AFB_1 and covalent binding to macromolecular targets may occur at levels of exposure anticipated by ingestion of contaminated food. The incidence of AFB_1-induced hepatic tumors in both rats and trout was a linear function of the concentration of DNA adducts. Even though rats and trout have very different biotransformation pathways for AFB_1, the correlation between DNA adducts and tumor incidence was similar [98]. Thus, if human cancer risk was similarly linked to DNA adduct formation, the impact of AFB_1 exposure could be estimated.

C. Biotransformation and Comparative Carcinogenicity of Aflatoxin

The relative susceptibility of different species to AFB_1-induced tumors has been estimated (Table 2). For the mouse and rat, differences in DNA binding in vivo reflect the relative susceptibility of these two rodent species to hepatocarcinogenesis by AFB_1 [73, 102]. DNA binding is also closely linked to the relative capacities for detoxification of the epoxide via GSH conjugation [19, 72, 73]. Rainbow trout (*Oncorhyncus mykis*) and coho salmon (*Oncorhyncus kisutch*) have significantly different rates of tumors induced by AFB_1. Although conjugative biotransformation rates of the two species are essentially the same, trout liver appears to have more efficient microsomal activation of AFB_1 [103, 104].

Human liver microsomes efficiently activate AFB_1 (Table 1; Fig. 3), whereas human liver cytosol has very low GST activity toward AFB_1-8,9-epoxide (Table 1); these observations suggest that humans may be relatively susceptible to hepatocarcinogenesis by AFB_1. However, postinitiation events, such as DNA repair, immune surveillance, and promotional factors as important in the ultimate expression of AFB_1-induced DNA damage as the tumor initiation event, and thus substantive conclusions about the relative susceptibility of humans to AFB carcinogenesis should not be made on in vitro studies alone.

IV. DIETARY MODULATION OF AFLATOXIN TOXICITY AND BIOTRANSFORMATION

Both acute hepatotoxicity and hepatocarcinogenicity of AFB_1 may be affected by dietary factors, and relevant research has been carried out with the goal of developing chemoprotective regimens for humans exposed to AFB_1. Both nutrient and nonnutrient substances have been found to alter AFB_1 disposition and hence its toxicity and carcinogenicity.

A. Dietary Components

Induction of preneoplastic foci in the livers of rats given AFB_1 is enhanced by levels of dietary protein exceeding requirements for normal growth [105]; this effect does not appear to be related to aflatoxin biotransformation, however, because increased foci formation was observed when high protein levels were fed after AFB_1 dosing [106]. High dietary protein levels also increased the incidence of hepatocellular carcinoma in trout fed AFB_1 [107].

Vitamin deficiency can alter the biotransformation and DNA binding of AFB_1. Bioactivation of AFB_1 and binding of AFB_1 to DNA were impaired in riboflavin-deficient rats [108]. However, riboflavin deficiency did not alter the cytosolic conjugation of AFB_1-8,9-epoxide with GSH. Thus, it is likely that decreased microsomal oxidation of AFB_1 due to

TABLE 2. Species Differences in Hepatocarcinogenicity of Aflatoxin B_1

Species	Initial age (or weight)	Dosing regimen	Treatment period (mo.)	Response[a] (percentage)	Ref.
Rainbow trout	Fingerling	20 ppb in diet	12	62/80 (78%)	27
Coho salmon	Fingerling	20 ppb in diet	20	0/100 (0%)	99
Duck	7 days	35 ppb in diet	14	8/11 (73%)	100
Rat	80 g	15 ppb in diet	17–20	25/25 (100%)	30
Monkey					
cynomolgus	9–10 days	53–1369 mg total dose, i.p. and/or p.o.	45–145	0/8[b] (0%)	101
rhesus	9–10 days	99–1354 mg total dose, i.p. and/or p.o.	39–147	2/13[c] (15%)	101
Hamster (Syrian)	4 wk	2 mg/(kg/day) (p.o.) 5 days/wk	4–19	1/25[d] (4%)	35
Mouse	Postweaning	1 ppm in diet	20	0/60 (0%)	32

[a]Incidence of hepatocellular carcinomas (tumor-bearing animals/total treated).
[b]Hemangioendothelial sarcoma of the liver or bile duct tumors were found in three of eight animals (38%).
[c]Hemangioendothelial sarcoma of the liver or bile duct tumors occurred in six of 13 animals (46%).
[d]Cholangiocarcinomas were observed in eight of 25 animals (32%).

lowered NADPH-cytochrome P-450 reductase levels was responsible for the results [108]. In contrast, vitamin A deficiency resulted in greater in vitro binding of AFB_1 to DNA mediated by rat liver fractions; this apparently was due to a combination of greater microsomal activation and diminished cytosolic conjugation [109]. Although vitamin E supplementation of rat diet has been reported to enhance in vitro biotransformation of AFB_1 [110], no differences in macromolecular binding were observed between rats deficient in both vitamin E and selenium and those given a vitamin E supplement [111].

Dietary treatments that result in metabolic stress on the liver may affect the response to AFB_1. A diet marginally deficient in lipotropes increased the sensitivity of rats to the carcinogenic effects of AFB_1 [112]; the dietary treatment protected rats against the lethality of a single dose of AFB_1 but enhanced the hepatic hyperplastic response to repeated low doses of AFB_1 [113]. In vitro studies have identified dietary components with protective activity against the effects of AFB_1. Cupric sulfate was particularly effective at reducing the mutagenicity of AFB_1 to *Salmonella typhimurium* strains, and selenite also had some antimutagenic activity [114]. Some naturally occurring carotenoids were shown to reduce the rat liver microsomal-mediated binding of AFB_1 to exogenous DNA [115].

Nonnutrient compounds that occur in plant foods may also modulate the effects of AFB_1 in vitro. The phenolic compounds eugenol and chlorogenic acid inhibited the mutagenicity of AFB_1 in *Salmonella typhimurium* strain TA100 [116]. Ellagic acid, a commonly occurring plant phenol, reduced the mutagenicity of AFB_1 in the same bacterial strain and was also found to reduce DNA adduct formation in both rat and human tracheal explant cultures [117]. Microsomal activation and binding of AFB_1 to exogenous DNA were both reduced by a variety of plant flavonoids [118]. Some flavonoids, in particular the hydroxylated flavonols morin, rutin, fisetin, and kaempferol, had antimutagenic activity against AFB_1 in the *Salmonella typhimurium* reverse mutation assay [119].

B. Cruciferous Vegetables

Cruciferous vegetables have attracted attention because of human epidemiological evidence that consumption is associated with decreased cancer rates [120–122]. Studies in both rats [123, 124] and humans [125, 126] demonstrated that consumption of cruciferous vegetables alters the activities of xenobiotic-metabolizing enzymes. Feeding cauliflower (freeze-dried, 20% in the diet) to rats protected against the hepatotoxic effects of dietary AFB_1 [127] and reduced the appearance of the hepatic tumor indicator α-fetoprotein in the serum [128]. Dietary cabbage (10–25%, by dry weight) had similar protective effects and also reduced the multiplicity of liver tumors in rats given AFB_1 in the diet [129, 130]. Brussels sprouts (25% of the dry weight of the diet) inhibited the induction of enzyme-altered hepatic foci in AFB_1-treated rats [131].

The antitumor effects of cruciferous vegetables appear to be mediated via modulation of biotransformation enzyme activities. Dietary cabbage [132] and brussels sprouts [133] (freeze-dried, 25% of diet) reduced in vivo binding of AFB_1 to hepatic DNA in rats and induced hepatic GST activity toward 1-chloro-2,4-dinitrobenzene (CDNB), but microsomal aryl hydrocarbon hydroxylase and ethoxycoumarin *O*-deethylase activities were unaffected. Dietary, freeze-dried broccoli (25% by dry weight) also reduced the binding of AFB_1 to hepatic DNA in rats [134]. The broccoli diet resulted in a 2.8-fold increase in the specific activity of liver cytosolic GST toward AFB_1-8,9-epoxide, which exceeded the level of induction of CDNB activity (1.6-fold). The microsomal formation of AFB_1-8,9-epoxide was unaffected by the broccoli diet.

The identity of the active principle(s) in cruciferous vegetables responsible for modulation of AFB_1 toxicity and biotransformation has been investigated. The indole products of glucosinolates, especially indole-3-carbinol, have been shown to be effective as enzyme inducers [135]. Induction of aryl hydrocarbon hydroxylase and the related cytochrome P-450-dependent activities (e.g., ethoxycoumarin *O*-deethylase and ethoxyresorufin *O*-deethylase) by indoles has been demonstrated in rats [130, 136,

137]. However, no effect of dietary indole-3-carbinol on microsomally catalyzed binding of AFB_1 to exogenous DNA in vitro was observed [133]. Indole-3-carbinol was also ineffective as an inducer of rat hepatic GST activity toward CDNB; the results suggest that additional substance(s) must be involved in the anticarcinogenic effects of cruciferous vegetables in rats.

Another glucosinolate breakdown product obtained from cruciferous vegetables, goitrin (5-vinyloxazolidine-2-thione), has been found to be a potent inducer of GST activity in rat liver [138]. Rats fed the compound had lower levels of AFB_1 binding to DNA in vivo, excreted more radioactivity in bile after a dose of labeled AFB_1, and higher GST activity toward CDNB in vitro [139]. These results suggest that goitrin may play a major role in alterations of AFB_1 biotransformation that results from feeding cruciferous vegetables. The chemistry of the *Cruciferae* is complex [140, 141], and additional compounds may be involved in antitumor activities.

C. Antioxidants and Drugs

Dietary pretreatment of animals with synthetic antioxidants such as butylated hydroxyanisole (BHA), butylated hydroxytoluene, and ethoxyquin has been found to protect against the toxic and carcinogenic effects of a variety of compounds [142, 143]. The protective effects are usually unrelated to the direct antioxidant properties but instead are linked to the ability of the chemicals to induce drug-metabolizing enzymes [144–147].

Administration of both butylated hydroxyanisole [147] and ethoxyquin [148] to rats protects against AFB_1-induced hepatocarcinogenesis. Modulation of AFB_1-induced hepatocarcinogenesis in rats by antioxidants is also reflected in changes in the amount of AFB_1 covalently bound to DNA after in vivo treatment with radiolabeled AFB_1 [73, 149, 150]. A similar reduction in AFB_1-DNA adduct formation was observed following in vitro incubation of isolated hepatocytes obtained from rats pretreated with BHA [151, 152]. Alterations of AFB_1 biotransformation activities have been closely linked to the protective effects of antioxidants. Pretreatment of rats with BHA increased GSH conjugation of AFB_1, enhanced biliary excretion of the AFB_1-GSH conjugate, and reduced in vivo binding to DNA [73, 146]. The same results were obtained with ethoxyquin, which was also shown to largely prevent the occurrence of enzyme-altered foci induced by AFB_1 [150, 153].

In contrast to the effects observed in rats, pretreatment of mice with dietary BHA did not result in significant reduction in binding of AFB_1 to hepatic DNA either in vivo [73, 154] or in vitro [155]. Determination of AFB_1 biotransformation enzyme activities in rats and mice explained the observed species difference [73]. Pretreatment of rats with BHA did not significantly affect AFB_1 oxidation by rat microsomes, whereas AFB_1 epoxidation activity was increased over threefold in mouse microsomes. GST activity toward AFB_1-8,9-epoxide in rat liver cytosol was induced threefold by BHA, but similar treatment of mice did not significantly increase mouse cytosolic GST activity toward AFB_1-8,9-epoxide. Treatment with BHA induced GST in mice, as measured by mRNA expression [156, 157] and activity toward CDNB [145]. The lack of significant induction of

AFB_1-8,9-epoxide conjugation activity in mouse liver cytosol by BHA apparently was the result of a relatively small increase in the amount of the one isoenzyme with very high activity toward AFB_1-8,9-epoxide [158].

Treatment of trout with BHA did not reduce AFB_1-induced hepatocarcinogenesis or binding of AFB_1 to DNA in vivo [159]. Little effect of BHA pretreatment was observed on biotransformation of AFB_1 by trout hepatocytes. The low basal GST activity toward AFB_1-8,9-epoxide in trout liver thus appears not to be inducible by BHA.

In addition to antioxidants, drugs that increase hepatic glutathione *S*-transferase activity have been investigated for potential use as chemoprotective agents. Dithiolthiones, including the antischistosomal drug Oltipraz, induce GST activities and increase hepatic GSH levels [160, 161]. Oltipraz protected rats agatinst acute and subchronic AFB_1 hepatotoxicity [162], and reduced both the binding of AFB_1 to rat liver DNA in vivo and the appearance of enzyme-altered foci in the livers of rats treated with AFB_1 [163, 164]. The potential for Oltipraz to protect humans against AFB_1 hepatotoxicity may be limited, however, by the extent and inducibility of human GST with activity toward AFB_1-8,9-epoxide.

V. SUMMARY

Wide species differences exist in both the hepatotoxic and carcinogenic effects of aflatoxin B_1, although species differences in carcinogenicity are most remarkable. Carcinogenicity of AFB appears to be closely related to the extent of DNA adduct formation, which in turn is determined largely by the relative rates of biotransformation to active versus inactive metabolites. Because of this relationship and the relative ease of determining enzyme activities toward AFB_1 in liver tissue from humans, comparative in vitro studies with human and rodent liver may prove useful in estimating the relative sensitivity of humans to the carcinogenic effects of AFB. However, more data are needed to determine the relative extent of AFB-DNA adduct formation in human, rat, and mouse liver under identical conditions. Studies with precision liver slices from these species may provide these important data.

Because the expression and activity of the enzymes principally responsible for both activation (certain cytochromes P-450) and inactivation (glutathione *S*-transferases and certain cytochromes P-450) of AFB_1 can be modulated by diet and other environmental factors, there is substantial promise in the development of chemoprotective interventions that could significantly reduce the carcinogenicity of AFB in high-risk populations. However, before such widespread efforts are implemented, additional studies are necessary to more fully understand (1) the molecular basis of the differences in biotransformation kinetics between species, especially humans; (2) the relationship between changes in specific human biotransformation pathway activities and AFB-DNA adduct formation; and (3) how such dietary and chemointervention factors could influence later stages in the carcinogenic process. Such studies will be valuable in improving the biochemical basis for risk extrapolations from laboratory animals to humans, as well as providing a solid rationale for dietary and chemopreventive measures in humans at high risk for aflatoxin exposure.

ACKNOWLEDGMENTS

The research described in this chapter was supported by National Institutes of Health grants T32ES-07032, ES-03933, CA-47561, and ES-04696.

REFERENCES

1. Asplin, F. D., and Carnaghan, R. B. A. The toxicity of certain groundnut meals for poultry with special reference to their effect on ducklings and chickens. *Vet. Rec.*, *73*:1215, 1961.
2. Sargeant, K., O'Kelly, J., Carnaghan, R. B. A., and Allcroft, R. The assay of a toxic principle in certain groundnut meals. *Vet. Rec.*, *73*:1219, 1961.
3. Carnaghan, R. B. A., Hartley, R. D., and O'Kelly, J. Toxicity and fluorescence properties of the aflatoxins. *Nature*, *200*:1101, 1963.
4. Bauer, D. H., Lee, D. J., and Sinnhuber, R. O. Acute toxicity of aflatoxins B_1 and G_1 in the rainbow trout (*Salmo gairdneri*). *Toxicol. Appl. Pharmacol.*, *15*:415, 1969.
5. Butler, W. H. Acute toxicity of aflatoxin B_1 in rats. *Br. J. Cancer*, *18*:756, 1964.
6. Hayes, A. W., Cain, J. A., and Moore, B. G. Effect of aflatoxin B_1, ochratoxin A and rubratoxin B on infant rats. *Food Cosmet. Toxicol.*, *15*:23, 1977.
7. Busby, W. F., and Wogan, G. N. Aflatoxins. *Chem. Carcinogen.*, *2*:945, 1984.
8. Hanigan, H. M., and Laishes, B. A. Toxicity of aflatoxin B_1 in rat and mouse hepatocytes in vivo and in vitro. *Toxicology*, *30*:185, 1984.
9. Ngindu, A., Johnson, B. K., Kenya, P. R., Ngira, J. A., Ocheng, D. M., Nandwa, H., Omondi, T. N., Jansen, A. J., Ngare, W., Kaviti, J. N., Gatei, D., and Siongok, T. A. Outbreak of acute hepatitis caused by aflatoxin poisoning in Kenya. *Lancet*, *i*:1346, 1982.
10. Newberne, P. M., and Butler, W. H. Acute and chronic effects of aflatoxin on the liver of domestic and laboratory animals: a review. *Cancer Res.*, *29*:236, 1969.
11. Gelboin, H. V., Wortham, J. S., Wilson, R. G., Friedman, M., and Wogan, G. N. Rapid and marked inhibition of rat-liver RNA polymerase by aflatoxin B_1. *Science*, *154*:1205, 1966.
12. Lafarge, C., and Frayssinet, C. The reversibility of inhibition of RNA and DNA synthesis induced by aflatoxin in rat liver. A tentative explanation for carcinogenic mechanism. *Int. J. Cancer*, *6*:74, 1970.
13. Sarasin, A., and Moule, Y. In vivo effect of aflatoxin B_1 on protein synthesis. *FEBS Lett.*, *29*:329, 1973.
14 Wagner, G., and Unterreiner, A. M. Inhibition of rat liver aminoacyl-tRNA synthetases in vitro after acute and chronic aflatoxin B_1 administration in vivo. *Chem.-Biol. Interact.*, *37*:233, 1981.
15. Wagner, G., and Unterreiner, A. M. Synthesis of transfer RNA in rat liver after acute and chronic aflatoxin B_1 administration. *Chem.-Biol. Interact.*, *41*:353, 1982.

16. Toskulkao, C., and Glinsukon, T. Hepatic lipid peroxidation and intracellular calcium accumulation in ethanol potentiated aflatoxin B_1 toxicity. *J. Pharmacobio-Dynam.*, *11*:191, 1988.
17. Ch'ih, J. J., Lin, T., and Devlin, T. M. Effect of inhibitors of microsomal enzymes on aflatoxin B_1-induced cytotoxicity and inhibition of RNA synthesis in isolated rat hepatocytes. *Biochem. Biophys. Res. Commun.*, *115*:15, 1983.
18. Godoy, H. M., and Neal, G. E. Some studies of the effects of aflatoxin B_1 in vivo and in vitro on nucleic acid synthesis in rat and mouse. *Chem.-Biol. Interact.*, *13*:257, 1976.
19. O'Brien, K., Moss, E., Judah, D., and Neal, G. Metabolic basis of the species differences to aflatoxin B_1 induced hepatotoxicity. *Biochem. Biophys. Res. Commun.*, *114*:813, 1983.
20. Clark, J. D., Hatch, R. C., Jain, A. V., and Weiss, R. Effect of enzyme inducers and inhibitors and glutathione precursor and depleter on induced acute aflatoxicosis in rabbits. *Am. J. Vet. Res.*, *43*:1027, 1982.
21. Nixon, J. E., Luebke, J. P., Shelton, D. W., Hendricks, J. D., Pawlowski, N. E., and Bailey, G. S. Reduced acute aflatoxin B_1 (AFB_1) toxicity by inhibitors of AFB_1 carcinogenesis in rats. *Fed. Proc.*, *44*:769, 1985.
22. Chentanez, T., Patradilok, P., Glinsukon, T., and Piyachaturawat, P. Effects of cortisol pretreatment on the acute hepatotoxicity of aflatoxin B_1. *Toxicol. Lett.*, *42*:237, 1988.
23. Metcalfe, S. A., and Neal, G. E. Some studies on the relationship between the cytotoxicity of aflatoxin B_1 to rat hepatocytes and metabolism of the toxin. *Carcinogenesis*, *4*:1013, 1983.
24. Karenlampi, S. O. Mechanism of cytotoxicity of aflatoxin B_1: role of cytochrome P_1-450. *Biochem. Biophys. Res. Commun.*, *145*:854, 1987.
25. Butler, W. H., Greenblatt, M., and Lijinsky, W. Carcinogenesis in rats by aflatoxins B_1, G_1 and B_2. *Cancer Res.*, *29*:2206, 1969.
26. Epstein, S. M., Bartus, B., and Farber, E. Renal epithelial neoplasms induced in male Wistar rats by oral aflatoxin B_1. *Cancer Res.*, *29*:1045, 1969.
27. Sinnhuber, R. O., Hendricks, J. D., Wales, J. H., and Putnam, G. B. Neoplasms in rainbow trout, a sensitive animal model for environmental carcinogenesis. *Ann. N.Y. Acad. Sci.*, *298*:389, 1977.
28. Hendricks, J. D., Wales, J. H., Sinnhuber, R. O., Nixon, J. E., Loveland, P. M., and Scanlan, R. A. Rainbow trout (*Salmo gairdneri*) embryos: a sensitive animal model for experimental carcinogenesis. *Fed. Proc.*, *39*:3222, 1980.
29. Wogan, G. N., Paglialunga, S., and Newberne, P. M. Carcinogenic effects of low dietary levels of aflatoxin B_1 in rats. *Food Cosmet. Toxicol.*, *12*:681, 1974.
30. Wogan, G. N., and Newberne, P. M. Dose-response characteristics of aflatoxin B_1 carcinogenesis in the rat. *Cancer Res.*, *27*:2370, 1967.
31. Carnaghan, R. B. A. Hepatic tumours and other chronic liver changes in rats following a single oral administration of aflatoxin. *Br. J. Cancer*, *21*:811, 1967.

32. Wogan, G. Aflatoxin carcinogenesis. In *Methods in Cancer Research* (H. Busch, Ed.), Academic, New York, 1973, p. 309.
33. Vesselinovitch, S. D., Mihailovich, N., Wogan, G. N., Lombard, L. S., and Rao, K. V. N. Aflatoxin B_1, a hepatocarcinogen in the infant mouse. *Cancer Res.*, *32*:2289, 1972.
34. Herrold, K. M. Aflatoxin induced lesions in Syrian hamsters. *Br. J. Cancer*, *23*:655, 1969.
35. Moore, R. M., Pitot, H. C., Miller, E. C., and Miller, J. A. Cholangiocellular carcinomas induced in Syrian golden hamsters administered aflatoxin B_1 in large doses. *J. Natl. Cancer Inst.*, *68*:271, 1982.
36. Peers, F. G., and Linsell, C. A. Dietary aflatoxins and liver cancer—a population based study in Kenya. *Br. J. Cancer*, *27*:473, 1973.
37. Bulatao-Jayme, J., Almero, E. M., Castro, M. C. A., Jardeleza, M. T. R., and Salamat, L. A. A case-control dietary study of primary liver cancer risk from aflatoxin exposure. *Int. J. Epidemiol.*, *11*:112, 1982.
38. Groopman, J. D., Cain, L. G., and Kensler, T. Aflatoxin exposure in human populations: measurements and relationship to cancer. *CRC Crit. Rev. Toxicol.*, *19*:113, 1988.
39. Stoloff, L. Aflatoxin is not a probable human carcinogen: the published evidence is sufficient. *Reg. Toxicol. Pharmacol.*, *10*:272, 1989.
40. Peers, F., Bosch, X., Kaldor, J., Linsell, A., and Pluuman, M. Aflatoxin exposure, hepatitis B virus infection and liver cancer in Swaziland. *Int. J. Cancer*, *39*:545, 1987.
41. Yeh, F.-S., Yu, M. C., Mo, C.-C., Luo, S., Tong, M. J., and Henderson, B. E. Hepatitis B virus, aflatoxins, and hepatocellular carcinoma in Southern Guangxi, China. *Cancer Res.*, *49*:2506, 1989.
42. Wogan, G. N. Chemical nature and biological effects of the aflatoxins. *Bacteriol. Rev.*, *30*:460, 1966.
43. Roebuck, B. D., Lu, Y.-L., Rogers, A. E., Groopman, J. D., and Kensler, T. W. Inhibition of aflatoxin-induced liver cancer in the rat by Oltipraz, 5-(2-pyrazinyl)-4-methyl-1,2-dithiole-3-thione. *Proc. Am. Assoc. Cancer Res.*, *31*:125, 1990.
44. Swenson, D. H., Lin, J.-K., Miller, E. C., and Miller, J. A. Aflatoxin B_1-2,3-oxide as a probable intermediate in the covalent binding of aflatoxins B_1 and B_2 to rat liver DNA and ribosomal RNA *in vivo*. *Cancer Res.*, *37*:172, 1977.
45. Battista, J. R., and Marnett, L. J. Prostaglandin H synthase-dependent epoxidation of aflatoxin B_1. *Carcinogenesis*, *6*:1227, 1985.
46. Garner, R. C., and Wright, C. M. Binding of [^{14}C]aflatoxin B_1 to cellular macromolecules in the rat and hamster. *Chem.-Biol. Interact.*, *11*:123, 1975.
47. Portman, R. S., Plowman, K. M., and Campbell, T. C. Aflatoxin metabolism by liver microsomal preparations of two different species. *Biochem. Biophys. Res. Commun.*, *33*:711, 1968.
48. Masri, M. S., Booth, A. N., and Hsieh, D. P. H. Comparative metabolic conversion of aflatoxin B_1 to M_1 and Q_1 by monkey, rat and chicken liver. *Life Sci.*, *15*:203, 1974.

49. Eaton, D. L., Monroe, D. H., Bellamy, G., and Kalman, D. A. Identification of a novel dihydroxy metabolite of aflatoxin B_1 produced in vitro and in vivo in rats and mice. *Chem. Res. Toxicol.*, *1*:108, 1988.
50. Ramsdell, H. S., and Eaton, D. L. Species susceptibility to aflatoxin B_1 carcinogenesis: comparative kinetics of microsomal biotransformation. *Cancer Res.*, *50*:615, 1990.
51. Moss, E. J., and Neal, G. E. The metabolism of aflatoxin B_1 by human liver. *Biochem. Pharmacol.*, *34*:3193, 1985.
52. Purchase, I. F. H. Acute toxicity of aflatoxins M_1 and M_2 in one-day-old ducklings. *Food Cosmet. Toxicol.*, *5*:339, 1967.
53. Pong, R. S., and Wogan, G. N. Toxicity and biochemical and fine structural effects of synthetic aflatoxins M_1 and B_1 in rat liver. *J. Natl. Cancer Inst.*, *47*:585, 1971.
54. Stoloff, L., Verrett, M. J., Dantzman, J., and Reynaldo, E. F. Toxicological study of aflatoxin P_1 using the fertile chicken egg. *Toxicol. Appl. Pharmacol.*, *23*:528, 1972.
55. Hsieh, D. P. H., Salhab, A. S., Wong, J. J., and Yang, S. L. Toxicity of aflatoxin Q_1 as evaluated with the chicken embryo and bacterial auxotrophs. *Toxicol. Appl. Pharmacol.*, *30*:237, 1974.
56. Wong, J. J., and Hsieh, D. P. H. Mutagenicity of aflatoxins related to their metabolism and carcinogenic potential. *Proc. Natl. Acad. Sci. USA*, *73*:2241, 1976.
57. Coulombe, R. A., Shelton, D. W., Sinnhuber, R. O., and Nixon, J. E. Comparative mutagenicity of aflatoxins using a *Salmonella*/trout hepatic enzyme activation system. *Carcinogenesis*, *3*:1261, 1982.
58. Sinnhuber, R. O., Lee, D. J., Wales, J. H., and Landers, M. K. Hepatic carcinogenesis of aflatoxin M_1 in rainbow trout (*Salmo gairdneri*) and its enhancement by cyclopropene fatty acids. *J. Natl. Cancer Inst.*, *53*:1285, 1974.
59. Hsieh, D. P. H., Cullen, J. M., and Ruebner, B. H. Comparative hepatocarcinogenicity of aflatoxins B_1 and M_1 in the rat. *Food Chem. Toxicol.*, *22*:1027, 1984.
60. Hendricks, J. D., Sinnhuber, R. O., Nixon, J. E., Wales, J. H., Masri, M. S., and Hsieh, D. P. H. Carcinogenic response of rainbow trout (*Salmo gairdneri*) to aflatoxin Q_1 and synergistic effect of cyclopropenoid fatty acids. *J. Natl. Cancer Inst.*, *64*:523, 1980.
61. Loveland, P. M., Nixon, J. E., Pawlowski, N. E., Eisele, T. A., Libbey, L. M., and Sinnhuber, R. O. Aflatoxin B and aflatoxicol metabolism in rainbow trout (*Salmo gairdneri*) and the effects of dietary cyclopropene. *J. Environ. Pathol. Toxicol.*, *2*:707, 1979.
62. Loveland, P. M., Nixon, J. E., and Bailey, G. S. Glucuronides in bile of rainbow trout (*Salmo gairdneri*) infected with [^{3}H]aflatoxin B_1 and the effects of dietary β-naphthoflavone. *Comp. Biochem. Physiol.*, *78C*:13, 1984.
63. Wong, Z. A., and Hsieh, D. P. H. The comparative metabolism and toxicokinetics of aflatoxin B_1 in the monkey, rat and mouse. *Toxicol. Appl. Pharmacol.*, 55:115, 1980.
64. Chen, S. C. G., Wei, R. D., and Hsieh, D. P. H. Purification and some properties of chicken-liver aflatoxin B_1 reductase. *Food Cosmet. Toxicol.*, *19*:19, 1981.

65. Detroy, R. W., and Hesseltine, C. W. Isolation and biological activity of a microbial conversion product of aflatoxin B_1. *Nature, 219*: 967, 1968.
66. Detroy, R. W., and Hesseltine, C. W. Aflatoxicol: structure of a new transformation product of aflatoxin B_1. *Can. J. Biochem., 48*: 830, 1970.
67. Schoenhard, G. L., Hendricks, J. D., Nixon, J. E., Lee, D. J., Wales, J. H., Sinnhuber, R. O., and Pawlowski, N. E. Aflatoxicol-induced hepatocellular carcinoma in rainbow trout (*Salmo gairdneri*) and the synergistic effects of cyclopropenoid fatty acids. *Cancer Res., 41*:1011, 1981.
68. Loveland, P. M., Sinnhuber, R. O., Berggren, K. E., Libbey, L. M., Nixon, J. E., and Pawlowski, N. E. Formation of aflatoxin B_1 from aflatoxicol by rainbow trout (*Salmo gairdneri*) liver in vitro. *Res. Commun. Chem. Pathol. Pharmacol., 16*:167, 1977.
69. Salhab, A. S., and Edwards, G. S. Comparative in vitro metabolism of aflatoxicol by liver preparations from animals and humans. *Cancer Res., 37*:1016, 1977.
70. Wong, Z. A., Decad, G. M., Byard, J. L., and Hsieh, D. P. H. Conversion of aflatoxicol to aflatoxin B_1 in rats in vivo and in primary hepatocyte culture. *Food Cosmet. Toxicol., 17*:481, 1979.
71. Neal, G. E., and Colley, P. J. Some high-performance liquid-chromatographic studies of the metabolism of aflatoxins by rat liver microsomal preparations. *Biochem. J., 174*:839, 1978.
72. Degen, G. H., and Neumann, H.-G. Differences in aflatoxin B_1-susceptibility of rat and mouse are correlated with the capability in vitro to inactivate aflatoxin B_1-epoxide. *Carcinogenesis, 2*:299, 1981.
73. Monroe, D. H., and Eaton, D. L. Comparative effects of butylated hydroxyanisole on hepatic in vivo DNA binding and in vitro biotransformation of aflatoxin B_1 in the rat and mouse. *Toxicol. Appl. Pharmacol., 90*:401, 1987.
74. Buchi, G. H., Muller, P. M., Roebuck, B. D., and Wogan, G. N. Aflatoxin Q_1: a major metabolite of aflatoxin B_1 produced by human liver. *Res. Commun. Chem. Pathol. Pharmacol., 8*:585, 1974.
75. Roebuck, B. D., and Wogan, G. N. Species comparison of in vitro metabolism of aflatoxin B_1. *Cancer Res., 37*:1649, 1977.
76. Aoyama, T. S., Yamano, S., Guzelian, P. S., Gelboin, H. V., and Gonzalez, F. J. Five of 12 forms of vaccinia virus-expressed human hepatic cytochrome P450 metabolically activate aflatoxin B_1. *Proc. Natl. Acad. Sci. USA, 87*:4790, 1990.
77. Decad, G. M., Dougherty, K. K., Hsieh, D. P. H., and Byard, J. L. Metabolism of aflatoxin B in cultured mouse hepatocytes: comparison with rat and effects of cyclohexene oxide and diethylmaleate. *Toxicol. Appl. Pharmacol., 50*:429, 1979.
78. Gurtoo, H. L., and Bejba, N. Hepatic microsomal mixed function oxygenase: enzyme multiplicity of the metabolism of carcinogens to DNA-binding metabolites. *Biochem. Biophys. Res. Commun., 61*: 685, 1974.

79. Lotlikar, P. D., Jhee, E. C., Insetta, S. M., and Clearfield, M. S. Modulation of microsome-mediated aflatoxin B_1 binding to exogenous and endogenous DNA by cytosolic glutathione *S*-transferases in rat and hamster livers. *Carcinogenesis, 5*:269, 1984.
80. Degen, G. H., and Neumann, H. G. The major metabolite of aflatoxin B_1 in the rat is a glutathione conjugate. *Chem.-Biol. Interact., 22*:239, 1978.
81. Moss, E. J., Judah, D. J., Przybylski, M., and Neal, G. E. Some mass-spectral and n.m.r. analytical studies of a glutathione conjugate of aflatoxin B_1. *Biochem. J., 210*:227, 1983.
82. Holeski, C. J., Eaton, D. L., Monroe, D. H., and Bellamy, G. M. Effects of phenobarbital on the biliary excretion of aflatoxin P_1-glucuronide and aflatoxin B_1-*S*-glutathione in the rat. *Xenobiotica, 17*:139, 1987.
83. Baertschi, S. W., Raney, K. D., Shimada, T., Harris, T., and Guengerich, F. P. Comparison of rates of enzymatic oxidation of aflatoxin B_1, aflatoxin G_1, and sterigmatocystin and activities of the epoxides in forming guanyl-N^7 adducts and inducing different genetic responses. *Chem. Res. Toxicol., 2*:114, 1989.
84. Swenson, D. H., Miller, J. A., and Miller, E. C. The reactivity and carcinogenicity of aflatoxin B_1-2,3-dichloride, a model for the putative 2,3-oxide metabolite of aflatoxin B_1. *Cancer Res., 35*:3811, 1975.
85. Lin, J.-K., Kennan, K. A., Miller, E. C., and Miller, J. A. Reduced nicotinamide adenine dinucleotide phosphate-dependent formation of 2,3-dihydro-2,3-dihydroxyaflatoxin B_1 from aflatoxin B_1 by hepatic microsomes. *Cancer Res., 38*:2424, 1978.
86. Essigmann, J. M., Croy, R. G., Nadzan, A. M., Busby, W. F., Reinhold, V. N., Buchi, G., and Wogan, G. N. Structural identification of the major DNA adduct formed by aflatoxin B_1 in vitro. *Proc. Natl. Acad. Sci. USA, 74*:1870, 1977.
87. Lin, J.-K., Miller, J. A., and Miller, E. C. 2,3-Dihydro-2-(guan-7-yl)-3-hydroxy-aflatoxin B_1, a major acid hydrolysis product of aflatoxin B_1-DNA or -ribosomal RNA adducts formed in hepatic microsome-mediated reactions and in rat liver in vivo. *Cancer Res., 37*:4430, 1977.
88. Croy, R. G., Essigmann, J. M., Reinhold, V. N., and Wogan, G. N. Identification of the principal aflatoxin B_1-DNA adduct formed in vivo in rat liver. *Proc. Natl. Acad. Sci. USA, 75*:1745, 1978.
89. Croy, R. G., and Wogan, G. N. Temporal patterns of covalent DNA adducts in rat liver after single and multiple doses of aflatoxin B_1. *Cancer Res., 41*:197, 1981.
90. Essigmann, J. M., Croy, R. G., Bennett, R. A., and Wogan, G. N. Metabolic activation of aflatoxin B_1: patterns of DNA adduct formation, removal, and excretion in relation to carcinogenesis. *Drug Metab. Rev., 13*:581, 1982.
91. Gopalakrishnan, S., Byrd, S., Stone, M. P., and Harris, T. M. Carcinogen—nucleic acid interactions: equilibrium binding studies of aflatoxin B_1 with the oligodeoxynucleotide $d(ATGCAT)_2$ and with plasmid pBR322 support intercalative association with the β-DNA helix. *Biochemistry, 28*:726, 1989.

92. Raney, K. D., Gopalakrishnan, S., Byrd, S., Stone, M. P., and Harris, T. M. Alteration of the aflatoxin cyclopentenone ring to a δ-lactone reduces intercalation with DNA and decreases formation of guanine N^7 adducts by aflatoxin epoxides. *Chem. Res. Toxicol.*, *3*:254, 1990.
93. Wild, C. P., Montesano, R., VanBenthem, J., Scherer, E., and DenEngelse, L. Intercellular variation in levels of adducts of aflatoxin B_1 and G_1 in DNA from rat tissues: a quantitative immunocytochemical study. *J. Cancer Res. Clin. Oncol.*, *116*:134, 1990.
94. Ueno, I., Friedman, L., and Stone, C. L. Species difference in the binding of aflatoxin B_1 to hepatic macromolecules. *Toxicol. Appl. Pharmacol.*, *52*:177, 1980.
95. Lutz, W. K., Jaggi, W., Luthy, J., Sagelsdorff, P., and Schlatter, C. In vivo covalent binding of aflatoxin B_1 and aflatoxin M_1 to liver DNA of rat, mouse and pig. *Chem.-Biol. Interact*, *32*:249, 1980.
96. Cole, K. E., Jones, T. W., Lipsky, M. M., Trump, B. F., and Hsu, I.-C. In vitro binding of aflatoxin B_1 and 2-acetylaminofluorene to rat, mouse and human hepatocyte DNA: the relationship of DNA to carcinogenicity. *Carcinogenesis*, *9*:711, 1988.
97. Appleton, B. S., Goetchius, M. P., and Campbell, T. C. Linear dose-response curve for the hepatic macromolecular binding of aflatoxin B_1 in rats at very low exposures. *Cancer Res.*, *42*:3659, 1982.
98. Bechtel, D. H. Molecular dosimetry of hepatic aflatoxin B_1-DNA adducts: linear correlation with hepatic cancer risk. *Reg. Toxicol. Pharmacol.*, *10*:74, 1989.
99. Halver, J. E., Ashley, L. M., and Smith, R. R. Aflatoxicosis in coho salmon. *Natl. Cancer Inst. Monogr.*, *31*:141, 1969.
100. Carnaghan, R. B. A. Hepatic tumors in ducks fed a low level of toxic groundnut meal. *Nature*, *208*:308, 1965.
101. Seiber, S. M., Correa, P., Dalgard, D. W., and Adamson, R. H. Induction of osteogenic sarcomas and tumors of the hepatobiliary system in nonhuman primates with aflatoxin B_1. *Cancer Res.*, *39*:4545, 1979.
102. Croy, R. G., and Wogan, C. N. Quantitative comparison of covalent aflatoxin-DNA adducts formed in rat and mouse livers and kidneys. *J. Natl. Cancer Inst.*, *66*:761, 1981.
103. Bailey, G. S., Williams, D. E., Wilcox, J. S., Loveland, P. M., Coulombe, R. A., and Hendricks, J. D. Aflatoxin B_1 carcinogenesis and its relation to DNA adduct formation and adduct persistence in sensitive and resistant salmonid fish. *Carcinogenesis*, *9*:1919, 1988.
104. Valsta, L. M., Hendricks, J. D., and Bailey, G. S. The significance of glutathione conjugation for aflatoxin B_1 metabolism in rainbow trout and coho salmon. *Food Chem. Toxicol.*, *26*:129, 1988.
105. Dunaif, G. E., and Campbell, T. C. Dietary protein level and aflatoxin B_1-induced preneoplastic hepatic lesions in the rat. *J. Nutr.*, *117*:1298, 1987.
106. Appleton, B. S., and Campbell, T. C. Dietary protein intervention during the postdosing phase of aflatoxin B_1-induced hepatic preneoplastic lesion development. *J. Natl. Cancer Inst.*, *70*:547, 1983.

107. Bailey, G., Taylor, M., Selivonchick, D., Eisele, T., Hendricks, J., Nixon, J., Pawlowski, N., and Sinnhuber, R. Mechanisms of dietary modification of aflatoxin B_1 carcinogenesis. *Basic Life Sci., 21*:149, 1982.
108. Prabhu, A. L., Aboobaker, V. S., and Bhattacharya, R. K. In vivo effect of dietary factors on the molecular action of aflatoxin B_1: role of riboflavin on the catalytic activity of liver fractions. *Cancer Lett., 48*:89, 1989.
109. Bhattacharya, R. K., Prabhu, A. L., and Aboobaker, V. S. In vivo effect of dietary factors on the molecular action of aflatoxin B_1: role of vitamin A on the catalytic activity of liver fractions. *Cancer Lett., 44*:83, 1989.
110. Emerole, G. O., Thabrew, M. I., and Kwanashie, H. O. Effect of dietary vitamin E (α-tocopherol) on aflatoxin B metabolism. *Eur. J. Drug Metab. Pharmacokin., 9*:295, 1984.
111. Chen, J., Goetchius, M. P., Campbell, T. C., and Combs, J. G. F. Effects of dietary selenium and vitamin E on hepatic mixed-function oxidase activities and in vivo covalent binding of aflatoxin B_1 in rats. *J. Nutr., 112*:324, 1982.
112. Newberne, P. M., Rogers, A. E., and Wogan, G. N. Hepatorenal lesions in rats fed a low lipotrope diet and exposed to aflatoxin. *J. Nutr., 94*:331, 1968.
113. Rogers, A. E., and Newberne, P. M. Diet and aflatoxin B_1 toxicity in rats. *Toxicol. Appl. Pharmacol., 20*:113, 1971.
114. Francis, A. R., Shetty, T. K., and Bhattacharya, R. K. Modifying role of dietary factors on the mutagenicity of aflatoxin B_1: in vitro effect of trace elements. *Mutat. Res., 199*:85, 1988.
115. Goswami, U. C., Saloi, T. N., Firozi, P. F., and Bhattacharya, R. K. Modulation of some natural carotenoids of DNA adduct formation by aflatoxin B_1 in vitro. *Cancer Lett., 47*:127, 1989.
116. Francis, A. R., Shetty, T. K., and Bhattacharya, R. K. Modification of the mutagenicity of aflatoxin B_1 and *N*-methyl-*N'*-nitro-*N*-nitrosoguanidine by certain phenolic compounds. *Cancer Lett., 45*:177, 1989.
117. Mandal, S., Ahuja, A., Shivapurkar, N. M., Cheng, S.-J., Groopman, J. D., and Stoner, G. D. Inhibition of aflatoxin B_1 mutagenesis in *Salmonella typhimurium* and DNA damage in cultured rat and human tracheobronchial tissues by ellagic acid. *Carcinogenesis, 8*: 1651, 1987.
118. Bhattacharya, R. K., and Firozi, P. F. Effect of plant flavonoids on microsome catalyzed reactions of aflatoxin B_1 leading to activation and DNA adduct formation. *Cancer Lett., 39*:85, 1988.
119. Francis, A. R., Shetty, T. K., and Bhattacharya, R. K. Modifying role of dietary factors on the mutagenicity of aflatoxin B_1: in vitro effect of plant flavonoids. *Mutat. Res., 222*:393, 1989.
120. Graham, S., Dayal, H., Swanson, M. A. M., and Wilkinson, G. Diet in the epidemiology of cancer of the colon and rectum. *J. Natl. Cancer Inst., 61*:709, 1978.
121. Kune, S., Kune, G. A., and Watson, L. F. Case-control study of dietary etiological factors: the Melbourne colorectal cancer study. *Nutr. Cancer, 9*:21, 1987.

122. Young, T. B., and Wolf, D. A. Case-control study of proximal and distal colon cancer and diet in Wisconsin. *Int. J. Cancer, 42*:167, 1988.
123. Wattenberg, L. W. Studies of polycyclic hydrocarbon hydroxylases of the intestine possibly related to cancer. Effect of diet on benz-pyrene hydroxylase activity. *Cancer, 28*:99, 1971.
124. Pantuck, E. J., Hsiao, K.-C., Loub, W. D., Wattenberg, L. W., Kuntzman, R., and Conney, A. H. Stimulatory effect of vegetables on intestinal drug metabolism in the rat. *J. Pharmacol. Exp. Ther., 198*:278, 1976.
125. Pantuck, E. J., Pantuck, C. B., Garland, W. A., Min, B. H., Wattenberg, L. W., Anderson, K. E., Kappas, A., and Conney, A. H. Stimulatory effect of brussels sprouts and cabbage on human drug metabolism. *Clin. Pharmacol. Ther., 25*:88, 1979.
126. Pantuck, E. J., Pantuck, C. B., Anderson, K. E., Wattenberg, L. W., Conney, A. H., and Kappas, A. Effect of brussels sprouts and cabbage on drug conjugation. *Clin. Pharmacol. Ther., 35*:161, 1984.
127. Stoewsand, G. S., and Babish, J. B. Inhibition of hepatic toxicities from polybrominated biphenyls and aflatoxin B_1 in rats fed cauliflower. *J. Environ. Pathol. Toxicol., 2*:399, 1978.
128. Boyd, J. N., Sell, S., and Stoewsand, G. S. Inhibition of aflatoxin-induced serum α-fetoprotein in rats fed cauliflower. *Proc. Soc. Exp. Biol. Med., 161*:473, 1979.
129. Boyd, J. N., and Stoewsand, G. S. Blood α-fetoprotein changes in rats fed aflatoxin B_1 and various levels of cabbage. *J. Food Sci., 46*:1923, 1981.
130. Boyd, J. N., Babish, J. G., and Stoewsand, G. S. Modification by beet and cabbage diets of aflatoxin B_1-induced rat plasma α-fetoprotein elevation, hepatic tumorigenesis, and mutagenicity of urine. *Food Chem. Toxicol., 20*:47, 1982.
131. Godlewski, C. F., Boyd, J. N., Sherman, W. K., Anderson, J. L., and Stoewsand, G. S. Hepatic glutathione *S*-transferase activity and aflatoxin B_1-induced enzyme altered foci in rats fed fractions of brussels sprouts. *Cancer Lett., 28*:151, 1985.
132. Whitty, J. P., and Bjeldanes, L. F. The effects of dietary cabbage on xenobiotic metabolizing enzymes and the binding of aflatoxin B_1 to hepatic DNA in rats. *Food Chem. Toxicol., 25*:581, 1987.
133. Salbe, A. D., and Bjeldanes, F. Effect of diet and route of administration on the DNA binding of aflatoxin B_1 in the rat. *Carcinogenesis, 10*:629, 1989.
134. Ramsdell, H. S., and Eaton, D. L. Modification of aflatoxin B_1 biotransformation in vitro and DNA binding in vivo by dietary broccoli in rats. *J. Toxicol. Environ. Health, 25*:269, 1988.
135. Loub, W. D., Wattenberg, L. W., and Davis, D. W. Aryl hydrocarbon hydroxylase induction in rat tissues by naturally occurring indoles of cruciferous plants. *J. Natl. Cancer Inst., 54*:985, 1975.
136. Bradfield, C. A., and Bjeldanes, L. F. Effect of dietary indole-3-carbinol on intestinal and hepatic monooxygenase, glutathione *S*-transferase and epoxide hydrolase activities in the rat. *Food Chem. Toxicol., 22*:977, 1984.

137. Bradfield, C. A., and Bjeldanes, L. F. Structure—activity relationships of dietary indoles: a proposed mechanism of action as modifiers of xenobiotic metabolism. *J. Toxicol. Environ. Health, 21*:311, 1987.
138. Chang, Y., and Bjeldanes, L. F. Effects of dietary R-goitrin on hepatic and intestinal glutathione *S*-transferase, microsomal epoxide hydratase, and ethoxycoumarin *O*-deethylase activities in the rat. *Food Chem. Toxicol., 23*:905, 1985.
139. Chang, Y., and Bjeldanes, L. F. R-Goitrin- and BHA-induced modulation of aflatoxin B_1 binding to DNA and biliary excretion of thiol conjugates in rats. *Carcinogenesis, 8*:585, 1987.
140. Buttery, R. G., Guadagni, D. G., Ling, L. C., Seifert, R. M., and Lipton, W. Additional volatile components of cabbage, broccoli and cauliflower. *J. Agric. Food Chem., 24*:829, 1976.
141. Fenwick, G. R., Heaney, R. K., and Mullin, W. J. Glucosinolates and their breakdown products in food and food plants. *CRC Crit. Rev. Food Sci. Nutr., 18*:123, 1983.
142. Wattenberg, L. W. Inhibition of carcinogenic and toxic effects of polycyclic hydrocarbons by phenolic antioxidants and ethoxyquin. *J. Natl. Cancer Inst., 48*:1425, 1972.
143. Miranda, C. L., Henderson, M. C., Schmitz, J. A., and Buhler, D. R. Protective role of dietary butylated hydroxyanisole against chemical-induced acute liver damage in mice. *Toxicol. Appl. Pharmacol., 69*:73, 1983.
144. Rahimtula, A. D., Jernstrom, B., Dock, L., and Moldeus, P. Effects of dietary and in vitro 2(3)-*t*-butyl-4-hydroxy anisole and other phenols on hepatic enzyme activities in mice. *Br. J. Cancer, 45*:935, 1982.
145. Benson, A. M., Batzinger, R. P., Ou, S.-Y. L., Bueding, E., Cha, Y. N., and Talalay, P. Elevation of hepatic glutathione *S*-transferase activities and protection against mutagenic metabolites of benzo[*a*]pyrene by dietary antioxidants. *Cancer Res., 38*:4486, 1978.
146. Monroe, D. H., Holeski, C. J., and Eaton, D. L. Effects of single-dose and repeated-dose pretreatment with 2(3)-*tert*-butyl-4-hydroxyanisole (BHA) on the hepatobiliary disposition and covalent binding to DNA of aflatoxin B_1 in the rat. *Food Chem. Toxicol., 24*:1273, 1986.
147. Williams, G. M., Tanaka, T., and Macura, Y. Dose-related inhibition of aflatoxin B_1 induced hepatocarcinogenesis by the phenolic antioxidants, butylated hydroxyanisole and butylated hydroxytoluene. *Carcinogenesis, 7*:1043, 1986.
148. Cabral, J. R. P., and Neal, G. E. The inhibitory effects of ethoxyquin on the carcinogenic action of aflatoxin B_1 in rats. *Cancer Lett., 19*:126, 1983.
149. Lotlikar, P. D., Clearfield, M. S., and Jhee, E. C. Effect of butylated hydroxyanisole on in vivo and in vitro hepatic aflatoxin B_1-DNA binding in rats. *Cancer Lett., 24*:241, 1984.
150. Kensler, T. W., Egner, P. A., Davidson, N. E., Roebuck, B. D., Pikul, A., and Groopman, J. D. Modulation of aflatoxin metabolism,

aflatoxin-N^7-guanine formation and hepatic tumorigenesis in rats fed ethoxyquin: role of induction of glutathione *S*-transferases. *Cancer Res.*, *46*:3924, 1986.
151. Salocks, C. B., Hsieh, D. P. H., and Byard, J. L. Effects of butylated hydroxytoluene pretreatment on the metabolism and genotoxicity of aflatoxin B_1 in primary cultures of adult rat hepatocyes: selective reduction of nucleic acid binding. *Toxicol. Appl. Pharmaccol.*, *76*:498, 1981.
152. Jhee, E.-C., Ho, L. L., Tsuji, K., Gopalan, P., and Lotlikar, P. D. Effect of butylated hydroxyanisole pretreatment on aflatoxin B_1-DNA binding and aflatoxin B_1-glutathione conjugation in isolated hepatocyes from rats. *Cancer Res.*, *49*:1357, 1989.
153. Mandel, H. G., Manson, M. M., Judah, D. J., Simpson, J. L., Green, J. A., Forrester, L. M., Wolf, C. R., and Neal, G. E. Metabolic basis for the protective effect of the antioxidant ethoxyquin on aflatoxin B_1 hepatocarcinogenesis in the rat. *Cancer Res.*, *47*:5218, 1987.
154. Monroe, D. H., and Eaton, D. L. Effects of modulation of hepatic glutathione on biotransformation and covalent binding of aflatoxin B_1 to DNA in the mouse. *Toxicol. Appl. Pharmacol.*, *94*:118, 1988.
155. Rahimtula, A. K., and Martin, M. Dietary administration of 2(3)-*t*-butyl-4-hydroxyanisole elevates mouse liver microsome-mediated DNA binding and mutagenicity of aflatoxin B_1. *Chem.-Biol. Interact.*, *48*:207, 1984.
156. Pearson, W. R. A., Windle, J. H., Morrow, J. F., Benson, A. M., and Talalay, P. Increased synthesis of glutathione-*S*-transferases in response to anticarcinogenic antioxidants. Cloning and measurement of messenger RNA. *J. Biol. Chem.*, *258*:2052, 1983.
157. Pearson, W. R., Reinhart, J., Sisk, S. C., and Anderson, K. S. Tissue-specific induction of murine glutathione transferase mRNAs by butylated hydroxyanisole. *J. Biol. Chem.*, *263*:13324, 1988.
158. Ramsdell, H. S., and Eaton, D. L. Mouse liver glutathione-*S*-transferase isoenzyme activity toward aflatoxin B_1-8,9-epoxide and benzo-[*a*]pyrene-7,8-dihydrodiol-9,10-epoxide. *Toxicol. Appl. Pharmacol.*, 1990.
159. Goeger, D. E., Shelton, D. W., Hendricks, J. D., Pereira, C., and Bailey, G. S. Comparative effect of dietary butylated hydroxyanisole and β-naphthoflavone on aflatoxin B_1 metabolism, DNA adduct formation and carcinogenesis in rainbow trout. *Carcinogenesis*, *9*:1793, 1988.
160. Ansher, S. S., Dolan, P., and Bueding, E. Chemoprotective effects of two dithiolthiones and of butylated hydroxyanisole against carbon tetrachloride and acetaminophen toxicity. *Hepatology*, *3*:932, 1983.
161. Davies, M. H., Blacker, A. M., and Schnell, R. C. Dithiolthione-induced alterations in hepatic glutathione and related enzymes in male mice. *Biochem. Pharmacol.*, *36*:568, 1987.
162. Liu, Y. L., Roebuck, B. D., Yager, J. D., Groopman, J. D., and Kensler, T. W. Protection by 5-(2-pyrazinyl)-4-methyl-1,2-dithiol-3-thione (Oltipraz) against the hepatoxicity of aflatoxin B_1 in the rat. *Toxicol. Appl. Pharmacol.*, *93*:442, 1988.

163. Kensler, T. W., Egner, P. A., Trush, M. A., Bueding, E., and Groopman, J. D. Modification of aflatoxin B_1 binding to DNA in vivo in rats fed phenolic antioxidants, ethoxyquin and a dithiothione. *Carcinogenesis, 6*:759, 1985.
164. Kensler, T. W., Egner, P. A., Dolan, P. M., Groopman, J. D., and Roebuck, B. D. Mechanism of protection against aflatoxin tumorigenicity in rats fed 5-(2-pyrazinyl)-4-methyl-1,2-dithiol-3-thione (Oltipraz) and related 1,2-dithiol-3-ones. *Cancer Res., 47*:4271, 1987.
165. Ramsdell, H. S., Parkinson, A., Eddy, A. C., and Eaton, D. L. Bioactivation of aflatoxin B_1 by human liver microsomes. Role of cytochrome P450IIIA isoenzymes. *Toxicol. Appl. Pharmacol., 108*: 436, 1991.

7

FORMATION OF AFLATOXIN-DNA ADDUCTS IN TROUT AND THEIR USE AS MOLECULAR DOSIMETERS FOR TUMOR PREDICTION

RODERICK H. DASHWOOD,* KOENRAAD MARIËN,† PATRICIA M. LOVELAND, DAVID E. WILLIAMS, J. D. HENDRICKS, and GEORGE S. BAILEY
Oregon State University, Corvallis, Oregon

I. INTRODUCTION

The rainbow trout (*Oncorhyncus mykiss*) tumor model has been used extensively to study the mechanisms of action of a number of chemical carcinogens, including several structurally related aflatoxins (AFB_1, AFM_1, AFL, and AFL-M_1) [1–3]. Test agents have been administered to whole trout at various stages in the life cycle including embryo, fingerling, and yearling [4–8]. Several routes of exposure have been employed including dietary administration, intraperitoneal (i.p.) injection, bath exposure (gill uptake), and embryo microinjection [8–12]. In addition to in vivo studies, isolated intact hepatocytes and tissue fractions (liver microsomes, S9, S20) from trout have been used to investigate aflatoxin metabolism and covalent DNA binding [13–18]. The latter studies have included both quantitative analyses of total aflatoxin–DNA binding levels and qualitative investigations to identify the type of DNA adducts formed and their relative persistence in DNA. As a result of these in vitro and in vivo studies, an extensive database exists on trout with respect to mechanisms of aflatoxin carcinogenesis.

It is well documented that rainbow trout are extremely sensitive to AFB_1-induced tumorigenesis [1, 3, 19]. To illustrate this point, the carcinogenic potency of AFB_1 may be compared in trout and rodent models, where dietary levels as low as 4 ppb produce a 60% incidence of hepatic tumors in trout [20] compared with only a 4.5% incidence in rats fed 5 ppb AFB_1 for a comparable period of time [21]. These differences in susceptibility to AFB_1 could be due to a number of factors in rats including

Present affiliations:
*University of Hawaii, Honolulu, Hawaii.
†Department of Health, State of Washington, Olympia, Washington.

reduced activation of AFB_1 to its ultimate carcinogen (AFB_1-8,9-epoxide), enhanced rates of detoxification, and efficient rates of DNA repair activity. The interplay between these various factors may result in a net reduction in AFB_1-DNA adduct formation and persistence and thereby decrease the probability of fixing a mutation in DNA.

In this chapter we discuss various factors that relate to aflatoxin-DNA adduct formation in rainbow trout. We initially review AFB_1 metabolism (Section II) and DNA adduct formation in vivo (Section III). Subsequent sections focus on comparative aspects of aflatoxin–DNA binding in vivo and in vitro and include studies with isolated trout hepatocytes and subcellular fractions. First, with trout as the test species, the activities of AFB_1 and several other aflatoxins will be compared with respect to metabolism, DNA binding, molecular dosimetry, and DNA sequence interaction specificities (Sections IV.A–IV.C). Second, interspecies differences in susceptibility to AFB_1 will be illustrated by comparing trout with other species, where differences in AFB_1 metabolism, DNA adduct formation, and persistence are emphasized. The coho salmon (Section V.A) is included as an example of a second salmonid species that is extremely resistant to AFB_1-induced carcinogenesis. The final section (Section V.B) discusses the sensitivity of trout versus rodents and includes a comparison of AFB_1–DNA binding and tumor data from trout and Fischer rats using the "molecular dosimetry" concept.

II. METABOLISM OF AFB_1 IN TROUT

A. Phase I Metabolism of AFB_1 In Vivo

Metabolism of AFB_1 by the cytochrome P-450–dependent mixed-function oxidase (MFO) system converts the parent molecule to a variety of metabolites of increased polarity. In trout the principal metabolites include AFM_1, aflatoxicol (AFL), AFL-M_1, and AFB_1-8,9-epoxide (Fig. 1) [17, 18]. The latter metabolite is regarded as the putative "ultimate carcinogen" [22, 23], an activated species that binds covalently to cellular macromolecules such as DNA (Section III) and elicits much of the genotoxic activity both in vitro and in vivo. The formation of AFM_1 represents metabolic detoxification because the carcinogenic potency of AFM_1 in trout is considerably lower than that of parent AFB_1. Carcinogenicity data for AFL indicate that the potency of the metabolite is approximately equivalent to that of AFB_1 in trout, although the route of exposure has been observed to influence the relative potencies (Section IV.B). The carcinogenic status and potency of AFL-M_1 is presently under investigation in trout.

Several MFO pathways produce both activated and detoxified products of AFB_1; differences are a product of multiple forms of P-450 with distinct regioselectivity toward the substrate. In the case of AFB_1, hydroxylation at the 4-position is preferentially catalyzed by a β-naphthoflavone (BNF)-inducible form of P-450 (P-448), whereas formation of AFB_1-8,9-epoxide is mediated by one or more forms of phenobarbital-induced P-450. In trout [24], the two principal cytochrome P-450 monooxygenase pathways involved in the metabolic activation and detoxification of AFB_1 are termed, respectively, cytochrome P-450 LM_2 and cytochrome P-450 LM_4b pathways

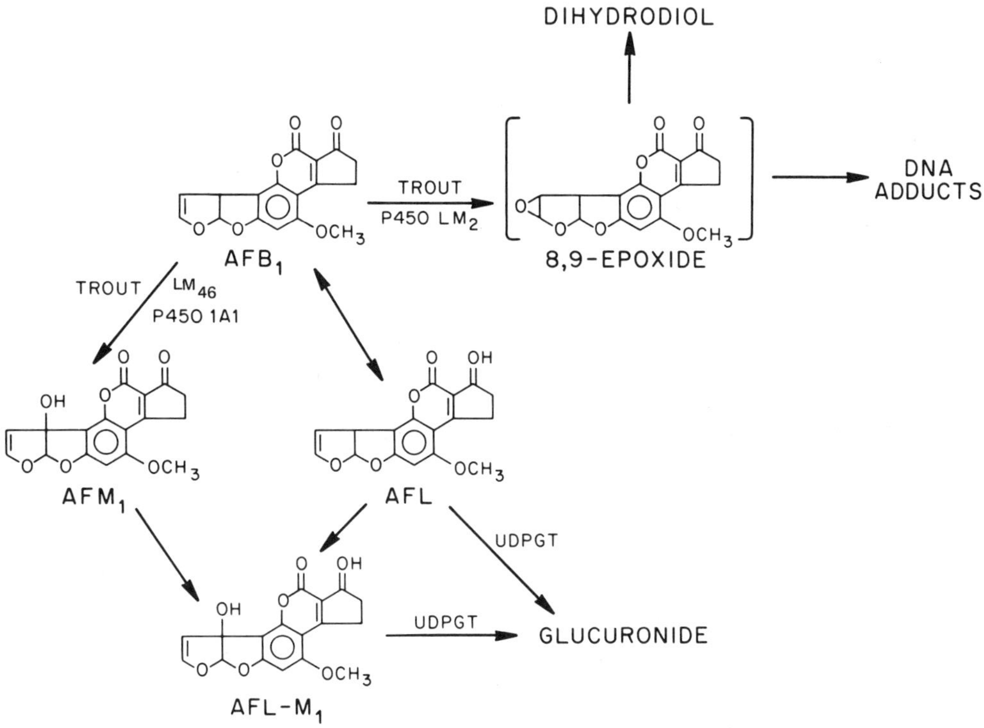

FIGURE 1. Pathways of AFB_1 metabolism in rainbow trout.

(Fig. 1). Thus, cytochrome P-450 LM_2, a constitutive isoenzyme with high activity toward AFB_1, forms the putative ultimate carcinogen AFB_1-8,9-epoxide, whereas cytochrome P-450 LM_{4b} (BNF-inducible P4501A1) produces AFM_1 as the major metabolite [24]. In addition to cytochrome P-450—dependent pathways, cytosolic enzymes in trout liver reduce AFB_1 to AFL; that is, they convert the cyclopentenone ring to a cyclopentenol (Fig. 1). Metabolism studies using isolated trout hepatocytes have provided evidence for the interconversion of the aflatoxins (Section IV.A).

B. Phase II Metabolism of AFB_1

Loveland et al. [25] studied AFB_1 conjugation reactions in vivo in trout and found that the major phase II metabolites formed and excreted in bile were AFL and AFL-M_1 glucuronides (Fig. 1). The findings are consistent with previous studies in trout, which indicate that the metabolites of compounds such as tetrachlorobiphenyl, di(2-ethylhexyl) phthalate, 1-naphthyl-*N*-methylcarbamate, 2',5-dichloro-4'-nitrosalicylanilide, and indole-3-carbinol are excreted in bile predominantly as glucuronides [26–29]. Current evidence does not suggest a major role for glutathione (GSH) in the detoxification of AFB_1 in trout. Recent studies from this laboratory determined that AFB_1-glutathione (AFB_1-SG) conjugates in trout bile represent less than 0.2% of the total administered dose 24 h after i.p. injection of ^{3}H-AFB_1 [30]. In studies using isolated liver cell fractions from trout, less than 0.5% of the original AFB_1 dose was recovered as AFB_1-SG. The findings are contrary to results in rodents, which indicate that AFB_1-SG conjugation plays a major role in some species in the phase II detoxification and elimination of AFB_1 (Section V.B).

III. AFB_1—DNA BINDING STUDIES IN TROUT

A. In Vivo Studies—An Overview

Early AFB_1—DNA binding studies from the Department of Food Science and Technology, Oregon State University, investigated the formation of covalent aflatoxin-DNA adducts in embryonic and yearling trout [4]. A linear relationship was found between the amount of AFB_1 absorbed into 21-day-old eggs and the level of covalent modification of embryo DNA. The principal AFB_1-DNA adduct was identified as 8,9-dihydro-8-(N^7-guanyl)-9-hydroxy aflatoxin (AFB_1-N^7-Gua); a time-dependent conversion of this adduct to the ring-opened major and minor species (the N^7-FAPyr major and minor formamidopyrimidine adducts) was observed. Although the trout eggs were exposed for a short time (1 h, 0.25 and 0.5 ppm AFB_1), covalent modification of DNA was evident over an extended period of time, with the highest levels being detected approximately 24 h after exposure to 0.5 ppm AFB_1. A similar pattern of covalent DNA adducts was found in the livers of yearling trout 10 h after i.p. administration of 0.3 mg/kg AFB_1. Subsequent investigations aimed at providing a more detailed view of aflatoxin—DNA binding in trout by studying (1) several structurally related aflatoxins (Section IV), (2) alternative routes of carcinogen exposure, including isolated hepatocytes and subcellular fractions (Section IV.A), and (3) a range of aflatoxin doses (dose—response studies).

Dose—response studies were initially conducted in trout given 5, 25, 100, and 300 μg AFB_1/kg body weight by single i.p. injection [6]. With the denoted dosing regime, an approximately linear increase in AFB_1—DNA binding in liver occurred with doses of AFB_1. In the same study, time-course parameters for AFB_1—DNA binding were investigated; although binding values reached a peak at 24 h (263 ± 55.5 pmol AFB_1/mg DNA), the relative binding levels were remarkably similar 4, 12, 24, and 48 h after dosing. Collectively, the results indicated that AFB_1—DNA adducts in trout liver are more persistent than those in rat liver; the latter adducts are rapidly removed by DNA repair enzymes. The AFB_1—DNA binding levels were also studied in trout pretreated with BNF and in a second salmonid species (coho salmon) that is more resistant to AFB_1-induced carcinogenesis. Compared to positive controls in trout, AFB_1—DNA binding was twofold lower in animals pretreated with BNF and 20-fold lower in coho salmon [6]; these levels also reflected the relative tumor incidence in each treatment group and suggested that AFB_1—DNA binding levels in trout may correlate with the incidence of hepatocellular carcinoma.

In subsequent, more detailed studies, the possible correlation was investigated between AFB_1—DNA binding and tumor dose response [10, 31].

Four [^{3}H]AFB_1 doses (10, 20, 40, and 80 ppb) were administered in the diet to separate groups of fingerling trout on a daily basis for 2 weeks. Each AFB_1 dose was tested using triplicate tanks of 150 animals; after 7 and 14 days of dosing, 15 animals were selected at random from each tank, livers were combined, and DNA was extracted to assess AFB_1–DNA binding. The remaining 120 animals in each tank were returned to control diets, and the tumor incidences were evaluated after 1 year. AFB_1–DNA binding increased linearly with dose of AFB_1 after both 7 days (open symbols) and 14 days (filled symbols) of carcinogen treatment (Fig. 2). When the slopes of both lines in Fig. 2 are plotted versus time of AFB_1 treatment (inset, Fig. 2a), it also is apparent that AFB_1–DNA binding increased linearly with time of carcinogen administration.

At the termination of study (12 months), animals were killed, and the livers were removed and analyzed for the presence of gross tumors. The

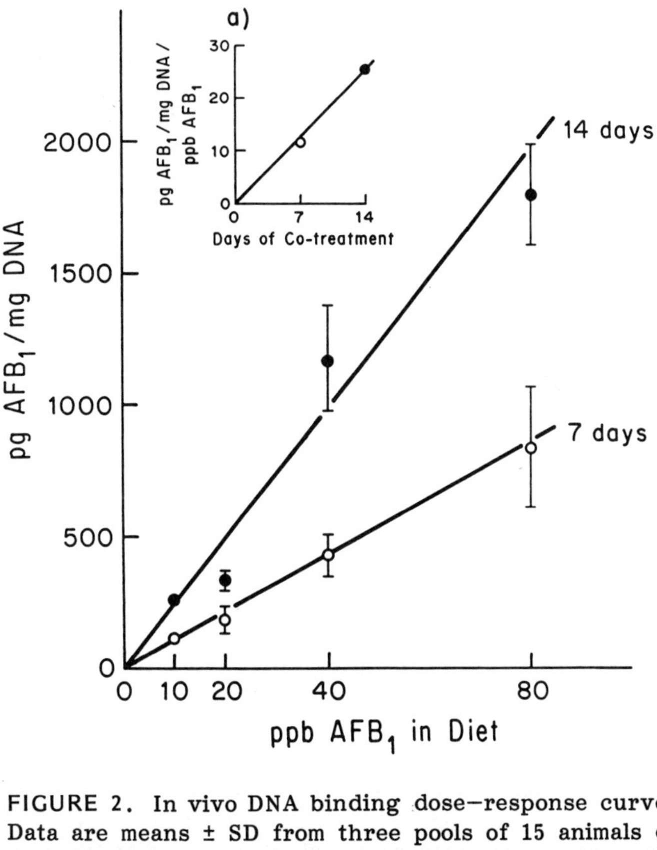

FIGURE 2. In vivo DNA binding dose–response curves for AFB_1 in trout. Data are means ± SD from three pools of 15 animals each. The slope of each line was determined and plotted versus time of AFB_1 administration (see inset, Fig. 2a). All lines were analyzed by statistical linear regression analysis; correlation coefficient >0.98, all cases.

number of animals with one or more liver tumors was recorded and expressed on a percent tumor incidence basis. For the four AFB_1 doses tested, the mean percent tumor incidence values were 10 ppb AFB_1, 10.7%; 20 ppb AFB_1, 32.2%; 40 ppb AFB_1, 55.6%; 80 ppb AFB_1, 68.4%. A linear dose-related increase in tumor response was observed when these data were plotted as a function of ppb AFB_1 in the diet (logit incidence versus log dose plot, Fig. 3). To determine the degree of correlation between tumor and DNA binding data from this study, the molecular dosimetry approach was adopted (Section III.B).

B. Molecular Dosimetry of AFB_1 in Trout

A number of pharmacokinetic factors modulate the final dose of a carcinogen achieved at the target site in vivo, and as a result it can be misleading when tumor data are expressed in terms of the chemical's exposure concentration (carcinogen dose). One measure of dose effectively achieved at the target site is the concentration of carcinogen-DNA adducts, which may represent the biological outcome after integrating various pharmacokinetic parameters [32]. The results have led to the suggested use of DNA adducts as an improved exposure index for quantitative risk assessment, interspecies extrapolations, comparison of data from in vivo and in vitro genotoxicity assays, and studies on mechanisms of carcinogen action [32–37]. The "molecular dosimetry" concept has become particularly

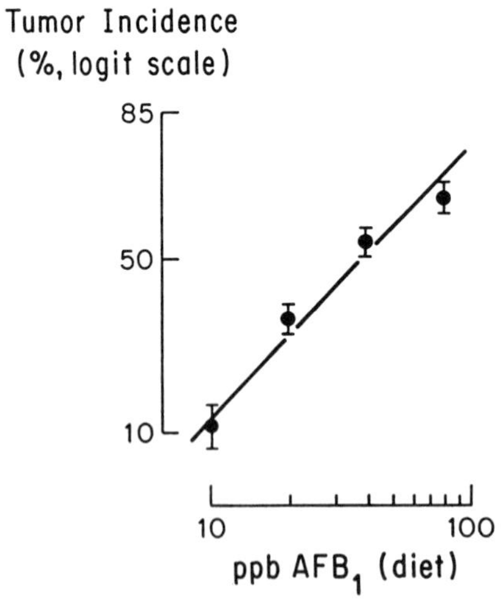

FIGURE 3. Tumor dose–response curves for trout given AFB_1 in the diet. Fingerling trout received 10, 20, 40, or 80 ppb AFB_1 in diet on a daily basis for 2 weeks, after which they were switched to control diet for 1 year. Each data point represents pooled data from three tanks, each containing ~100 animals at necropsy. Data are given as means ± SD.

appealing with the advent of new methodologies such as ^{32}P-postlabeling and immunoassays using specific antibodies [38–40], which enable detection of carcinogen-DNA adducts without the prerequisite for labeled test compounds.

The molecular dosimetry concept has been used to investigate the possible correlation between AFB_1–DNA binding and tumorigenesis in trout. A molecular dosimetry plot was produced when each data point from the tumor dose–response analyses (Fig. 3) was presented as a function of its corresponding DNA binding data point from Fig. 2. A straight-line relationship was evident when either the 7- or 14-day DNA binding data were used (Fig. 4). The line for the 14-day data was displaced in a parallel fashion toward higher carcinogen dose; this displacement is predicted from the time-course binding data, which increased linearly over the 2-week dosing period (Fig. 2a). It should be emphasized that trout were switched to control diet after 14 days in order to assess tumor incidences, and it is the 14-day DNA adduct data that represent the true final dose. In summary, the molecular dosimetry approach provides good evidence that, under specific exposure conditions, a direct quantitative correlation exists between tumor formation and the dose actually achieved in the target organ.

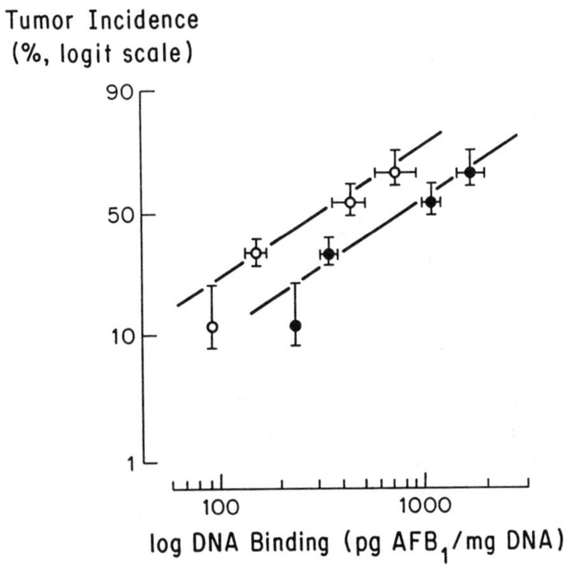

FIGURE 4. Molecular dosimetry of AFB_1–DNA adducts in trout. Tumor incidence data from Fig. 3 were plotted as a function of 7-day (open symbols) and 14-day AFB_1–DNA (filled symbols) binding data from Fig. 2. Data points are means; vertical bars represent the upper and lower tumor incidences (range from three tanks of ~100 animals); horizontal bars indicate standard deviation.

Relevant information has been used to gain insight into three important areas:

1. Molecular dosimetry can provide mechanistic information on structurally related aflatoxins.
2. Interspecies extrapolations are facilitated through the use of molecular dosimetry.
3. In studies with tumor modulators, molecular dosimetry can provide important information on mechanisms of anticarcinogenesis. Specifically, by comparing inhibitor effects on tumor formation and DNA adduction, it is possible to ascertain whether a compound operates as a pure "blocking agent" [41, 42]—that is, solely through inhibition of carcinogen—DNA binding—or whether other mechanisms are involved.

Recent studies of a tumor modulator, indole-3-carbinol (I3C), demonstrated that this natural constituent of cruciferous vegetables (cauliflower, cabbage, broccoli) inhibits chemically induced carcinogenesis and carcinogen—DNA binding in a number of animal models, including rats, mice, and rainbow trout [8—10, 43—45]. When trout were pretreated with dietary I3C doses up to 2000 ppm and a range of $[^3H]AFB_1$ levels, the data points described a single line when plotted as logit tumor incidence versus log AFB_1—DNA binding (Fig. 5). The excellent correlation in the molecular dosimetry plot provides good evidence that I3C is a pure blocking agent in trout because the reduced tumor incidence is accounted for entirely by the degree of inhibition of AFB_1—DNA binding. A detailed account of this study has been published by Dashwood et al. [31].

IV. COMPARATIVE STUDIES WITH STRUCTURALLY RELATED AFLATOXINS

A. Studies with Isolated Trout Hepatocytes and Subcellular Fractions

As a corollary to studies conducted in vivo, the mutagenicity, metabolism, and covalent DNA binding of AFB_1 has been examined ex vivo using isolated trout hepatocytes and subcellular fractions (microsomes, S9, S20). Results from in vitro mutagenicity assays using trout liver S9 and S20 mixes for metabolic activation indicated that the relative mutagenic potencies of the aflatoxins shown in Fig. 1 were $AFB_1 > AFL > AFL\text{-}M_1 > AFM_1$ [13, 16]. Qualitatively similar mutagenicity results were obtained using rat liver activation systems [46]. For the denoted aflatoxins, a general concordance appears to exist between carcinogenic potency in trout and mutagenic potency in the *Salmonella* assay. Isolated hepatocytes from trout readily activate AFB_1 to mutagens detectable by *Salmonella typhimurium* strain TA98 [16]. Analysis of supernatants for unbound $[^3H]AFB_1$ metabolites revealed that presence of bacteria during incubation did not perturb the various pathways of AFB_1 metabolism within hepatocytes. Under defined conditions, isolated trout hepatocytes reflect in vivo cell capabilities for AFB_1 metabolism that include formation of polar conjugates and the major unbound metabolites AFL and AFM_1 [14, 17, 18]. DNA-binding studies with isolated trout hepatocytes [14] established that intracellular

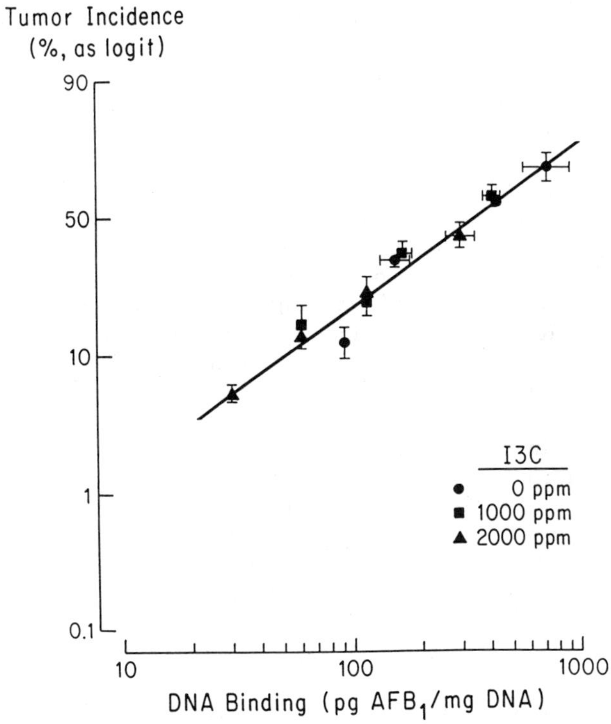

FIGURE 5. Molecular dosimetry data for AFB_1 in trout, illustrating the modulatory effect of indole-3-carbinol, a natural anticarcinogen. Data points and bars are means and standard errors of observations from triplicate tanks (~150 animals) for each AFB_1-I3C dose point.

DNA adduct formation is linearly related to AFB_1 dose and that the AFB_1–DNA adducts are qualitatively similar to those formed in vivo.

Isolated trout hepatocytes were employed to compare AFB_1 with AFL, AFM_1, and AFL-M_1 with respect to patterns of metabolism and DNA binding [17, 18]. Tritium-labeled aflatoxins were incubated with hepatocytes for up to 1 h; cellular DNA was isolated to determine the specific activities, and the cellular supernatants from each of the incubations were analyzed by HPLC to follow the metabolism of the test compounds. In DNA binding studies, a linear relationship was observed for each aflatoxin when data were plotted as pmol aflatoxin bound/μg DNA/μmol dose versus time. After 1 h, the DNA binding levels relative to AFB_1 were AFL-M_1 (0.83) > AFM_1 (0.81) > AFL (0.53). Current mutagenicity data and limited carcinogenicity studies with the four aflatoxins indicate that AFM_1 and AFL-M_1 have DNA binding levels in hepatocytes that are higher than expected compared to those of AFB_1 and AFL. The discrepancy is possibly related to pharmacokinetic factors that differ between isolated hepatocytes and the in vivo situation (hepatocytes in situ); the difference could influence the doses achieved at the target site [47].

Because the four aflatoxins possess an 8,9 double bond (Fig. 1), DNA binding could be due to activation of AFB_1 phase I metabolites (AFL, AFM_1, AFL-M_1) as well as the parent compound. To investigate activation, AFL was ^3H-labeled in the 1-position, that is, on the carbon atom containing the cyclopentenol function [17]. After incubation of [1-^3H]AFL with isolated hepatocytes, the DNA was hydrolyzed and analyzed by HPLC. The primary adduct was found to be nonradioactively labeled AFB_1-N^7-Gua. A minor adduct, possibly AFL-N^7-Gua, was also observed, but only at levels of about 3% those for AFB_1-N^7-Gua. The results suggest that direct interaction of AFL-8,9-epoxide with DNA is of minor quantitative importance in trout hepatocytes and that the major adduct results from conversion of AFL to AFB_1 prior to epoxide formation [17].

In support of the conclusions, supernatants from trout hepatocyte incubations were found to contain, in addition to parent substrates, hydroxylated metabolites and glucuronides (Table 1). Thus, in isolated trout hepatocytes, the dominant pathway producing stable phase I products involves direct interconversion of cyclopentenones with the corresponding cyclopentenols [18]. Recent molecular dosimetry studies support the view that interconversion of aflatoxins also may occur in vivo.

B. Molecular Dosimetry Studies with AFB_1 and AFL

The molecular dosimetry concept has been applied to the analysis of several different aflatoxins including AFB_1, AFL, AFM_1, and AFL-M_1. In rainbow trout and other animal models, the compounds appear from limited data to possess widely differing carcinogenic potencies; it is of interest to determine on a systematic basis the extent of carcinogenic potency of each compound in terms of the covalent DNA binding levels in various species. Preliminary data are presented for rainbow trout treated with a range of

TABLE 1. Interconversion of Aflatoxins in Isolated Trout Hepatocytes[a]

Substrate	Major metabolite
AFB_1	AFL
AFL	AFB_1
AFM_1	AFL-M_1
AFL-M_1	AFM_1

[a]Isolated trout hepatocytes were incubated with parent aflatoxin for 1 h, and the metabolites from supernatants were analyzed by HPLC (see Ref. 18 for method).

AFB_1 and AFL doses by embryo bath exposure. Two distinct tumor dose–response curves were observed when data were plotted as tumors versus dose given (Fig. 6a), but when plotted in terms of dose *achieved* (DNA adducts formed from each aflatoxin), the data yielded a single line (Fig. 6b). The results suggest that the differing carcinogenic potencies of AFL and AFB_1 reflect divergent pharmacokinetics rather than differences in their relative DNA adduct initiating potencies. Studies with AFB_1 and AFL with the dietary route of exposure [37] produced a single line not only for the molecular dosimetry plot but also in the classical plot of

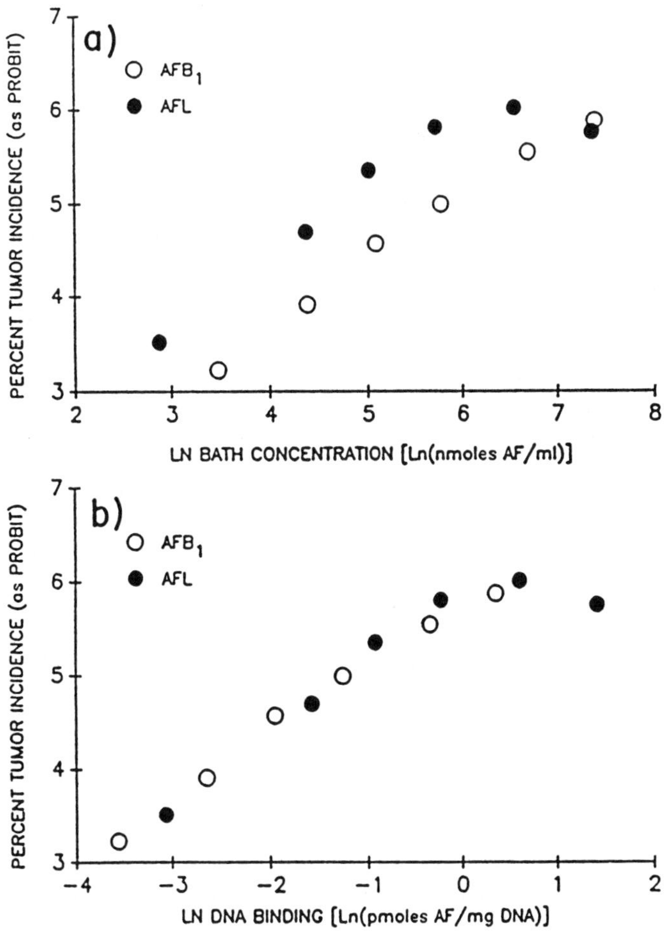

FIGURE 6. Tumor dose–response curves for AFB_1 and AFL in trout following embryo water-bath exposure. (a) "Classical" approach of plotting tumor incidences versus dose given. (b) Molecular dosimetry curves for the same tumor data (tumor incidences versus dose achieved, i.e., DNA adducts in liver).

tumors versus dose given (ppb aflatoxin in diet). Thus, the carcinogenic potency of AFB_1 and AFL is influenced by route of exposure in the classical approach, but molecular dosimetry effectively integrates the various pharmacokinetic parameters to produce a single line regardless of exposure route. When expressed on a "per DNA adduct" basis, AFB_1 and AFL are found to be of equivalent potency with respect to inducing liver tumors in trout. The observation is consistent with data obtained from studies using isolated trout hepatocytes (previous section); the results indicate that AFL is primarily converted to AFB_1 with subsequent activation to the 8,9-epoxide and the production of a DNA adduct profile similar to that following direct AFB_1 treatment. The molecular dosimetry approach is currently being employed to investigate the underlying reasons for higher than expected DNA binding values with AFM_1 and AFL-M_1 in trout relative to carcinogenicity data for the denoted compounds and parent AFB_1.

C. DNA Sequence Interaction Specificities of Structurally Related Aflatoxins

In addition to comparing the metabolism, DNA binding, and molecular dosimetry of various aflatoxins, the interaction specificities of guanyl—aflatoxin binding have been considered. Previous work demonstrated that the 8,9-dichloride derivative and the naturally produced epoxide of AFB_1 react predominantly with the N-7 of guanine and that both produce adducts with similar frequency at identical sites [48—51]. The investigation suggested that 8,9-dihalo derivatives of AFB_1 can be suitable electronic analogs for the ultimate in vivo carcinogen, AFB_1-8,9-oxide. The DNA sequence interaction selectivities of AFB_1-8,9-dibromide and AFB_1-8,9-dichloride have been compared using a cloned fragment of the human c-Ha-*ras* proto-oncogene. The method takes advantage of the fact that DNA fragments containing N-7 guanyl adducts can be cleaved quantitatively under alkaline treatment [52, 53]. Radiolabeled fragments subsequently can be separated on a denaturing polyacrylamide gel; by scanning the autoradiogram with a densitometer, a relative indication of the degree of alkylation at each site can be obtained.

Densitometric scans of the two AFB_1 dihalides indicated that their site reactivity profiles are similar (Fig. 7). A more complete comparison of the results from the scans illustrates that the frequency spectrum of adduct formation among the various guanine sites for the two compounds is very similar (Table 2). Each guanyl site was assigned a reference number from the 5' terminus, and each site was prioritized according to susceptibility, with the smallest ranked numbers indicating sites of highest AFB_1-dichloride and AFB_1-dibromide binding. Standard deviations were ranked in order for each site to number 15; beyond 15 the values are not prioritized because they reflect very low binding affinities that preclude precise quantitation. For both dihalides, site number 17 exhibited the greatest susceptibility to aflatoxin binding. With the exception of site 19, all sites were similar in reactivity for the dichloride and dibromide (within one standard deviation). The results indicate that the two dihalides interact with double-stranded DNA in a similar sequence-selected manner and that AFB_1-8,9-dichloride and AFB_1-8,9-dibromide can be used equally well as electronic analogs of the putative "ultimate carcinogen" AFB_1-8,9-oxide.

Misra and co-workers [49] provided a set of empirical rules to generalize the affinity of activated AFB_1 for guanines based on nearest-neighbor nucleotides. Benasutti and co-workers [54] made an extensive study of various DNA sequences to ascertain a set of values determining activated AFB_1 reaction intensities at a particular guanine based on nearest neighbors. Many positions in the DNA fragment used in the studies alluded to above have binding frequencies that are not in agreement with the values predicted by the algorithms. In addition, values predicted by various workers are not always in agreement [49, 54–59]. Disparity between results must indicate a present lack in understanding of all the variables required to correctly predict reaction affinities. Although the basis for variation in AFB–guanyl reactivity is not understood, it should be stressed that the pattern of variation along a given gene sequence is relatively reproducible. If one wishes to know the reaction intensities of selected sites within any gene sequence, they cannot currently be predicted but must be determined empirically.

In a previous study, evidence was obtained that activated AFM_1 and AFB_1 had significantly (>1 SD) different reactivities at several sites within two sequences examined [55]. The results suggested that the two aflatoxins could possibly exhibit different relative reactivities at the codon 12 guanines in *ras* protooncogenes. However, no major differences between AFB_1 and AFM_1 were found in relative reactivity at any of the 26 guanyl sites examined (Table 3) with a less rigorous ranking of reaction intensity (+, -, =) than was used in other tests (Table 2). Failure to observe differences at any site may simply reflect the small number of sites examined in the study; a conceivable alternative is that differences previously observed might have been a consequence of the methodology by which the aflatoxins were activated with mild peroxy acid.

Densitometric scans of the proximate carcinogen of benzo[*a*]pyrene, anti-BPDE, produced a reaction profile similar to those of the AFB_1 dihalides. The results indicate that many of the factors determining the target affinity of the AFB_1 dihalides for the N-7 of guanine also hold true for anti-BPDE. No mechanistic explanation for reactivity preference at particular sites is available. Local variations in helix twist, propeller twist, base role angle, and torsion angle [60] were calculated for the DNA fragment, but no obvious correlation was observed between relative site reactivity and local neighboring variation in any of these parameters. Also, electrostatic potential in the neighborhood of each guanyl residue was calculated using an algorithm provided by Pullman and Pullman [61] (Fig. 7). The electrostatic potentials created by the adjacent base pairs appear to have some effect, though not a dominant one, in influencing the selectivity of the aflatoxins or anti-BPDE for a particular N-7 of guanine. Further, since anti-BPDE, AFB_1, and AFM_1 analogs all react with similar selectivity profiles with the sequence examined, the determinants cannot be exquisitely sensitive to electrophile configuration.

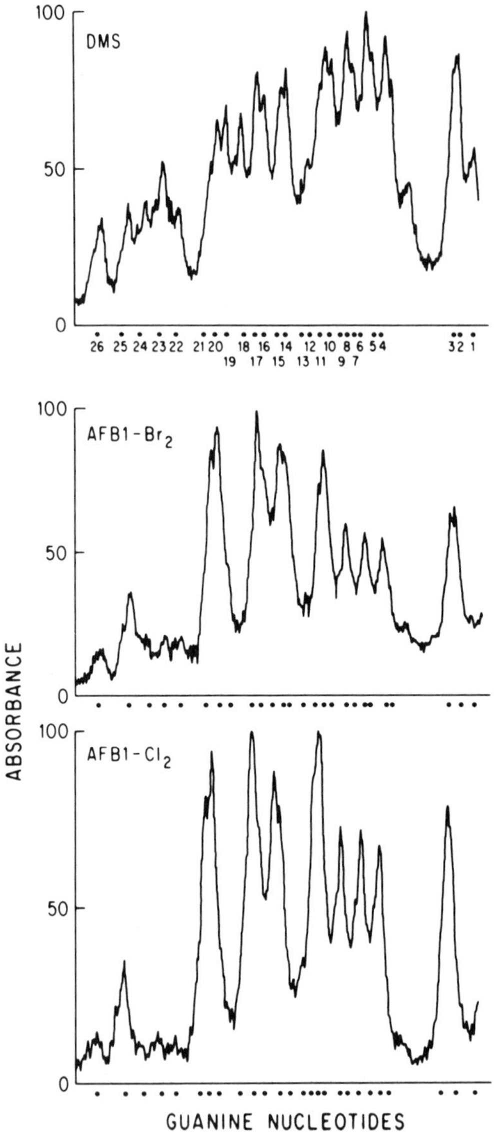

FIGURE 7. Densitometric scans of guanine N-7 alkylation patterns produced by dimethyl sulfate, AFB_1-2,3-dibromide, AFB_1-2,3-dichloride, benzo[*a*]pyrene-7,8-diol-9,10-epoxide, and AFM_1-2,3-dibromide. Numbers below peaks in the dimethyl sulfate (DMS) scan at top indicate the positions of guanines. Direction of sequence is 3' to 5'. The DMS reacts specifically with the N-7 of guanine and is used as a positive control. The graph at the bottom represents the electrostatic potential (kcal/mol) at the guanine N-7 for each guanine nucleotide within the DNA sequence studied. Codon 12 corresponds to guanines 14 and 15.

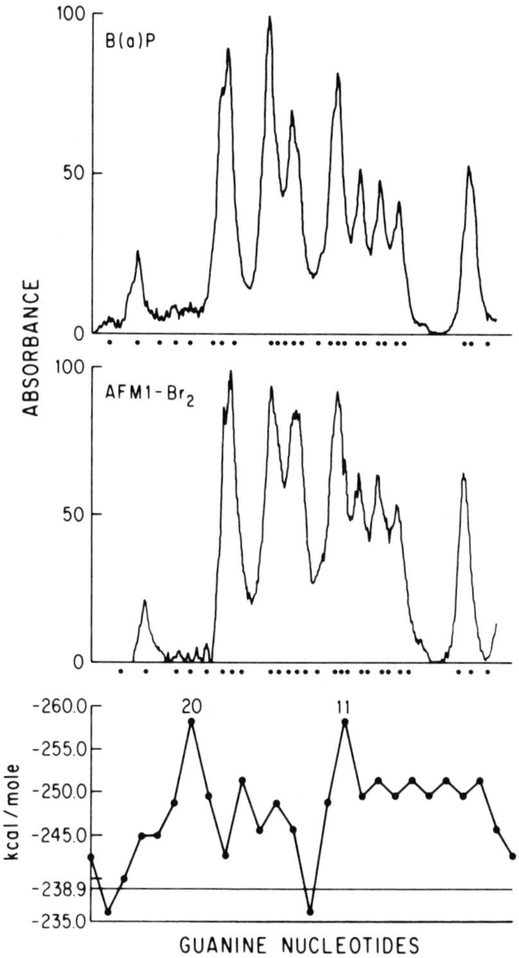

V. INTERSPECIES COMPARISONS—SUSCEPTIBILITY TO AFB_1

A. Sensitivity of Trout Versus Coho Salmon

It is well documented that different salmonid species can exhibit widely divergent susceptibilities to AFB_1-induced carcinogenesis. A classical example of this interspecies variation is provided by the coho salmon (*Onchorynus kisutch*), an animal model that is considerably more resistant than trout with respect to the induction of hepatocarcinoma by AFB_1 [6, 30,62]. A recent publication described a detailed study of AFB_1 carcinogenesis in the two species that included parallel analyses of AFB_1 tumorigenesis, metabolism, DNA binding, and adduct persistence in vivo and in

TABLE 2. Site Reactivity Comparison of AFB_1-2,3-Dichloride with AFB_1-2,3-Dibromide

Guanine ref. no.	Tri-N seq	AFB_1-Cl_2 (± SD)	AFB_1-Br_2 (± SD)
1	TGC	>15	>15
2	CGG	12 (2)	11 (1)
3	GGA	13 (1)	14 (1)
4	TGG	>15	>15
5	GGT	8 (2)	9 (3)
6	TGG	>15	>15
7	GGT	9 (2)	9 (3)
8	TGG	>15	>15
9	GGT	10 (2)	11 (2)
10	TGG	15 (1)	15 (1)
11	GGG	4 (3)	6 (2)
12	GGC	11 (2)	13 (3)
13	CGC	>15	>15
14	CGG	3 (2)	4 (2)
15	GGC	7 (3)	7 (2)
16	CGG	5 (2)	3 (3)
17	GGT	1 (1)	1 (1)
18	TGT	>15	>15
19	TGG	14 (2)	10 (1)
20	GGG	2 (2)	2 (2)
21	GGC	6 (3)	4 (2)
22	AGA	>15	>15
23	AGT	>15	>15
24	TGC	>15	>15
25	CGC	>15	>15
26	TGA	>15	>15

For each guanine position the reaction intensities were determined and prioritized with the value of 1 (guanine No. 17) representing the guanine most susceptible to adduct formation. Positions of low reactivity were not categorized and are depicted as >15. All other positions list standard deviations in ranking order assignment (three replicates per genotoxin with each replicate scanned twice) to indicate possible variation in priority number. The trinucleotide sequence is given for each position in the second column.

isolated hepatocytes [62]. Trout exposed as embryos to an aqueous solution of 0.5 ppm AFB_1 (15 min) showed a 62% tumor incidence after 12 months, whereas coho salmon exposed to this concentration for twice the period of time (30 min) showed only a 9% incidence (Table 4). The difference between these two salmonid species as even greater when AFB_1

TABLE 3. Sequence of In Vitro DNA Segment

													*	*	*											
Test sequence 5'	ATGACG		GAATATTAACTG		GTG		GTG		GT	G	G	GC	G	CCG	GC	G	G	TGT	G	G	GCAAGA		GT	GC	GCTGACC	3'
Reference no.	1	2	3	4	5	6	7	8	9	10	11	12	13	14	15	16	17	18	19	20	21	22	23	24	25	26
Reaction intensity																										
DMS	+	+	=	=	=	=	=	=	=	=	=	=	=	+	+	+	+	=	=	=	=	=	=	=	=	=
AFB_1-Br_2	–	=	=	–	=	–	=	–	=	=	+	=	–	+	+	+	+	–	=	+	+	–	–	–	–	–
AFB_1-Cl_2	–	=	=	–	=	–	=	–	=	=	+	=	–	+	+	+	+	–	=	+	+	–	–	–	–	–
AFM_1-Br_2	–	=	=	–	=	–	=	–	=	=	+	=	–	+	+	+	+	–	=	+	+	–	–	–	–	–
B(*a*)P	–	=	=	–	=	–	=	–	=	=	+	=	–	+	+	+	+	–	=	+	+	–	–	–	–	–

Sequence is listed 5' to 3' with individual guanines numbered. Reaction intensities for the individual sites by the mutagens as indicated: –, poor target; +, strong target; and = depicting the wide range of targets that fall inbetween. Codon 12 is represented by *.

TABLE 4. Carcinogenic Response of Trout (RT) and Coho Salmon (CS) to AFB_1

Species	AFB_1 treatment regimen	Total tumors	Percent incidence
RT	Embryo bath (0.5 ppm; 15 min)	37/60	62%
CS	Embryo bath (0.5 ppm; 30 min)	5/54	9%
RT	Diet (20 ppb; 4 wk)	74/120	62%
RT	Diet (control)	1/121	0.8%
CS	Diet (40 ppb; 4 wk)	0/116	0%
CS	Diet (control)	0/118	0%
CS	Diet (5000 ppb; 3 wk)	9/176	5%
CS	Diet (control)	0/178	0%

Tumors in trout were diagnosed as predominantly hepatocellular carcinomas; those in coho salmon were hepatic adenomas.

was administered in the diet; trout exposed for 4 weeks to 20 ppb AFB_1 had a 62% tumor response at 12 months, while coho salmon exposed to 40 ppb AFB_1 for 4 weeks failed to develop tumors. Moreover, only a 5% tumor incidence was observed in salmon given 5000 ppb AFB_1 in the diet for 3 weeks. Of particular interest, the neoplastic response of salmon also differed from that of trout in producing benign hepatic adenomas versus malignant hepatocellular carcinomas, respectively [62].

To understand the mechanistic reasons for the interspecies differences, various biochemical endpoints were studied [30, 62]. The role of metabolism and initiation in the test species was analyzed by studying phase I and phase II pathways of AFB_1 metabolism; the results included analyses of liver P-450 isoenzyme composition. In studies of phase II metabolism, the rates of bile elimination of AFB_1 detoxification products and the total excretion of aflatoxins into the water, which reflect combined gill and urinary elimination, were similar between trout and salmon. Thus, the different susceptibility of the two species to AFB_1 apparently is not related to an ability of salmon to detoxify AFB_1 more efficiently than trout [30, 62].

Comparative DNA binding studies were also conducted in vivo with several routes of AFB_1 exposure. Aflatoxin B_1–DNA binding was 7–26 times greater in trout than in salmon at various times after i.p. injection (Fig. 8). Qualitatively similar results also were obtained after 3 weeks of dietary treatment and after 1 h exposure of embryos and isolated hepatocytes (Table 5). Differences in initial adduct level appear to reflect relatively low production of AFB_1-8,9-epoxide by salmon liver enzymes. Trout liver microsomes displayed a K_m for AFB_1–DNA adduction in vitro that was sevenfold lower than for salmon (Fig. 9), and immunoquantitation

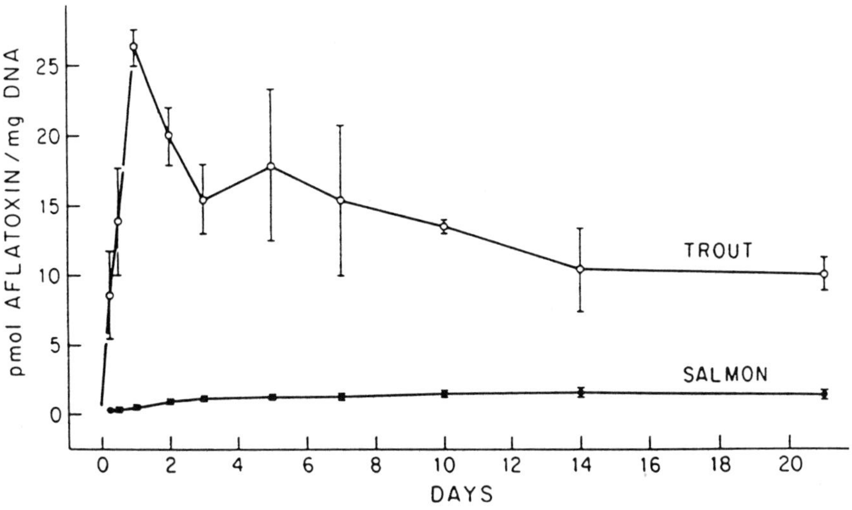

FIGURE 8. AFB_1–DNA binding and persistence in coho salmon and trout. Thirty fish were given $[^3H]AFB_1$ by i.p. injection and were killed in groups of three at various times after exposure. Data points are means ± SD, $n = 3$. Selected data from this figure are included in Table 5. Figure reproduced with permission from Ref. 62.

TABLE 5. Sensitivity of Trout Versus Coho Salmon to AFB_1–DNA Binding Using Various Routes of AFB_1 Exposure

Route of AFB_1 exposure	Dosing schedule	Time point	Relative AFB_1–DNA binding (trout:salmon)
Injection (i.p.)	10 μg AFB_1/kg	(6 h)	7:1
		(12 h)	14:1
		(24 h)	26:1
		(3 days)	15:1
		(21 days)	10:1
Diet	80 ppb 3 wk	(24 h)	18:1
Embryo	0.5 ppm 15–30 min	(24 h)	20:1
Hepatocyte	0.64 μM	(1 h)	20:1

Numbers in parentheses are time points at which DNA binding was determined after the final AFB_1 administration. For time course in i.p. studies, see Fig. 8.

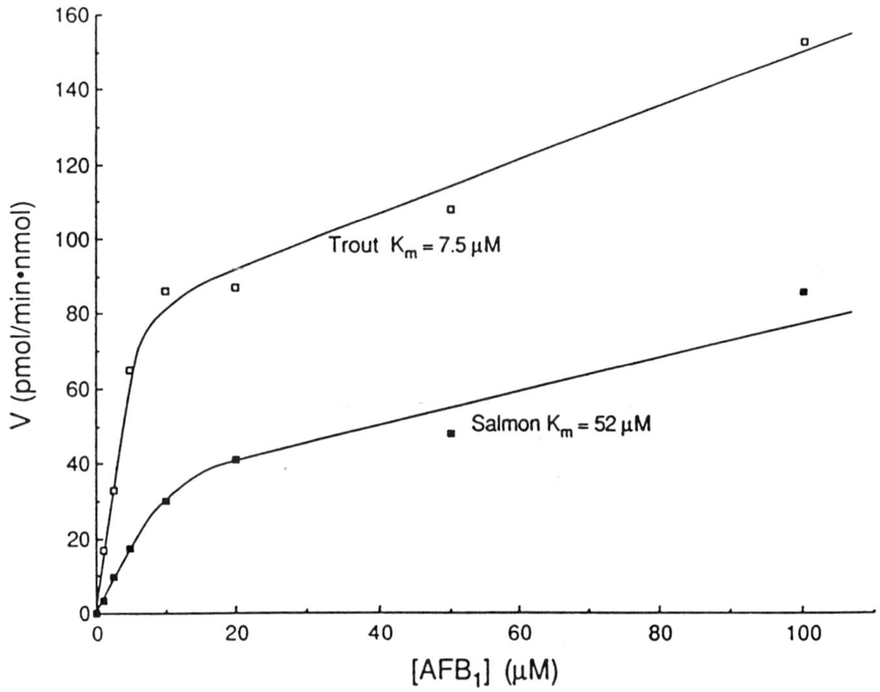

FIGURE 9. Enzyme kinetic studies of trout and salmon microsomal [^{3}H]-AFB_1–DNA adduct formation in vitro. Liver microsomes (0.2 mg protein, pooled from six fish) from trout and salmon were incubated with 1–100 µM tritiated AFB_1 in the presence of 0.4 mg of salmon testes DNA, and the AFB_1–DNA adducts were quantitated as described before [62]. The K_m values were calculated from double reciprocal plots (not shown). Figure reproduced with permission from Ref. 62.

of specific P-450 isoenzymes in the two species indicated that salmon liver microsomes have much lower concentrations of the major isoenzyme that activates AFB_1 to the epoxide (LM_2 in trout, Fig. 10). The major DNA adduct from both species was the N^7-Gua adduct, and the persistence of adducts in vivo was high in both salmon and trout compared to mammalian systems. The findings imply that the greater sensitivity of trout versus salmon is related to quantitative, but not qualitative, differences in AFB_1–DNA binding in the target organ, caused by higher concentrations and higher activity of the P-450 LM_2 isoenzyme in trout [62]. Since DNA repair activity is low in both salmonid species [62], it would appear that the enzymatic removal of bulky adducts is not a major factor in their differential response; the results contrast with data from rodent species, which indicate that different rates of DNA repair activity may be an important factor in interspecies variability to AFB_1.

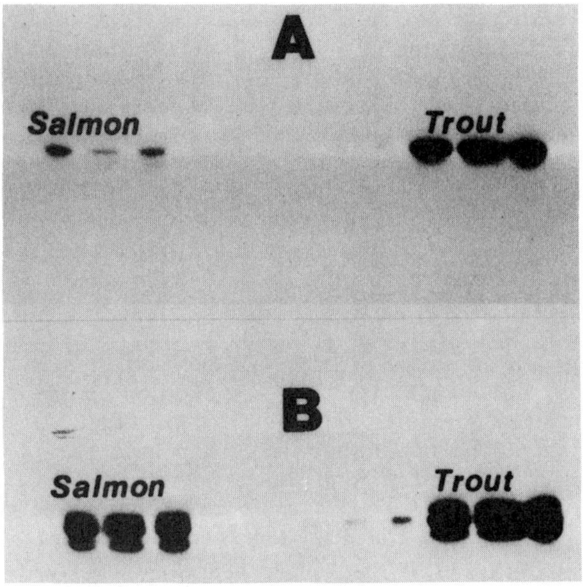

FIGURE 10. Western blots of trout and salmon liver microsomes. (A) Ten µg of microsomal protein from trout and salmon were separated by SDS-PAGE and electrophoretically transferred to nitrocellulose probed with rabbit anti-trout LM_2 IgG (20 µg/mL) followed by ^{125}I-labeled protein A (0.1 µCi/mL). The autoradiogram was developed at -85°C for 4 h. Figure (B) was obtained by reincubating the blot with a mixture of rabbit antibodies directed against four other trout cytochromes P-450 (LMC_1, LMC_3, LMC_4, and LMC_5). Reproduced with permission from Ref. 62.

B. Sensitivity of Trout Compared with Rodents

Species differences in response to AFB_1 have been studied extensively in rodents; studies have focused primarily on the sensitivity of rats and the relative resistance of mice and guinea pigs. Various biochemical endpoints studied include (1) detoxification via GSH conjugation and (2) rates of AFB_1–DNA adduction [63–67]. In rodents, conjugation with GSH and elimination via the bile represents the major detoxification pathway [65, 66]; an inverse relationship exists between GSH conjugation and AFB_1–DNA binding [67]. Thus, comparative metabolic studies in rats and mice have provided evidence that the much lower susceptibility of the latter species to AFB_1 may be related to higher rates of AFB_1-SG production and correspondingly lower AFB_1–DNA adduction. Although the rate of AFB_1-SG formation is lower in trout than in rats and mice [30], the extreme sensitivity of trout to AFB_1 is related to factors in addition to poor GSH detoxification, including small body size at the time of AFB_1 treatment and poor rates of DNA repair activity. Rodents possess active DNA repair systems [33, 68] and can exhibit high rates of activity in in vivo and in vitro unscheduled DNA synthesis (UDS) assays [69, 70]. DNA repair synthesis has been observed in isolated trout hepatocytes in response to DNA strand-breaking (clastogenic) agents such as bleomycin [71]. However, trout exhibit poor DNA repair activity and low rates of UDS after treatment with potent alkylating carcinogens such as AFB_1 [72, 73]. In the case of removal of thymine dimers, evidence suggests that the repair mechanism in fish cells involves a direct reversal process using a photoreactive protein [74–76], whereas repair systems in rodents proceed by mechanisms involving excision repair enzymes.

Another factor to consider in the sensitivity of trout to AFB_1 is the body size at the time of carcinogen exposure. Trout are exposed to AFB_1 at various developmental stages, but animals typically are exposed as embryos (microinjection or bath uptake) or as fingerlings (i.p. injection, dietary or gill exposure). It should be noted that trout have been reported as showing "high sensitivity" based on the minimal total dose of AFB_1 administered per individual required to achieve a certain tumor response. A more informative comparison is based on the dose required to achieve *equal tumor response* in rats and trout, where dose is defined as amount of test compound per kilogram of body weight (dose given) or number of carcinogen-DNA adducts formed (dose achieved). Molecular dosimetry studies in trout and other species have been undertaken to compare tumor and DNA binding data following long-term administration of the carcinogen. For example, the AFB_1 data presented in this chapter for trout (Fig. 5) were converted to risk versus adducts per 10^8 nucleotides and plotted together with tumor and steady-state AFB_1 adduct data from Fischer rats [77]. The interspecies molecular dosimetry plot (Fig. 11) exhibited a remarkably high correlation coefficient ($r^2 = 0.985$). In an interesting and somewhat provocative extension to this study, Bechtel [77] suggested that the excellent correlation in the molecular dosimetry plot provided a possible basis for use of adduct measurements for estimating human cancer risk. Thus, with Fig. 11 as a calibration curve, the minimum human virtual safe dose for aflatoxin was estimated to be 0.264 ng/kg/day. The "provocative" aspect relates, in part, to the fact that rats were treated continuously for 2 years, whereas trout were treated on a daily basis for 2 weeks and analyzed for tumors after 1 year. Previous data indicate that tumor incidences would have been considerably higher had trout been examined, as in the rat studies, after 2 years instead of 1 year. Clearly, the experimental time course would be predicted to adversely affect the correlation in Fig. 11. However, the fact that 1-year tumor data from trout correlated with 2-year results from rodents using molecular dosimetry provides perhaps the most definitive evidence for the relative "sensitivity" of trout versus rodents to AFB_1.

In addition to reviewing aflatoxin-DNA adduct formation, this chapter has highlighted the unique features of the trout tumor model that facilitate molecular dosimetry comparisons. Pertinent features include (1) poor DNA repair activity, which facilitates DNA adduct persistence, and (2) relatively low animal rearing costs, which make it practical to use sufficient numbers of animals in combined binding/tumor studies. By taking advantage of

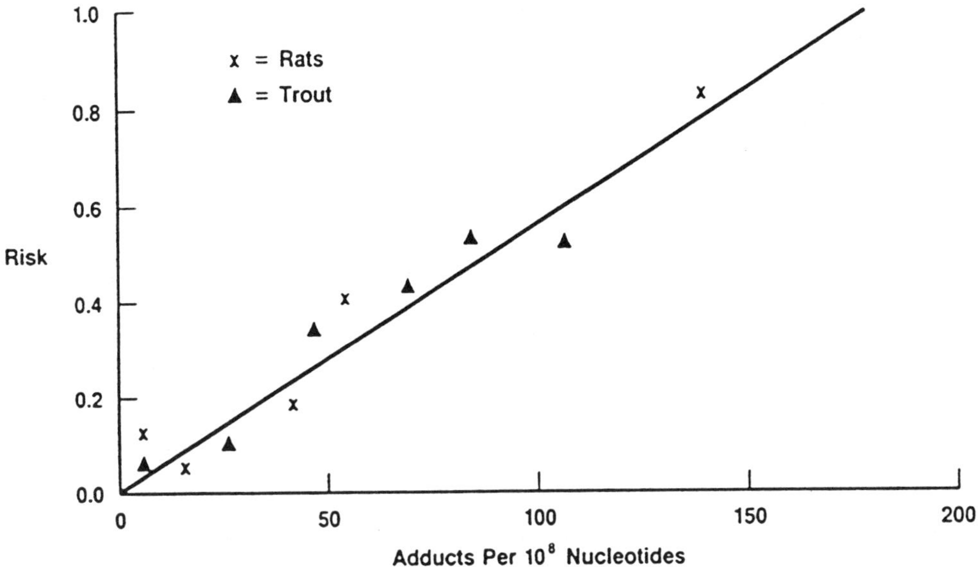

FIGURE 11. Molecular dosimetry of AFB_1 in trout and Fischer rats. Trout data from Fig. 5 were combined with tumor and steady-state AFB_1–DNA adduct data from rats. The data were fitted with a regression line assuming that risk is zero when the adduct concentration is zero. Reproduced from Ref. 77 with permission.

these features of the trout tumor model, current research is directed toward clarifying pertinent issues that relate to molecular dosimetry and its use in studies with aflatoxins and other carcinogens and anticarcinogens.

ACKNOWLEDGMENTS

We thank Wan Mohaiza for assistance in typing the manuscript. The original studies from our laboratory discussed in the text and reported in Figs. 2–10 and Table 1–5 were supported in part by U.S. Public Health Service grants ES03850, ES00210, CA34732, and CA39398.

REFERENCES

1. Sinnhuber, R. O., Hendricks, J. D., Wales, J. H., and Putnam, G. B. Neoplasms in rainbow trout, a sensitive animal model for environmental carcinogens. *Ann. N.Y. Acad. Sci.*, *298*:389–408, 1977.
2. Sinnhuber, R. O., Wales, J. H., Hendricks, J. D., Putnam, G. B., Nixon, J. E., and Pawlowski, N. E. Trout bioassay of mycotoxins.

In *Mycotoxins in Human and Animal Health* (J. D. Rodricks, Ed.), Pathotox, Chicago, Ill., 1977, pp. 731–744.

3. Hendricks, J. D. Chemical carcinogenesis in fish. In *Aquatic Toxicology* (L. J. Weber, Ed.), Raven, New York, 1982, pp. 149–211.
4. Croy, R. G., Nixon, J. E., Sinnhuber, R. O., and Wogan, G. N. Investigation of covalent aflatoxin B_1-DNA adducts formed *in vivo* in rainbow trout (*Salmo gairdneri*) embryos and liver. *Carcinogenesis, 1*:903, 1980.
5. Hendricks, J. D., Nixon, J. E., Loveland, P. M., Casteel, J. L., and Sinnhuber, R. O. High dietary protein promotes embryo-induced aflatoxin carcinogenesis in rainbow trout. *Fed. Proc., 40*:948, 1981.
6. Whitham, M., Nixon, J. E., and Sinnhuber, R. O. Liver DNA bound *in vivo* with aflatoxin B_1 as a measure of hepatocarcinoma initiation in rainbow trout. *J. Natl. Cancer Inst., 68*:623, 1982.
7. Nixon, J. E., Hendricks, J. D., Pawlowski, N. E., Pereira, C. B., Sinnhuber, R. O., and Bailey, G. S. Inhibition of aflatoxin B_1 carcinogenesis in rainbow trout by flavone and indole compounds. *Carcinogenesis, 5*:615–619, 1984.
8. Goeger, D. E., Shelton, D. W., Hendricks, J. D., and Bailey, G. S. Mechanisms of anti-carcinogenesis by indole-3-carbinol: effect on the distribution and metabolism of aflatoxin B_1 in rainbow trout. *Carcinogenesis, 7*:2025–2031, 1986.
9. Goeger, D. E., Shelton, D. W., Hendricks, J. D., Pereira, C., and Bailey, G. S. Comparative effect of dietary butylated hydroxyanisole and β-naphthoflavone on aflatoxin B_1 metabolism, DNA adduct formation, and carcinogenesis in rainbow trout. *Carcinogenesis, 9*:1793–1800, 1988.
10. Dashwood, R. H., Arbogast, D. N., Fong, A. T., Hendricks, J. D., and Bailey, G. S. Mechanisms of anti-carcinogenesis by indole-3-carbinol: detailed in vivo DNA binding dose-response studies after dietary administration with aflatoxin B_1. *Carcinogenesis, 9*:427–432, 1988.
11. Fong, A. T., Hendricks, J. D., Dashwood, R. H., Van Winkle, S., and Bailey, G. S. Formation and persistence of ethylguanines in liver DNA of rainbow trout (*Salmo gairdneri*) treated with diethylnitrosamine by water exposure. *Food Chem. Toxicol., 26*:699–704, 1988.
12. Dashwood, R. H., Fong, A. T., Hendricks, J. D., and Bailey, G. S. Inhibition by indole-3-carbinol acid products of AFB_1–DNA binding and tumorigenesis in trout following embryo coinjection. (In preparation.)
13. Coulombe, R. A., Shelton, D. W., Sinnhuber, R. O., and Nixon, J. E. Comparative mutagenicity of aflatoxins using a *Salmonella*/trout hepatic enzyme activation system. *Carcinogenesis, 3*:1261–1264, 1982.
14. Bailey, G. S., Taylor, M. J., and Selivonchik, D. P. Aflatoxin B_1 metabolism and DNA binding in isolated hepatocytes from rainbow trout (*Salmo gairdneri*). *Carcinogenesis, 3*:511–518, 1982.
15. Shelton, D. W., Coulombe, R. A., Pereira, C. B., Casteel, J. L., and Hendricks, J. D. Inhibitory effect of Aroclor 1254 on aflatoxin-initiated carcinogenesis in rainbow trout and mutagenesis using a

Salmonella/trout hepatic activation system. *Aquatic Toxicol.*, *3*:229–238, 1983.

16. Coulombe, R. A., Jr., Bailey, G. S., and Nixon, J. E. Comparative activation of aflatoxin B_1 to mutagens by isolated hepatocytes from rainbow trout (*Salmo gairdneri*) and coho salmon (*Oncorhynchus kisutch*). *Carcinogenesis*, *5*:29–33, 1984.
17. Loveland, P. M., Wilcox, J. S., Pawlowski, N. E., and Bailey, G. S. Metabolism and DNA binding of aflatoxicol and aflatoxin B_1 *in vivo* and in isolated hepatocytes from rainbow trout (*Salmo gairdneri*). *Carcinogenesis*, *8*:1065–1070, 1987.
18. Loveland, P. M., Wilcox, J. S., Hendricks, J. D., and Bailey, G. S. Comparative metabolism and DNA binding of aflatoxin B_1, aflatoxin M_1, aflatoxicol and aflatoxicol-M_1 in hepatocytes from rainbow trout (*Salmo gairdneri*). *Carcinogenesis*, *9*:441–446, 1988.
19. Wales, J. H., Sinnhuber, R. O., Hendricks, J. D., Nixon, J. E., and Eisele, T. A. Aflatoxin B_1 induction of hepatocellular carcinoma in the embryos of rainbow trout (*Salmo gairdneri*). *J. Natl. Cancer Inst.*, *60*:1133–1139, 1978.
20. Sinnhuber, R. O., Lee, D. J., Wales, J. H., Landers, M. K., and Keyl, A. C. Hepatic carcinogenesis of aflatoxin M_1 in rainbow trout (*Salmo gairdneri*) and its enhancement by cyclopropene fatty acids. *J. Natl. Cancer Inst.*, *53*:1285–1288, 1974.
21. Wogan, G. N., Paglialunga, S., and Newberne, P. M. Carcinogenic effects of low dietary levels of aflatoxin B_1 in rats. *Food Cosmet. Toxicol.*, *12*:681–685, 1974.
22. Swenson, D. H., Miller, E. C., and Miller, J. A. Aflatoxin B_1-2,3-oxide: evidence for its formation in rat liver *in vivo* and by human liver microsomes *in vitro*. *Biochem. Biophys. Res. Commun.*, *60*:1036–1043, 1974.
23. Swenson, D. H., Lin, J. K., Miller, E. C., and Miller, J. A. Aflatoxin B_1-2,3-oxide as a probable intermediate in the covalent binding of aflatoxins B_1 and B_2 to rat liver DNA and ribosomal RNA *in vivo*. *Cancer Res.*, *37*:172–181, 1977.
24. Williams, D. E., and Buhler, D. R. Purified form of cytochrome P-450 from rainbow trout with high activity toward conversion of aflatoxin B_1 to aflatoxin B_1-2,3-epoxide. *Cancer Res.*, *43*:4752–4756, 1983.
25. Loveland, P. M., Nixon, J. E., and Bailey, G. S. Glucuronides in bile of rainbow trout (*Salmo gairdneri*) injected with ^3H-aflatoxin B_1 and the effects of dietary β-naphthoflavone. *Comp. Biochem. Physiol.*, *78C*:13, 1984.
26. Melancon, M. J., and Lech, J. J. Isolation and identification of a polar metabolite of tetrachlorobiphenyl from bile of rainbow trout exposed to ^{14}C-tetrachlorobiphenyl. *Bull. Environ. Contam. Toxicol.*, *15*:181, 1976.
27. Melancon, M. J., and Lech, J. J. Distribution and biliary excretion products of di-2-ethylhexyl phthalate in rainbow trout. *Drug Metab. Dispos.*, *4*:112, 1976.
28. Statham, C. N., and Lech, J. J. Metabolism of 2',5-dichloro-4'-nitrosalicylanilide (Bayer 73) in rainbow trout (*Salmo gairdneri*). *J. Fish. Res. Bd. Can.*, *32*:515, 1975.

29. Dashwood, R. H., Uyetake, L., Fong, A. T., Hendricks, J. D., and Bailey, G. S. *In vivo* disposition of the natural anti-carcinogen indole-3-carbinol after po administration to rainbow trout. *Food Chem. Toxicol.*, *27*:385–392, 1989.
30. Valsta, L. M., Hendricks, J. D., and Bailey, G. S. The significance of glutathione conjugation for aflatoxin B_1 metabolism in rainbow trout and coho salmon. *Food Chem. Toxicol.*, *26*:129–135, 1988.
31. Dashwood, R. H., Arbogast, D. N., Fong, A. T., Pereira, C., Hendricks, J. D., and Bailey, G. S. Quantitative inter-relationships between aflatoxin B_1 carcinogen dose, indole-3-carbinol anti-carcinogen dose, target organ DNA adduction and final tumor response. *Carcinogenesis*, *10*:175–181, 1989.
32. Swenberg, J. A. Quantitative relationships between DNA adducts and carcinogenesis. *Proceedings*, American Association for Cancer Research '89, San Francisco, Calif., Vol. 30, p. 658, 1989.
33. Swenberg, J. A., Richardson, F. C., Tyeryar, L., Deal, F., and Boucheron, J. The molecular dosimetry of DNA adducts formed by continuous exposure of rats to alkylating hepatocarcinogens. *Prog. Exp. Tumor Res.*, *31*:42–51, 1987.
34. Aaron, C. S. Molecular dosimetry as a tool in genetic risk assessment. In *Genetic Risk Assessment* (A. D. Bloom and P. K. F. Poskitt, Eds.), Monograph No. 1, Environmental Health Institute, March of Dimes, New York, 1988, pp. 53–78.
35. van Zeeland, A. A. Molecular dosimetry of alkylating agents: quantitative comparisons of genetic effects on the basis of DNA adduct formation. *Mutagenesis*, *3*:179–191, 1988.
36. Dashwood, R. H., Loveland, P. M., Fong, A. T., Hendricks, J. D., and Bailey, G. S. Combined *in vivo* DNA binding and tumor-dose response studies to investigate the molecular dosimetry concept. *Environ. Mol. Mutagen.*, *14*:44, 1989.
37. Dashwood, R. H., Loveland, P. M., Fong, A. T., Hendricks, J. D., and Bailey, G. S. Combined *in vivo* DNA binding and tumor-dose response studies to investigate the molecular dosimetry concept. In *Mutation and the Environment* — Part D: Carcinogenesis (M. Mendelsohn and R. J. Albertini, Eds.), Wiley-Liss, New York, pp. 335–344, 1990.
38. Hsieh, L. L., Hsu, S. W., Chen, D. S., and Santella, R. M. Immunological detection of aflatoxin B_1-DNA adducts formed *in vivo*. *Cancer Res.*, *48*:6328–6331, 1988.
39. Santella, R. M. Application of new techniques for the detection of carcinogen adducts to human population monitoring. *Mutat. Res.*, *205*:271–282, 1988.
40. Randerath, K., Randerath, E., Danna, T. F., van Golen, K. L., and Putnam, K. L. A new sensitive ^{32}P-postlabeling assay based on the specific enzymatic conversion of bulky DNA lesions to radiolabeled dinucleotides and nucleoside 5'-monophosphates. *Carcinogenesis*, *10*: 1231–1239, 1989.
41. Williams, D. E., Dashwood, R. H., Hendricks, J. D., and Bailey, G. S. Anticarcinogens and tumor promoters in food. In *Food Toxicology: A perspective on the Relative Risks* (S. L. Taylor and R. A. Scanlan, Eds.), Marcel Dekker, New York, 1989, pp. 101–150.

42. Wattenberg, L. W. Chemoprevention of cancer. *Cancer Res.*, *45*:1–8, 1985.
43. Wattenberg, L. W., and Loub, W. D. Inhibition of polycyclic aromatic hydrocarbon-induced neoplasia by naturally occurring indoles. *Cancer Res.*, *38*:1410–1413, 1978.
44. Shertzer, H. G. Protection by indole-3-carbinol against covalent binding of benzo[*a*]pyrene metabolites to mouse liver DNA and protein. *Food Chem. Toxicol.*, *21*:31–35, 1983.
45. Shertzer, H. G. Indole-3-carbinol protects against covalent binding of benzo[*a*]pyrene and *N*-nitrosodimethylamine metabolites to mouse liver macromolecules. *Chem.-Biol. Interactions*, *48*:81–90, 1984.
46. Wong, J. J., and Hsieh, D. P. H. Mutagenicity of aflatoxins related to their metabolism and carcinogenic potential. *Proc. Natl. Acad. Sci. USA*, *73*:2241–2244, 1976.
47. Dashwood, R. H., and Combes, R. D. Comparisons between *in vivo* and *in vitro* systems: the *in vivo* equivalent dose concept. *Mutagenesis*, *3*:527–531, 1988.
48. Fahmy, M. J., Fahmy, O. G., and Swenson, D. H. Aflatoxin B_1-2,3-dichloride as a model of the active metabolite of aflatoxin B_1 in mutagenesis and carcinogenesis. *Cancer Res.*, *38*:2608–2161, 1978.
49. Misra, R. P., Muench, K. F., and Humayun, M. Z. Covalent and non-covalent interactions of aflatoxin with defined deoxyribonucleic acid sequences. *Biochemistry*, *22*:3351–3359, 1983.
50. Muench, K. F., Misra, R. P., and Humayun, M. Z. Sequence specificity in aflatoxin B_1–DNA interactions. *Proc. Natl. Acad. Sci. USA*, *80*:6–10, 1983.
51. Refolo, L. M., Conley, M. P., Sambamurti, K., Jacobsen, J. S., and Humayun, M. Z. Sequence context effects in DNA replication blocks induced by aflatoxin B_1. *Proc. Natl. Acad. Sci. USA*, *82*:3096–3100, 1985.
52. Clewell, D. B., and Helinsky, D. R. Properties of a supercoiled deoxyribonucleic acid-protein relaxation complex and strand specificity of the relaxation event. *Biochemistry*, *9*:4428–4440, 1970.
53. Mattes, W. B., Hartley, J. A., and Kohn, K. W. Mechanism of DNA strand breakage by piperidine at sites on N^7-alkylguanine. *Biochim. Biophys. Acta*, *868*:71–76, 1986.
54. Benasutti, M., Ejadi, S., Whitlow, M. D., and Loechler, E. L. Mapping the binding site of aflatoxin B_1 in DNA: systematic analysis of the reactivity of aflatoxin B_1 with guanines in different DNA sequences. *Biochemistry*, *27*:472–481, 1988.
55. Marien, K., Moyer, R., Loveland, P. M., van Holde, K., and Bailey, G. S. Comparative binding and sequence interaction specificities of aflatoxin B_1, aflatoxicol, aflatoxin M_1 and aflatoxicol M_1 with purified DNA. *J. Biol. Chem.*, *262*:7455–7462, 1987.
56. Boles, T. C., and Hogan, M. E. Site specific carcinogen binding to DNA. *Proc. Natl. Acad. Sci. USA*, *81*:5623–5627, 1984.
57. Boles, T. C., and Hogan, M. E. High resolution mapping of carcinogen binding sites on DNA. *Biochemistry*, *25*:3039–3043, 1986.
58. Lobanenkov, V. V., Plumb, M., Goodwin, G. H., and Grover, P. L. The effect of neighbouring bases on G-specific DNA clevage mediated

by treatment with the anti-diol epoxide of benzo[*a*]pyrene *in vitro*. *Carcinogenesis*, *7*:1689–1695, 1986.

59. Reardon, D., Bigger, A., Strandberg, J., Haruhiko, Y., Jerina, D., and Dipple, A. Sequence selectivity in the reaction of optically active hydrocarbon dihydrodiol epoxides with rat H-*ras* DNA. *Chem. Res. Toxicol.*, *2*:12–14, 1989.
60. Dickerson, R. E. Base sequence and helix structure variation in B- and A-DNA. *J. Mol. Biol.*, *166*:419–441, 1983.
61. Pullman, A., and Pullman, B. Molecular electrostatic potential of the nucleic acids. *Quart. Rev. Biophys.*, *14*:289–380, 1981.
62. Bailey, G. S., Williams, D. E., Wilcox, J. S., Loveland, P. M., Coulombe, R. A., and Hendricks, J. D. Aflatoxin B_1 carcinogenesis and its relation to DNA adduct formation and adduct persistence in sensitive and resistant salmonid fish. *Carcinogenesis*, *9*:1919–1926, 1988.
63. Lotlikar, P. D., Jhee, E. C., Insetta, S. M., and Clearfield, M. S. Modulation of microsome-mediated aflatoxin B_1 binding to exogenous and endogenous DNA by cystolic glutathione *S*-transferases in rat and hamster livers. *Carcinogenesis*, *5*:269–276, 1984.
64. Ueno, I., Friedman, L., and Stone, C. L. Species differences in the binding of aflatoxin B_1 to hepatic macromolecules. *Toxicol. Appl. Pharmacol.*, *52*:177–180, 1980.
65. Degen, G. H., and Neumann, H.-G. Differences in aflatoxin B_1 susceptibility of rat and mouse are correlated with the capacity *in vitro* to inactivate aflatoxin B_1-epoxide. *Carcinogenesis*, *2*:299–306, 1981.
66. Degen, G. H., and Neumann, H.-G. The major metabolite of aflatoxin B_1 in the rat is a glutathione conjugate. *Chem.-Biol. Interactions*, *22*:239–255, 1978.
67. Raj, H. G., Clearfield, M. S., and Lotlikar, P. D. Comparative kinetic studies on aflatoxin B_1-DNA binding and aflatoxin B_1-glutathione conjugation with rat and hamster livers *in vitro*. *Carcinogenesis*, *5*: 879–884, 1984.
68. Kensler, T. W., Egner, P. A., Davidson, N. E., Roebuck, B. D., Pikul, A., and Groopman, J. D. Modulation of aflatoxin-N^7-guanine formation and hepatic tumorigenesis in rats fed ethoxyquin: role of induction of glutathione *S*-transferases. *Cancer Res.*, *46*:3924–3931, 1986.
69. Ashby, J., Lefevre, P. A., Burlinson, B., and Penman, M. G. An assessment of the in vivo rat hepatocyte DNA-repair assay. *Mutat. Res.*, *156*:1–18, 1985.
70. Waters, R., Ashby, J., Burlinson, B., Lefevre, P., Barrett, R., and Martin, C. Unscheduled DNA synthesis. In *Report of the UKEMS Sub-Committee on Guidelines for Mutagenicity Testing Part II* (B. J. Dean, Ed.), U.K. Environmental Mutagen Society, Swansea, U.K., 1984, pp. 63–87.
71. Miller, M. R., Blair, J. B., and Hinton, D. E. DNA repair synthesis in isolated rainbow trout liver cells. *Carcinogenesis*, *10*:995–1001, 1989.
72. Walton, D. G., Acton, A. B., and Stich, H. F. DNA repair synthesis following exposure to chemical mutagens in primary liver, stomach,

and intestinal cells isolated from rainbow trout. *Cancer Res.*, *44*: 1120–1121, 1984.
73. Walton, D. G., Acton, A. B., and Stich, H. F. DNA repair synthesis in cultured mammalian and fish cells following exposure to chemical mutagens. *Mutat. Res.*, *124*:153–161, 1983.
74. Regan, J. D., Carrier, W. L., Samet, C., and Olla, B. L. Photoreactivation in two closely related marine fishes having different longevities. *Mech. Ageing Dev.*, *18*:59–66, 1982.
75. Yew, F. H., and Chang, L. M. DNA replication and repair in Tilapia cells. I. The effect of UV radiation. *J. Cell Sci.*, *72*:213–226, 1984.
76. Mano, Y., Kator, K., and Egami, N. Photoreactivation and excision repair of thymidine dimers in UV-cultured fish cells. *Radiat. Res.*, *90*:501–508, 1982.
77. Bechtel, D. H. Molecular dosimetry of hepatic aflatoxin B_1-DNA adducts: linear correlation with hepatic cancer risk. *Req. Toxicol. Pharmacol.*, *10*:74–81, 1989.

8

HUMAN EXPOSURE TO AFLATOXINS—BIOLOGICAL MONITORING

HERMAN AUTRUP* and JUDITH L. AUTRUP *Fibiger Institute, Copenhagen, Denmark*

I. AFLATOXINS AND HUMAN HEALTH

Aflatoxin B_1 (AFB_1) is one of the most potent hepatocarcinogens known; it has been classified by IARC as a group I carcinogen [1]. Aflatoxin B_1 has also been found to induce cancer in other organs in experimental animals, depending on the species and route of administration [1]. A positive correlation has been established between estimated aflatoxin intake determined from the level of aflatoxin of either market food samples or cooked food samples and the incidence of liver cancer in a number of studies in African and Asian countries [2–4].

Inhalation of aflatoxin-contaminated dust in the working environment may be another route of exposure to aflatoxin. An increased incidence of liver cancer has been reported among Danish male workers in livestock feed processing companies, and aflatoxin was suggested as the causative agent [5]. A causative association of aflatoxin exposure and human cancer has also been established in Dutch workers in an oil mill processing peanut and linseed oils. In contrast to the Danish study, an increase in respiratory cancers was evident [6]. However, the assessment of aflatoxin exposure was not based upon environmental or biological monitoring. A single case report has associated aflatoxin exposure with lung cancer. In this case, a blue fluorescent material was detected in the lung tissue [7].

Exposure to high levels of aflatoxins may result in an acute toxic effect. Although outbreak of aflatoxicosis is quite common in livestock, there have only been a few reports on aflatoxicosis in humans [8].

In developing countries, a relationship between exposure to aflatoxins and kwashiorkor has been established. Aflatoxins have been detected more frequently and at higher concentrations in the sera of kwashiorkor patients than in other children, and autopsy liver samples from individuals afflicted with kwashiorkor contained AFB_1 [9]. An association between

**Present affiliation*: Institute of Environmental and Occupational Medicine, University of Aarhus, Aarhus, Denmark.

AFB_1 exposure and Reye's syndrome has been suggested, but AFB_1 was not recoverable from the livers of all the diseased children [10].

It has previously been suggested that the role of aflatoxin in human liver carcinogenesis was an immunosuppressive effect rather than a genotoxic effect. The AFB-exposed individuals may be more susceptible to infection with hepatitis B virus, which is considered another major etiological factor in liver carcinogenesis [11]. Animal studies support this hypothesis, as aflatoxins have been shown to be immunomodulators in both experimental and domestic animals [12, 13].

II. METABOLISM OF AFLATOXIN B_1

Aflatoxin B_1 is metabolized by the mixed-function oxidase systems to a number of reduced and hydroxylated metabolites (Fig. 1) [14]. The hydroxylated metabolites, for example, aflatoxin M_1 (AFM_1) and aflatoxin P_1 (AFP_1), can be further metabolized by conjugating enzymes, forming glucuronides and sulfates. These metabolites are more hydrophilic and are excreted into the urine of monkeys, mice, and rats treated with AFB_1 [15]. Aflatoxin M_1 was the first metabolite detected in urine from humans exposed to AFB_1 [16].

Epoxidation of the 8,9 double bond in the furan moiety of AFB_1 creates the ultimate carcinogenic metabolite. This metabolite is very reactive, and the principal DNA adduct is formed by reaction between this epoxide and the N-7 position of guanine. The major adduct has been identified as

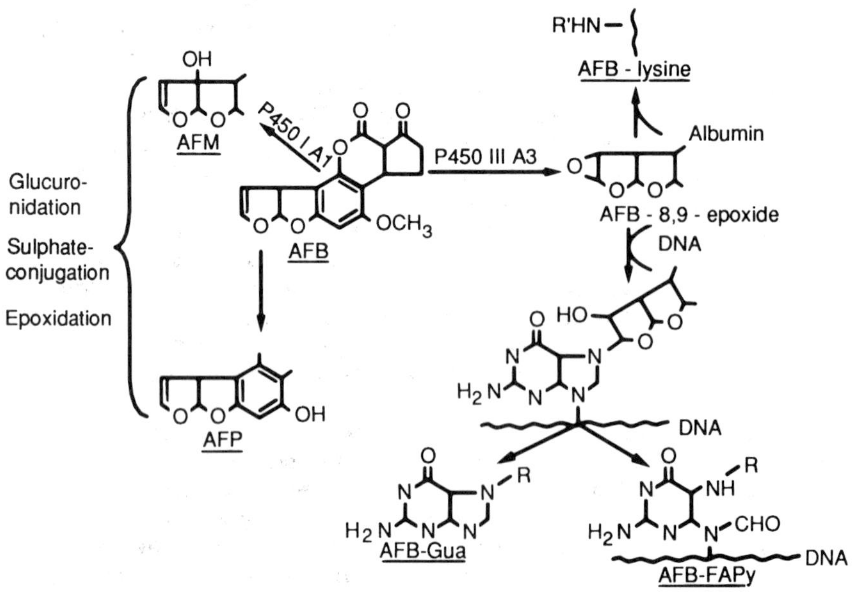

FIGURE 1. Metabolic pathways of AFB_1 to metabolites relevant for biological monitoring.

8,9-dihydro-8-(guan-7-yl)-9-hydroxy-AFB_1 [17]. In addition to the cytochrome P-450-mediated metabolism, the AFB_1-8,9-epoxide and the same AFB_1-DNA adduct can also be formed via the prostaglandin H synthase oxidative pathway, for example, when AFB_1 is incubated with ram seminal vesicle microsomes [18].

The amount of the ultimate carcinogen, AFB_1-8,9-epoxide, in the target tissues may be modified by many endogenous and exogenous factors, and subsequently a modification of the biological effect may occur. The 8,9-epoxide is a good substrate for glutathione transferases, and it has been postulated that a high level of glutathione transferases protects against the liver carcinogenic activity of AFB by diminishing the amount of the active metabolite available for reaction with DNA [19]. Thiol conjugates of aflatoxin have been detected in urine of rats and hamsters treated with AFB_1 [20].

The major hydroxylated metabolites retain the 8,9 double bond, and an AFB_1-8,9-epoxide can be formed by further metabolism. However, the biological activity of these hydroxylated products, for example, aflatoxin M_1, is less than that of the parent compound [21].

Cytochrome P-450 IIIA3 is the principal P-450 enzyme involved in the bioactivation of AFB_1 [22], whereas P-450 IA1 and IA2, which are important in the metabolism of polycyclic aromatic hydrocarbons, do not play any significant role in the activation of AFB_1 to mutagenic metabolites [23]. Aflatoxin B_1 has been transformed to a mutagenic metabolite in Chinese hamster V79 cells constitutively expressing P-450 IIB1 [24].

Several potential target tissues for the carcinogenic action of AFB_1, including the liver, are able to metabolize AFB_1 to DNA-binding metabolites. Metabolic studies have been conducted in intact animals, in cells, and in subcellular fractions isolated from human tissues and experimental animals.

It is well recognized that the liver is the major target organ for the carcinogenic action of AFB_1, and AFB_1 is also extensively metabolized to the ultimate carcinogenic form in the liver. Using human liver microsomes, AFM_1 and aflatoxin Q_1 (AFQ_1) were identified as the major metabolites extractable in organic solvent [25]. It is generally assumed that an association exists between the amount of specific DNA adducts in the target and the carcinogenic response. When the binding level of AFB_1 to DNA was compared in hepatocytes isolated from male and female mice, rats, and humans, the highest level was observed in hepatocytes from male rats, which are the most susceptible to AFB_1 liver carcinogenesis [26].

Aflatoxins are also metabolized by extrahepatic tissues, such as cells in the upper airways, including olfactory cells from animals [27, 28]. These cells may be targets for AFB_1 toxicity as they are the first tissues to come into contact with AFB_1 present in airborne particulates [29]. Furthermore, it has been suggested that AFB_1-contaminated feed may be an important etiological factor in the development of nasal tumors in cattle [27]. Aflatoxin M_1 is the major hydroxylated metabolite formed by respiratory tissues. In addition, aflatoxicol (AFL) and AFQ_1 were detected when tracheal cells were incubated with AFB_1, whereas AFL could not be detected in olfactory tissues. Explant cultures of human bronchus and trachea tissue also metabolize AFB_1, and the major DNA adduct is identical to the adduct detected in rat liver DNA of rats treated with AFB_1

[30, 31]. In a comparative study, the binding levels of AFB_1 to tracheal DNA did not correlate with the relative susceptibilities of different species to the carcinogenic action of AFB_1. The highest binding level was seen in dogs and hamsters, which are insensitive to the carcinogenic action of AFB_1, whereas rats and monkeys had lower binding levels [31].

Aflatoxin B_1 is metabolized to DNA-binding metabolites by human bronchus, bladder, colon, and esophagus explant cultures [30] and by human fetal liver explant. A detectable level of AFB_1-DNA adducts was seen in 3-month-old fetuses [32]. Very few studies on the formation of hydroxylated metabolites have been conducted in human tissues. Aflatoxin M_1 and AFP_1 are the major metabolites formed by cultured human colon [33].

No qualitative differences could be detected between human tissues and animal tissues from species in which AFB_1 is carcinogenic.

III. METHODS TO DETECT HUMAN EXPOSURE TO AFLATOXIN

Human populations are exposed to aflatoxins by the consumption of commodities that are contaminated by aflatoxins or by occupational exposure. Exposure assessment has mostly been done by analysis of the commodities or by environmental analysis. The development of new, sensitive analytical procedures has made it possible to assess an individual's exposure to aflatoxin [34, 35]. Several of the primary metabolites of AFB_1 have been detected in human body fluids, urine, blood, and mother's milk, and AFB_1-macromolecule adducts have been detected in both blood cells and liver tissue (Table 1).

TABLE 1. Methods to Detect Human Exposure to Aflatoxins

Biological material	Method	Ref.
Urine	TLC	16, 36
	ELISA	35, 37, 38
	HPLC, TLC	39
Blood	ELISA	40–42
	HPLC	43–45
Mother's milk	HPLC	44
	ELISA	46
Serum albumin	ELISA	47, 48
Liver DNA adduct	ELISA	49, 50
AFB_1-Gua (in urine)	HPLC	51–53

Quantitation of AFB_1-macromolecule adducts is considered the biologically most relevant measure of exposure, as it is an indirect measure of the dose of the ultimate carcinogenic form, AFB_1-8,9-epoxide. However, the amount of AFB_1-DNA adducts in lymphocytes and AFB_1-serum protein adducts is not a measure of the amount of the active metabolite reaching the target organ. An association between the target dose in experimental animals and these binding parameters has to be established before binding to serum proteins can be used in risk assessment.

A. Urine

Common methods for detection and quantitation of AFB_1 in urine include thin-layer chromatography (TLC) or high-pressure liquid chromatography (HPLC) analyses of organic solvent extracts of urine samples. Identification of the various metabolites is based upon cochromatography of known aflatoxin standards. The chromatographic methods are quite time-consuming, and their use in routine monitoring of human populations has been limited. In the last decade, competitive enzyme-linked immunosorbent assay (ELISA) and radioimmunoassay (RIA) methods using monoclonal or polyclonal antibodies against AFM_1 or AFB_1 have been developed and used in a number of human monitoring studies [54]. Recently, a modification of the ELISA procedure, time-resolved fluoroimmunoassay, has been developed for quantitation of AFB_1. This method has not been used on biological material, but it is assumed to be simpler and quicker than the traditional ELISA assays. However, the europium-labeled antibody is reported to have slightly lower affinity [55].

Metabolites

In addition to AFB_1, many of the primary AFB metabolites such as AFM_1 and AFL_1, have been detected in human urine. Quantitation of these metabolites as a measure of exposure may be an underestimate, as most of the metabolites excreted in the urine are conjugated as sulfates, glucuronides, or glutathione derivatives. The conjugated metabolites may not be detected by the conventional HPLC/TLC procedures, but they may be recognized by the antibodies used in competitive ELISA. However, a higher concentration of the conjugated metabolites may be required to achieve the same level of competition as the primary metabolite itself. Therefore, the urine samples should be treated with deconjugating enzymes, such as sulfatase and glucuronidase, to release the primary metabolites. Unfortunately, this step has not been included in many of the studies.

The advantage of the chromatographic methods is that the identity of the metabolites can be verified by cochromatography of standards, while the advantage of the ELISA is that the urine samples can be used directly in a competitive assay. When the results in the competitive RIA assay were compared with the results obtained by the classical TLC method, the TLC method gave an overestimate by 11%. The sensitivity of the ELISA assay has been increased by concentrating the urine on a Sep-Pak cartridge or on an AFB affinity column prior to the quantitative analysis by ELISA. A number of monoclonal and polyclonal antibodies recognizing AFB_1,

AFM_1, and/or AFQ_1 have been developed [54], and these antibodies have been used to monitor human urine samples for AFB_1 or its metabolites. Although the antibodies are raised against specific aflatoxins, some of the antibodies also recognize the other metabolites, but the affinity is much less. Consequently, the results should be expressed as aflatoxin B_1 equivalents, the amount of AFB_1 giving the same level of inhibition.

A number of studies to detect AFB_1 and its metabolites in human urine have been performed in different countries (Table 2). The level of exposure to AFB_1 was highest in African and Asian countries. A comprehensive study was conducted in the Quangxi region of China, an area with a high incidence of liver cancer [37]. In this study, food samples were collected and analyzed for AFB_1 in order to assess the exposure to AFB_1 by the traditional analytical procedures and to obtain some information on the pharmacokinetics of AFB_1 in humans. A highly specific polyclonal antiserum against AFM_1 was used in a RIA assay. Aflatoxin M_1 was the major urinary metabolite, and the excreted amount of AFM_1 accounted for 1–4% of the consumed amount of AFB_1 [38]. Fifteen percent of the dietary intake

TABLE 2. Detection of Aflatoxin Metabolites in Urine Samples

Country (no. samples)	Percent samples positive	Metabolite	Amount (ng/mL)	Method	Ref.
Nigeria (161)	42	AFB2A	0.70	TLC & HPLC	39
	11	AFM	0.89		
	3	AFB	7.24		
	9	AFL	0.64		
China (252)	n.r.	AFM	0.03–0.3	ELISA	38
Zimbabwe (1228)	4.9	AFM	4.2	TLC	36
Gambia (63)	50	AFB-eq.	0–145	ELISA	35
Philippines (33)	37	AFB-eq.	0–4.25	ELISA	35
Singapore (50)	12	AFB-eq.	0–2.5	ELISA	35
Lyon, France (32)	15	AFB-eq.	0–0.6	ELISA	35
Denmark (80)	90	AFB-eq.	0–3.0	ELISA	56, 57

of aflatoxin is excreted per day [35]. A good correlation between the dietary intake of AFB_1 and the urinary excretion of AFM_1 was established. The largest study on human exposure to AFB_1 to date was conducted in Zimbabwe, and the TLC method was used for quantitation. The average level of exposure on the national level was 4.2 ng/mL urine, but large regional differences could be observed, ranging from 0.7% to 8.0% of the urine samples being positive for AFB_1. The lowest rate of exposure was in an area located at an altitude greater than 2500 m and having an average temperature of 17°C, that is, an area not conducive to fungal growth and toxin production. Aflatoxin M_1 was also detected in the urine of individuals living in an urban area, where most of the consumed food had been processed [36]. In an occupational study to assess the exposure to AFB_1 in longshoremen in Denmark, it was discovered that the control group also excreted some material that was recognized by the monoclonal antibody against AFB_1 [56].

Aflatoxinlike Antigenic Material

Urine samples from 78 individuals on a normal Danish diet were analyzed for AFB_1 by ELISA. The expected exposure to AFB_1 in Denmark is low, and imported rice, corn, dried fruit, and nuts are considered the major known sources. However, most of the individuals on a normal diet excreted aflatoxinlike antigenic material (ALAM). The level of ALAM ranged from 0 to 6 ng AFB-equiv/mg creatinine, which corresponds to 0–150 ng AFB-equiv/kg body weight. The results of a follow-up study indicate that the source of ALAM has been partly removed (unpublished results). This may be ascribed to stricter control of the raw material used in the production of various types of animal feed.

A dietary restriction study suggested that the high levels of ALAM were caused by the consumption of dairy products and meat, whereas low levels of ALAM were seen following the consumption of fresh seafood, fruits, and vegetables. The antibody used in this study recognized the AFB_1 molecules with possible modifications in either the 8,9 positions or at the carbon atom in the α position to the cyclopentone group. On the basis of the epitope mapping and UV spectrum of the antigenic material, it was proposed that ALAM contains the AFB_1 moiety conjugated to amino acid residues [57]. However, the molecular structure of ALAM has not been verified by chemical analysis. HPLC analysis of the antigenic material revealed the presence of several compounds. The biological activity of the major fraction was tested in the Ames system without an activation system, and it was cytotoxic even at low concentrations (3 ng AFB-equiv/plate). An aflatoxinlike antigenic substance has also been detected in urine samples collected in other Western societies [58].

DNA Repair Product

Aflatoxin B_1 is metabolized to the ultimate carcinogen, AFB_1-8,9-epoxide. This metabolite reacts with the N-7 position of guanine in rats, and the adduct 8,9-dihydro-8-(7'-guanyl)-9-hydroxy-AFB (AFB_1-Gua) accounts for more than 80% of the adducts in rat liver 2 h after administration of AFB_1 [59]. Adducts at the N-7 position of guanine lead to destabilization of the imidazole ring structure of guanine owing to the delocalization of a positive charge. The major pathways of stabilization are (1) the release of AFB_1-Gua, leaving an apurinic site, and (2) ring opening of the imidazole structure to form a stable adduct identified as 8,9-dihydro-9-(N^5-formyl-2',5',6'-triamino-4'-oxo-N^5-pyrimidyl)-9-hydroxy-AFB_1 (AFB_1-NPy) [60]. After a single dose of AFB_1, 10–15% of the initial amount of AFB_1 bound to DNA persisted as the ring-opened adduct. The released AFB_1-Gua has been detected in the urine of rats dosed with AFB_1 [61] and in human urine samples collected in an area of Kenya with high suspected exposure to AFB_1 [51]. Detection of AFB_1-Gua in urine is an indirect proof that formation of a carcinogen-DNA adduct has occurred following exposure to AFB_1. The AFB_1-Gua was detected by cochromatography with a synthetic marker by HPLC, and the chemical structure was verified by synchronous fluorescence spectrophotometry [51]. A good correlation between dose of AFB_1, initial AFB_1-DNA level, and amount of AFB_1-Gua in urine within 48 h was established in male rats [62]. In rats, approximately 1% of the administered AFB_1 was excreted as AFB_1-Gua within 24 h [61]. Assuming the same rate of excretion in humans, the range of exposure in Kenya was estimated at 360–3600 ng AFB_1/day [52]. Urine samples were collected in different districts, and 12.6% of all the tested urine samples were positive for AFB_1-Gua. The highest rate of exposure was found in the areas with the highest incidence of liver cancer. This level corresponded to the estimated level of AFB_1 exposure based on food sampling [63].

The sensitivity of the assay for AFB_1-Gua has been increased by including an affinity column prepurification step prior to analysis by HPLC [53]. Aflatoxin B_1-guanine adduct was detected in urine samples collected in the Quangxi region of China using a direct competitive ELISA assay and a polyclonal antiserum against AFB_1 [37]. The sensitivity of the competitive ELISA was less than the combination of affinity-chromatography and HPLC, and AFB_1-Gua could be detected only in the urine of individuals exposed to a dose higher than 87 μg/day. A correlation between the amount of AFB_1-Gua and AFB_1-serum albumin adduct was observed in the same population group [47].

The advantage of the DNA repair product assay is that the presence of AFB_1–Gua in urine is an indication of consumption of AFB_1 and shows that the individual is able to transform AFB_1 into the biologically active form. Conversely, the presence of AFB_1 metabolites is an indication that a person can transform AFB_1 to less toxic metabolites. The disadvantages of this assay are that it is time-consuming and the amount of AFB_1-Gua in urine does not represent the biologically active dose of AFB_1 in the liver. Contributions of AFB_1-DNA formed in other tissues and from AFB_1-RNA adducts that are similar to the DNA adducts cannot be excluded. Furthermore, AFB_1-Gua may originate from meat obtained from animals that have consumed an AFB_1-contaminated diet.

B. Blood

The uptake and vascular transport of AFB_1 in humans has not been investigated. It is assumed that ingested AFB_1 is absorbed into the intestinal vascular drainage, reaching the liver via the hepatic portal system, which has been suggested to explain the organ selectivity of AFB_1 carcinogenicity.

Aflatoxin B_1 is not associated with any circulating lipoprotein, in contrast to other lipophilic carcinogens, which mostly associate with low-density lipoprotein [64]. In rats, AFB_1 was primarily associated with plasma albumin [65].

Metabolites

Aflatoxin B_1 and its metabolites have been detected in human serum from individuals living in industrial as well as developing countries. The levels of AFB_1 in blood in Western societies are generally low or undetectable. The same methodological approach as for the detection of urinary metabolites has been used—ELISA and HPLC techniques.

Using a combined RIA-HPLC technique, AFB_1 was detected in sera from normal healthy persons eating a standard Japanese diet and living in Kobe, Japan. The chemical identity of the antigenic material was verified as AFB_1 by mass spectrometry. The level of AFB_1 was 0.02–1.17 ng/mL, and 34% of all the samples were positive for aflatoxin, but a possible origin of AFB_1 was not suggested [40]. AFB_1 was detected by HPLC analysis in the serum of a patient with primary hepatic carcinoma in the United States [43] and in patients with Reye's syndrome. A survey in the United Kingdom using ELISA indicated that exposure in the general population is low [41]. A significant level of AFB_1 was detected in serum from normal subjects living in African countries (Table 3).

TABLE 3. Detection of Aflatoxin B_1 in Human Sera

Country (no. samples)	Method	Percent positive	Level (ng/mL)	Ref.
Nigeria (78)	ELISA	76	0.02–3.1	42
Nepal (29)	ELISA	100	0.06–10	42
Japan (34)	HPLC	34	0.02–1.17	40
Sudan (41)	HPLC	32	0.08–0.7	45
Kenya (125)	HPLC	43	0.01–1.69	44, 66
U.S.A (6)	HPLC	66	3.4–12	43
U.K.	ELISA	n.r.	0.02	41

Aflatoxins have been detected in cord blood samples collected in African countries. Aflatoxin M_1 was the most frequently observed metabolite, and it was detected in 50% of the samples positive for aflatoxins, while AFB_1 was detected in 15% of the samples. A seasonal variation in the detection rate was observed in the Kenyan cord blood samples, and the relative frequency was higher in the warm months than in the "cold" dry months [44]. Maxwell et al. suggested that the higher concentration of aflatoxins in cord blood compared to the maternal blood collected simultaneously is an indication that the aflatoxins may accumulate in the fetus. The birth weights of females born to AFB_1-positive mothers was significantly lower than those born to AFB_1-free mothers [66]. A number of studies in experimental animals have shown that AFB_1 exposure in utero influences development [67]. Treatment of pregnant rats with AFB_1 increased the incidence of tumors at various sites in the offspring, while the dams were not affected [68]. The occurrence of primary hepatocellular carcinoma in young children in Kenya may be a result of exposure to AFB_1 in utero, although exposure to AFB_1 and AFM_1 through mother's milk cannot be excluded as the causative agent [69]. Human fetal liver from the age of 3 months is able to metabolize AFB_1 to the epoxide [32], and exposure to AFB_1 in utero may therefore be responsible for the early development of liver cancer in people living in areas with high AFB_1 exposure.

Serum Protein Adducts

Aflatoxin B_1-8,9-epoxide binds covalently to serum albumin, and the major AFB_1-serum albumin adduct is formed by reaction with lysine. The adduct formed in rats treated with AFB_1 has been fully characterized. It was proposed that the initial adduct is formed by condensation of the dialdehyde tautomer of AFB_1-8,9-dihydro-8,9-dihydroxy and the ε-amino group of lysine. This adduct is rearranged to an α-amino ketone [70].

Twenty-four hours after a single dose of AFB_1 to rats, 1–2% of the administered dose was covalently bound to serum albumin, and a dose response was observed. Binding of AFB_1 to rat hemoglobin could not be detected [71]. Aflatoxin-albumin adducts were also detected in sera from individuals living in Quangxi province, China. Some of the food in this region is highly contaminated with AFB_1, and, as in the rat experiment, a positive association between the intake and level of binding to serum albumin was reported [37].

Two different approaches have been used to isolate albumin from serum either by ammonium sulfate precipitation or by binding of albumin to Reactive Blue-2 Sepharose. For quantitative analysis, albumin has to be degraded by pronase treatment and concentrated on an immunoaffinity column prior to quantitation by immunological analysis or by HPLC with fluorescence detection. Various modifications to the assay procedure have been proposed to increase the sensitivity and hence make the AFB-albumin adduct assays applicable for monitoring humans exposed to AFB_1 [48]. Using the "hydrolysis"-ELISA assay, no AFB_1-lysine could be detected in albumin isolated from individuals living in France, whereas the levels in Gambians were in the range of 5–200 pg AFB_1-lysine/mg albumin.

The advantage of using serum albumin adducts to assess exposure to AFB_1 is that this measure is more biologically relevant than quantitation of AFB_1 or its metabolites. Animal experiments suggests an association between AFB_1 binding to serum albumin and to liver DNA in rats treated with AFB_1 [71]. This suggests that AFB_1-serum albumin adduct is a reliable indirect measure of the target genotoxic dose of AFB_1. Furthermore, albumin has a half-life of 20 days in humans, allowing an assessment of exposure during the half-life of the proteins rather than the exposure within the last 24 h.

Lymphocyte DNA Adducts

The binding of AFB_1 to lymphocyte DNA is much less than the binding to serum albumin. No binding could be found to lymphocyte DNA isolated from rats treated with a single dose of AFB_1 [71], and AFB_1-DNA adducts could not be detected in lymphocytes from humans exposed to AFB_1. Although different types of carcinogen-DNA adducts have been detected in human lymphocytes, these cells appear not to be applicable to assess exposure to AFB_1 with the present technology. The ^{32}P postlabeling that is presently the most sensitive technique to detect carcinogen-DNA adducts has not been used successfully on AFB_1-DNA adducts.

C. Other Body Fluids and Tissues

Mother's Milk

Aflatoxin B_1 is metabolized to different hydroxylated metabolites such as AFM_1, which is excreted in the milk of both animals and humans. Although AFM_1 is far less toxic and mutagenic than the parent compound AFB_1, exposure to AFM_1 through mother's milk may pose a health hazard, as newborns are more sensitive to the toxic action. In most Western societies the guidelines for AFB_1 in animal feed for milk cattle are based upon the risk associated with the amount of AFM_1 expected to be excreted in the milk.

Aflatoxin M_1 could not be detected in human milk samples collected in France, whereas 11% of the milk samples collected in Zimbabwe were positive for AFM_1 using ELISA and an antiserum against AFM_1. The level of AFM_1 in the positive samples was up to 50 pg AFM_1/mL milk [46]. A higher percentage of milk samples collected in Ghana, Sudan, and Kenya were positive for AFM_1 using HPLC with fluorescent detection [44]. As expected, a higher percentage of samples collected in the wet season were positive for AFM_1 than of those collected in the dry season. The concentration of AFM_1 in the positive samples ranged from 5 to 1379 pg/mL. Aflatoxin B_1 and AFL could be detected in some of the positive samples in addition to AFM_1 [44].

Administration of AFB_1 to lactating rats resulted in a transfer of AFB_1 metabolites to the suckling offspring, and the metabolites were distributed to many different tissues. Binding to DNA in these tissues could not be detected, but binding to proteins and RNA was observed [72]. In a small carcinogenesis study, hepatocellular carcinomas developed predominantly in male offspring of rats that had received AFM_1 or other AFB_1 metabolites in their mother's milk [73].

Liver DNA Adducts

Determination of AFB_1-DNA adducts in liver cell DNA is the most relevant measure for the biological active dose, the amount of the ultimate carcinogen formed in the target organ. A number of studies in rats have established a relationship between dose and the adduct level in liver DNA [59, 62], but no quantitative associations have been found between the amount of the adducts and development of liver carcinomas. However, a quantitative relationship between AFB_1-DNA binding and the development of liver tumors in rainbow trout has been established [74].

The major adduct of AFB_1 formed in vivo is unstable and disappears rapidly from DNA by spontaneous hydrolysis of the glycosyl bond, or the imidazole ring of guanine opens to form AFB_1-FAPy. This adduct is stable in DNA for a prolonged period and is therefore the most obvious choice for monitoring AFB_1 exposure at the DNA level. Monoclonal antibodies have been developed against AFB_1-FAPy, and the antibody has been used to detect AFB_1 adducts in liver tumor DNA from people in Taiwan [49]. The adduct level was in the range of 1.2–3.5 μmol/mol DNA. A slightly lower level (0–1.1) was detected in DNA isolated from primary hepatocellular carcinomas from Czechoslovakians using a monoclonal antibody against AFB_1 [50]. A number of other antibodies against AFB_1-modified DNA have been developed, but these have not been used successfully in human monitoring studies [75].

Many dietary components as well as nutritional status influence the activity of enzymes involved in the metabolism of AFB_1 and hence the level of AFB_1-DNA adducts [34, 76–78] and subsequently the tumor initiation.

IV. OTHER METHODS

Animals immunized with AFB_1 conjugated with BSA or other proteins produced antibodies against AFB_1 [54]. Aflatoxin-serum albumin has been detected in highly exposed individuals, but the level of modification was less than the AFB_1-BSA used for antibody production. However, rats treated with AFB_1 do develop an antigenic activity against an AFB_1 epitope. Antibody activity against an AFB_1 epitope has been demonstrated in the sera from individuals with known high exposures to AFB_1 [79]. The titers of sera collected in Kenya were significantly higher than in sera collected in Denmark. The highest titers were found in sera from individuals living in the western part of Kenya, an area with high exposures to AFB_1. Furthermore, individuals who had recently been exposed to AFB_1 as measured by the detection of urinary AFB_1-Gua had a high titer. This observation is interesting, as immunization of rabbits with AFB_1-BSA confers some protection against the carcinogenic effect of AFB_1, as the covalent binding of radiolabeled AFB_1 to liver DNA was reduced in male rabbits [80].

V. CONCLUSION

Aflatoxins are a human health hazard, and continuous exposure to even small amounts may be detrimental. Sensitive analytical procedures, some requiring sophisticated instrumentation, have been developed to assess human exposure to aflatoxins. The application of these methods has made it possible to establish a valid association between the incidence of liver cancer and exposure to AFB_1. No clear relationship between exposure to aflatoxins and other diseases has been established. The immunosuppressive effect of AFB_1 may also be of concern for human health, especially for malnourished people in less developed countries.

An unequivocal association between the sources of exposure and disease is essential in disease prevention. Establishment of quantitative relationships between individual exposure and risk is an important step in establishing the scientific basis for regulatory guidelines to protect the population from the health hazards associated with exposure to aflatoxins.

The sensitivity and complexity of the present methods to assess human exposure to AFB_1 still limit comprehensive studies on the exposure to AFB_1 in western societies. In these countries, aflatoxins are generally not considered a risk factor. However, in Denmark a recent risk calculation estimated that consumption of an average diet will contribute three to four cases of liver cancer a year in a population of 5 million. This is based upon the current level of aflatoxins, 10 μg/kg AFB_1 in all food items potentially contaminated and 0.02 μg/L AFM_1 in milk [81].

REFERENCES

1. IARC Monographs. Evaluation of carcinogenic risk of chemicals to humans. *IARC Monogr. Suppl.*, *7*:83, 1987.
2. van Rensburg, S. J., Cook-Mozaffari, P., van Schalkwyk, D. J., et al. Hepatocellular carcinoma and dietary aflatoxin in Mozambique and Transkei. *Br. J. Cancer*, *51*:713, 1985.
3. Bulatao-Jayme, J., Almero, E. M., Castro, M. C. A., et al. A case-control dietary study of primary liver cancer risk from aflatoxin exposure. *Int. J. Cancer*, *11*:112, 1982.
4. Sun, T.-T., and Chu, Y.-Y. Carcinogenesis and prevention strategy of liver cancer in areas of prevalence. *J. Cell. Physiol.*, *Suppl. 3*: 39, 1984
5. Olsen, J. H., Dragsted, L., and Autrup, H. Cancer risk and occupational exposure to aflatoxins in Denmark. *Br. J. Cancer*, *58*:392, 1988.
6. Hayes, R. B., van Nieuwenhuize, J. P., Raatgever, J. W., and ten Kate, F. J. W. Aflatoxin exposures in the industrial setting: an epidemiological study of mortality. *Food Chem. Toxicol.*, *22*:39, 1984.
7. Dvorackova, I., Stora, C., and Ayraud, N. Evidence for aflatoxin B_1 in two cases of lung cancer in man. *J. Cancer Res. Clin. Oncol.*, *100*: 221, 1981.
8. Ngindu, A., Johnson, B. K., Kenya, P. R., et al. Outbreak of acute hepatitis caused by aflatoxin poisoning in Kenya. *Lancet*, *i*:1346, 1982.
9. Hendrickse, R. G., and Maxwell, S. M. Aflatoxins and child health in the tropics. *J. Toxicol. Toxin Rev.*, *8*:30, 1989.

10. Rogan, W. J., Yang, G. C., and Kimbrough, R. D. Aflatoxin and Reye's syndrome: a study of livers from deceased cases. *Arch. Environ. Health, 40*:91, 1985.
11. Lutwick, L. I. Relationship between aflatoxin, hepatitis-B virus and hepatocellular carcinoma. *Lancet, i*:755, 1979.
12. Padmanaban, V. D. Role of aflatoxins as immunomodulators. *J. Toxicol. Toxin Rev., 8*:242, 1989.
13. Pier, A. C., Richard, J. L., and Thurston, J. R. Effects of aflatoxin on the mechanisms of immunity and native resistance. *Med. Mycol. Zbl. Bakt., Suppl. 8*:301, 1980.
14. Busby, W. F., and Wogan, G. N. Aflatoxins. In *Chemical Carcinogens* (C. D. Searle, Ed.), ACS Monograph 182, American Chemical Society, Washington, D.C., 1984, p. 945.
15. Wei, C. I., Marshall, M. R., and Hsieh, D. P. H. Characterization of water-soluble glucuronide and sulphate conjugates of aflatoxin B_1. 1. Urinary excretion in monkey, rat and mouse. *Food Chem. Toxicol., 23*:809, 1985.
16. Campbell, T. C., Caedo, J. P., and Bulato-Jayme, J. Aflatoxin M_1 in human urine. *Nature, 227*:403, 1970.
17. Essigmann, J. M., Croy, R. G., Nadzan, A. M., et al. Structural identification of the major DNA adduct formed by aflatoxin B_1 *in vitro. Proc. Natl. Acad. Sci. USA, 74*:1870, 1977.
18. Battista, J. R., and Marnett, L. J. Prostaglandin H synthase-dependent epoxidation of aflatoxin B_1. *Carcinogenesis, 6*:1227, 1985.
19. Jhee, E.-C., Ho, L. L., Tsuji, K., et al. Effect of butylated hydroxyanisol pretreatment on aflatoxin B_1–DNA binding and aflatoxin B_1-glutathione conjugation in isolated hepatocytes from rats. *Cancer Res., 49*:1357, 1989.
20. Raj, H. G., and Lotlikar, P. D. Urinary excretion of thiol conjugates of aflatoxin B_1 in rats and hamsters. *Cancer Lett., 22*:125, 1984.
21. Wong, J. J., and Hsieh, D. P. H. Mutagenicity of aflatoxins related to their metabolism and carcinogenic potential. *Proc. Natl. Acad. Sci. USA, 73*:2241, 1976.
22. Shimada, T., and Guengerich, F. P. Evidence for cytochrome P-450NF, the nifedipine oxidase, being the principal enzymes involved in the bioactivation of aflatoxins in human liver. *Proc. Natl. Acad. Sci. USA, 86*:462, 1989.
23. Aoyama, T., Gonzales, F. J., and Gelboin, H. V. Mutagen activation by cDNA-expressed P1-450, P3-450, and P-450a. *Mol. Carcinogen., 1*:253, 1989.
24. Doehmer, J., Dogra, S., Friedberg, T., et al. Stable expression of rat cytochrome P450IIB1 cDNA in Chinese hamster cells (V79) and metabolic activation of aflatoxin B_1. *Proc. Natl. Acad. Sci. USA, 85*:5769, 1988.
25. Moss, E. J., and Neal, G. E. The metabolism of aflatoxin B_1 by human liver. *Biochem. Pharmacol., 34*:3193, 1985.
26. Cole, K. E., Jones, T. W., Lipsky, M. M., et al. *In vitro* binding of aflatoxin B_1 and 2-acetylaminofluorene to rat, mouse and human

hepatocyte DNA: the relationship of DNA binding to carcinogenicity. *Carcinogenesis, 9*:711, 1988

27. Larsen, P., Pettersson, H., and Tjävle, H. Metabolism of aflatoxin B_1 in the bovine olfactory mucosa. *Carcinogenesis, 10*:1113, 1989.
28. Coulombe, R. A., Wilson, D. W., Hsieh, D. P. H., Plopper, C. G., and Serabjit-Singh, C. Metabolism of aflatoxin B_1 in the upper airways of the rabbit: role of the nonciliated tracheal epithelial cells. *Cancer Res., 46*:4091, 1986.
29. Sorensen, W. G., Jones, W., Simpson, J., and Davidson, J. L. Aflatoxin in respirable airborne peanut dust. *J. Toxicol. Environ. Health, 14*:525, 1984.
30. Autrup, H., and Harris, C. C. Metabolism of chemical carcinogens by human tissues. In *Human Carcinogenesis* (C. C. Harris and H. Autrup, Eds.), Academic, New York, 1983, p. 169.
31. Stoner, G. D., Daniel, F. B., Schenck, K. M., et al. DNA binding and adduct formation of aflatoxin B_1 in cultured human and animal tracheobronchial and bladder tissues. *Carcinogenesis, 3*:1345, 1982.
32. Autrup, H., Harris, C. C., Wu, S.-M., et al. Activation of chemical carcinogens by cultured human fetal liver, esophagus and stomach. *Chem.-Biol. Interactions, 50*:15, 1984.
33. Autrup, H., Schwartz, R. D., and Essigmann, J. M. Metabolism of aflatoxin B_1, benzo[*a*]pyrene, and 1,2-dimethylhydrazine by cultured rat and human colon. *Teratogen. Carcinogen. Mutagen., 1*:3, 1980.
34. Groopman, J. D., Cain, L. G., and Kensler, T. W. Aflatoxin exposure in human populations: measurements and relationship to cancer. *CRC Crit. Rev. Toxicol., 19*:113, 1988.
35. Wild, C. P., Chapot, B., Scherer, E., Den Engelse, L., and Montesano, R. Application of antibody methods to the detection of aflatoxin in human body fluids. *IARC Sci. Publ., 89*:67, 1988.
36. Nyathi, C. B., Mutiro, C. F., Hasler, J. A., and Chetsanga, C. J. A survey of urinary aflatoxin in Zimbabwe. *Int. J. Cancer, 16*:516, 1987.
37. Gan, L.-S., Skipper, P. L., Peng, X., et al. Serum albumin adducts in the molecular epidemiology of aflatoxin carcinogenesis: correlation with aflatoxin B_1 intake and urinary excretion of aflatoxin M_1. *Carcinogenesis, 9*:1323, 1988.
38. Zhu, J.-Q., Zhang, L.-S., Xiao, Y., et al. Correlation of dietary aflatoxin B_1 levels with excretion of aflatoxin M_1 in human urine. *Cancer Res., 47*:1848, 1987.
39. Bean, T. A., Yourtee, D. M., Akande, B., and Ogunlewe, J. Aflatoxin metabolites in the urine of Nigerians—comparison of chromatographic methods. *J. Toxicol. Toxin Rev., 8*:41, 1989.
40. Tsuboi, S., Nakagawa, T., Tomita, M., et al. Detection of aflatoxin B_1 in serum samples of male Japanese subjects by radioimmunoassay and high pressure liquid chromatography. *Cancer Res., 44*:1231, 1984.
41. Wilkinson, A. P., Denning, D. W., and Moran, M. R. Analysis of UK sera for aflatoxin by enzyme-linked immunosorbent assay. *Hum. Toxicol., 7*:353, 1988.

42. Wilkinson, A. P., Denning, D. W., and Morgan, M. R. Immunoassay of aflatoxin in food and human tissues. *J. Toxicol. Toxin Rev., 8*: 41, 1989.
43. Siraj, M. Y., Hayes, A. W., Unger, P. D., et al. Analysis of aflatoxin B_1 in human tissues with high pressure liquid chromatography. *Toxicol. Appl. Pharmacol., 58*:422, 1981.
44. Maxwell, S. M., Apeagyei, F., de Vries, H. R., et al. Aflatoxins in breast milk, neonatal cord blood and sera of pregnant women. *J. Toxicol. Toxin Rev., 8*:18, 1989.
45. Hendrickse, R. G., Coulter, J. B. S., Lamplugh, S. M., et al. Aflatoxins and kwashiorkor: a study in Sudanese children. *Br. Med. J., 285*:843, 1984.
46. Wild, C. P., Pionneau, F., Montesano, R., Mutiro, C. F., and Chetsanga, C. J. Aflatoxin detected in human breast milk by immunoassay. *Int. J. Cancer, 40*:328, 1987.
47. Groopman, J. D., Gan, L., Skipper, P., et al. Dosimetry of aflatoxin DNA and serum albumin adducts in people living in Guangxi province, P.R.C. *Proc. Am. Assoc. Cancer Res., 30*:317, 1989.
48. Wild, C. P., Jiang, Y.-Z., Sabbioni, G., Chapot, B., and Montesano, R. Evaluation of methods for quantitation of aflatoxin-albumin adducts and their application to human exposure assessment. *Cancer Res., 50*:245, 1990.
49. Hsieh, L.-L., Hsu, S.-W., Chen, D.-S., and Santella, R. M. Immunological detection of aflatoxin B_1-DNA adducts formed *in vivo*. *Cancer Res., 48*:6328, 1988.
50. Garner, R. C., Dvorackova, I., and Tursi, F. Immunoassay procedures to detect exposure to aflatoxin B_1 and benzo[*a*]pyrene in animals and man at the DNA level. *Int. Arch. Occup. Environ. Health, 60*:145, 1988.
51. Autrup, H., Bradley, K. A., Shamsuddin, A. K. M., et al. Detection of a putative adduct with fluorescence characteristics identical to 2,3-dihydro-(7'-guanyl)-3-hydroxyaflatoxin B_1 in human urine collected in Muranga district, Kenya. *Carcinogenesis, 4*:1193, 1983.
52. Autrup, H., Seremet, T., Wakhisi, J., and Wasunna, A. Aflatoxin exposure measured by urinary excretion of aflatoxin B_1-guanine adduct and hepatitis B virus infection in areas with different liver cancer incidence in Kenya. *Cancer Res., 47*:3430, 1987.
53. Groopman, J. D., Donahue, P. R., Zhu, J. Q., et al. Aflatoxin metabolism in human: detection of metabolites and nucleic acid adduct in urine by affinity chromatography. *Proc. Natl. Acad. Sci. USA, 82*:6492, 1986.
54. Martin, C. N., Garner, R. C., Garner, J. V., et al. An enzyme-linked immunosorbent procedure for assaying aflatoxin B_1. *IARC Sci. Publ., 59*:313, 1985.
55. Degan, P., Montagnoli, G., and Wild, C. P. Time-resolved fluoroimmunoassay of aflatoxins. *Clin. Chem., 35*:2308, 1989.
56. Dragsted, L. O., Bull, I., Larsen, A., and Autrup, H. Aflatoksinudskillelse hos erhverseksponerede havnearbejdere. *Ugskr. Laeger, 147*:30, 1985.

57. Dragsted, L. O., Bull, I., and Autrup, H. Substances with affinity to a monoclonal aflatoxin B_1 antibody in Danish urine samples. *Food Chem. Toxicol., 26*:233, 1988.
58. Groopman, J. D., Trudel, L. J., Donahue, P. R., and Wogan, G. N. High-affinity monoclonal antibodies for aflatoxins and their applications to solid-phase immunoassays. *Proc. Natl. Acad. Sci. USA, 81*: 1984, 1984.
59. Croy, R. G., Essigmann, J. M., Reinhold, V. N., and Wogan, G. N. Identification of the principal aflatoxin B_1-DNA adducts formed *in vivo* in rat liver. *Proc. Natl. Acad. Sci. USA, 75*:1745, 1978.
60. Hertzog, P. J., Smith, J. R. L., and Garner, R. C. A high pressure liquid chromatography study on the removal of DNA-bound aflatoxin B_1 in rat liver and in vitro. *Carcinogenesis, 1*:787, 1980.
61. Bennett, R. A., Essigmann, J. M., and Wogan, G. N. Excretion of an aflatoxin-guanine adduct in the urine of aflatoxin treated rats. *Cancer Res., 41*:650, 1981.
62. Essigmann, J. M., Croy, R. C., Bennett, R. A., and Wogan, G. N. Metabolic activation of aflatoxin B_1: patterns of DNA adduct formation, removal, and excretion in relationship to carcinogenesis. *Drug Metab. Rev., 13*:581, 1982.
63. Peers, F. G., and Linsell, C. A. Dietary aflatoxins and liver cancer—a population based study in Kenya. *Br. J. Cancer, 27*:473, 1973.
64. Wilson, R., Ziprin, R., Ragsdale, S., and Busbee, D. Uptake and vascular transport of ingested aflatoxin. *Toxicol. Lett., 29*:169, 1985.
65. Dirr, H. W., and Schabort, J. C. Aflatoxin B_1 transport in rat blood plasma. Binding to albumin *in vivo* and *in vitro* and spectrofluorimetric studies into the nature of the interaction. *Biochim. Biophys. Acta, 81*:383, 1986.
66. De Vries, H. R., Maxwell, S. M., and Hendrickse, R. G. Foetal and neonatal exposure to aflatoxins. *Acta Paediatr. Scand., 78*: 373, 1989.
67. Dietert, R. R., Qureshi, M. A., Nanna, U. C., and Bloom, S. E. Embryonic exposure to aflatoxin B_1: mutagenicity and influence on development and immunity. *Environ. Mutagen., 7*:715, 1985.
68. Goerttler, K., Löhrke, H., Schweizer, H.-J., and Hesse, B. Effects of aflatoxin B_1 on pregnant inbred Sprague-Dawley rats and their F1 generation. A contribution to transplacental carcinogenesis. *J. Natl. Cancer Inst., 64*:1349, 1980.
69. Cameron, H. M., and Warwick, G. P. Primary cancer of the liver in Kenyan children. *Br. J. Cancer, 36*:793, 1978.
70. Sabbioni, G., Skipper, P. L., Büchi, G., and Tannenbaum, S. R. Isolation and characterization of the major serum albumin adduct formed by aflatoxin B_1 *in vivo* in rats. *Carcinogenesis, 8*:819, 1987.
71. Wild, C. P., Garner, R. C., Montesano, R., and Tursi, F. Aflatoxin B_1 binding to plasma albumin and liver DNA upon chronic administration to rats. *Carcinogenesis, 7*:853, 1986.

72. Allameh, A., Saxena, M., and Raj, H. G. Interaction of aflatoxin B_1 metabolites with cellular macromolecules in neonatal rats receiving carcinogen through their mothers milk. *Carcinogenesis, 10*:2131, 1989.
73. Wakhisi, J. Exposure of rat offsprings to aflatoxin risks through suckling mothers dosed with aflatoxin B_1. *J. Toxicol. Toxin Rev., 8*:275, 1989.
74. Dashwood, R. H., Arbogast, D. N., Fong, A. T., et al. Quantitative interrelationship between aflatoxin B_1 carcinogen dose, indole-3-carbinol anti-carcinogen dose, target organ DNA adduction and final tumor response. *Carcinogenesis, 10*:175, 1989.
75. Haugen, A., Groopman, J. D., Hsu, I. C., et al. Monoclonal antibody to AFB_1 modified DNA detected by immunoassay. *Proc. Natl. Acad. Sci. USA, 78*:4124, 1981.
76. Prabhu, A. L., Aboobaker, V. S., and Bhattacharya, R. K. *In vivo* effect of dietary factors on the molecular action of aflatoxin B_1: role of riboflavin on the catalytic activity of liver fractions. *Cancer Lett., 48*:89, 1989.
77. Salbe, A. D., and Bjeldanes, L. F. Effect of diet and route of administration on the DNA binding of aflatoxin B_1 in the rat. *Carcinogenesis, 10*:629, 1989.
78. Schrager, T. F., Newberne, P. M., Pikul, A. H., and Groopman, J. D. Aflatoxin-DNA adduct formation in chronically dosed rats fed a choline-deficient diet. *Carcinogenesis, 11*:177, 1990.
79. Autrup, H., Seremet, T., and Wakhisi, J. Evidence for human antibodies that recognize an aflatoxin epitope in groups with high- and low-exposure to aflatoxins. *Arch. Environ. Health, 43*:31, 1990.
80. Caviezel, M., Aeschbach, A. P., Lutz, W. K., and Schlatter, C. Reduction of covalent binding of aflatoxin B_1 to rabbit liver DNA after immunization against this carcinogen. *Arch. Toxicol. Suppl., 7*:249, 1984.
81. Danish Institute of Toxicology. Report from a working group—health evaluation of aflatoxins in foodstuffs, 1989, p. 31.

9

GENOTOXICITY OF NATURALLY OCCURRING METABOLITES: STRUCTURAL ANALOGS OF AFLATOXIN AND RELATED CHEMICALS

HIDEKI MORI *Gifu University School of Medicine, Gifu, Japan*

I. INTRODUCTION

A variety of naturally occurring fungal products have been isolated, and a number of associated metabolites have been found to be toxic and carcinogenic in animals [1–4]. Aflatoxin B_1 (AFB_1) was isolated from *Aspergillus flavus* [3]. Studies on the metabolic products of *Aspergillus versicolor* resulted in the isolation of sterigmatocystin [4]. Aflatoxins and sterigmatocystin are foodborne carcinogens that have been implicated in the etiology of human liver neoplasms [1, 2]. In addition to AFB_1, closely related compounds such as aflatoxicol, aflatoxin G_1 (AFG_1), and aflatoxin M_1 (AFM_1) have also been identified as hepatotumorigenic [5–7]. Biochemical modification of aflatoxins has been studied intensively, and biotransformation of related mycotoxins has been elucidated [8].

Ideally, naturally occurring mycotoxins should be tested for carcinogenicity in long-term bioassays. However, such evaluations are time-consuming, and difficulty in the isolation of sufficient amounts of pertinent fungal metabolites has hampered long-term bioassays. Meanwhile, recent technological developments have produced remarkable changes in genetic toxicology; significant progress has been made in short-term screening systems for mutagens and carcinogens. For example, the *Salmonella* microsome test established by Ames and Yanofsky [9] has been validated as a promising routine method; employing several tester strains, a number of mycotoxins have been found to be mutagenic in the assay [10–13]. However, the bacterial mutagenicity test has some limitations in the assessment of certain substances, including naturally occurring compounds such as mycotoxins.

Direct mutagenicity induced by chemical agents was initially identified as a genotoxic property. Subsequently, short-term assays have been established to detect genotoxicity of fungal metabolites and related substances.

The hepatocyte primary culture (HPC)/DNA repair test measures unscheduled DNA synthesis (UDS) [14]. It is a relevant test system for identification of the genotoxicity of naturally occurring carcinogens as well as of a variety of chemically synthesized carcinogens [15–17]. The Department of Pathology group of Gifu University, Gifu, Japan, has examined the genotoxicity of mycotoxins and has demonstrated the genotoxicity of a number of compounds [18–20]. Several UDS-positive compounds were found to be carcinogenic in rodents [21–23]. Some mycotoxins exhibit carcinogenic properties in in vivo long-term assays, although their genotoxicities in short-term systems are not apparent [12, 18]. Thus, the results of genotoxicity tests such as the HPC/DNA repair system are important for understanding the mode of action of certain mycotoxins.

In this chapter, mutagenicity in the *Salmonella* microsome assay and other assays and genotoxicity in the HPC/DNA repair test and other genotoxic properties of structural analogs of aflatoxins and related chemicals are reviewed. Observations of the structure–activity relationship are also considered.

II. STRUCTURAL CLASSIFICATION OF MYCOTOXINS INVOLVED IN THE METABOLIC PATHWAY OF AFLATOXINS

A number of fungal metabolites are intermediates in the aflatoxin biosynthetic pathway. Anthraquinones such as versicolorin A, versicolorin B, averufin, norsolorinic acid, and nidurufin are representative of the intermediates. Another aflatoxin intermediate, sterigmatocystin, is a compound in which a substituted anthraquinone is fused to a bisfuran ring. Chemicals structurally related to the sterigmatocystin include dihydrosterigmatocystin, 5-methoxysterigmatocystin, 5-methoxydihydrosterigmatocystin, demethylsterigmatocystin, sterigmatin, *O*-methylsterigmatocystin, and *O*-acetylsterigmatocystin. Anthraquinones that do not directly relate to aflatoxin biosynthesis include emodin, chrysophanol, luteoskyrin, rugulosin, islandicin, skyrin, and iridoskyrin. Naphthoquinone is the principal structure of a group of fungal metabolites that includes xanthomegnin, luteosporin, and fluccosin. Trichothecenes are a family of closely related sesquiterpenoids including T-2 toxin, nivalenol, fusarenon-X, crotocin, roridin A, and verrucarin A. Other mycotoxins include groups of lactones, halogens, and indoles. Mycotoxins such as patulin, penicillic acid, and austidiol belong to certain lactones. Griseofulvin, ochratoxin A, and chloropeptide are representatives of the halogen group, whereas fumitremorgen B and TR2 toxin are substituted indoles. Mycotoxins with unique individual structures include austocystins A and C, mitomycin C, actinomycin D, cytochalasin B, and zearalenone.

III. MUTAGENICITY OF MYCOTOXINS INCLUDING THOSE RELATED TO AFLATOXIN BIOSYNTHESIS AND OTHERS

A. Mutagenicity in the *Salmonella* Microsome Assay

Current advances in genetic toxicology have produced a variety of short-term tests for screening environmental mutagens and carcinogens [13]. Of these assays, the *Salmonella* microsome mutagenicity test developed by Ames and co-workers [9] has been regarded as the most practical method and is widely used for detecting environmental mutagens. The test system has identified the mutagenicity of numerous mycotoxins and related compounds, and further studies have demonstrated the presence of a number of fungal metabolites with carcinogenic potentials [10–13]. Ames et al. [24] revealed the mutagenicity of AFB_1 in the tester strains of TA 1537 and TA 1538 and that of sterigmatocystin in TA 1538 of *Salmonella typhimurium*. Using TA 1538, Kubota et al. [25] and McCann and Ames [26] also identified the mutagenicity of several mycotoxins in TA 1538. Wong and Hsieh [10] compared the mutagenicity of aflatoxins and their biotransformation products in TA 98, a new strain established by McCann et al. [27]. In this study, aflatoxicol, aflatoxin G_1 (AFG_1), aflatoxin M_1 (AFM_1), aflatoxin H_1 (AFH_1), aflatoxin Q_1 (AFQ_1), aflatoxin B_2 (AFB_2), aflatoxin P_1 (AFP_1), aflatoxin G_2 (AFG_2), aflatoxin $B_{2}a$ ($AFB_{2}a$), and aflatoxin $G_{2}a$ ($AFG_{2}a$) were all less active than AFB_1. Recently, mutagenicity of AFB_1 to *S. typhimurium* was characterized as a forward mutation by Stark et al. [28], who determined forward mutation by AFB_1 to 8-azaguanine resistance in treated cultures. Other investigations showed that mutagenicity of AFB_1 in *S. typhimurium* is expressed to a greater degree in the frameshift tester strains (TA 1538 and TA 98) than in the base-pair tester strains (TA 1535 and TA 100) [29]. Wong et al. [11] examined mutagenicity of fungal metabolites related to aflatoxin biosynthesis in TA 98; versicolorin A and sterigmatocystin were significantly mutagenic, although norsolonic acid, averufin, and versiconal acetate showed only marginal mutagenicity. The results suggested that mutagenic activity was related to the bisfuran structure and not to the anthraquinone moiety [11]. Ueno et al. [12] examined the mutagenicity of 15 mycotoxins and two chemically modified toxins, employing tester strains of TA 98 and TA 100. In the study, bisfuranoid mycotoxins such as AFB_1, AFG_1, sterigmatocystin, and *O*-acetylsterigmatocystin were positive in the routine test method, and epoxide mycotoxins such as PR-toxin and crotin were positive only when the tester strains were preincubated with the mycotoxins in the presence of the fortified, microsomal S-9 mixture [12]. Bjeldanes et al. [30] examined mutagenic activity of 41 fungal isolates with TA 100; extracts of two aflatoxin-producing isolates, *Aspergillus flavus* and *A. parasiticus*, and five other isolates (*A. heterothallicus*, *A. nidulans*, *A. terricola*, *Alternaria tenuis*, and *Fusarium moniliforme*) were clearly mutagenic. Mutagenic activity of *Aspergillus* and *Penicillium* mycotoxins were also examined by Wehner et al. [31] using TA 98, TA 100, TA 1535, and TA 1537; of 17 mycotoxins, austdiol, austocystins A and D, kojic acid, and viridicatumtoxin were found to be mutagenic after metabolic activation, while austdiol was mutagenic per se. No mutagenicity was evident in the cases of citrinin, cyclopiazonic acid, fumitremorgen B, griseofulvin, luteoskyrin, *O*-methylsterigmatocystin, mycophenolic acid, ochratoxin A, patulin, penicillic acid, secalonic acid D, and TR2 toxin. Furthermore, the mutagenic potential of veriscolorin A, an intermediate of aflatoxin biosynthesis in *A. flavus* and *A. parasiticus*, was identified with half the activity of sterigmatocystin. Nagao et al. [32] reported that dimethylsterigmatocystin, 5-methylsterigmatocystin, sterigmatin, and 6,8-dimethylversicolorin A were all strong or relatively strong mutagens in TA 98 and TA 100 with metabolic activation.

In other studies, versicolorins A and B, dimethylversicolorins A and B, and deoxyversicolorin A exhibited mutagenic effects in all tester strains of TA 97, TA 100, and TA 2637 [19]. TA 97 was the strain most sensitive to these compounds, and mutagenicity was enhanced with supplementation of microsomal S-9 mixture. Nidurufin and norsolonic acid did not exhibit mutagenicity in the tests [19]. Further study using *Salmonella* microsome assay was applied to 11 fungal metabolites structurally related to sterigmatocystin [20]. 5-Methoxysterigmatocystin, 5,6-dimethoxysterigmatocystin, and *O*-acetylsterigmatocystin exhibited mutagenic effects equivalent to those of sterigmatocystin and AFB_1 in TA 100. Other fungal metabolites such as dihydrosterigmatocystin, 5-methoxydihydrosterigmatocystin, and demethyldihydrosterigmatocystin were negative in the tester strain. No mutagenic activity was confirmed for sterigmatocystin and related compounds on TA 98 [20]. Liberman et al. [33] examined bacterial mutagenicity of a number of anthraquinone mycotoxins using five tester strains of TA 98, TA 100, TA 1535, TA 1537, and TA 1538. Chrysophanol and islandicin, which are structurally related to emodin, were positive in TA 1537 with microsomal S-9 mixture, but emodin was negative in all five tester strains. Similar studies were done by Tikkanen et al. [34]; they examined the bacterial mutagenicity of 15 naturally occurring anthraquinones using the preincubation method with three tester strains, TA 98, TA 100, and TA 2637. Seven compounds (chrysazin, emodin, islandicin, alizalin, chrysophanol, 2-hydroxyanthraquinone, and emodic acid) were strongly mutagenic in TA 2637. 2-Hydroxyanthraquinone, alizalin, emodin, islandicin, and chrysazin were also mutagenic in TA 100. All of the bisanthraquinones [skyrin, (+)-rugulosin, (-)-luteoskyrin, (-)-rubroskyrin, and sennoside A] were not mutagenic in the test system with or without metabolic activation. The results suggested that hydroxy substitutes are necessary for the expression of mutagenicity for the bisanthraquinones. Brown [35] studied bacterial mutagenicity in the *Salmonella* microsome assay with several phenolic anthraquinones derived from fungi. In this study, rubrofusarin was positive in TA 100 and TA 1537, and toralactone was positive in TA 100 and TA 1535. Torachrysone in TA 1537 was not mutagenic [35, 36]. Recently, the bacterial mutagenicity of some quinone compounds from pathogenic fungi was examined; aurosporin, floccosin, rugulosin, and xanthomegnin were not mutagenic in the tester strains of TA 98 and TA 100 with or without microsomal S-9 mixtures [37].

In Table 1, mutagenic activities of a variety of mycotoxins related to aflatoxin biosynthesis and others are summarized. There are several discrepancies in the mutagenicity data that may be related to the inconsistency of microsomal S-9 mixtures among the studies. Some species-dependent effects of dietary substances on the microsomal activation of AFB_1 into mutagenic derivatives have been reported [38]. Baker et al. [39] showed that the use of 3-methylcholanthrene-induced guinea pig liver microsomal S-9 was valuable for detecting weak mutagenicity of some chemicals but did not apply to AFB_1. Previously, Ueno et al. [8, 12] reported that there is no significant correlation between the mutagenic conversion of AFB_1 and the qualitative and quantitative behaviors of the

TABLE 1. Mutagenicity of Mycotoxins in the *Salmonella* Microsome Assay

Chemical	Tester strains						
	TA 97	TA 98	TA 100	TA 1535	TA 1537	TA 1538	TA 2637
Aflatoxins							
AFB_1		+[10,20], +*[12]	+*[12], +[20]	-[31]	+[24]		+[24]
AFB_2		+[10]					
AFG_1		+[10,12]					
AFG_2		+[10]					
AFH_1		+[10]					
AFM_1		+[10]					
AFQ_1		+[10]					
AFB_{2a}		-[10]					
AFG_{2a}		-[10]					
Aflatoxicol		+[10]					
Sterigmatocystins							
sterigmatocystin		+*[12], +[31,29]	+[20,31]	-[31]	+[24], -[31]	+[24]	
dihydrosterigmatocystin		-[20]	-[20]				
5-methoxysterigmatocystin		-[20]	+[20]				
5-methoxydihydrosterig-matocystin		-[20]	-[20]				
5,6-dimethoxysterigmato-cystin		-[20]	+[20]				
5,6-dimethoxydihydro-sterigmatocystin		-[20]	-[20]				
demethylsterigmatocystin		+*[32], -[20]	+[20]				
demethyldihydrosterig-matocystin		-[20]	-[20]				
sterigmatin		+[32], -[20]	+[32], -[20]				
dihydrosterigmatin		-[20]	-[20]				

(continued)

TABLE 1. (Continued)

Chemical	Tester strains						
	TA 97	TA 98	TA 100	TA 1535	TA 1537	TA 1538	TA 2637
[Sterigmatocystins]							
O-acetylsterigmatocystin		-[20]	+[20]				
O-methylsterigmatocystin		-[31]	-[31]	-[31]	-[31]		
Anthraquinones related to the aflatoxin biosynthesis							
versicolorin A	+*[19]	+*[31]	+† [19], +*[31]	+*[31]			+*[19]
deoxyversicolorin A	+[19]		+† [19]				+*[19]
dimethylversicolorin A	+[19]	+[32]	+† [19], +[32]				+*[19]
versicolorin B	+[19]		+† [19]				+*[19]
dimethylversicolorin B	+*[19]		+† [19]				+*[19]
versiconal acetate		-[29]					
averufin		-[29]					
norsolorinic acid		-[29]	-[19]				
nidurufin			-[19]				
Other anthraquinones							
chrysophanol		-[33,34]	-[33]	-[33]	+[33]	-[33]	+[34]
emodin		-[33,34]	+[34], -[33]	-[33]	+[33]		+[34]
emodic acid		-[34]	-[34]				+[34]
islandicin		-[33]	-[33]	-[33]			+[34]
chrysazin		-[34]					+[34]
luteoskyrin		-[12,31,34]	-[12,31,34]	-[31]	-[31]		-[34]
rugulosin		-[34,37]	-[34,37]				-[34]
skyrin		-[34]	-[34]				-[34]
rubroskyrin		-[35]	-[35]				-[34]
rubrofusarin		-[35]	+[35]	-[35]	+[35]	-[35]	
toralactone		-[35]	+[35]	+[35]	-[35]	-[35]	
torachrysone						-[35]	

Naphthoquinones				
aurosporin	-[37]	-[37]		
xanthomegnin	-[37]	-[37]		
floccosin	-[37]	-[37]		
Epoxides				
PR toxin	-[12]	+[12]		
fusarenon-X	-[12]	-[12]		
crotocin	-[12]	+[12]		
Lactones				
patulin	-[31]	-[31]	-[31]	-[31]
penicillic acid	-[31]	-[31]	-[31]	-[31]
mycophenolic acid	-[31]	-[31]	-[31]	-[31]
austidol	+*[31]	-[31]	-[31]	-[31]
kojic acid	-[31]	-[31]	-[31]	-[31]
citrinin	-[31]	-[31]	-[31]	-[31]
Halogens				
griseofulvin	-[31]	-[31]	-[31]	-[31]
ochratoxin	-[31]	-[31]	-[31]	-[31]
Indoles				
fumitremorgen B	-[31]	-[31]	-[31]	-[31]
TR2 toxin	-[31]	-[31]	-[31]	-[31]
Miscellaneous types				
secalonic acid D	-[31]	-[31]	-[31]	-[31]
autocystin A	+[31]	+[31]	-[31]	+[31]
autocystin D	-[31]	+[31]	-[31]	+[31]
cyclopiazonic acid	-[31]	-[31]	-[31]	-[31]
viridicatumtoxin	+[31]	-[31]	-[31]	-[31]

Reference numbers are shown in brackets.
*Positive with or without S-9.
†Positive only without S-9.

microsomal drug-metabolizing system of rat liver. Ueno and Ueno emphasized that several other factors such as epoxide transferase, free GSH, and ligandin may control the exact nature of AFB_1 mutagenicity [8]. Liver microsomal metabolism of AFB_1 to a reactive derivative toxic to TA 1530 was studied by Garner et al. [40]. The livers from different species of rodents and a single autopsy sample of human liver were active in the microsome-mediated AFB_1 toxicity assay. Vaught and Bresnick [41] reported that carcinogen binding to macromolecules presented a possible activation of carcinogen by a nuclear P-450 system. These evidences also suggest that the organ and species differences of AFB_1 mutagenicity may correlate with variations of nuclear enzyme systems.

B. Mutagenic Evidence in Other Systems

Aflatoxin B_1 is mutagenic to some microorganisms other than *Salmonella*. Gose et al. [42] observed induction of prophage λ and mutagenicity of AFB_1 in *Escherichia coli* k12. Ueno et al. [43] reported that fusarenon-X, a trichothecene mycotoxin of *Fusarium nivale* Fn 2B, *Fusarium episharia* Fn M, and *Gibberella zeae* Ohoita-II, induced a respiratory-deficient mutant on yeast cells, *Saccharomyces cerevisiae*. Ueno and Nakajima confirmed a high frequency of mutation by (−)-luteoskyrin and (+)-rugulosin in yeast cells [44]. Further studies using *S. cerevisiae* with a variety of mycotoxins have been done by Kuczak et al. [45]. In this study, AFB_1 and sterigmatocystin were positive with microsomal S-9 activation. However, fungal metabolites of citrinin, diacetoxyscipenol, griseofulvin, ochratoxin A, oosporin, patulin, penicillic acid, penitrem A, rubratoxin B, T-2 toxin, verruculogen, viriditoxin, and zearalenone did not elicit mutagenicity with or without S-9 activation. Nestmann et al. [46] also reported positive response in *S. cerevisiae* of actinomycin D, which did not generate mutagenicity in *E. coli*. Meanwhile, the mutagenicity of several compounds known to be carcinogenic in various animals has been tested by determining induction of recessive lethal mutations in the adenine-3 region of a two-component heterokaryon of *Neurospora crassa* [47]. In this system, mutagenicity of aflatoxins was examined by Ong and de Serres [48, 49]; the assay showed that AFB_1 and AFG_1 were mutagenic although AFB_2 and AFG_2 had negative responses [48, 49]. Mutagenicity of actinomycin D in *N. crassa* was also confirmed by Fisher et al. [50].

Some mycotoxins exhibit mutagenic effects in mammalian cells. Kuroki et al. [51] reported mutagenic activity of AFB_1 in Chinese hamster cells. Aflatoxin B_1 showed clear mutagenicity and cytotoxicity in V79 Chinese hamster cells with supplementation of 2% liver S15 microsomal fractions from aroclor-pretreated rats, but AFG_2 did not induce cytotoxicity or mutagenicity. Mutagenicity of aflatoxins in mice in vivo was reported by Epstein and Shafner [52]. In the dominant lethal test, AFB_1 and AFG_1 were identified as mutagens [52]. The mutagenicity of various mycotoxins and their efficiency in producing DNA single-strand breaks and chromosome aberrations were evaluated by Umeda et al. [53] in a mammalian cell line (FM3A) derived from C3H mouse mammary carcinoma. In the study, AFB_1, mycophenolic acid, patulin, penicillic acid, and sterigmatocystin induced 8-azaguanine-resistant mutations. AFB_1, mycophenolic acid, and sterigmatocystin had little effect on single-strand DNA at high concentrations. In the treatment with patulin and penicillic acid, severe breaks were obtained at higher concentration than those required to induce mutation. The incidence of chromosome aberrations by the treatment correlated with their mutagenic activities. Chaetoglobosin B, fusarenon-X, (−)-luteoskyrin, and ochratoxin A did not induce 8-azaguanine-resistant mutations.

It has been known that the fruit fly, *Drosophila melanogaster*, provides for assessment of the total spectrum of genetic effects [54]. The wealth of specific tester strains, special markers, useful inversions, or other rearrangements make this insect a desirable agent for mutagenicity tests. *Drosophila* also has its own metabolic activation system as does the mammalian liver and provides an opportunity for screening both gene and chromosome effects [54, 55]. The mutagenicity of AFB_1 in *Drosophila* was tested by Vogel and Sobels [55]. Aflatoxin B_1 was an effective mutagen in IInd chromosome recessive lethal mutants of *Drosophila,* with a peak in the third brood (mating schedule one male per two females with 3-day brood changes) that was produced by AFB_1 treatment of spermatocytes or early spermatids. The biological activities of AFB_1 2,3-dichloride (AFB_1-C12) and the parent mycotoxin AFB_1 in *Drosophila* were compared by Fahmy et al. [56]; general toxicity was observed in males treated by injection, and cytotoxic and mutagenic effects of toxins on the various stages of testicular tissues were characterized. The cytotoxic and mutagenic efficiencies of AFB_1-C12 were considerably higher than those of AFB_1, and the differences in activities were increased in the metabolizing earlier germ cell stages. The results indicated that AFB_1-C12 is a model for the active metabolite of AFB_1 in mutagenesis and carcinogenesis. Actinomycin D has also been reported to be mutagenic in *D. melanogaster* [57]. Some mycotoxins have mutagenic effects in mammalian epithelial cells. Tong and Williams [58] reported induction of purine analog-resistant mutants in adult rat liver epithelial lines by AFB_1.

IV. GENOTOXICITY OF MYCOTOXINS INCLUDING METABOLITES RELATED TO AFLATOXIN BIOSYNTHESIS

A. Genotoxicity in the HPC/DNA Repair Assay

Many mycotoxins are toxic and carcinogenic in the liver [1, 2]. Consequently, the HPC/DNA repair assay [14] is a highly relevant system for identification of the genotoxicity of mycotoxins. This method offers two distinct advantages over the traditional UDS assay employing human fibroblasts (WI-38 cells) [59]. The technique does not require chemical blockage of scheduled DNA synthesis, since the hepatocytes are nonreplicating over 48 h after collection. In addition, the primary hepatocytes carry out all microsomal enzyme reactions and ensure reliable detection of metabolic reaction products. The HPC/DNA repair assay appears to be the most reliable initial test for primary DNA damage in mammalian cells. Williams [60] examined the genotoxicity of aflatoxins in the HPC/DNA repair test. Three aflatoxins (AFB_1, AFB_2, and AFG_1) showed genotoxicity; the results indicated a correlation between genotoxicity and in vivo carcinogenicity. Genotoxicity was determined for seven fungal metabolites related to AFB_1 biosynthesis: versicolorin A, deoxyversicolorin A, dimethylversicolorin A, versicolorin B, dimethylversicolorin B, norsolorinic acid, and nidurufin [19]. Genotoxicity was observed for dimethylversicolorins A and B and deoxyversicolorin A in addition to versicolorins A and B. The toxicity of versicolorin A, deoxyversicolorin A, and dimethylversicolorin A was apparently greater than that of versicolorin B and dimethylversicolorin B. The results indicated that the dihydrobisfuran ring was more favorable than the tetrahydrobisfuran ring for induction of genotoxicity of versicolorins. Nidurufin and norsolorinic acid did not elicit clear DNA repair. Genotoxicity examinations were also carried out on a variety of sterigmatocystin-related compounds: sterigmatocystin, dihydrosterigmatocystin, 5-methoxysterigmatocystin, 5,6-dimethoxysterigmatocystin, 5,6-dimethoxydihydrosterigmatocystin, demethylsterigmatocystin, demethyldihydrosterigmatocystin, sterigmatin, dihydrosterigmatocystin, *O*-methylsterigmatocystin, and *O*-acetylsterigmatocystin [20]. Of these chemicals, dihydrosterigmatocystin, 5-methoxydihydrosterigmatocystin, 5,6-dimethoxysterigmatocystin, 5,6-dimethoxydihydrosterigmatocystin, demethylsterigmatocystin, demethyldihydrosterigmatocystin, sterigmatin, *O*-methylsterigmatocystin, and *O*-acetylsterigmatocystin elicited DNA repair. The UDS levels of metabolites such as dihydrosterigmatocystin, demethyldihydrosterigmatocystin, and sterigmatin were relatively lower than responses produced by other active compounds. Dihydrosterigmatin did not generate a positive UDS response (Fig. 1 and Table 2). Several quinone pigments from pathogenic fungi such as *Microsporum cookei* and *Epidermophyton floccosum* were also examined for genotoxicity in this test [37]. Of the four metabolites xanthomegnin, luteosporin, floccosin, and rugulosin, the first two produced positive responses. The data suggested that these two compounds could represent a potential hazard as genotoxic carcinogens. Negative results of floccosin and (+)-rugulosin were in agreement with data from *Salmonella* microsome mutagenicity assays [37]. Further studies of genotoxicity in the assay have been applied to naturally occurring anthraquinones and related compounds. Luteoskyrin, chrysazin (1,8-dihydroxyanthraquinone), and 1-hydroxyanthraquinone were genotoxic. However, emodin, chrysophanol, alizalin, quinizalin, anthrarufin, islandicin, iridoskyrin, (−)-rubroskyrin, carminic acid, and skyrin were negative. Other mycotoxins that exhibited genotoxicity in this assay included mitomycin C, actinomycin D, and azaserine, whereas griseofulvin, penicillic acid, xenoclauxin, duclauxin, *N*-methylduclauxin, *N*-ethylduclauxin, duclauxamide, protoduclauxamide, flavoglaucin, averufin, echinurin, secaronic acid D, cytochalacin B, flavoskyrin, and violaceol I were negative [61-66]. In vivo long-term assays for carcinogenicity were carried out with 5,6-dimethoxysterigmatocystin, chrysazin (1,8-dihydroxyanthraquinone), and 1-hydroxyanthraquinone. 5,6-Dimethoxysterigmatocystin was found to be as potent a hepatocarcinogen as sterigmatocystin and also induced bone neoplasms in rats [21]. Chrysazin induced carcinogenic activities in the large bowel of rats and mice [22, 23], and 1-hydroxyanthraquinone exhibited carcinogenicity in the large bowel, liver, and stomach of rats [67].

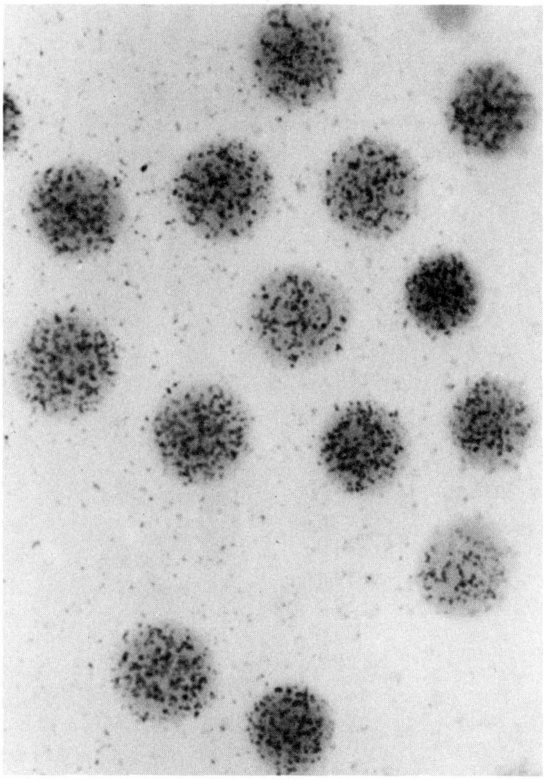

FIGURE 1. A photograph showing cultured rat hepatocytes in the DNA repair assay. Significant incorporation of [^{3}H]thymidine is recognized in the nuclei of 5,6-dimethoxysterigmatocystin-treated cells.

Some HPC/DNA repair-positive mycotoxins display species differences in their carcinogenic activities. For example, AFB_1 is carcinogenic for rat liver but not for mouse liver [1, 2]. In contrast, luteoskyrin shows a stronger carcinogenic activity in mouse liver than in rat liver [1, 2]. However, species differences are generally not apparent in the genotoxicity of mycotoxins in the HPC/DNA assay with hepatocytes of rats and mice [18]. Genotoxicity assays like the HPC/DNA repair test are useful in understanding the modes of action of carcinogens; the test also provides for screening of carcinogens to assess chemical hazard to humans. Recently, epigenetic mechanisms of liver carcinogenesis have been proposed for nongenotoxic agents [68]. Thus, it may be that negative results in the HPC/DNA repair test with some mycotoxins such as griseofulvin and rugulosin of known hepatotumorigenicity represent a new class of nongenotoxic carcinogens.

TABLE 2. Genotoxicity of Mycotoxins in the HPC/DNA Repair Assay

Chemical	DNA repair
Aflatoxins	
aflatoxin B_1	+[18,60]
aflatoxin B_2	+[60]
aflatoxin G_1	+[60]
aflatoxin G_2	-[60]
Sterigmatocystins	
sterigmatocystin	+[20,60]
dihydrosterigmatocystin	+[20]
5-methoxysterigmatocystin	+[20]
5-methoxydihydrosterigmatocystin	+[20]
5,6-dimethoxysterigmatocystin	+[20]
5,6-dimethoxydihydrosterigmatocystin	+[20]
demethylsterigmatocystin	+[20]
demethyldihydrosterigmatocystin	+[20]
sterigmatin	+[20]
dihydrosterigmatin	-[20]
O-methylsterigmatocystin	+[20]
O-acetylsterigmatocystin	+[20]
Anthraquinones related to aflatoxin biosynthesis	
versicolorin A	+[18,19]
deoxyversicolorin A	+[19]
dimethylversicolorin A	+[19]
versicolorin B	+[18,19]
dimethylversicolorin B	+[18,19]
averufin	+[18]
norsolorinic acid	-[19]
nidurufin	-[19]
Other anthraquinones	
1-hydroxyanthraquinone	+[61]
chrysazin	+[18,61]
chrysophanol	-[18]
emodin	-[18]
islandicin	-[62]
luteoskyrin	+[18]
rugulosin	-[18]
skyrin	-[18]
flavoskyrin	-[63]
iridoskyrin	-[62]
rubroskyrin	-[62]

TABLE 2. (Continued)

Chemical	DNA repair
Naphthoquinones	
xanthomegnin	+[38]
luteosporin	+[38]
floccosin	-[38]
Miscellaneous	
mitomycin C	+[18]
actinomycin D	+[18]
azaserine	+[18]
penicillic acid	-[18]
griseofulvin	-[18]
patulin	-[18]
duclauxin	-[18]
desacetylduclauxin	-[64]
secalonic acid	-[18]
violaceol I	-[18]
xenoclauxin	-[64]
echinurin	-[18]
cytochalasin B	-[18]

B. Genotoxic Evidence in Other Systems

Ueno and Kubota [69] examined DNA-attacking ability in recombination-deficient mutant cells of *Bacillus subtilis* [70] with 30 mycotoxins and five chemically modofied toxins. Six *Penicillium* toxins [citrinin, penicillic acid, patulin, (−)-luteoskyrin, (+)-rugulosin, and PR-toxin], five *Aspergillus* toxins (AFB_1, AFG_1, sterigmatocystin, *O*-acetylsterigmatocystin, and *O*-acetyldihydrosterigmatocystin), and two *Fusarium* toxins (zearalenon and zearalenol-b) exhibited positive responses in the assay. Aflatoxins have also been studied in in vitro bioassay using morphological transformation of early passage Golden Syrian hamster embryo cells derived from cryopreserved primary cultures (Pienta et al. [71]). In the assay, AFB_1 and AFG_1 showed positive responses, indicating good correlation between morphological transformation and carcinogenicity. Aflatoxin B_1 is also known to transform some other cells; neoplastic transformation of guinea pig fetal cells in culture by AFB_1 was reported by Evans and DiPaolo [72]. In the fetal cell study, cells morphologically transformed by AFB_1 developed concurrently with the potential for neoplastic growth. Furthermore, DiPaolo et al. [73] reported neoplastic transformation of Balb/3T3 cloned cell lines by AFB_1. The carcinogen-sensitive rodent cell lines are recognized as reliable, in vitro, quantitative bioassay models for screening of chemical carcinogens. Haugen et al. [74] recently examined enzyme immunoassay of monoclonal antibody to AFB_1-modified DNA in mammalian cells. Competitive ELISA with the monoclonal antibodies reliably quantitiated AFB_1 adducted in vitro to rat liver DNA at adduct levels of one AFB_1 residue per 250,000 nucleotides.

Some mycotoxins have been known to induce chromosome abnormalities. Iskandar and Vijayalaxmi [75] observed AFB_1 induction of chromosome abnormality in cultured human lymphocytes with an S-9-NADPH activation system. Friedman and Staub [76] reported positive activity of AFB_1 in the micronucleus assay [77, 78] with mouse or hamster bone marrow cells. Trzos et al. [79] also showed positive results with AFB_1 in the micronucleus assay employing bone marrow cells of rats. Heddle and Bruce [80] studied the effects of AFB_1, mytomycin C, griseofulvin, and actinomycin D; only mitomycin C was positive. Jenssen and Ramel [81] observed no activity of actinomycin D and griseofulvin in mouse bone marrow cells, but AFB_1 and mitomycin C had positive activities. Sister chromatid exchanges and chromosome aberrations in Chinese hamster V79 cells by aflatoxins were studied by Batt et al. [82]. With metabolic activation by microsomal S-9 mixtures, AFB_1, AFB_2, AFG_1, and AFG_2 caused significant increases in frequency of sister chromatid exchanges. The relative order of potency of the sister chromatid exchange induction was established as $AFB_1 > AFG_1 > AFG_2 > AFB_2$. Without microsomal S-9, the aflatoxins had essentially no effect. A dose- and time-dependent increase of chromosome aberrations was observed for AFB_1 and AFG_1. Induction of sister chromatid exchanges by aflatoxins through cultured rat liver epithelial cells was observed by Ray-Chaudhuri et al. [83]. In that study, AFB_1 induced sister chromatid exchanges in Chinese hamster ovary cells, and in the presence of liver cells the exchange frequency was greatly increased. Aflatoxin G_2 also produced a response in the cocultivation system. Furthermore, it is known that AFB_1 induced DNA-repair synthesis in human peripheral blood monocytes [84].

Terao and Ito [85] investigated the cytotoxicity of 16 naturally occurring mycotoxins possessing bisdihydrofuran by electron microscopy and radioautography in a primary culture of chick embryo hepatocytes. In tested cells, these authors observed a series of changes of nucleolar structure, formation of macrosegregation, fragmentation of nucleoli, and the appearance of homogeneously granular nucleoli; the results suggested that genotoxicity was linked to compounds containing a 2,3 double bond in the bisdihydrofuran ring such as AFB_1, sterigmatocystin, and versicolorin A. Wyrobek and Bruce [86] found that abnormal sperm were produced as a result of mutagenic treatment during spermatogenesis; this phenomenon could be useful for screening carcinogens. Although AFB_1 did not generate clear abnormal sperm in the assays, actinomycin D, griseofulvin, and mitomycin C had positive responses. Engelbrecht and Altenkirk [87] investigated toxic effects of 15 analogous compounds related to either sterigmatocystin or AFB_1 in a primary culture system evaluating nucleolar morphology and incorporation of [^{3}H]thymidine and [^{3}H]uridine. They concluded that some compounds such as *O*-methylsterigmatocystin may be carcinogenic on the basis of their effects on the mitotic rate and nuclear changes. Kawasaki et al. [88] examined cytotoxic effects of penicillic acid and patulin in HeLa cells. Penicillic acid was slow in action and associated with accumulation of metaphasic cells and elongation of the whole cell cycle. Patulin, on the other hand, acted rapidly; the whole cell cycle stopped immediately after toxin administration. Fusarenon-X exhibits inhibitory effects on the protein and DNA synthesis in rabbit reticulocytes, protozoa, and cultured cells [89–91]. Another genotoxic effect of the chemicals is teratogenicity in rodents. For instance, T-2 toxin and deoxynivalenol induce malformations in mice [92, 93]. However, teratogenicity has not been linked to fusarenon-X.

Recently, the capability of tumor-promoting activity of some mycotoxins such as the trichothecenes has been suggested [95]. Watanabe et al. [96] examined the tumor-promoting activity of a variety of trichothecenes; palytoxin, roridin, verrucarin, and diacetoxyscirpenol did not exhibit tumor-promoting activity.

V. CORRELATION BETWEEN EXPRESSION OF GENOTOXICITY AND CHEMICAL STRUCTURES OF MYCOTOXINS

Aflatoxin B_1 has a molecular structure optimal for acute toxicity, mutagenicity, and carcinogenicity; any changes in the bisfuran ring, cyclopentenone ring, and methoxy structures of AFB_1 result in reduction of biological activities. The active form of AFB_1 has been proposed as the 2,3-epoxide generated metabolically from the 2,3 vinyl ether double bond in the terminal furan ring. 2,3-Epoxides are considered to be formed by epoxidation of the 2,3 double bond [8, 95]. Comparison of the mutagenic activity of the aflatoxins and the presence of the 2,3 vinyl ether double bond in the terminal furan ring support the involvement of this functional group in toxicologic action. In fact, there is evidence that the 2,3-dihydro compounds (AFB_2 and AFG_2) that are saturated in their 2,3-positions exhibit little or no activity of mutagenicity; the hydrated forms AFB_2 and AFG_2 are less acutely toxic and carcinogenic than the unsaturated analogs [10]. Fahmy et al. [56] showed that AFB_1-2,3-dichloride (AFB_1-C12) was a model for the active metabolite of AFB_1 in mutagenesis and carcinogenesis by comparison of AFB_1-C12 and AFB_1 in *Drosophila*. In the study, genetic results were consistent with the chemical deductions that AFB_1-C12 was a model of the oxidative metabolite of AFB_1, AFB_1-2,3-dioxide; both AFB_1 and the oxide induced mutagenicity through reactions between C-2 of the AFB_1 residues and the bases of DNA, especially N-7 of guanine moieties. Stark et al. [28] reported that approximately 70% of the AFB_1 bound to DNA was chromatographically identical to 2,3-dihydro-2-(N^7-guanyl)-3-hydroxy AFB_1. A mutagenicity assay with fungal metabolities identified as the intermediates in the aflatoxin biosynthetic pathway was carried out by Wong and Hsieh [10]. In their study, norsolorinic acid, averufin, and versiconal acetate were negative in the Ames assay, but versicolorin A and sterigmatocystin were mutagenic. The results indicate that mutagenic activity is related to the bisfuran and not to the anthraquinone moieties. Nidurufin and norsolonic acid, which do not possess a bisfuran ring, exhibited questionable activities in the bacterial mutagenicity assay and no genotoxicity in the HPC/DNA repair test [19]. In another study, dihydrosterigmatocystin was negative in the DNA repair test; 5-methoxydihydrosterigmatocystin and 5,6-dimethoxydihydrosterigmatocystin were not mutagenic in the bacterial mutagenicity assay [20]. Genotoxicity of versicolorin A with the double bond at the 2,3-position, was stronger than that of versicolorin B with the saturated bond [19]; these results confirm the significance of the 2,3 bond in the bisfuran ring. In genotoxicity assays with rodent hepatocytes, 5-methoxydihydrosterigmatocystin and 5,6-dimethoxydihydrosterigmatocystin exhibited positive responses although both chemicals contain a saturated 2,3 bond in the bisfuran ring. The results appear to suggest that methoxylation may also yield genotoxic capacity. *O*-Methylsterigmatocystin and *O*-acetylsterigmatocystin were positive in the HPC/DNA repair test; these compounds are also mutagenic in the bacterial mutagenicity assay. It is assumed that *O*-acetylation or *O*-methylation does not reduce the genotoxic capacity of these chemicals. Englebrecht and Altenkirk [87] examined the toxicity of 15 analogous compounds that were related to either sterigmatocystin or AFB_1 in a primary cell culture system. In their study, a saturated 1,2-furobenzofuran system and the position of methoxy and hydroxy groups on the xanthine position of the sterigmatocystin analogs affected cytotoxicity; toxicity was determined by incorporation of [^{3}H]thymidine and [^{3}H]-uridine and nucleolar morphology. Correlations between expression of genotoxicity and chemical structures of the mycotoxins of AFB_1 biosynthesis and related fungal metabolites somewhat resemble those of other naturally occurring chemicals such as pyrrolizidine alkaloids. In the case of pyrrolizidine alkaloids, compounds that lack the unsaturated double bond at the 1,2-position of the pyrrolizidine ring generally do not elicit clear genotoxicity [15].

Current studies have clarified that there is a remarkable difference between the mycotoxins of angular types and those of linear types for the biological actions. Terao and Ito [85] reported that mycotoxins containing a 2,3 double bond in an angular-type bisdihydrofuran ring were far more toxic than those with a linear-type ring. Results of our genotoxicity studies on the metabolites related to aflatoxin biosynthesis and sterigmatocystin-related compounds [19, 20] are also in agreement with these reports.

In addition to the mycotoxins of aflatoxins and related compounds, there is a body of evidence showing correlation between chemical structures and biological activities [8, 95]. Recently, the genotoxicity of anthraquinones in the HPC/DNA repair assay was examined [61]. In the test, of the dihydroxyanthraquinones, chrysazin (1,8-dihydroxyanthraquinone) was genotoxic. However, alizalin (1,2-dihydroxyanthraquinone), quinizalin (1,4-dihydroxyanthraquinone), and anthrarufin (1,5-dihydroxyanthraquinone) were negative. Furthermore, emodin, a derivative of 1,8-dihydroxyanthraquinone, was also negative in the assay. The reasons are not clear why 1,2-, 1,4-, and 1,5-dihydroxyanthraquinones are negative. An analysis of quantum chemical properties of 19 anthraquinones has been carried out; the results indicate that electrophilic frontier electron densities in the β-position may be associated with genotoxic or carcinogenic capability [97]. Emodin has been found to be transformed, mainly by hydroxylation at the β-position of the anthraquinone nucleus. The presence of a hydroxy group at the 2-, 4-, or 5-position may hinder enzymatic activation at the β-position. Accordingly, the genotoxicities of naturally occurring chemicals of this class may also be mutually related to their chemical structures.

REFERENCES

1. Enomoto, M. Carcinogenicity of mycotoxins. In *Toxicology Biochemistry and Pathology of Mycotoxins* (K. Uraguchi and M. Yamazaki, Eds.), Kodansha Press, Tokyo and Wiley, New York, 1978, p. 239.
2. Hirono, I. Natural carcinogenic products of plant origin. *CRC Crit. Rev. Toxicol., 11*:235, 1981.
3. Worgan, G. N., and Newberne, P. M. Dose-response characteristics of aflatoxin B_1 carcinogenesis in the rat. *Cancer Res., 27*:2370, 1967.
4. Van der Watt, J. J. Sterigmatocystin. In *Mycotoxins* (I. F. H. Purchase, Ed.), Elsevier, New York, 1974, p. 369.
5. Butler, W. H., Greenblatt, B., and Lijinsky, W. Carcinogenesis in rats by aflatoxin B_1, G_1, and B_2. *Cancer Res., 29*:2206, 1969.
6. Wogan, G. N., Edwards, G. S., and Newberne, P. M. Structure–activity relationships in toxicity and carcinogenicity of aflatoxins and analogs. *Cancer Res., 31*:1936, 1971.
7. Wogan, G. N., and Paglialunga, S. Carcinogenicity of synthetic aflatoxin M_1 in rats. *Food Cosmet. Toxicol., 12*:381, 1974.
8. Ueno, Y., and Ueno, I. Toxicology and biochemistry of mycotoxins. In *Toxicology, Biochemistry, and Pathology of Mycotoxins* (K. Uraguchi and M. Yamazaki, Eds.), Kodansha Press, Tokyo, and Wiley, New York, 1978, p. 107.
9. Ames, B. N., and Yanofsky, C. The detection of chemical mutagens with enteric bacteria. In *Chemical Mutagens: Principles and Methods for Their Detection* (A. Hollaender, Ed.), Plenum, New York, 1971, p. 267.
10. Wong, J. J., and Hsieh, D. P. H. Mutagenicity of aflatoxins related to their metabolism and carcinogenic potential. *Proc. Natl. Acad. Sci. USA, 73*:2241, 1976.
11. Wong, J. J., Singh, R., and Hsieh, D. P. H. Mutagenicity of fungal metabolites related to aflatoxin biosynthesis. *Mutat. Res., 44*:447, 1977.
12. Ueno, Y., Kubota, K., and Nakamura, Y. Mutagenicity of carcinogenic mycotoxins in *Salmonella typhimurium*. *Cancer Res., 38*:536, 1978.
13. Nagao, M., Sugimura, T., and Matsushima, T. Environmental mutagens and carcinogens. *Ann. Rev. Genet., 12*:117, 1978.
14. Williams, G. M. Detection of chemical carcinogens by unscheduled DNA synthesis in rat liver primary cell cultures. *Cancer Res., 37*:1845, 1977.
15. Mori, H., Sugie, S., Yoshimi, N., Asada, Y., Furuya, T., and Williams, G. M. Genotoxicity of a variety of pyrrolizidine alkaloids in the hepatocyte primary culture-DNA repair test using rat, mouse, and hamster hepatocytes. *Cancer Res., 45*:3125, 1985.

16. Mori, H., Sugie, S., Hirono, I., Yamada, K., Niwa, H., and Ojika, M. Genotoxicity of ptaquiloside, a bracken carcinogen, in the hepatocyte of primary culture/DNA-repair test. *Mutat. Res.*, *143*:75, 1985.
17. Williams, G. M., Laspia, M. F., Mori, H., and Hirono, I. Genotoxicity of cycasin in the hepatocyte primary culture/DNA repair test supplemented with beta-glucosidase. *Cancer Lett.*, *12*:329, 1981.
18. Mori, H., Kawai, K., Ohbayashi, F., Kuniyasu, T., Yamazaki, M., Hamasaki, T., and Williams, G. M. Genotoxicity of a variety of mycotoxins in the hepatocyte primary culture/DNA repair test using rat and mouse hepatocytes. *Cancer Res.*, *44*:2918, 1984.
19. Mori, H., Kitamura, J., Sugie, S., Kawai, K., and Hamasaki, T. Genotoxicity of fungal metabolites related to aflatoxin B_1 biosynthesis. *Mutat. Res.*, *143*:121, 1985.
20. Mori, H., Sugie, S., Yoshimi, N., Kitamura, J., Niwa, M., Hamasaki, T., and Kawai, K. Genotoxic effects of a variety of sterigmatocystin-related compounds in the hepatocyte/DNA-repair test and the *Salmonella* microsome assay. *Mutat. Res.*, *173*:217, 1986.
21. Mori, H., Sugie, S., Yoshimi, N., Kuniyasu, T., Iwata, H., Kawai, K., and Hamasaki, T. Potential carcinogenicity of 5,6-dimethoxysterigmatocystin in rats. *Carcinogenesis*, *9*:1039, 1988.
22. Mori, H., Sugie, S., Niwa, K., Takahashi, M., and Kawai, K. Induction of intestinal tumours in rats by chrysazin. *Br. J. Cancer*, *52*:781, 1985.
23. Mori, H., Sugie, S., Niwa, K., Yoshimi, N., Tanaka, T., and Hirono, I. Carcinogenicity of chrysazin in large intestine and liver of mice. *Jpn. J. Cancer Res. (Gann)*, *77*:871, 1986.
24. Ames, B. N., Durston, W. E., Yamasaki, E., and Lee, F. D. Carcinogens are mutagens; a simple test system combining liver homogenates for activation and bacteria for detection. *Proc. Natl. Acad. Sci. USA*, *70*:2281, 1973.
25. Kubota, K., Asahara, Y., and Ueno, Y. Mutagenicity of toxic fungal metabolites. *Abstracts*, 95th Annual Meeting Jpn. Soc. Pharmaceut. Sci., 1975, p. 111.
26. McCann, J., and Ames, B. N. Detection of carcinogens as mutagens in the *Salmonella*/microsome test; assay of 300 chemicals: discussion. *Proc. Natl. Acad. Sci. USA*, *73*:950, 1976.
27. McCann, J., Spingarn, N. E., Kobari, J., and Ames, B. N. Detection of carcinogens as mutagens; bacterial tester strains with R factor plasmids. *Proc. Natl. Acad. Sci. USA*, *72*:979, 1975.
28. Stark, A. A., Essigmann, J. A., Demain, A. L., Skopek, T. R., and Wogan, G. N. Aflatoxin B_1 mutagenesis, DNA binding, and adduct formation in *Salmonella typhimurium*. *Proc. Natl. Acad. Sci. USA*, *76*:1343, 1979.
29. Wheeler, L., Malula, M., and Demeo, M. A comparison of aflatoxin B_1-induced cytotoxicity, mutagenicity and prophage induction in *Salmonella typhimurium* mutagen tester strains TA 1535, TA 1538, TA 98 and TA 100. *Mutat. Res.*, *83*:39, 1981.
30. Bjeldanes, L. F., Chang, G. W., and Thomson, S. V. Detection of mutagens produced by fungi with the *Salmonella typhimurium* assay. *Appl. Environ. Microbiol.*, *35*:1150, 1978.

31. Wehner, F. C., Thiel, P. G., Van Rensburg, S. J., and Demasius, I. P. C. Mutagenicity to *Salmonella typhimurium* of some *Aspergillus* and *Penicillium* mycotoxins. *Mutat. Res., 58*:193, 1978.
32. Nagao, M., Honda, M., Hamasaki, T., Natori, S., Ueno, Y., Yamazaki, M., Seino, Y., Yahagi, T., and Sugimura, T. Mutagenicities of mycotoxins on *Salmonella*. *Proc. Jpn. Assoc. Mycotoxicol., 3/4*: 41, 1976.
33. Liberman, D. F., Schaefer, F. L., Fink, R. C., Ramgopal, M., Ghosh, A. C., and Mulcahy, R. Mutagenicity of islandicin and chrysophanol in the *Salmonella*/microsome system. *Appl. Environ. Microbiol., 40*:476, 1980.
34. Tikkanen, L., Matsushima, T., and Natori, S. Mutagenicity of anthraquinones in the *Salmonella* preincubation test. *Mutat. Res., 116*: 297, 1983.
35. Brown, J. P. A review of the genetic effects of naturally occurring flavonoids, anthraquinones and related compounds. *Mutat. Res., 75*: 243, 1980.
36. Stark, A. A., Kobbe, B., Marsuo, K., Buchi, G., Wogan, G. N., and Demain, A. L. Mollicellins: mutagenic and antibacterial mycotoxins. *Appl. Environ. Microbiol., 36*:412, 1978.
37. Mori, H., Kawai, K., Ohbayashi, F., Kitamura, J., and Nozawa, Y. Genotoxicity of quinone pigments from pathogenic fungi. *Mutat. Res., 122*:29, 1983.
38. Decloitre, P., and Hamon, G. Species-dependent effects of dietary lindane and/or zineb on the activation of aflatoxin B_1 into mutagenic derivatives. *Mutat. Res., 79*:185, 1980.
39. Baker, R. S. U., Bonin, A. M., Stupans, I., and Holder, G. M. Comparison of rat and guinea pig as sources of the S9 fraction in the *Salmonella*/mammalian microsome mutagenicity test. *Mutat. Res., 71*:43, 1980.
40. Garner, R. C., Miller, E. C., and Miller, J. A. Liver microsomal metabolism of aflatoxin B_1 to a reactive derivative toxic to *Salmonella typhimurium* TA 1530. *Cancer Res., 32*:2058, 1972.
41. Vaught, J., and Bresnick, E. Binding of polycyclic hydrocarbons to nuclear components in vitro. *Biochem. Biophys. Res. Commun., 69*: 587, 1976.
42. Gose, A., Sarasin, A., Moule, Y., and Devoret, R. Induction and mutagenesis of prophage λ in *Escherichia coli* k12 by metabolites of aflatoxin B_1. *Mutat. Res., 28*:1, 1975.
43. Ueno, Y., Ueno, I., Iitoi, Y., Tsunoda, H., Enomoto, M., and Ohtsubo, K. Toxicological approaches to the metabolites of *Fusaria*. III. Acute toxicity of fusarenon-X. *Jpn. J. Exp. Med., 41*:521, 1971.
44. Ueno, Y., and Nakajima, M. Production of respiratory-deficient mutants of *Saccharomyces cerevisae* by (−)-luteoskyrin and (+)-rugulosin. *Chem. Pharm. Bull., 22*:2258, 1974.
45. Kuczak, M. H., Benson, P. H., Heath, H., and Hayes, A. W. Evaluation of the mutagenic potential of mycotoxins using *Salmonella typhimurium* and *Saccharomyces cerevisiae*. *Mutat. Res., 53*:11, 1978.

46. Nestmann, E. R., Nasim, A., Haynes, R. H., and Kowbel, D. J. Genetic activity of actinomycin D in *Saccharomyces cerevisiae* but not in *Escherichia coli*. *Mutat. Res.*, *89*:229, 1981.
47. Malling, H. V., and de Serres, F. G. Correlation between base-pair transition and complementation pattern in nitrous acid-induced ad-3B mutant of *Neurospora crassa*. *Mutat. Res.*, *5*:359, 1968.
48. Ong, T. Mutagenicity of aflatoxins in *Neurospora crassa*. *Mutat. Res.*, *9*:615, 1970.
49. Ong, T., and de Serres, F. J. Mutagenicity of chemical carcinogens in *Neurospora crassa*. *Cancer Res.*, *32*:1890, 1972.
50. Fisher, C. R., Malling, H. V., de Serres, F. J., and Snyder, S. Mutagenicity of actinomycin D in *Neurospora crassa*. *Mutat. Res.*, *33*:187, 1975.
51. Kuroki, T., Malaveille, C., Drevon, C., Piccoli, C., Macleod, M., and Selkirk, J. K. Critical importance of microsome concentration in mutagenesis assay with V79 Chinese hamster cells. *Mutat. Res.*, *63*:259, 1979.
52. Epstein, S. S., and Shafner, H. Chemical mutagens in the human environment. *Nature*, *219*:385, 1968.
53. Umeda, M., Tsutsui, T., and Saito, M. Mutagenicity and inducibility of DNA single-strand breaks and chromosome aberrations by various mycotoxins. *Gann*, *68*:619, 1977.
54. Vogel, E. The relation between mutational pattern and concentration by chemical mutagens. In *Screening Tests in Chemical Carcinogens* (R. Montesano, Ed.), Publ. 12, Int. Agency Res. Cancer, Lyons, 1976, p. 117.
55. Vogel, E., and Sobels, F. H. The function of *Drosophila* in genetic toxicology. In *Chemical Mutagens: Principles and Methods for Their Detection*, Vol. 4 (A. Hollaender, Ed.), Plenum, New York, 1976, p. 93.
56. Fahmy, M. J., Fahmy, O. G., and Swenson, D. H. Aflatoxin B_1-2,3-dichloride as a model of the active metabolite of AFB_1 in mutagenesis and carcinogenesis. *Cancer Res.*, *38*:2608, 1976.
57. Proust, J. P., Sankaranarayanan, K., and Sobels, F. H. The effects of treating *Drosophila* females with actinomycin-D on the yields of dominant lethals, translocations and recessive lethals recovered from X-irradiated spermatozoa. *Mutat. Res.*, *16*:65, 1972.
58. Tong, C., and Williams, G. M. Induction of purine analog-resistant mutants in adult rat liver epithelial lines by metabolic activation-dependent and independent carcinogens. *Mutat. Res.*, *58*:339, 1978.
59. Stich, H. F., and Laishes, B. A. DNA repair and chemical carcinogens. *Pathobiol. Ann.*, *3*:341, 1973.
60. Williams, G. M. The detection of genotoxic chemicals in the hepatocyte primary culture/DNA repair test. *Gann Monogr. Cancer Res.*, *27*:45, 1981.
61. Kawai, K., Mori, H., Sugie, S., Yoshimi, N., Inoue, T., Nakamaru, T., Nozawa, Y., and Matsushima, T. Genotoxicity in the hepatocyte/DNA repair test and toxicity to liver mitochondria of 1-hydroxyanthraquinone and several dihydroxyanthraquinones. *Cell. Biol. Toxicol.*, *2*:457, 1986.

62. Mori, H., Yoshimi, N., Iwata, H., Tanaka, T., Kawai, K., and Sankawa, U. Additional survey on genotoxicity of natural anthraquinones in the hepatocyte primary culture/DNA repair assay. *J. Toxicol. Sci.*, *13*:161, 1988.
63. Kawai, K., Nozawa, Y., Mori, H., and Ogihara, Y. Inhibition of mitochondrial respiration by flavoskyrin, a toxic metabolite of *Penicillium islandicum* Sopp. *Toxicol. Lett.*, *30*:105, 1986.
64. Kawai, K., Shiojiri, H., Nakamaru, T., Nozawa, Y., Sugie, S., Mori, H., Kato, T., and Ogihara, Y. Cytotoxicity and genotoxicity of xenoclauxin and desacetylduclauxin from *Penicillium declauxii* (Delacroix). *Cell Biol. Toxicol.*, *1*:1, 1985.
65. Kawai, K., Mori, H., and Kitamura, J. The uncoupling effect of flavoglaucin, a quinol pigment from *Aspergillus chevalieri* (Mangin), on mitochondrial respiration. *Toxicol. Lett.*, *19*:321, 1983.
66. Mori, H., and Kawai, K. Genotoxicity in rodent hepatocytes and carcinogenicity of mycotoxins and related chemicals. In *Mycotoxins and Phycotoxins '88* (S. Natori, K. Hashimoto, and Y. Ueno, Eds.), Elsevier, New York, 1989, p. 81.
67. Mori, H., Yoshimi, N., Iwata, H., Mori, Y., Hara, A., Tanaka, T., and Kawai, K. Carcinogenicity of naturally occurring 1-hydroxyanthraquinone in rats: induction of large bowel, liver and stomach neoplasms. *Carcinogenesis*, *11*:799, 1990.
68. Williams, G. M., and Mori, H. Liver carcinogenesis by chemicals through promotion mechanism involving modification of the cell membrane. *J. Pesticide Sci.*, *8*:619, 1983.
69. Ueno, Y., and Kubota, K. DNA-attacking ability of carcinogenic mycotoxins on recombination-deficient mutant cells of *Bacillus subtilis*. *Cancer Res.*, *36*:445, 1976.
70. Kada, K., Tsuchikawa, K., and Sadaie, Y. In vitro and host mediated Rec-assay procedures for screening chemical mutagens, and phloxine, mutagenic red dye detected. *Mutat. Res.*, *16*:165, 1972.
71. Pienta, R. J., Poiley, J. A., and Lebherz, W. B. Morphological transformation of early passage golden Syrian hamster embryo cells derived from cryopreserved primary cultures as a reliable *in vitro* bioassay for identifying diverse carcinogens. *Int. J. Cancer*, *19*:642, 1977.
72. Evans, C. H., and DiPaolo, J. A. Neoplastic transformation of guinea pig fetal cells in culture induced by chemical carcinogens. *Cancer Res.*, *35*:1035, 1975.
73. DiPaolo, J. A., Takano, K., and Popescu, N. C. Quantitation of chemically induced neoplastic transformation of Balb/3T3 cloned cell lines. *Cancer Res.*, *32*:2686, 1972.
74. Haugen, A., Groopman, J. D., Hsu, I.-C., Goodrich, G. R., Wogan, G., and Harris, C. C. Monoclonal antibody to aflatoxin B_1-modified DNA detected by enzyme immunoassay. *Proc. Natl. Acad. Sci. USA*, *78*:4124, 1981.
75. Iskandar, O., and Vijayalaxmi. The enhancement of the aflatoxin B_1 by metabolic activation with rat-liver microsomes on human lymphocytes associated with the micronucleus test. *Mutat. Res.*, *91*:63, 1981.

76. Friedman, M. A., and Staub, J. Induction of micronuclei in mouse and hamster bone-marrow by chemical carcinogens. *Mutat. Res.*, *43*:255, 1977.
77. Miller, R. C. The micronucleus test as an *in vivo* cytogenetic method. *Environ. Health Perspect.*, *6*:167, 1973.
78. Schmid, W. The micronucleus test. *Mutat. Res.*, *31*:9, 1975.
79. Trzos, R. J., Petzold, G. L., Brunden, M. N., and Swenberg, J. A. The evaluation of sixteen carcinogens in the rat using the micronucleus test. *Mutat. Res.*, *58*:79, 1978.
80. Heddle, J. A., and Bruce, W. R. Comparison of tests for mutagenicity or carcinogenicity using assays for sperm abnormalities, formation of micronuclei and mutations in *Salmonella*. In *Origins of Human Cancer* (H. H. Hiatt, J. D. Watson, and J. A. Winsten, Eds.), Cold Spring Harbor Conferences on Cell Proliferation, 4, Cold Spring Harbor Lab., Cold Spring Harbor, N.Y., 1977, p. 1549.
81. Jenssen, D., and Ramel, C. The micronucleus test as part of a short-term mutagenicity test program for the prediction of carcinogenicity evaluated by 143 agents tested. *Mutat. Res.*, *75*:191, 1980.
82. Batt, T. R., Hsueh, J. L., Chen, H. H., and Huang, C. C. Sister chromatid exchanges and chromosome aberrations in V79 cells induced by aflatoxin B_1, B_2, G_1 and G_2 with or without metabolic activation. *Carcinogenesis*, *1*:759, 1980.
83. Ray-Chaudhuri, R., Kelley, S., and Iype, P. T. Induction of sister chromatid exchanges by carcinogens mediated through cultured rat liver epithelial cells. *Carcinogenesis*, *1*:779, 1980.
84. Lake, R. S., Kropko, M. L., McLachlan, S., Pezzutti, M. R., Shoemaker, R. H., and Igel, H. J. Chemical carcinogen induction of DNA-repair synthesis in human peripheral blood monocytes. *Mutat. Res.*, *74*:357, 1980.
85. Terao, K., and Ito, E. The effects of naturally occurring bisdihydrofuran ring-containing mycotoxins on cultured chick embryonal hepatocytes. *Gann*, *72*:81, 1981.
86. Wyrobek, A. J., and Bruce, W. R. Chemical induction of sperm abnormalities in mice. *Proc. Natl. Acad. Sci. USA*, *72*:4425, 1975.
87. Engelbrecht, J. C., and Altenkirk, B. Comparison of some biological effects of sterigmatocystin and aflatoxin analogues on primary cell cultures. *J. Natl. Cancer Inst.*, *48*:1647, 1972.
88. Kawasaki, I., Oki, T., Umeda, M., and Saito, M. Cytotoxic effects of penicillic acid and patulin on HeLa cells. *Jpn. J. Exp. Med.*, *42*:327, 1970.
89. Ueno, Y., and Hosaya, M. Bioassay of toxic principles of rice infested with *Fusarium nivale* employing rabbit reticulocytes. *Proceedings*, 1st U.S.-Japan Conf. on Toxic Micro-Organisms, 1968, p. 80.
90. Ueno, Y., and Yamakawa, H. Antiprotozoal activity of scirpene mycotoxins of *Fusarium nivale* Fn 2B. *Jpn. J. Exp. Med.*, *40*:385, 1970.
91. Ohtsubo, K., and Saito, M. Cytotoxic effects of scirpine compounds, fusarenon-X produced by *Fusarium nivale*, dihydronivalenol and dihydrofusarenon-X on HeLa cells. *Jpn. J. Med. Sci. Biol.*, *23*:217, 1970.

92. Stanford, G. K., Hood, R. D., and Hayes, A. W. Effect of prenatal administration of T-2 toxin to mice. *Res. Commun. Chem. Pathol. Pharmacol.*, *10*:743, 1975.
93. Khera, K. S., Whalen, C., Angers, G., Vesonder, R. F., and Kuiper-Goodman, T. Embryotoxicity of 4-deoxynivalenol (vomitoxin) in mice. *Bull. Environ. Contam. Toxicol.*, *29*:487, 1982.
94. Ito, Y., Ohtsubo, K., and Saito, M. Effects of fusarenon-X, a trichothecene produced by *Fusarium nivale*, on pregnant mice and their fetuses. *Jpn. J. Exp. Med.*, *50*:167, 1980.
95. Ueno, Y. The toxicology of mycotoxins. *CRC Crit. Rev. Toxicol.*, *14*:99, 1985.
96. Watanabe, T., Taguchi, Y., Kitamura, Y., Tsuyama, K., Fujiki, H., Tanooka, H., and Sugimura, T. Induction of histidine decarboxylase activity in mouse skin after application of indole alkaloids, a new class of tumor promoter. *Biochem. Biophys. Res. Commun.*, *109*:478, 1982.
97. Kuzuya, M., Noguchi, A., Kawai, K., and Mori, H. Quantum chemical study for genotoxic and antitumor activities of hydroxyanthraquinones. *Regul. Toxicol. Pharmacol.*, *13*:185, 1991.

10

OXIDATION-REDUCTION REACTIONS IN BIOSYNTHESIS OF SECONDARY METABOLITES

DEEPAK BHATNAGAR, KENNETH C. EHRLICH, and THOMAS E. CLEVELAND
Southern Regional Research Center, Agricultural Research Service, U.S. Department of Agriculture, New Orleans, Louisiana

I. INTRODUCTION

Aflatoxins are extremely biologically active, secondary metabolites produced by the fungi *Aspergillus flavus* Link ex. Fries and *A. parasiticus* Speare [1]. Aflatoxins are receiving increased attention from researchers, the food industry, and the general public for two main reasons: (1) the metabolites (particularly aflatoxin B_1) are not only toxic to animals as well as humans but are also the most carcinogenic of all known natural compounds [2, 3]; and (2) there is a high incidence of the compounds in food and feed throughout the world [4]. The major aflatoxins are B_1, B_2, G_1, and G_2; synthesis of these compounds serves as a model for the basic description of biosynthesis of polyketide-derived secondary metabolites.

Aflatoxin synthesis has no obvious physiological role in primary growth and metabolism of the organism and therefore is considered to be a "secondary" process [5]. As yet, there is no confirmed biological role of aflatoxins in the ecological survival of the fungal organism. However, since aflatoxins are toxic to certain potential competitor microbes in the ecosystem [1], a survival benefit to the producing fungi is implied. Theories have also been proposed about a possible biological role of aflatoxins or related compounds as deterrents to insect feeding activity on fungal overwintering bodies (sclerotia) [6, 7]. It should be noted, however, that aflatoxin per se is a poor antibiotic [8].

Aflatoxins are polyketide-derived secondary metabolites [9]. The mode of action, metabolism, and biosynthesis of the toxins have been extensively studied [3, 5, 10–15]. Mutants of *A. parasiticus* and *A. flavus* that are impaired in aflatoxin synthesis have been useful in studying the aflatoxin biosynthetic pathway [5, 9, 10, 16–18]. Bioconversion experiments using blocked mutants, metabolic inhibitors, and radiolabeled precursors have produced an accepted general scheme of precursors in aflatoxin B_1 biosynthesis: acetate → polyketide → anthraquinones → xanthones → aflatoxins [13, 19, 20]. The enzymes required for the production of aflatoxins have not been characterized in as much detail as have the enzymes involved in primary metabolism [21]. Several efforts have, however, been made toward developing a cell-free system for identifying and purifying the enzymes involved in aflatoxin biosynthesis [11, 22–26].

A solution to the aflatoxin problem through biotechnology could involve acquisition of information on the molecular regulation of biosynthesis of the toxin. Relevant knowledge is being acquired about the biosynthetic intermediates and enzymes associated with the aflatoxin pathway. In this chapter a comprehensive analysis of the molecular mechanism of aflatoxin biosynthesis will be discussed that involves identification and characterization of pathway intermediates and enzymes involved in the synthesis of aflatoxins B_1, B_2, G_1, and G_2. An understanding of the complex steps of aflatoxin biosynthesis will enable investigators to characterize genes responsible for the pathway as well as to understand the regulation of pertinent genes involved in secondary metabolism.

II. EFFECT OF PRIMARY METABOLISM ON SECONDARY METABOLISM

Drew and Demain [27] have elegantly discussed the fact that primary and secondary metabolism are separate and distinct processes. The interdependent catabolic and biosynthetic activities of primary metabolism give structure and biological function to organisms. On the other hand, secondary metabolism produces compounds whose function in the welfare, development, and reproduction of the producing organisms is not apparent. The metabolic pathway of primary metabolism supplies the precursors of secondary metabolism. Therefore, factors that influence primary metabolism affect secondary metabolism. Secondary metabolites fail to appear during the growth phase (trophophase) because the enzymes necessary for their formation are absent [28]. Secondary metabolism is initiated only in the production phase (idiophase). Drew and Demain [27] have postulated that primary and secondary metabolism may be separated only by differences in pathway regulation; the poorly defined role of secondary metabolites in the welfare of the producing organism probably reflects an absence of relevant information on the role of these metabolites.

Biosynthesis of secondary metabolites in fungi shares a common lipid and protein pool with primary metabolites [27, 29–32]. The relationship between fatty acids and secondary biosynthesis has been reported in several species of fungi [33–36]. A close association between aflatoxins and lipid biosynthesis has also been established [37–40], and it is known that protein synthesis declines during the aflatoxin-producing phase (idiophase) [41–43].

III. POLYKETIDE SYNTHESIS

The first step in formation of most acetate-derived secondary metabolites is chain elongation by polyketide synthesis (Fig. 1) [44]. Polyketide biosynthesis represents the initial steps in which the organism is committed

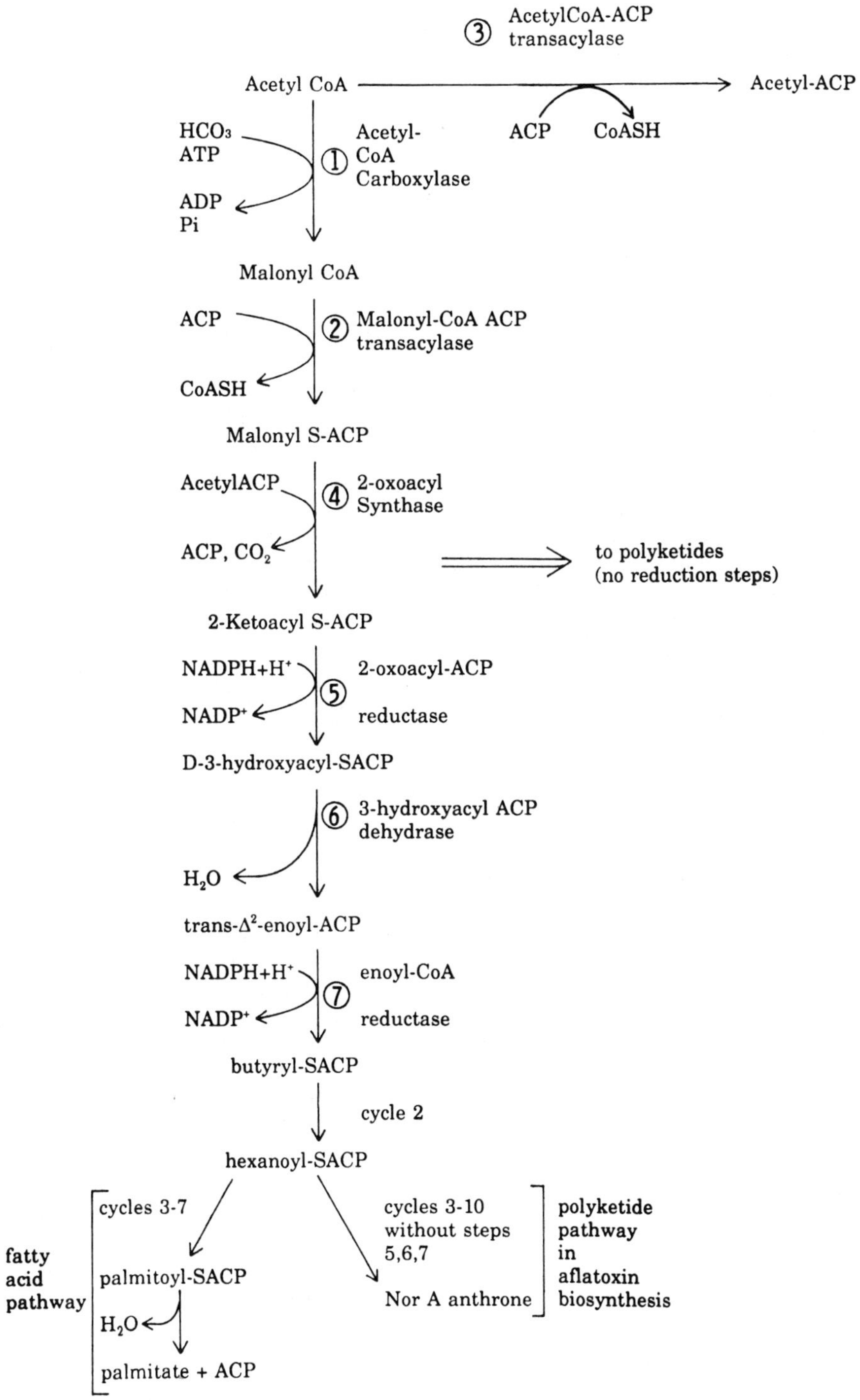

FIGURE 1. Pathway for the conversion of acetyl CoA and malonyl CoA to polyketides and fatty acids. (Adapted from Ref. 74.)

to the production of secondary metabolic products [45–47]. Therefore, similar to fatty acid biosynthesis, many of the enzymatic steps in polyketide biosynthesis are essentially irreversible.

Polyketides are a large, structurally diverse class of natural products arising from self-condensation of acetate units by head-to-tail linkage via poly-β-carbonyl intermediate compounds [48]. The origin and current status of the specificity by which the simple polyketide chain is folded, condensed, oxidized, reduced, cleaved and rearranged is of significance for understanding the biosynthesis of aflatoxins as well as other secondary metabolites of polyacetate origin. The key reaction in the polyacetyl chain formation is cyclization [49].

Table 1 illustrates several examples of polyketides ranging from tetra- to decaketides [58]. Although a great variety of secondary metabolites have been identified, the initial process of polyketide synthesis is remarkably simple and involves steps similar to those of fatty acid biosynthesis. It is not known whether cyclization is catalized by specific enzymes or is a consequence of carbon chain conformational folding in a multienzyme synthase complex [45, 46]. In most cases, although a variety of products can be found (indicating a flexibility in the allowed conformations), the major product is the one expected based on the principal of maximal aromatization [59]. Thus, with the hepta- and octaketides in which no initial reduction steps occur, the cyclization products have the greatest number of condensed aromatic rings [58]. The final step in the polyketide biosynthetic process is assumed to be hydrolysis of the polyketide acyl–acyl carrier protein bond, which releases the product from the synthase. It is probable that cyclization occurs prior to product release because it is unlikely that a structurally unstable polyketide would be capable of folding correctly after it has dissociated from the enzyme. Therefore, no uncyclized precursor polyketides have been identified as fungal metabolites.

After hydrolysis to release the aromatized polyketide, the molecule must remain accessible to further chemical transformations by a variety of different enzymes and cofactors. Since after cyclization the molecule is relatively more hydrophobic than the polyketide, it may be transported and processed within the lipid matrix of the fungal cell. Pertinent enzymes could be bound to cell membrane components or, if free, have lipophilic portions that facilitate tethering the hydrophobic end of the metabolite.

Polyketide synthesis is strikingly similar to fatty acid biosynthesis [60], without some of the intermediate reduction steps. Schemes for both fatty acid and polyketide synthesis are shown in Fig. 1. A polyketide synthase, 6-methylsalicylic acid synthase (MSAS), has been purified recently from *Penicillium patulum*, and the gene coding for the protein has been cloned and sequenced [61]. Although similar to fatty acid synthase in the sequential order of catalytic domains, MSAS has only a low degree of sequence homology to the fatty acid synthase from *P. patulum*.

The first step in both fatty acid and polyketide biosynthesis is the formation of malonyl CoA from acetyl CoA (Fig. 1); for fatty acid synthesis this conversion is catalyzed by the enzyme acetyl CoA carboxylase, which from yeast contains a biotin prosthetic group [62]. The carboxylation step is catalyzed in two stages. The first stage involves

TABLE 1. Examples of Polyketide-Derived Products from Various Fungi

Ketide /Structure		Probable primary product in biosynthesis	Number of reduction steps to primary product	Final product	Organ-ism	Ref.
Tetra	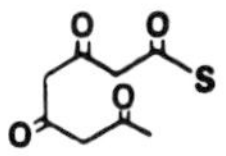	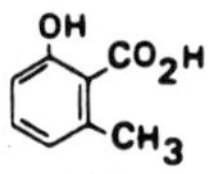 6MSA	1	Patulin	P. patulum	67
		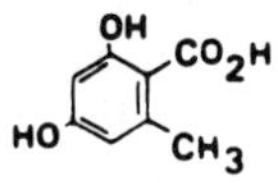 Orsellinic acid	0	Penicillic acid	P. cyclopium	67
penta	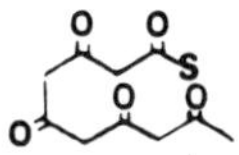	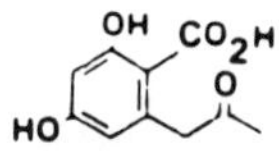	0	Citrinin	P. citrinum	119
		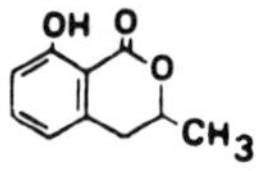	1	Ochratoxin	A. ochraceus	74
hexa	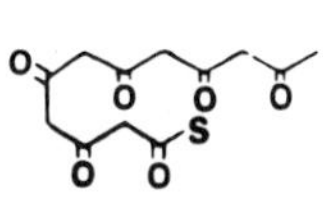	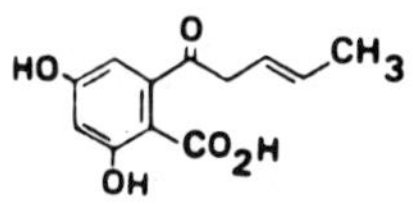	1	Maltoryzine	A. oryzae	120
hepta	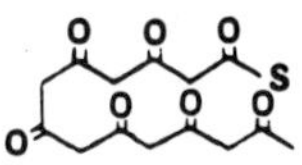	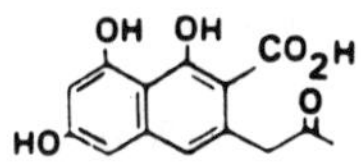	0	Viriditoxin	A. viridinutans	121
			0	Xanthomegnins	A. melleus	122
Octa	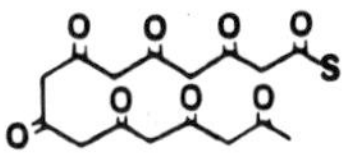	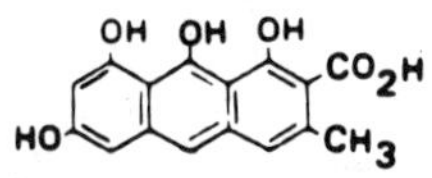	0	Ergochromes	A. ochraceus P. oxalicum P. islandicum	123
			0	Anthraquinones		123
Nona	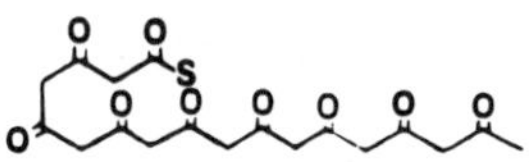	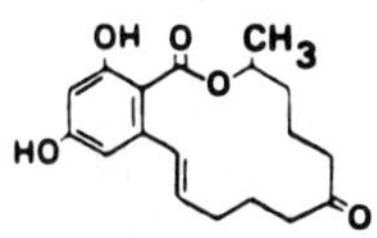	3	Zearalenone	Fusarium sp.	124
Deca	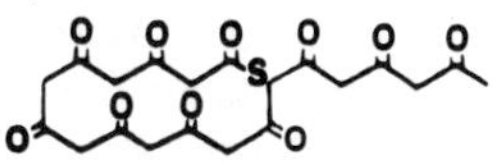	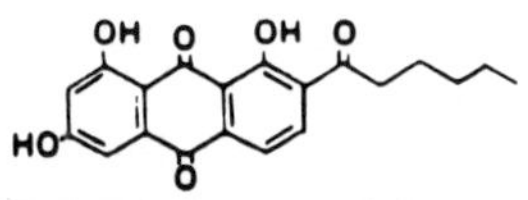	2	Aflatoxin	A. flavus	124

formation of a carboxybiotin intermediate, which is transferred to acetyl CoA. In eukaryotes, the carboxylase is subject to allosteric control in which citrate activates the enzyme while the end product, palmitoyl CoA, inactivates it; however, this pathway may not be the source of malonyl

CoA used in polyketide synthesis. The malonyl CoA used in secondary metabolism may arise from oxaloacetate via decarboxylation catalyzed by oxaloacetate dehydrogenase [63]; oxaloacetate is generated by carboxylation of pyruvate or phosphoenolpyruvate during glycolysis [36, 64]. Polyketide biosynthesis, in aflatoxin formation, has been related to pyruvate kinase activity, which was found to be high in toxigenic strains of *Aspergillus parasiticus* and low in nontoxigenic strains [65]. Asparagine, like pyruvate, has also been shown to stimulate aflatoxin formation [66]. Asparagine or aspartate by transamination to oxaloacetate could also be a plentiful source of the large amounts of malonyl CoA needed for polyketide biosynthesis [63].

Any factors that suppress tricarboxylic acid cycle activity would be expected to enhance polyketide biosynthesis [67]; the stimulatory effect of glucose and sucrose on aflatoxin formation by carbon catabolite induction represents such an effect [68, 69], just as the synthesis of fatty acids is enhanced when carbohydrate is abundant. The balance between fatty acid biosynthesis and polyketide biosynthesis may be affected by factors that influence NADPH formation. For example, the requirement for zinc in aflatoxin formation may result from its ability to inhibit certain NADPH-generating enzymes [70]. However, NADPH is required for polyketide biosynthesis as well as many of the subsequent oxidation steps.

In both fatty acid biosynthesis and polyketide biosynthesis, chain elongation begins with the condensation of acetyl CoA and malonyl CoA [71]; this reaction is catalyzed by a transferase activity (the fatty acid or polyketide synthase) fixed into a multienzyme complex in the form of a flexible protein "arm" [72]. The condensation steps are driven by release of carbon dioxide. The acyl group of both acetyl CoA and malonyl CoA are first transferred to a protein called acyl carrier protein (ACP), which is part of the multienzyme complex. The transfers are catalyzed by acetyl CoA-ACP transacylase and malonyl CoA-ACP transacylase, respectively; for MSAS the same site catalyzes both transfers [61]. Malonyl CoA and acetyl CoA are linked to the sulfhydryl terminus of a phosphopantethienyl group attached to a serine residue of ACP (the "flexible arm") [72, 73]. MSAS from *P. patulum* has a subunit mass of 191 kDa and functions in polyketide synthesis as a homomultimer (0.8–1.1 MDa). The single polypeptide chain contains the acyl carrier protein, the synthase, the 2-oxoacyl reductase, and the transacylase activity for the synthesis of a tetraketide [61]. In the biosynthesis of longer chain ketides, where reduction to hydrocarbon must occur, the additional activities of β-hydroxyacyl dehydratase and enoyl reductase may be part of the enzyme complex. In fatty acid synthesis, multiple activities occur on two polypeptide chains, each encoded by a single gene. Within the enzyme complex, the growing polyketide chain is transferred from the ACP to the synthase and then back again, involving a series of translocations in which the first reaction makes room for the incoming malonyl unit and the second is involved in the condensation step [73, 74]. The flexibility needed for the process is provided by the phosphopantetheinyl group, which allows the growing chain to remain bound to the complex but still accessible to the active sites of the enzymes [50].

It is unlikely that the polyketide synthase complex evolved from the fatty acid synthase by gene duplication as has been proposed [9], since, at least in *P. patulum*, MSAS has only low sequence homology to its own fatty acid synthase. The MSAS gene shows considerable homology to other fatty acid synthase genes in a variety of organisms [61]. The most homology occurs in the amino acid region around the substrate binding sites. MSAS differs from most fatty acid synthases in that a single transacylase is capable of accepting both the malonyl CoA and acetyl CoA for transacylations as well as performing the thioesterase function to release the polyketide from the complex. Several other regions of the MSAS gene have no analogy to fatty acid synthases and may be involved in the final condensation and cyclization steps. Oligonucleotide probes prepared to identify the most conserved regions of *P. patulum* MSAS [61] are currently being used to screen cDNA and genomic libraries prepared from *A. parasiticus* and *A. flavus* mRNA and DNA to identify the polyketide synthase gene involved in aflatoxin biosynthesis (Bhatnagar, unpublished observations).

IV. TRANSFORMATION OF POLYKETIDE TO AFLATOXINS

With almost all polyketide-derived secondary metabolites, the condensation product undergoes extensive modification, a process that is dictated by two features of the metabolite: (1) the presence of resorcinol hydroxyls, which make the aromatic portion highly electron-rich and readily susceptible to oxidation, and (2) the presence of hydrocarbon side chains that can be oxidized.

What purpose does modification of the aflatoxin precursor serve? It has been estimated that 16 further enzyme-catalyzed steps are needed to complete the biosynthesis of the toxin. It is unlikely that the later modification enzymes are present in the cytosol for general oxidative or reductive processes, since the enzyme reactions have recently been shown to be highly substrate-specific [13, 112] and synthesis of the pertinent enzymes occurs exclusively during idiophase [24, 28, 75]. The primary driving force for further biosynthetic modification of the polyketide appears to be the synthesis of an end product that can be secreted by the mold. Although intermediate intracellular products are formed during aflatoxin biosynthesis, the main secondary metabolites are eventually excreted. Therefore, following polyketide biosynthesis, the subsequent steps in aflatoxin synthesis may be viewed as degradation stages of the polyketide product. The conversion process could be similar to fatty acid oxidation; the net result of oxidation steps after ring cyclization is a release of energy in a process independent of the fatty acid cycle. For illustration of the wide variety of enzymatic modification steps possible in secondary alterations of polyketide condensate, the aflatoxin B_1 pathway is presented as a model (Fig. 2). In aflatoxin synthesis, the initial decaketide condensate [58] has the most sites available for potential modification and rearrangement.

The timing of the reduction steps during formation of the hexanoate side chain of norsolorinic acid (NOR), the presumed first stable cyclization product following polyketide elongation, is uncertain (Fig. 3) [40]. It has been shown that hexanoic acid is incorporated intact into the

Decaketide

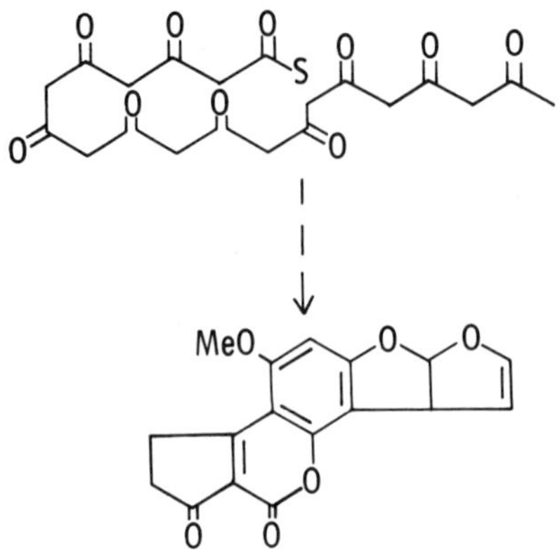

Aflatoxin B1

FIGURE 2. Aflatoxin B_1, a decaketide-derived secondary metabolite of *Aspergillus flavus* and *A. parasiticus*.

decaketide side chain, possibly by exchange with hexanoate [76] that is bound to the polyketide synthase; in an analogous process, thioesters can exchange in fatty acid biosynthesis [60]. The reductive steps occur prior to complete polyketide synthesis [77]. For example, MSAS reduction follows the addition of the second acetate unit [78]; in such cases, the polyketide synthase retains the reductive enzyme activities of fatty acid synthase, and it is the organization of the relevant activities within the complex that determines whether or not reduction occurs along the growing polyketide chain. In most, if not all, examples of metabolites derived from the polyketide pathway in which reduction has occurred [6-MSA (a tetraketide), maltoryzine (a hexaketide), NOR (a decaketide)] (Table 1), the reductive steps occur at the beginning of the growing chain rather than at the terminus, and the results indicate that the polyketide pathway can be entered via the fatty acid synthetic route [77]. Therefore, either the organization of subunits in the synthase determines when reduction occurs or the synthase accommodates a preformed short fatty acid CoA species prior to polyketide chain elongation.

A. Anthraquinone Synthesis

Norsolorinic Acid Synthesis

A schematic for the biosynthesis of aflatoxin B_1 starting from the decaketide condensation product Nor A is given in Fig. 3. The presumptive condensation product, an anthrone, is first oxidized to an anthraquinone norsoloronic acid (NOR) [79]; the oxidation step may take place before the polyketide leaves the synthase complex; no intermediate anthrone has ever been isolated, and nothing is known about the requisite oxygenase. Catalysis of the cyclization step may lead directly to NOR.

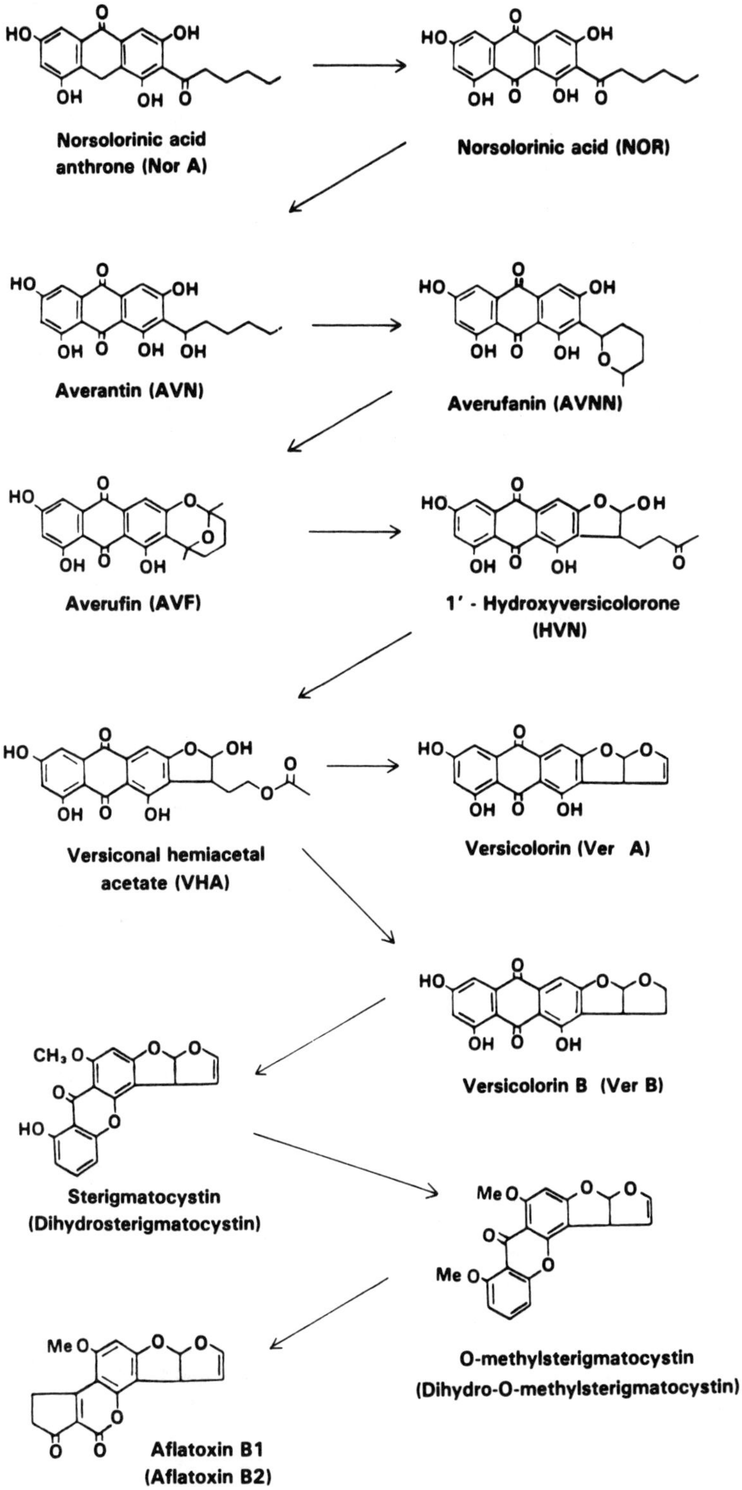

FIGURE 3. Schematic representation of the intermediates in aflatoxin B_1/B_2 biosynthesis from the first polyketide condensation product norsolorinic acid anthrone.

Dehydrogenation of Norsolorinic Acid to Averufin

The steps leading to averufin (AVF) have been investigated extensively using blocked mutants [19], [^{13}C]precursor uptake followed by nuclear magnetic resonance studies, and ^{14}C-radiolabeled precursor incorporation [52]. NOR is converted to averantin (AVN) by a simple reduction of the benzylic ketone to a hydroxyl in the side chain (Fig. 4); reaction from NOR has been demonstrated to require NADPH and an oxidoreductase (NOR reductase or a dehydrogenase). The reductase has been identified in cell-free extracts of fungal mycelia [75, 80] and purified to homogeneity [12]. The enzyme appears to be a single subunit protein with a molecular mass of approximately 40 kDa.

The initial reduction step is immediately followed by oxidation of AVN to averufin (AVF) with the intermediacy of averufanin (AVNN) (Figs. 3

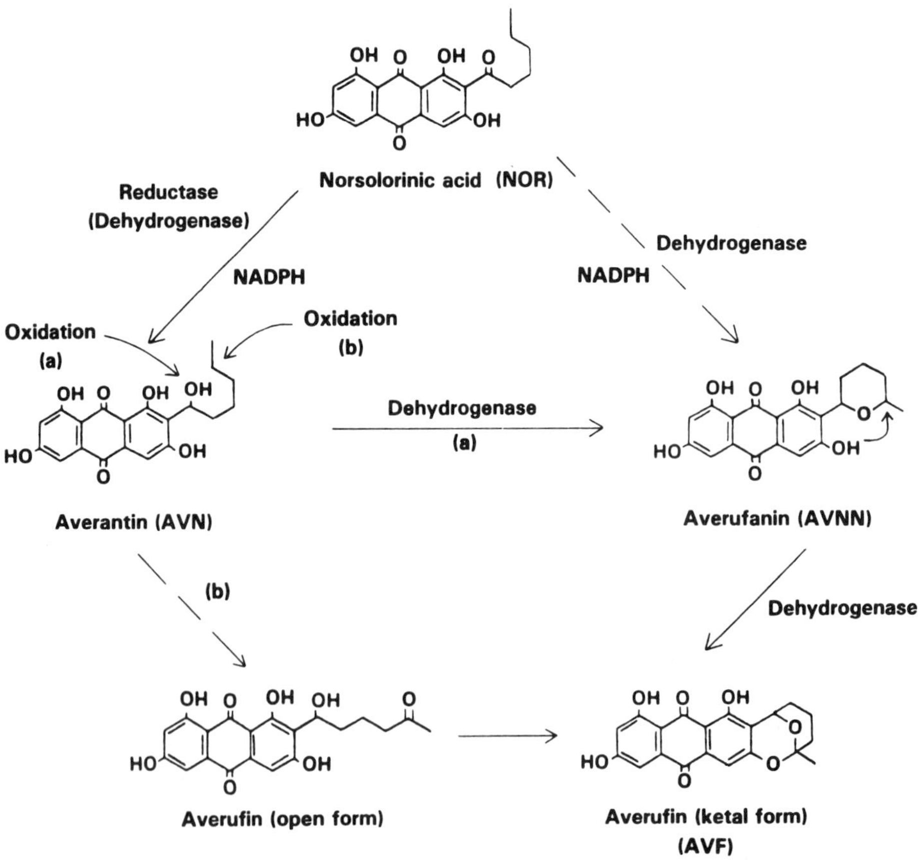

FIGURE 4. Proposed pathways for conversion of norsolorinic acid to averufin.

TABLE 2. Production of Aflatoxin by Various *A. parasiticus* Mutants

A. parasiticus culture	Primary metabolite accumulated	Aflatoxin produced[a] (ppb)			
		B_1	B_2	G_1	G_2
Wild-type (SRRC 143)	Aflatoxins	575,172	134,508	363,301	33,652
NOR-1 mutant (SRRC 162)	Norsolorinic acid	64,432	34,419	12,643	tr
AVN-1 mutant (SRRC 163)	Averantin	n.d.	n.d.	n.d.	n.d.
AVF-1 mutant (SRRC 165)	Averufin	967	11,209	tr	tr
VERA-1 mutant (SRRC 164)	Versicolorin A	n.d.	n.d.	n.d.	n.d.
OMST-1 mutant (SRRC 2043)	*O*-Methylsterigmatocystin	n.d.	n.d.	n.d.	n.d.

Abbreviations: tr = trace (< 100 ppb); n.d. = none detected.
[a]The mycelial cultures were grown for 7 days as described earlier [10] and analyzed for aflatoxin content and mycelial growth.

and 4) [16, 81]. In Fig. 4 the solid arrows represent the accepted pathway for the conversion; oxidation occurs at the penultimate carbon of the hydrocarbon side chain of AVN either to give averufanin (oxidation *a*) or possibly the ketone (oxidation *b*), which internally cyclizes to the ketal, averufin. Therefore, it is possible that multiple pathways to averufin can occur as indicated in Fig. 4. The *Aspergillus* strains used to deduce the structures of the aflatoxin precursors NOR and AVF are "leaky" mutants. Although NOR and AVF are the main products of the strains, a small amount of aflatoxin B_1 is formed, possibly by alternate pathways (Table 2).

It is possible that both steps in the oxidation of NOR to AVF are catalyzed by the same NADPH-dependent monooxygenase (reductase). Even if the same enzyme is used in the independent steps, the NOR reductase enzyme may favor the pathway leading from NOR to AVN but may be defective in further oxidation, whereas the less favored direct oxidation to averufanin may still occur. Until it is known whether a single dehydrogenase causes both steps in the NOR to AVF conversion or whether multiple activities prevail, the proposed scenario is unclear. However, any scheme must account for the co-occurrence of anthraquinone accumulation and aflatoxin synthesis.

The initial oxidation steps in aflatoxin biosynthesis appear to be similar to those of several other related (*A. versicolor, A. nidulans, A. ustes*) and unrelated fungi (*Solorina crocea, Bipolaris* sp., *Dothistroma* sp.,

TABLE 3. Decaketide Products Related to Aflatoxin Pathway Intermediates, Indicating Variation in Oxidation of the Hexanoate Side Chain

Decaketide	Possible routes from anthrone	Fungus	Ref.
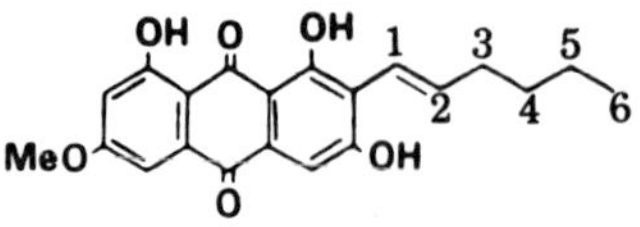 Averythrin	Dehydration at C-1	*Aspergillus versicolor*	83
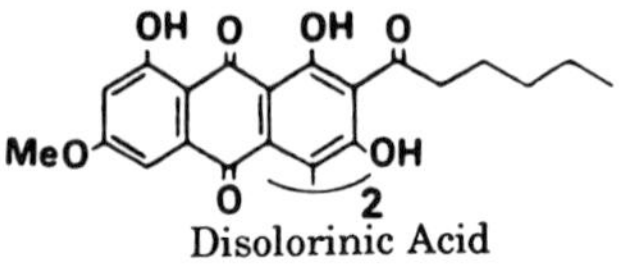 Disolorinic Acid	Ring oxidation	*Solorina crocea*	84
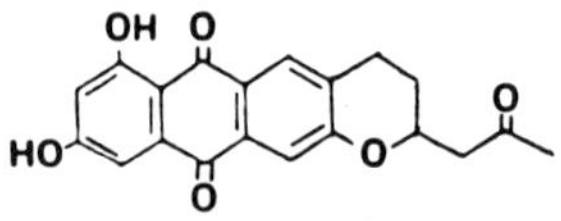 Deoxyaverufinone	Oxidation at C-3, C-5, C1 reduced	*A. versicolor*	85
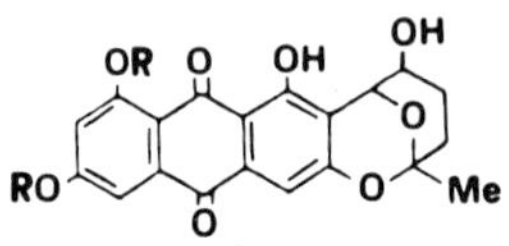 Nidurufin	Oxidation at C-4 and C-5 Ring closure at C-5	*A. nidulans* *Dothistroma pini* *A. versicolor*	86 87 88
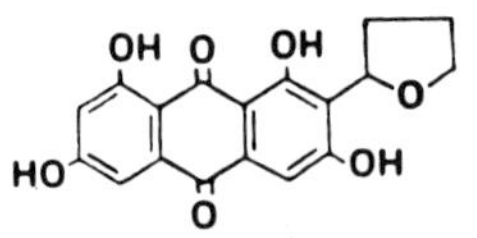 bipolarin	Oxidation at C-4 and C-5, ring closure at C-4 Oxidative decarboxylation	*Bipolaris* sp.	86

Cercospora sp.) [58, 82] that make decaketide products related to those of *A. parasiticus* and *A. flavus* (Table 3). Other metabolites can be formed by additional oxidative modification of the hexanoate side chain of NOR or of the anthraquinone ring.

Oxidative Rearrangement of Averufin

The ketal averufin (AVF) can exist in both an open-chain form and a closed form (Fig. 4). Further oxidation of the side chain of this anthraquinone is believed to involve three separate oxidation steps, the first two probably occurring as part of a multienzyme complex [21]. The end product of the oxidation steps is the dihydrobisfuran and tetrahydrobisfuran ring system characteristic of the sterigmatocystins and aflatoxins [82]; this ring system is not further modified by fungi, but it can be oxidized by eukaryotes (for example, epoxidation of the double bond in AFB_1) [89].

The stereochemical considerations and probable pathway for the oxidative rearrangement of averufin have been proposed by Townsend et al. [90]; both this scheme and previously described schemes are shown in Fig. 5. The product of the initial oxidation reaction of averufin is 1-hydroxy versicolorone (HVN) [90]. Mutant fungi have been isolated that accumulate

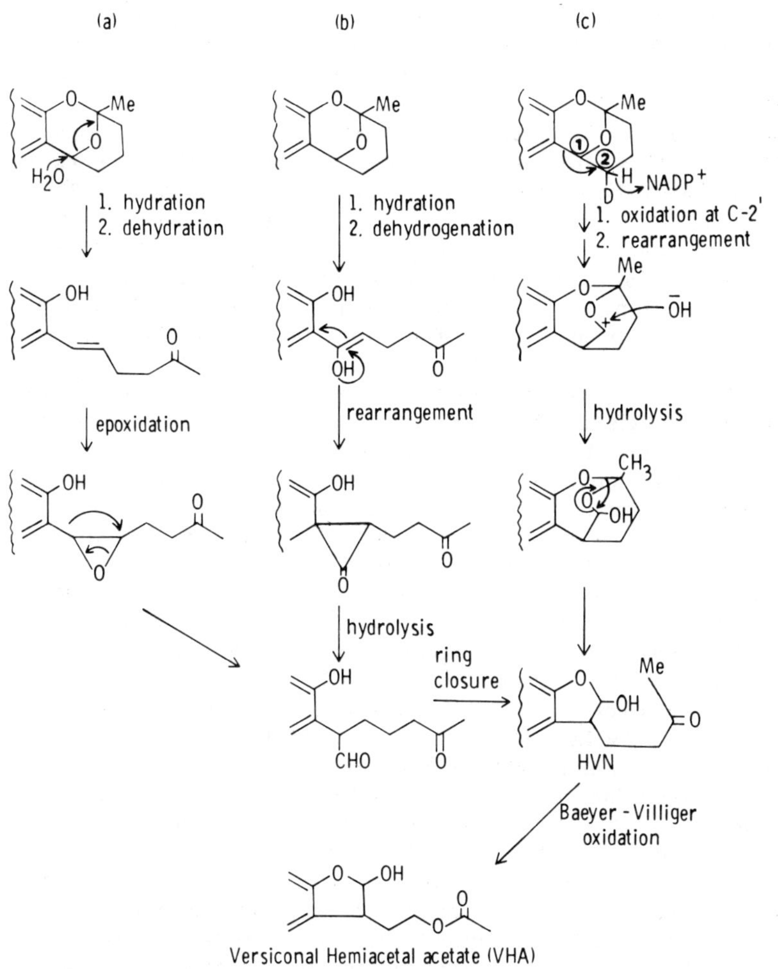

FIGURE 5. Proposed mechanisms for conversion of averufin to versiconal hemiacetal acetate.

HVN [91]; its further oxidation product, versiconal hemiacetal acetate (VHA), has been identified in various *A. parasiticus* culture extracts [92]. A Baeyer-Villiger (BV) type of oxidation converts HVN to VHA, probably as a separate oxidation reaction that is not mediated by the versicoloronal synthase complex.

In the three schemes for the conversion of AVF to HVN [93–95], only the third fits all the stereochemical requirements for the conversion. In the first scheme (Fig. 5a), after an initial hydroxylation reaction and subsequent dehydration to the enone, an epoxide is formed that could then rearrange to the open-chain form of HVN [93]. The second scheme (Fig. 5b) involves dehydrogenation with ring breakage followed by a rearrangement via a hypothetical cyclopropanone intermediate [94]. In the third scheme (Fig. 5c), the rearrangement is catalyzed by hydride removal, initial rearrangement to the intermediate ring-expanded ketal, followed by rearrangement to HVN [95] without formation of an open-chain intermediate. The intermediacy of nidurufin in the conversion of AVF to HVN was disproved because it could not be metabolized to aflatoxin by wild-type *A. parasiticus* [95].

Pathways *a* and *b* (Fig. 5) involve opening of the ketal ring. However, ^{18}O-induced isotope shift studies have shown that the bond between the 1'-oxygen and 5'-carbon remains intact in this rearrangement [90]. Therefore, pathways *a* and *b* are unlikely because this mechanism involves breakage of the denoted bond in the dehydrogenation step. Moreover, from deuterium labeling studies, equatorial deuteriums at C-2' and C-4' survive the oxidation step, and the 1'-hydrogen is retained in the oxidation. Therefore, only scheme (c) fits all of the stereochemical considerations demanded in the rearrangement.

The rearrangement may occur first by formation of 2'-hydroxyaverufin, which then rearranges via a carbonium ion at carbon 2' by loss of a hydroxyl group. Townsend et al. [90], in model studies, found that the intermediate, ring-expanded product of the 1,2-aryl shift could be trapped; the results provided evidence for a cationic reaction path. The most straightforward explanation for the rearrangement pathway has been proposed by Dutton [21] in which bifunctional catalysis occurs involving $NADP^+$-catalyzed abstraction of the 2'-trans hydrogen with simultaneous 1,2-aryl migration (Fig. 6). A hydroxyl on serine covalently tethers the C-1' carbon and catalyzes the rearrangement to the more stable, nonbridged, five-membered ring anthraquinone, HVN.

Further oxidation involves BV oxidation at the ketone to give versiconal hemiacetal acetate (VHA) [96]. VHA is converted to versiconal by an esterase [91, 96–99], which upon release from the enzyme complex with or without further oxidation condenses to the bis-tetrahydrofuran or dihydrofuran (Fig. 7). The esterase is inhibited by the pesticide dichlorvos [100], a known inhibitor of acetylcholine esterase [101]; these enzymatic activities must be intimately associated with the versicolorone synthase complex but apparently can be isolated as separate activities [98]. Following deesterification, the subsequent oxidation step to the dihydrobisfuran may involve oxidation of versiconal to an aldehyde that would condense to versicolorin A (Ver A) [102]. In the absence of oxidation, dehydration of VAL yields versicolorin C/B (Ver C/B) (Fig. 8) [97]. Versicolorin C is a racemate of Ver B [103]. The condensation reaction

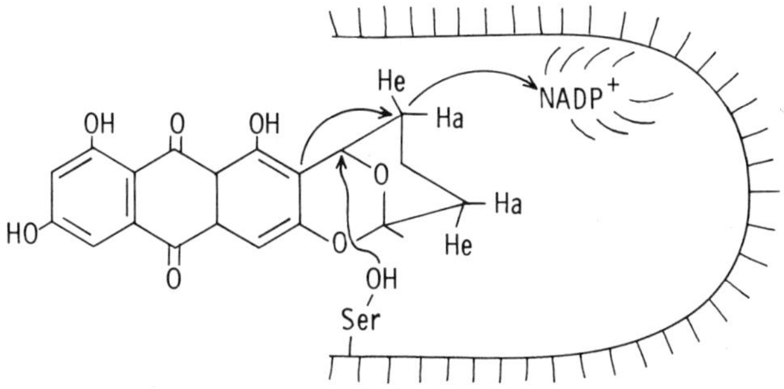

FIGURE 6. Model for the catalytic site of the hydroxyversicolorone synthase.

in the presence of the intact versicolorin synthase complex gives the stereochemically correct products for further conversion to aflatoxins B_1 (AFB_1) and B_2 (AFB_2). However, in the presence of dichlorvos a racemic versicoloronal and its oxidized product, versiconal acetate, are formed, presumably because the substrate is no longer properly tethered to the enzyme complex; because of this, the substrate may be subjected to nonspecific hydrolytic attack rather than attack of a serine hydroxyl within the complex (Fig. 6) [21].

McGuire et al. [103] have reported that cell-free systems gave time-dependent reactions of VHA to form Ver B but not Ver A by way of the diol corresponding to hydrolysis of the acetate in VHA. They hypothesized that Ver B was an intermediate in Ver A formation and that Ver B played a pivotal role in aflatoxin biosynthesis; the compound served as the point of partition between tetrahydrobisfuran- and dihydrobisfuran-containing products. McGuire et al. [103] further postulated that formation of the critical dihydrobisfuran is a three-step process from VHA involving an acetate hydrolysis/cyclization sequence to give Ver B and an apparent oxidative desaturation step to irreversibly form Ver A. Consequently, the strain of *A. flavus* (SRRC 141) that produces only AFB_2 [104] may lack the postulated desaturase activity (Fig. 7); Ver A but not Ver B (Bhatnagar, unpublished), ST, and OMST [13] are converted to AFB_1 by this strain in whole-cell feeding studies. The results of McGuire et al. [103], therefore, can also be explained by VAL being the intermediate in the formation of both Ver A and Ver B rather than Ver B being the requisite intermediate for Ver A if the oxidation of VAL is blocked. The mutant strain of *A. flavus* that produces only AFB_2 then would lack a functional dehydrogenase activity rather than a desaturase as proposed, and VAL rather than Ver B would be the point of partition between the tetrahydrobisfuran- and dihydrobisfuran-containing metabolites (Fig. 8). The results of McGuire et al. [103] also showed that Ver A hemiacetal cannot be an intermediate (proposed by Dutton [21]) in the conversion of Ver B to Ver A, since as with nidurufin [95], Ver A hemiacetal is reduced to Ver B rather than dehydrated by cell-free extracts of *A. parasiticus* [103].

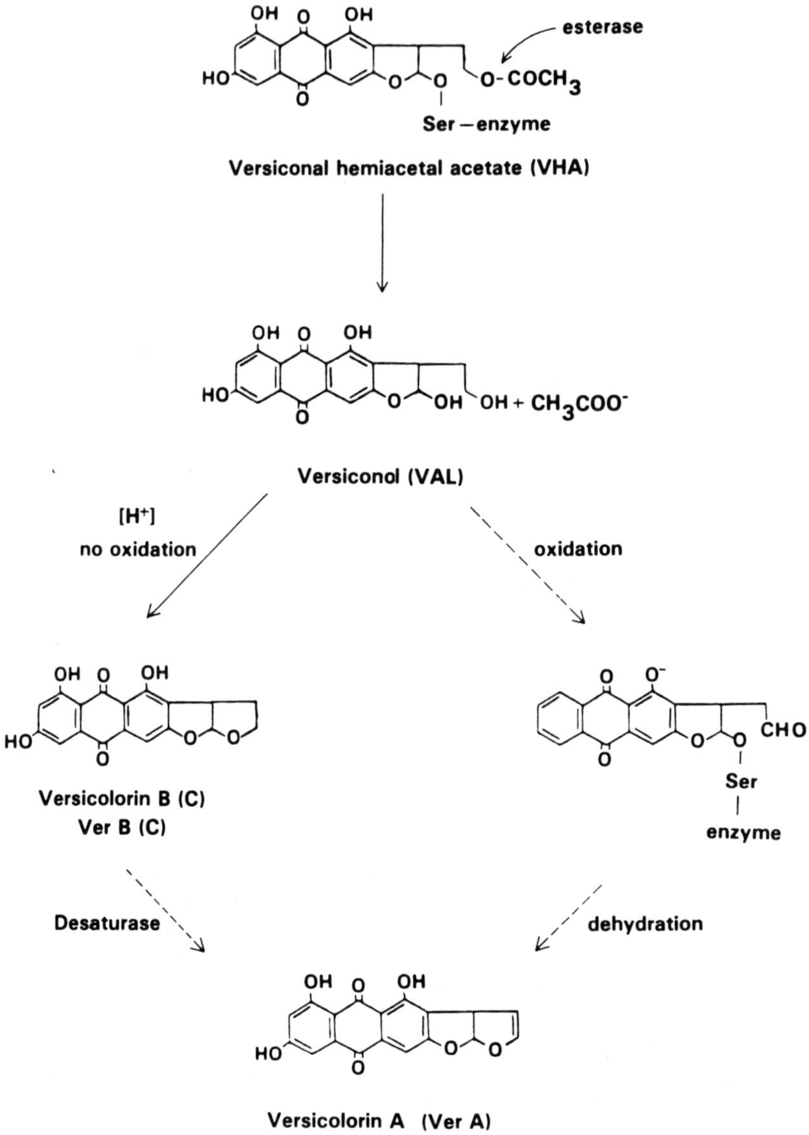

FIGURE 7. Schematic representation of the conversion of versiconal hemiacetal acetate to versicolorin A. (Dashed arrows indicate postulated pathways.)

B. Anthraquinone Ring Modification

Xanthone Synthesis

Once the rearrangement of AVF to Ver A and Ver B is complete, the dihydrobisfuran or tetrahydrobisfuran is probably released from the synthase complex and is not oxidized further during aflatoxin biosynthesis.

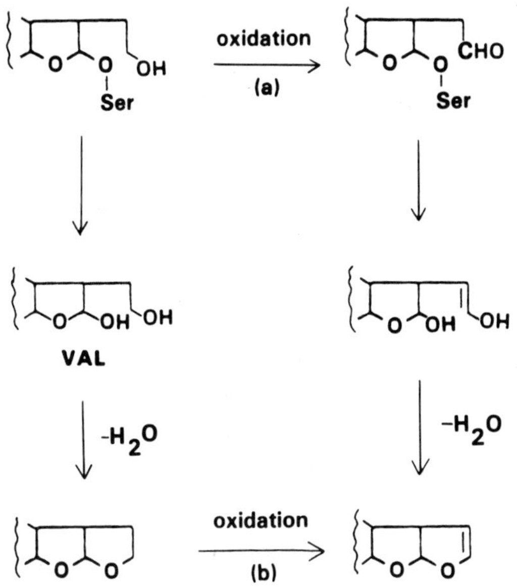

FIGURE 8. Oxidation steps leading from versiconal to versicolorin A and B (C).

Further modification involves cleavage of the quinone portion of the anthraquinone by a BV oxidation with subsequent rearrangement to produce the angular xanthone characteristic of sterigmatocystin. The exact timing of 6-hydroxyl loss is not clear, but 6-deoxyversicolorin A, a known metabolite of *A. versicolor*, is not isolated from blocked mutants that accumulate Ver A or Ver B [105]. Furthermore, Graybill et al. [105] found that 6-deoxy Ver A is not metabolized to aflatoxin by wild-type strains of *A. parasiticus*; the strains convert Ver A to aflatoxin B_1 with 20% efficiency according to isotope incorporation studies. Therefore, the deoxygenation of the 6-hydroxyl of the anthraquinone prior to rearrangement is highly unlikely.

Possible steps involved in the anthraquinone rearrangement to produce the xanthones are shown in Fig. 9; these involve BV oxidation of the quinone portion of the ring, rearrangement with decarboxylation and concomitant 6-deoxygenation, and methylation of one or both of the hydroxyls to give demethylsterigmatocystin, sterigmatocystin, and *O*-methylsterigmatocystin. Of the enzymes that catalyze the conversions, only the methyltransferases have been identified [24, 106], purified to homogeneity and partially characterized [107, 108]. In some cases, after oxidation of the quinone, linear xanthones are found; the results suggest that rearrangement is not a necessary consequence of the oxidation [109, 110]. In the formation of linear xanthones, the 6-hydroxyl of Ver A, the probable precursor, is also lost. Thus, deoxygenation at the denoted position may be a general phenomenon and a consequence of the oxidative decarboxylation pathway (Fig. 9).

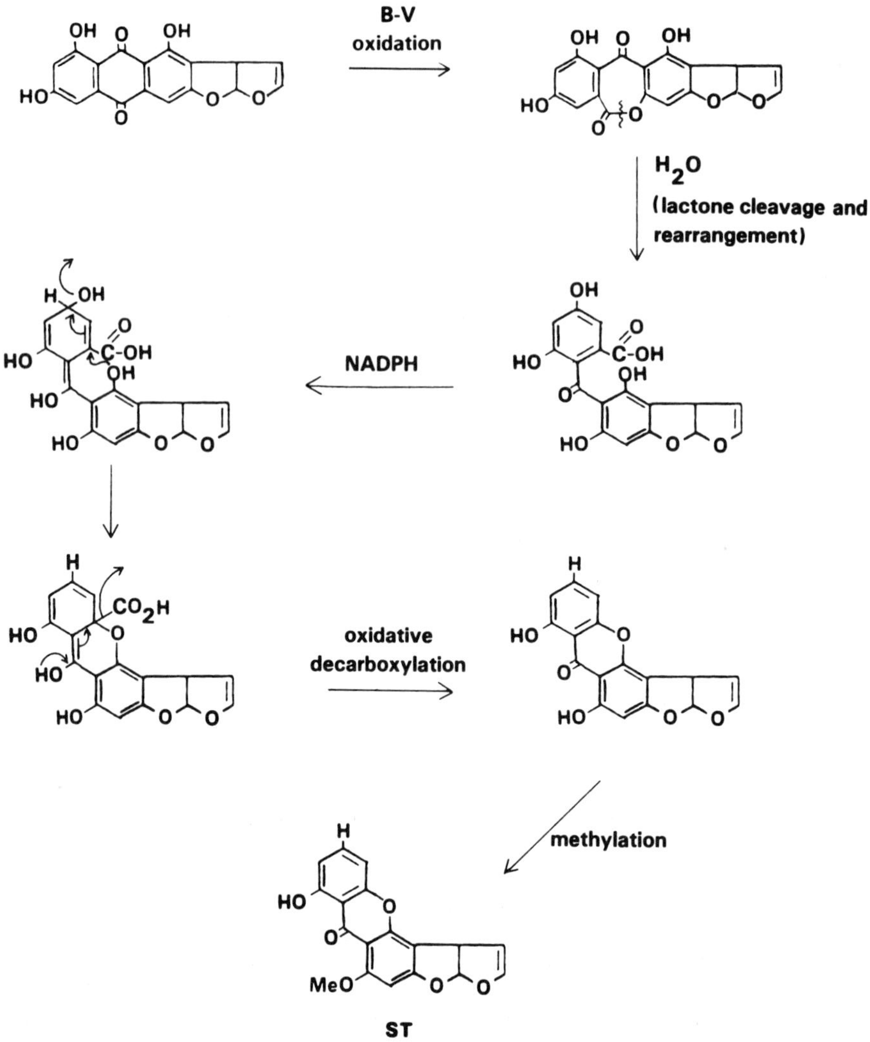

FIGURE 9. Possible steps involved in the rearrangement of anthraquinone to xanthones.

Examples of chemical reactions of anthraquinones are listed in Table 4. Consideration of these examples shows certain specific modifications to be most favorable; first, the ring is cleaved by oxidation at the C10—C12 or C10—C14 carbon bond followed by rearrangement to a xanthone, either with or without decarboxylation. Modification at C-6 (deoxygenation) or hydroxylation at C-5 or C-7 are the most common modifications of the A ring of the anthraquinones. The mycotoxin literature shows only a few examples of xanthones being end products of biosynthesis [58], for example, the heptaketide xanthones such as griseoxanthone

TABLE 4. Examples of Reactions of Anthraquinones

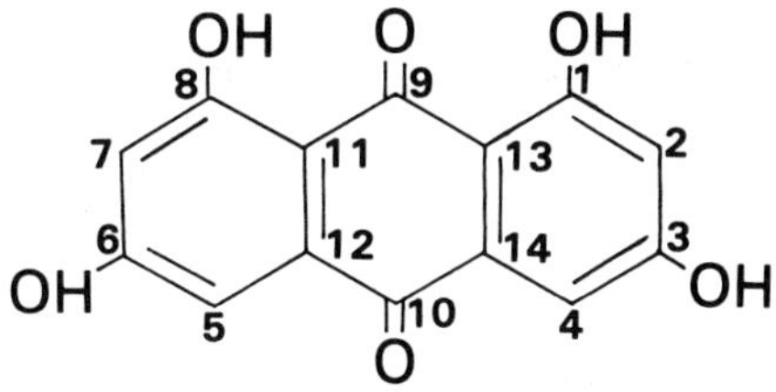

Reactions	Example
Dimerization at C-4	4,4'-Disolorinic acid
6-Deoxygenation (no rearrangement)	6-Deoxy Ver B
O-Methylation	6,8-*O*-Dimethyl Ver A
5-Hydroxylation, 6-deoxygenation	Bisdeoxydothistromin
Hydroxylation at C-7, 6-deoxygenation	1,6-Dihydroxy-3-methyl anthraquinone
Baeyer-Villiger oxidation at C-10	Xanthone rearrangement
decarboxylation	Variecoxanthones, shamixanthones, tajixanthones
ring expansions	Arugosins
no decarboxylation	Ergochromes
Xanthone rearrangement	
linear	Sterigmatin, austocystins, ergochromes
angular	Sterigmatocystin, aflatoxins

Source: Derived from Turner and Aldridge [58].

from *A. versicolor*. Only one natural isolate of *Aspergillus* spp. (*A. parasiticus* SRRC 2043, [10]) has been identified that does not produce aflatoxins but accumulates OMST.

Aflatoxin B_1, B_2 Synthesis

The oxidative rearrangement of the anisole ring of *O*-methylsterigmatocystin or *O*-methyldihydrosterigmatocystin is a process that is unique to *A. flavus* and *A. parasiticus*; many fungal species (e.g., *A. nidulans*) make sterigmatocystin and its precursors but do not produce aflatoxins B_1 or B_2. Also, no analogous rearrangements of other anthraquinones (e.g., ergochromes) serve as a precedent. A possible pathway utilizing a dioxygenase for the required oxidation is shown in Fig. 10.

It has now been established that AFB_1 and AFB_2 are produced by separate pathways [13, 106, 111, 130]; AFB_2 arises from dihydrosterigmatocystin (DHST) and dihydro-*O*-methylsterigmatocystin (DHOMST), whereas

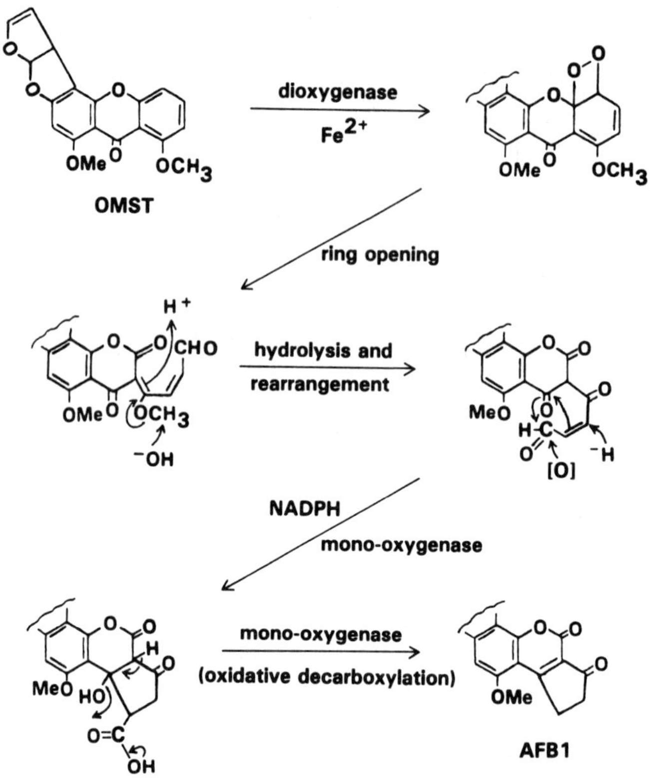

FIGURE 10. Postulated scheme for conversion of *O*-methylsterigmatocystin to aflatoxin B_1 involving a dioxygenase.

AFB_1 is produced from ST and OMST [10, 106]. The independent reactions in AFB_1 and AFB_2 synthesis are, however, catalyzed by common enzyme systems [13, 106, 112], and the AFB_1 precursors are the preferred substrates for the relevant enzymes [13]. The branch in the AFB_1/B_2 biosynthetic pathway has also been established recently [103]; the early precursors (NOR, AVN, AVNN, AFV, and VHA) are common to both pathways [113]. The following observations provide the basis for a metabolic scheme for AFB_1 and AFB_2 synthesis [106, 112, 114].

$$\text{VHA} \rightarrow \text{VAL} \begin{cases} \nearrow \text{Ver A} \rightarrow \text{DMST} \rightarrow \text{ST} \rightarrow \text{OMST} \rightarrow \text{AFB}_1 \\ \searrow \text{Ver B(C)} \rightarrow \text{DMDHST} \rightarrow \text{DHST} \rightarrow \text{DHOMST} \rightarrow \text{AFB}_2 \end{cases}$$

A scheme was proposed earlier by Maggon et al. [43] and Dutton et al. [113], who theorized that AFB_1 arose via Ver A and ST and that dihydrobisfuran aflatoxins developed from the pathway

VAL → Ver C → 5-hydrodihydroxysterigmatocystin → AFB_2

In the conversions OMST → AFB_1 and DHOMST → AFB_2, a dioxygenase first oxidizes the A ring, and then decarboxylation, possibly by a monooxygenase, completes the oxidative process. The enzymatic system requires NADPH [115]; the results indicate that the oxidation occurs as a two-step process. The OMST → aflatoxin B_1 catalyzing enzyme does not exhibit any cytochrome P-450 characteristics [23]. A complex mechanism for the conversion sterigmatocystin → OMST → AFB_1 was postulated earlier [116]. Unmethylated OMST (i.e., ST) cannot serve as a substrate in this oxidation reaction, possibly because it is not sufficiently electron-rich for the oxidation to occur; conversion to an anisolic group may provide the needed activation by election induction, making the *para* position particularly electron-rich. Oxidation is assumed to be catalyzed by a deoxygenase requiring Fe^{2+} for activity. An initial deoxetane ring is shown in Fig. 10 that opens to a ketone and an aldehyde. After hydrolysis (with loss of the methyl, possibly as methanol), the opened ring can rotate so that the correct cyclization can occur. During the cyclization, in the presence of a monooxygenase, NADPH serves as a hydride donor and is regenerated by oxidation of the aldehyde. NADPH has been demonstrated to be an optimal cofactor for the OMST → AFB_1 enzyme activity [117]. The cyclopentanone carboxylic acid is decarboxylated to the cyclopentanone group, the last modification in the AFB_1 and B_2 biosynthesis. It would be of interest to determine whether the austocystins, the linear xanthones related to ST, can be oxidized by the aflatoxin dioxygenase–monooxygenase system of *A. parasiticus*.

No intermediate metabolites between *O*-methylsterigmatocystin and AFB_1 or AFB_2 have been found, a fact suggesting that the substrate remains part of the oxidative enzyme complex until the cyclopentanone ring is formed. An enzyme complex (>200 kDa) that catalyzes the conversion of OMST to aflatoxin B_1 has been identified [11, 24] and partially purified [13, 118] from *A. parasiticus* cell-free extracts.

Aflatoxin G_1 and G_2 Synthesis

Aflatoxin B_1 is generally assumed to be a precursor of AFG_1, and AFB_2 a precursor of AFG_2 [119, 120]; several reviews include postulated schemes illustrating conversion of AFB_1 to AFG_1 [121–123]. The experimental role of AFB_1 as the precursor of AFG_1 is based on two studies. Maggon and Venkitasubramanian [124] used [^{14}C]AFB_1 as a substrate for in vitro incubation with a cell-free homogenate of *A. parasiticus* and reported recovery of low levels of ^{14}C label in AFB_2, AFG_1, and AFG_2; Heathcote et al. [125] reported similar results using an in vitro system with wild-type *A. flavus*. Other studies of in vivo feeding of precursors indicate that AFG_1 is synthesized from ST/OMST [10, 106] and AFG_2 from DHST/DHOMST [106, 111], namely the AFB_1 and AFB_2 precursors, respectively. Using mutant strains blocked in the aflatoxin pathway, Floyd et al. [126] and Henderberg et al. [127] did not observe any biotransformation of AFB_1 or AFB_2 into either AFG_1 or AFG_2; their results suggest independent pathways to the B and G aflatoxins. Aflatoxin B_1 is therefore probably not an intermediate, but rather an end product of the aflatoxin biosynthetic pathway. Furthermore, attempts to show that AFG_1 is formed by BV oxidation of AFB_1 have not been successful [128]. Other known metabolites of sterigmatocystin may be candidates for the denoted oxidation such as 5,6-dimethoxysterigmatocystin or 5-methoxysterigmatocystin [112]. A possible pathway is shown in Fig. 11 with 5-*O*-methylsterigmatocystin as precursor. The pertinent rearrangement involves oxidation by the same dioxygenase system, but instead of the 2 + 2 addition of oxygen, a 2 + 4 addition gives the required oxidation at C-4, a position both *para* to the xanthone ether and *ortho* to the methoxy and, therefore, specially electron-rich. In this scheme the amount of G_1 and G_2 formed would

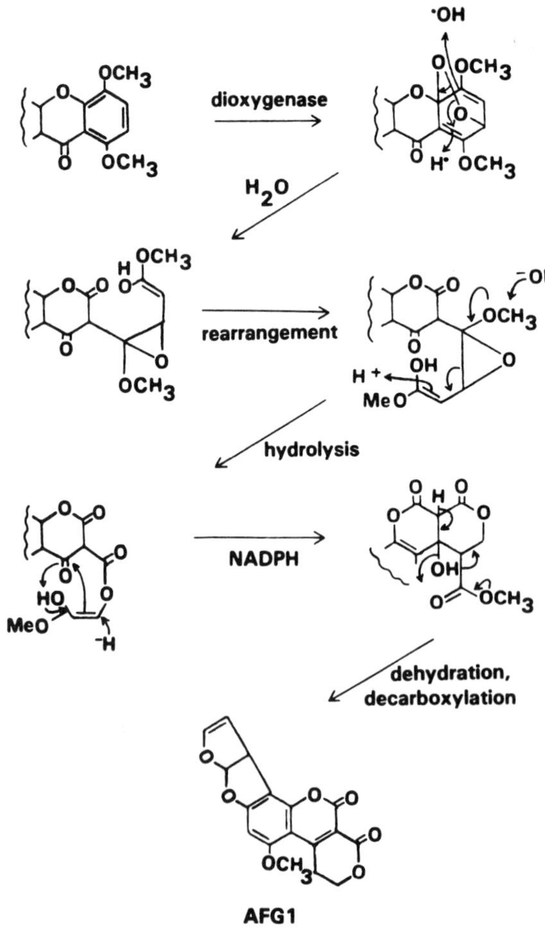

FIGURE 11. Aflatoxin G_1 synthesis from sterigmatocystins.

be dictated by the amount of the precursor sterigmatocystins rather than by further oxidation of AFB_1 and AFB_2. The conversion of 5-methoxy-OMST to AFG_1 by precursor feeding studies is currently being examined with fungi capable of converting sterigmatocystin to AFB_1 (Bhatnagar, unpublished observations).

Oxidation of Aflatoxins

Further oxidation of the aflatoxin bishydrofuran can occur at the double bond to give an epoxide [89] or at the allylic carbon to give the allylic alcohol (M series) [113]. Reduction of the cyclopentanone ring oxygen also can take place to yield the aflatoxicols [82]. The epoxide and allylic alcohol derivatives are produced during mammalian microsomal activation and the aflatoxicols by bacterial dehydrogenase reduction.

V. CONCLUDING REMARKS

In summary, the aflatoxins are the quintessential metabolites of the polyketide class and utilize the armamentarium of oxidation and reduction reactions available to fungi during secondary metabolism. The pathway appears to be extremely complicated, but since many enzymatic steps may be combined in multienzyme complexes, the conversions most likely occur without the intermediates being released into the cytosol or to storage sites within the cell. Such multifunctional complexes would require fewer genes for biosynthesis. Under efficient aflatoxin production in wild-type fungi, many of the pathway intermediates cannot be isolated even in small amounts, so for some, pools of these metabolites are probably not maintained by the cell. Most of the catalyzed steps in the pathway involve oxidations (Table 5). The monooxygenases act by incorporating one atom from molecular oxygen into the substrate with the other requiring reduction by NADPH [74]. Monooxygenases of two types are used in the biosynthesis of most mycotoxins, the first being a hydroxylase (oxygen insertion into a C—H bond), the second being a BV oxidase (oxygen insertion into an activated C—C bond).

Incorporation of oxygen is usually effected by a 1,2-cycloaddition pathway. However, in the presence of light (photooxidation) or other suitable catalysis, singlet oxygen could be the active species that would be able to react by a 1,4-cycloaddition pathway [129]. Since the biosynthetic pathway follows an orderly sequence of events with known stable and stereochemically correct precursors being necessary for successful continuation of the metabolic process, it is unlikely that the oxidases are maintained by the fungus for a separate purpose because they would catalyze nonspecific oxidations.

Although the cytochrome P-450 microsomal oxidase systems of most organisms are capable of accommodating a wide variety of analogous substrates, the difference between the latter systems and the mycotoxin machinery is probably that the toxin-synthesis enzymes are part of multienzyme systems with a strong requirement for correct stereochemical and spatial orientation of the substrate.

TABLE 5. Oxidative Enzymes Used in Aflatoxin Synthesis

Enzyme	Substrate	Product
Reductases		
β-Keto ACP-reductase	β-Ketoacyl S-ACP	D-3 hydroxyacyl S-ACP
Enoyl-CoA reductase	*trans*-Δ^2-Enoyl ACP	Butyryl S-ACP
Norsolorinic acid reductase	Norsolorinic acid	Averantin
Monooxygenases		
Hydroxylases		
anthrone oxidase	Hexanoyl anthrone	Norsolorinic acid
averantin hydroxylase	Averantin, averufanin	Averufin
hydroxyversicolorone synthase	Averufin	1'-Hydroxyversicolorone
versiconol oxidase	Versiconol	Versicolorin A, B
Baeyer-Villiger type		
versicolorone oxidase	1'-Hydroxyversicolorone	Versiconal acetate
versicolorin oxidase	Versicolorin A, B	Sterigmatocystin Dihydrosterigmatocystin
Dioxygenases		
O-Methylsterigmatocystin dioxygenase	*O*-Methylsterigmatocystin	AFB_1, AFB_2
	Dihydro-*O*-Methylsterigmatocystin	AFG_1, AFG_2

REFERENCES

1. Detroy, R. W., Lillehoj, E. B., and Ciegler, A. Aflatoxin and related compounds. In *Microbial Toxins*, Vol. 6 (A. Ciegler, S. Kadis, and S. J. Ajl, Eds.), Academic, New York, 1971, pp. 3–178.
2. Busby, W. F., and Wogan, G. N. Food-borne mycotoxins and alimentary mycotoxicosis. In *Food Borne Infections and Intoxications*, 2nd ed. (H. P. Reimann and F. L. Bryan, Eds.), Academic, New York, 1979, pp. 519–610.
3. Groopman, J. D., Croy, R. G., and Wogan, G. N. In vitro reactions of aflatoxin B_1-adducted DNA. *Proc. Natl. Acad. Sci. USA, 78*:5445–5449, 1981.
4. Jelinek, C. F., Pohland, A. E., and Wood, G. E. Review of mycotoxin contamination. Worldwide occurrence of mycotoxins in foods and feeds—an update. *J. Assoc. Off. Anal. Chem., 72*:223–230, 1989.
5. Malik, V. S. Genetics and biochemistry of secondary fungal metabolites. *Adv. Appl. Microbiol., 28*:27–116, 1982.

6. Wicklow, D. T., and Shotwell, O. L. Intrafungal distribution of aflatoxins among conidia and sclerotia of *Aspergillus flavus* and *Aspergillus parasiticus*. *Can. J. Microbiol.*, *29*:1–5, 1982.
7. Dowd, P. F. Insect interactions with mycotoxin producing fungi and their hosts. In *Mycotoxins in Ecological Systems* (D. Bhatnagar, E. B. Lillehoj, and D. K. Arora, Eds.), Marcel Dekker, New York, 1991, pp. 137–155.
8. Ciegler, A. Evolution, ecology, and mycotoxins: some musings. In *Secondary Metabolism and Differentiation in Fungi* (J. W. Bennett and A. Ciegler, Eds.), Marcel Dekker, New York, 1983, pp. 429–440.
9. Bennett, J. W., and Christensen, S. B. New perspectives on aflatoxin biosynthesis. *Adv. Appl. Microbiol.*, *19*:53–92, 1983.
10. Bhatnagar, D., McCormick, S. P., Lee, L. S., and Hill, R. A. Identification of *O*-methylsterigmatocystin as an aflatoxin B_1 and G_1 precursor in *Aspergillus parasiticus*. *Appl. Environ. Microbiol.*, *53*: 1028–1033, 1987.
11. Bhatnagar, D., Cleveland, T. E., and Lillehoj, E. B. Enzymes in late stages of aflatoxin B_1 biosynthesis: strategies for identifying pertinent genes. *Mycopathologia*, *107*:75–83, 1989.
12. Bhatnagar, D., and Cleveland, T. E. Purification and characterization of a reductase from *Aspergillus parasiticus* SRRC 2043 involved in aflatoxin biosynthesis. *FASEB J.*, *4*:A2164, 1990.
13. Bhatnagar, D., Cleveland, T. E., and Kingston, D. G. I. Enzymological evidence for separate pathways for aflatoxin B_1 and B_2 biosynthesis. *Biochemistry*, *30*:4343–4350, 1991.
14. Benasutti, M., Ejadi, S., Whitlow, M. D., and Loechler, E. L. Mapping the binding site of aflatoxin B_1 in DNA: system analysis of the reactivity of aflatoxin B_1 with guanines in different DNA sequences. *Biochemistry*, *27*:472–481, 1988.
15. Townsend, C. A. Progress towards a biosynthetic rationale of the aflatoxin pathway. *Pure Appl. Chem.*, *58*:227–238, 1986.
16. McCormick, S. P., Bhatnagar, D., and Lee, L. S. Averufanin is an aflatoxin B_1 precursor between averantin and averufin in the biosynthetic pathway. *Appl. Environ. Microbiol.*, *53*:14–16, 1987.
17. Papa, K. E. Genetics of *Aspergillus flavus*: complementation and mapping of aflatoxin mutants. *Gen. Res.*, *34*:1–9, 1979.
18. Papa, K. E. Genetics of *Aspergillus flavus*: linkage of aflatoxin mutants. *Can. J. Microbiol.*, *30*:68–73, 1984.
19. Bennett, J. W., Lee, L. S., Shoss, S. M., and Boudreaux, G. H. Identification of averantin as an aflatoxin B_1 precursor: placement in the biosynthetic pathway. *Appl. Environ. Microbiol.*, *39*:835–839, 1980.
20. Wan, N. C., and Hsieh, D. P. H. Enzymatic formation of the bisfuran structure in aflatoxin biosynthesis. *Appl. Environ. Microbiol.*, *39*:109–112, 1980.
21. Dutton, M. F. Enzymes and aflatoxin biosynthesis. *Microbiol. Rev.*, *52*:274–295, 1988.
22. Anderson, M. S., and Dutton, M. F. The use of cell-free extracts derived from fungal protoplasts in the study of aflatoxin biosynthesis. *Experientia*, *35*:21–22, 1979.

23. Bhatnagar, D., Cleveland, T. E., and Lax, A. R. Comparison of the enzymatic composition of cell-free extracts of non-aflatoxigenic *Aspergillus parasiticus* with respect to late stages of aflatoxin biosynthesis. *Arch. Environ. Contam. Toxicol.*, *18*:434–438, 1989.
24. Cleveland, T. E., Lax, A. R., Lee, L. S., and Bhatnagar, D. Appearance of enzyme activities catalyzing conversion of sterigmatocystin to aflatoxin B_1 in late-growth-phase *Aspergillus parasiticus* cultures. *Appl. Environ. Microbiol.*, *53*:1711–1713, 1987.
25. Jeenah, M. S., and Dutton, M. F. The conversion of sterigmatocystin to *O*-methylsterigmatocystin and aflatoxin B_1 by a cell-free preparation. *Biochem. Biophys. Res. Commun.*, *116*:1114–1118, 1983.
26. Hsieh, D. P. H., Lin, M. T., Raymond, C. Y., and Singh, R. Biosynthesis of aflatoxin. Conversion of norsolorinic acid and other hypothetical intermediates into aflatoxin B_1. *J. Agric. Food Chem.*, *24*:1170–1173, 1976.
27. Drew, S. W., and Demain, A. L. Effect of primary metabolites on secondary metabolism. *Ann. Rev. Microbiol.*, *31*:343–356, 1977.
28. Cleveland, T. E., and Bhatnagar, D. Evidence for *de novo* synthesis of an aflatoxin pathway methyltransferase near the cessation of active growth and the onset of aflatoxin biosynthesis by *Aspergillus flavus* mycelia. *Can. J. Microbiol.*, *36*:1–5, 1990.
29. Bu'lock, J. D. Secondary metabolism in fungi and its relationship to growth and development. In *The Filamentous Fungi: Industrial Mycology* (J. E. Smith and D. R. Berry, Eds.), Arnold, London, 1975, pp. 33–58.
30. Martin, J. F., and Demain, A. L. Fungal development and metabolite formation. In *The Filamentous Fungi: Developmental Mycology*, Vol. 3 (J. E. Smith and D. R. Berry, Eds.), Arnold, London, 1978, pp. 426–450.
31. Rao, V. M., Maggon, K. K., and Venkitasubramanian, T. A. Primary metabolism of *Aspergillus parasiticus* in relation to aflatoxin biosynthesis. *Eur. J. Appl. Microbiol. Biotechnol.*, *7*:307–312, 1979.
32. Turner, W. B. Polyketides and related metabolites. In *The Filamentous Fungi: Biosynthesis and Metabolism*, Vol. 2 (J. E. Smith and D. R. Berry, Eds.), Arnold, London, 1976, pp. 445–459.
33. Hopwood, D. A. Molecular genetics of polyketides and its comparison to fatty acid biosynthesis. *Ann. Rev. Gen.*, *24*:37–66, 1990.
34. Koman, V., and Betina, V. Diphasic production of secondary metabolites by *Penicillium notatum*. II. Biosynthesis of fatty acids and lipids. *Folia Microbiol (Prague)*, *18*:133–141, 1973.
35. Martin, J. F. Fatty acid biosynthesis and polyene macrolide antibiotic production in *Streptomyces*. *Dev. Ind. Microbiol.*, *17*:223–231, 1976.
36. Ward, A. C., and Packter, W. M. Relationship between fatty acid and phenol synthesis in *Aspergillus fumigatus*. *Eur. J. Biochem.*, *46*:323–333, 1974.
37. Detroy, R. W., and Hesseltine, C. W. Net synthesis of carbon-14-labeled lipids and aflatoxins in resting cells of *Aspergillus parasiticus*. *Dev. Ind. Microbiol.*, *10*:127–133, 1968.

38. Gupta, S. K., Viswanathan, L., and Venkitasubramanian, T. A. A comparative study of lipids of a toxigenic and a non-toxigenic strain of *Aspergillus flavus*. *Indian J. Biochem. Biophys.*, *7*:108–111, 1970.
39. Shih, C.-N., and Marth, E. H. Aflatoxin formation: lipid synthesis and glucose metabolism by *Aspergillus parasiticus* during incubation with and without agitation. *Biochim. Biophys. Acta*, *338*:286–296, 1976.
40. Townsend, C. A., Christensen, S. B., and Trautwein, K. Hexanoate as a starter unit in polyketide synthesis. *J. Am. Chem. Soc.*, *106*:3868–3869, 1984.
41. Detroy, R. W., and Ciegler, A. Aflatoxin biosynthesis in *Aspergillus parasiticus*: effect of methionine analogs. *Can. J. Microbiol.*, *17*:569–574, 1971.
42. Detroy, R. W., and Hesseltine, C. W. Secondary biosynthesis of aflatoxin B_1 in *Aspergillus parasiticus*. *Can. J. Microbiol.*, *16*: 959–963, 1970.
43. Maggon, K. K., Gupta, S. K., and Venkitasubramanian, T. A. Biosynthesis of aflatoxins. *Bacteriol. Rev.*, *41*:822–855, 1977.
44. Birch, A. J. Biosynthesis of polyketides and related compounds. *Science*, *156*:202–206, 1967.
45. Bu'Lock, J. D. *The Biogenesis of Natural Products*, McGraw-Hill, London, 1965, pp. 23–27.
46. Bu'Lock, J. D. Aspects of secondary metabolism in fungi. In *Biogenesis of Antibiotic Substances* (Z. Vanek and Z. Hostalek, Eds.), Academic, New York, 1965, pp. 61–71.
47. Bu'Lock, J. D. Mycotoxins as secondary metabolites. In *The Biosynthesis of Mycotoxins* (P. S. Steyn, Ed.), Academic, New York, 1980, pp. 13–15.
48. Applebaum, R. S., and Marth, E. H. Biogenesis of the C_{20} polyketide, aflatoxin. *Mycopathologia*, *76*:103–114, 1981.
49. Richards, J. H., and Hendrickson, J. B. *The Biosynthesis of Steroids, Terpenes, and Acetogenins*, Benjamin, New York, 1964, pp. 16–121.
50. Zamir, L. O. The biosynthesis of patulin and penicillic acid. In *The Biosynthesis of Mycotoxins* (P. S. Steyn, Ed.), Academic, New York, 1980, pp. 223–268.
51. Curtis, R. F., Hassal, C. H., and Nazar, M. Biosynthesis of phenols. XV. Some metabolites of *Penicillium citrinum* related to citrinin. *J. Chem. Soc. C*, *1968*:85–93.
52. Steyn, P. S., Vleggar, R., and Wessels, P. L. The biosynthesis of aflatoxin and its congeners. In *The Biogenesis of Mycotoxins* (P. S. Steyn, Ed.), Academic, New York, 1980, pp. 105–155.
53. Yamazaki, M. The biosynthesis of neurotropic mycotoxins. In *The Biosynthesis of Mycotoxins* (P. S. Steyn, Ed.), Academic, New York, 1980, pp. 193–222.
54. Weisleder, D., and Lillehoj, E. B. Structure of viriditoxin, a toxic metabolite of *Aspergillus viridinutans*. *Tetrahedron Lett.*, *48*:4705–4706, 1971.

55. Simpson, T. J. Carbon-13 nuclear magnetic resonance spectra and biosynthetic studies of xanthomegnin and related pigments from *Aspergillus sulphureus* and *melleus*. *J. Chem. Soc., Perkin Trans. 1*, *1977*:592–595.
56. Frank, C. B. The biosynthesis of the ergochromes. In *The Biosynthesis of Mycotoxins* (P. S. Steyn, Ed.), Academic, New York, 1980, pp. 157–191.
57. Steele, J. A., Lieberman, J. R., and Mirocha, C. J. Biogenesis of zearalenone (F2) by *Fusarium roseum* Graminearum. *Can. J. Microbiol.*, *20*:531–534, 1974.
58. Turner, W. B., and Aldridge, D. C. *Fungal Metabolites*, Vol. 2, Academic, London, 1983.
59. Herbert, R. B. *The Biosynthesis of Secondary Metabolites*, Chapman and Hall, London, 1981, pp. 28–49.
60. Wakil, S. J., Stoops, J. K., and Joshi, V. C. Fatty acid synthesis and its regulation. *Ann. Res. Biochem.*, *52*:537-579, 1983.
61. Beck, J., Ripka, S., Sienger, A., Schiltz, E., and Schweizer, E. The multifunctional 6-methylsalicylic acid synthase gene of *Penicillium patulum*. *Eur. J. Biochem.*, *192*:487–498, 1990.
62. Lane, M. D., Moss, J., and Polakis, S. E. Acetyl coenzyme A carboxylase. *Curr. Top. Cell. Regul.*, *8*:139–187, 1974.
63. Behal, V., Jehova, V., Vanek, Z., and Hostalek, Z. Alternate pathways of malonyl-CoA formation in *Streptomyces aureofaciens*. *Phytochemistry*, *16*:347–350, 1977.
64. Utter, M. F. Metabolic roles of oxaloacetate. In *Citric Acid Cycle: Control and Compartmentation* (J. M. Lowenstein, Ed.), Marcel Dekker, New York, 1969, pp. 249–296.
65. Tyagi, J. S., and Venkitasubramanian, T. A. The role of glycolysis in aflatoxin biosynthesis. *Can. J. Microbiol.*, *27*:1276–1282, 1981.
66. Reddy, T. V., Viswanathan, L., and Venkitasubramanian, T. A. High aflatoxin production on a chemically defined medium. *Appl. Microbiol.*, *22*:393–396, 1971.
67. Buchanan, R. L., Federowicz, D., and Stahl, H. G. Activities of tricarboxylic acid cycle enzymes in an aflatoxigenic strain of *Aspergillus parasiticus* after peptone to glucose carbon shift. *Trans. Br. Mycol. Soc.*, *84*:267–275, 1985.
68. Abdollahi, A., and R. L. Buchanan. Regulation of aflatoxin biosynthesis: induction of aflatoxin production by various carbohydrates. *J. Food Sci.*, *46*:633–635, 1981.
69. Buchanan, R. L., and Lewis, D. F. Regulation of aflatoxin biosynthesis: effect of glucose on the activities of various glycolytic enzymes. *Appl. Environ. Microbiol.*, *48*:306–310, 1984.
70. Bhatnagar, R. K., Ahmad, S., Mukerji, K. G., and Venkitasubramanian, T. A. Pyridine nucleotides and redox state regulation of aflatoxin biosynthesis in *Aspergillus parasiticus*. *J. Appl. Bacteriol.*, *60*:135–141, 1986.
71. Lynen, F. Biosynthetic pathways from acetate to natural products. *Pure Appl. Chem.*, *14*:137–167, 1967.
72. Dimroth, P., Ringelmann, E., and Lynen, F. 6-Methylsalicylic acid synthetase from *Penicillium patulum*. Some catalytic properties of

the enzyme and its relation to fatty acid synthetase. *Eur. J. Biochem.*, *68*:591–506, 1976.
73. Vagelos, P. R. Acyl group transfer (acyl carrier protein). In *The Enzymes*, 3rd ed., Vol. 8 (P. D. Boyer, Ed.), Academic, New York, 1973, pp. 155–199.
74. Mathews, C. K., and van Holde, K. E. *Biochemistry*, Benjamin/Cummings, New York, 1990, pp. 593–598.
75. Chuturgoon, A. A., and Dutton, M. F. The appearance of an enzyme activity catalyzing the conversion of norsolorinic acid to averantin in *Aspergillus parasiticus* cultures. *Mycopathologia, 113*:41–46, 1991.
76. Chandler, I. M., and Simpson, T. J. Studies on polyketide chain assembly processes: incorporation of (2-^{13}C)malonate into averufin in *Aspergillus parasiticus*. *J. Chem. Soc. Chem. Commun.*, *1987*: 17–18, 1987.
77. Townsend, C. A., Brobst, S. W., Ramer, S. E., and Vederas, J. C. Stereochemical features of enoyl thiol ester reductase in averufin and fatty acid biosynthesis in *Aspergillus parasiticus*. *J. Am. Chem. Soc.*, *110*:318–319, 1988.
78. Scott, A. J., Beadling, L. C., Georgopapadakou, N. H., and Subbarayan, C. R. Biosynthesis of polyketides. Purification and inhibition studies of 6-methylsalicylic acid synthase. *Bioorg. Chem.*, *3*:238–248, 1974.
79. Vederas, J. C., and Nakashima, T. T. Biosynthesis of averufin by *Aspergillus parasiticus*: detection of ^{18}O-label by ^{13}C NMR isotope shifts. *J. Chem. Soc. Chem. Commun.*, *1980*:183–185.
80. Chuturgoon, A. A., Dutton, M. F., and Berry, R. K. The preparation of an enzyme associated with aflatoxin biosynthesis by affinity chromatography. *Biochem. Biophys. Res. Commun.*, *166*:38–42, 1990.
81. Heathcote, J. G., and Dutton, M. F. New metabolites of *Aspergillus flavus*. *Tetrahedron*, *25*:1497–1500, 1969.
82. Cole, R. J., and Cox, R. H. *Handbook of Toxic Fungal Metabolites*, Academic, New York, 1981.
83. Roberts, J. C., and Roffey, P. Studies in mycological chemistry. Part XVII. Averythrin, an anthraquinonoid pigment from *Aspergillus versicolor*. *J. Chem. Soc.*, *1965*:3666–3672.
84. Steglich, W., and Jedtke, K. F. Novel anthraquinone pigments from *Solorina crocea*. *Z. Naturforsch. C*, *31*:197–198, 1976.
85. Berger, Y., Jadot, J., and Ramant, J. Isolement et structure de la deoxyaverufinone et de la dehydroverufine: deux metabolites de l'*Aspergillus versicolor* (Vuillemin) Triaboschi. *Bull. Soc. Chim. Belg.*, *85*:161–166, 1976.
86. Aucamp, P. J., and Holzapfel, C. W. Polyhydroxyanthraquinones from *Aspergillus versicolor*, *Aspergillus nidulans* and *Bipolaris* sp., their significance in relation to biogenetic theories on aflatoxin B_1. *J.S. Afr. Chem. Inst.*, *23*:40–56, 1970.
87. Danks, A. V., and Hodges, R. Group migrations in the mass spectra of some benzyl ethers and the acetals of some phenacetaldehydes. *Aust. J. Chem.*, *25*:2721–2724, 1972.

88. Kingston, D. G. I., Chen, P. N., and Vercellotti, J. R. Metabolites of *Aspergillus versicolor*: 6,8-di-*O*-methylnidurufin, griseofulvin, dechlorogriseofulvin, and 3,8-dihydroxy-6-methoxy-1-methylxanthone. *Phytochemistry, 15*:1037–1039, 1976.
89. Shimada, T., and Guengerich, F. P. Evidence for cytochrome P-450 NF, the nifedipine oxidase, being the principal enzyme involved in the bioactivation of aflatoxins in human liver. *Proc. Natl. Acad. Sci. USA, 86*:462–465, 1989.
90. Townsend, C. A., Isomura, Y., Davis, S. G., and Hodge, J. A. Reaction models of the oxidative rearrangement of averufin to 1'-hydroxyversicolorone: the first step in dihydrobisfuran formation in aflatoxin biosynthesis. *Tetrahedron, 45*:2263–2276, 1989.
91. Townsend, C. A., Plaucan, K. A., Pal, K., and Brobst, S. W. Hydroxyversicolorone: isolation and characterization of a potential intermediate in aflatoxin biosynthesis. *J. Org. Chem., 53*:2472–2477, 1988.
92. McCormick, S. P., Bowers, E., and Bhatnagar, D. High-performance liquid chromatographic procedure for determining the profiles of aflatoxin precursors in wildtype and mutant strains of *Aspergillus parasiticus*. *J. Chromatogr., 441*:400–405, 1988.
93. Gorst-Allman, C. P., Pachler, K. G. R., Steyn, P. S., Wessels, P. L., and Scott, D. B. Carbon-13 nuclear magnetic resonance assignments of some fungi C_{20} anthraquinones: their biosynthesis in relation to that of aflatoxin B_1. *J. Chem. Soc., Perkin Trans. 1, 1977*:2181–2188, 1977.
94. Tanabe, M., Uramoto, M., Hamasaki, T., and Cary, L. Biosynthetic studies with ^{13}C: a novel Favorsky rearrangement route to the fungal metabolite aspyrone. *Heterocycles, 5*:355–365, 1976.
95. Townsend, C. A., and Christensen, S. B. Concerning the role of nidurufin in aflatoxin biosynthesis. *J. Am. Chem. Soc., 107*:270–271, 1985.
96. Townsend, C. A., Christensen, S. B., and Davis, S. G. Bisfuran formation in aflatoxin biosynthesis: the role of versiconal acetate. *J. Am. Chem. Soc., 104*:6154–6155, 1982.
97. Hsieh, D. P. H., Yao, R. C., Fitzell, D. L., and Reece, C. A. Origin of the bisfuran ring structure in aflatoxin biosynthesis. *J. Am. Chem. Soc., 98*:1020–1021, 1976.
98. Hsieh, D. P. H., Wan, C. C., and Billington, J. A. A versiconal hemiacetal acetate converting enzyme in aflatoxin biosynthesis. *Mycopathologia, 107*:121–126, 1989.
99. Anderson, J. A., and Chung, C. H. Conversion of versiconal acetate to versiconal and versicolorin C in extracts from *Aspergillus parasiticus*. *Mycopathologia, 110*:31–35, 1990.
100. Yao, R. C., and Hsieh, D. P. H. Step of dichlorvos inhibition in the pathway of aflatoxin biosynthesis. *Appl. Microbiol., 28*:52–57, 1974.
101. Hassall, K. A. *The Chemistry of Pesticides*, Macmillan, London, 1982, pp. 73–78.
102. Anderson, M. S., and Dutton, M. F. Biosynthesis of versicolorin A. *Appl. Environ. Microbiol., 40*:706–709, 1980.

103. McGuire, S. M., Brobst, S. W., Graybill, T. L., Pal, K., and Townsend, C. A. Partitioning of tetrahydro- and dihydrobisfuran formation in aflatoxin biosynthesis defined by cell-free and direct incorporation experiments. *J. Am. Chem. Soc.*, *111*:8308–8309, 1989.
104. Schroeder, H. W., and Carlton, W. W. Accumulation of only aflatoxin B_2 by strain of *Aspergillus flavus*. *Appl. Microbiol.*, *25*: 146–148, 1973.
105. Graybill, T. L., Pal, K., McGuire, S. M., Brobst, S. W., and Townsend, C. A. The timing of aromatic deoxygenation in aflatoxin biosynthesis. *J. Am. Chem. Soc.*, *111*:8306–8308, 1989.
106. Yabe, K., Ando, Y., and Hamasaki, T. Biosynthetic relationship among aflatoxins B_1, B_2, G_1, G_2. *Appl. Environ. Microbiol.*, *54*: 2101–2106, 1988.
107. Bhatnagar, D., Ullah, A. H. J., and Cleveland, T. E. Purification and characterization of a methyltransferase from *Aspergillus parasiticus* SRRC 163 involved in aflatoxin biosynthetic pathway. *Prep. Biochem.*, *18*:321–349, 1988.
108. Keller, N., Dischinger, H. C., Jr., Bhatnagar, D., Cleveland, T. E., and Ullah, A. H. J. Purification of a second methyltransferase active in the aflatoxin biosynthetic pathway. *FASEB J.*, 5:A822, 1991.
109. Hamasaki, T., Matsui, K., Isono, K., and Hatsuda, Y. A new metabolite from *Aspergillus versicolor*. *Agric. Biol. Chem.*, *37*:1769–1770, 1973.
110. Steyn, P. S., and Vleggaar, R. Austocystins. Six novel dihydrofuro(3',2',4,5)furo(3,2-*b*)xanthones from *Aspergillus ustus*. *J. Chem. Soc., Perkin Trans. 1, 1974*:2250–2254.
111. Cleveland, T. E. Conversion of dihydro-*O*-methylsterigmatocystin to aflatoxin B_2 by *Aspergillus parasiticus*. *Arch. Environ. Contam. Toxicol.*, *18*:429–433, 1989.
112. Yabe, K., Ando, Y., Hashimoto, J., and Hamasaki, T. Two distinct *O*-methyltransferases in aflatoxin biosynthesis. *Appl. Environ. Microbiol.*, *55*:2172–2177, 1989.
113. Dutton, M. F., Ehrlich, K., and Bennett, J. W. Biosynthetic relationship among aflatoxin B_1, B_2, M_1, and M_2. *Appl. Environ. Microbiol.*, *49*:1392–1395, 1984.
114. Anderson, J. A., Chung, C. H., and Cho, S.-H. Versicolorin A hemiacetal, hydroxydihydrosterigmatocystin and aflatoxin G_{2a} reductase activity in extracts from *Aspergillus parasiticus*. *Mycopathologia*, *111*:39–45, 1990.
115. Singh, R., and Hsieh, D. P. H. Enzymatic conversion of sterigmatocystin to aflatoxin B_1 by cell-free extracts of *Aspergillus parasiticus*. *Appl. Environ. Microbiol.*, *31*:743–745, 1976.
116. Bhatnagar, D., and Cleveland, T. E. Fate of the methyl group during the conversion of sterigmatocystin into *O*-methylsterigmatocystin and aflatoxin B_1 by cell-free preparations of *Aspergillus parasiticus*. *Biochimie*, *70*:743–747, 1988.
117. Cleveland, T. E., and Bhatnagar, D. Individual reaction requirements of two enzyme activities, isolated from *Aspergillus parasiticus*, which together catalyze conversion of sterigmatocystin to aflatoxin B_1. *Can. J. Microbiol.*, *33*:1108–1112, 1987.

118. Bhatnagar, D., and Cleveland, T. E. Utilization of purified pertinent fungal enzymes for development of probes to identify genes responsible for aflatoxin biosynthesis. *J. Toxicol. Toxin Rev.*, *8*: 305–318, 1989.
119. Biollaz, M., Buchi, G., and Milne, G. The biosynthesis of aflatoxins. *J. Am. Chem. Soc.*, *92*:1035–1043, 1970.
120. Thomas, R. Biosynthetic pathway involving ring cleavage. In *Biogenesis of Antibiotic Substances* (Z. Vanek and A. Hostalek, Eds.), Academic, New York, 1965, pp. 155–167.
121. Berger, Y., and Jadot, J. Biosynthesis de l'averufine et de la versicolorine B: deux metabolites de l'*Aspergillus versicolor*. *Bull. Soc. Chim. Belg.*, *85*:271–276, 1976.
122. Bu'lock, J. D. Polyketide biosynthesis. *Comp. Biochem.*, 5:927–987, 1979.
123. Heathcote, J. G., and Hibbert, J. R. *Aflatoxins: Chemical and Biological Aspects*, Elsevier, New York, 1978.
124. Maggon, K. K. and Venkitasubramanian, T. A. Metabolism of aflatoxin B_1 and G_1 by *Aspergillus parasiticus*. *Experientia*, *29*:1211–1222, 1973.
125. Heathcote, J. G., Dutton, M. F., and Hibbert, J. R. Biosynthesis of aflatoxins. Part II. *Chem. Ind.*, *1976*:270–273, 1976.
126. Floyd, J. C., Mills, J., and Bennett, J. W. Biotransformation of sterigmatocystin and absence of aflatoxin biotransformation by blocked mutants of *Aspergillus parasiticus*. *Exp. Mycol.*, *11*:109–114, 1987.
127. Henderberg, A., Bennett, J. W., and Lee, L. S. Biosynthetic origin of aflatoxin G_1: confirmation of sterigmatocystin and lack of confirmation of aflatoxin B_1 as precursors. *J. Gen. Microbiol.*, *134*: 661–667, 1988.
128. Floyd, J. C., and Bennett, J. W. Preparation of ^{14}C labelled aflatoxins and incorporation of unlabelled aflatoxin in a blocked versicolorin A accumulating mutant of *Aspergillus parasiticus*. *J. Am. Oil Chem. Soc.*, *58*:956A–959A, 1981.
129. Gollnick, K., and Schenck, G. O. Oxygen as a dienophile. In *1,4-Cycloaddition Reactions* (J. Hamer, Ed.), Associated Press, New York, 1967, pp. 255–344.
130. Cleveland, T. E., Bhatnagar, D., Foell, C. J., and McCormick, S. P. Conversion of a new metabolite to aflatoxin B_2 by *Aspergillus parasiticus*. *Appl. Environ. Microbiol.*, *53*:2804–2807, 1987.

11

A MOLECULAR APPROACH TOWARDS UNDERSTANDING AFLATOXIN PRODUCTION

NANCY P. KELLER*, THOMAS E. CLEVELAND, and DEEPAK BHATNAGAR
Southern Regional Research Center, Agricultural Research Service, U.S. Department of Agriculture, New Orleans, Louisiana

I. INTRODUCTION

The imperfect fungi *Aspergillus flavus* Link and *A. parasiticus* Speare have achieved notoriety owing to their ability to produce putative carcinogenic secondary metabolites. These metabolites, aflatoxins, were first detected in *A. flavus*-infested peanut meal that killed thousands of poultry in hatcheries across England [1]. This incident was the impetus for intense research efforts to define and detect these toxins and to understand the biology of the causal fungi [1, 2]. In the last three decades, biochemical, epidemiological, and classic genetic studies have answered many of the questions asked about the aflatoxigenic aspergilli. Recently, attention has been focused on the power of molecular biology to unravel further secrets of metabolism through gene dissection. This chapter presents the current state of knowledge of the aflatoxin biosynthetic pathway and summarizes how modern techniques can be used to complement previous investigations in deciphering the pathway.

A. Aflatoxin: Mode of Action, Biosynthetic Pathway, Producer Strains

Aflatoxins have several structural forms. In general, *A. flavus* produces aflatoxin B_1 (AFB_1) and aflatoxin B_2 (AFB_2), whereas *A. parasiticus* also produces aflatoxin G_1 (AFG_1) and aflatoxin G_2 (AFG_2). Aflatoxin B_1 is the most toxic and most commonly occurring form in contaminated foodstuffs and has received the most attention in studies elucidating the mode of action. Aflatoxin B_1 is metabolically activated to a toxic form by liver-specific, cytochrome P-450 monooxygenases in mammals [3–5]. The activated form of AFB_1 induces virus expression and tumor formation associated with *ras* oncogenes [6, 7]. Studies on the mechanism of chemical carcinogenesis have focused on the chemical–DNA interaction; AFB_1 binds

Present affiliation: Texas A&M University, College Station, Texas.

to both chromosomal proteins [8] and DNA [9]. DNA adducts are formed when AFB_1 binds to guanine residues [10], preferentially to DNA with an alternating G-C sequence [11]. This can lead to frameshift mutations caused by intercalation of activated AFB_1 [11] and G:C to T:A transversions [6, 9]. The net effect of AFB_1 exposure is an inhibition of RNA synthesis [11, 12].

Whereas very little information is available on effects of AFB_1 on either of the aflatoxigenic aspergilli or on their host plants, fungi per se are not resistant to the effects of aflatoxin. Both AFB_1 and AFG_1, the two metabolites implicated in mammalian cancers, are highly mutagenic to the fungus *Neurospora crassa* [13]. It is not known how *A. flavus* and *A. parasiticus* avoid aflatoxin-mediated suicide. The mechanism of adduct formation should be uniform for DNA from any organism, although there are sequence preferences for AFB_1–DNA binding sites [11]. Other organisms that produce active metabolites (e.g., antibiotics, toxins) defend themselves by various means: (1) modification or replacement of the target site, (2) inactivation of the metabolite with enzymes, (3) sequestration of the metabolite, and (4) membrane permeability barriers or active afflux systems [14]. Antibiotic resistance genes are often linked to the respective antibiotic biosynthesis gene cluster [15]. Could a similar system be operating in the aflatoxigenic fungi? The molecular understanding and manipulation of a defense mechanism could provide clues to means of limiting the effects of aflatoxin on target organisms.

One area of progress in aflatoxin research has been the careful determination of the intermediate steps of the synthetic pathway. Using a combination of mutant strains, pathway inhibitors, and feeding studies with radioactively and isotopically labeled precursors [16], a scheme of the pathway intermediates leading to AFB_1 has been designed: polyketide precursor → norsolorinic acid [17] → averantin [18] → averufanin [19] → averufin [20] → versiconal hemiacetal acetate [21] → versicolorin A [22] → sterigmatocystin [23] → *O*-methylsterigmatocystin [24] → AFB_1. A decaketide is the immediate ancestor for the characterized anthraquinone precursors of aflatoxin. Anthraquinones are pigmented compounds [16] and include the precursors norsorolinic acid through versicolorin A. Sterigmatocystin, *O*-methylsterigmatocystin, and the aflatoxins are xanthones characterized by a bisdihydrofuran moiety; the aflatoxins are distinguished by a coumarin ring fused to this structure [16].

Aspergillus flavus and *A. parasiticus* are generally acknowledged to be the only aflatoxin-producing fungi, although reports exist of other *Aspergillus* and *Penicillium* species seemingly capable of producing aflatoxins [25–27]. Caution must be taken in analyzing for aflatoxin production; toxin synthesis has been mistakenly attributed to many genera and species other than *A. flavus* and *A. parasiticus* [28]. Other fungi produce precursors in the pathway such as sterigmatocystin and/or *O*-methylsterigmatocystin. These fungi include *Chaetomium* and *Farrowia* spp. [29], *Bipolaris* sp. [30–32], *Dothistroma pini* [33], and *Cercospora smilacis* [34] as well as several *Asergillus* species including the genetically well defined *A. nidulans* [32, 35–38]. Detection of the precursor compounds cannot always be repeated. Whether this is due to differential laboratory procedures, insufficient precautionary measures, or strain instability is unknown.

II. GENETICS OF AFLATOXIN BIOSYNTHESIS

A. Parasexual Cycle

The two aflatoxigenic fungi, *A. flavus* and *A. parasiticus*, are strictly anamorphic species (no known sexual stage) and thus lack the advantages of sexual analysis via Mendelian genetics. Classical genetic studies of anamorphic fungi have used the parasexual cycle since no meiotic products can be generated. The parasexual cycle was first described for the anamorphic stage of the fungus *Emericella nidulans* (= *A. nidulans*) in 1953 [39] and since then has been used for studies of several fungi [40, 41]. This cycle consists of hyphal anastomosis between two comparable isolates to form a heterokaryotic mycelium, karyogamy (between like or unlike nuclei), occasional recombination through mitotic crossing over, and haploidization. The recombination between unlike nuclei is the most important phase of this cycle as it confers some of the advantages of sexuality. Unlinked genes will segregate independently during haploidization, linked genes only rarely so (as a function of map distance and crossover frequency). Parasexuality, along with naturally occurring mutations, may be of importance for the evolution of the imperfect species [42–44].

Parasexual experimentation is inherently laborious. Genetic analyses of *A. flavus* and *A. parasiticus*, in particular, have proved painstaking because of several obstacles:

1. Haploid and diploid conidiospores are indistinguishable by size or morphology, making diploid recombinants difficult to identify [45].
2. Certain genotypes segregate nonrandomly during haploidization [45–47].
3. Vegetative incompatibility groups restrict heterokaryon compatibility [48, 49].

Nevertheless, several mutants have been induced and analyzed in parasexual studies leading to the establishment of linkage groups and elucidation of the aflatoxin biosynthetic pathway [47, 50–53].

B. Mutants and Linkage Groups

Mutants in both aflatoxin-producing species have been generated by mutagenic agents (ultraviolet light, nitrosoguanidine) or repeated serial transfer of isolates on artificial growth media. Ultraviolet- and chemically derived mutants include aberrant, toxin-producing strains with varied spore color and auxotrophic traits [16, 50, 51]. Since many of the precursors in the aflatoxin pathway are colored compounds, changes in mycelial pigment have often been associated with changes in toxin production [19, 52]. These mutants have been useful in establishing linkage groups for both fungi. Independent researchers using different strains have found evidence for one [18], four [47], six [53], and seven [54] linkage groups in *A. parasiticus*, none with a great number of loci. In two studies, one group showed that two aflatoxin mutants, *ver-1* and *nor-1*, were on the same linkage group [53], whereas another study put them on separate linkage groups [54]. *Aspergillus flavus* has over 30 genes assigned to eight linkage groups [35, 51, 55–57]. Of the aflatoxin mutants of *A. flavus*, 10 were assigned to linkage group VII, one to linkage group II, and one to linkage group VIII. All aflatoxin mutants of *A. parasiticus* and most in *A. flavus* are recessive; however, two *A. flavus* mutants, *afl-1* and *afl-b2*, are dominant [58].

C. Vegetative Incompatibility Groups

At present, vegetative incompatibility has prevented crosses between many strains of the aflatoxigenic aspergilli by either parasexual crosses [48, 49, 59] or protoplast fusions [60], the latter being a technique that has allowed interspecific hybrid fusions with other aspergilli [61–63]. Vegetative incompatibility, the inability of mycelia of two species to fuse and exchange nuclei, blocks heterokaryon formation between fungal strains and consequently restricts gene flow among fungal populations. This phenomenon has been well documented for several *Aspergillus* species [64]. *Aspergillus nidulans* has at least eight *het* loci; an allelic difference at any of the loci is sufficient to define a separate vegetative compatibility group (VCG) [61, 65, 66]. Papa [49] found 22 VCGs in *A. flavus* isolates from infested corn in Georgia, and Bayman and Cotty [48] found 28 VCGs in *A. flavus* isolates from cotton in Arizona. The molecular basis for heterokaryon incompatibility in any fungus is not well understood, although recently genes governing vegetative incompatibility have been cloned from *Neurospora crassa* [67] and *Podospora anserina* [68].

D. Subculturing Degeneracy

Repeated serial transfers of aflatoxigenic strains of both *A. flavus* and *A. parasiticus* can lead to reduction or elimination of toxin production [28, 50, 69–73]. This phenomenon is not unlike that observed with phytopathogenic fungi, which lose virulence upon extensive subculturing. Loss of virulence can often be reversed by inoculating the host plant with the degenerate strain. However, once aflatoxin production is attenuated in a toxigenic strain of *Aspergillus*, it rarely is fully expressed again [25, 69, 74, 75]. An exception to this was reported in an attenuated strain of *A. parasiticus* where transfer of single spores of the originally polysporous strain restored the strain to high toxin production [72].

Bennett et al. [69, 76, 77] have reported two distinct classes of atoxigenic strains of *A. parasiticus* isolated after serial subculturing, christened *fan* (complete block of pathway) and *fluff* (leaky block) based on colony morphology. Very similar results were obtained in an independent study with isolates of both *A. flavus* and *A. parasiticus* [73]. Both studies reported gross morphological changes in these strains: suppressed sporulation, increased vegetative mycelia, and decreased sclerotia formation in conjunction with a decrease in aflatoxin production. These changes may indicate the presence of a global regulatory system controlling vast tracts of genetic material.

III. MOLECULAR BIOLOGY

A. Genetic Transformation

In 1973 the first transformation of a fungus by DNA was reported. Intact cells of an auxotrophic mutant of *Neurospora crassa* were converted to prototrophy by DNA of the wild-type strain, presumably by uptake and incorporation of the transforming DNA into the genome of the auxotrophic strain [78]. The transformation procedure was improved considerably for *Neurospora* [79] and extended to *Saccharomyces cerevisiae* and *A. nidulans* [80, 81] when it was found that transformation of fungal protoplasts, rather than whole cells, was highly reproducible. The ability to transform fungi has made it possible to locate, isolate, and manipulate specific regions of the fungal genome (see Refs. 82–84).

Genetic transformation of *A. flavus* and *A. parasiticus* has been developed for the purpose of aiding in the study of the aflatoxin biosynthetic pathway [85–87]. In general, selectable genes used in transformation experiments fall into two categories: genes allowing for the return to prototrophy of an auxotrophic mutant [88–90] and genes conferring drug resistance [91–94]. The selectable genes may originate from either the fungus to be transformed (homologous genes) or from another organism (heterologous genes). Since aflatoxigenic aspergilli are resistant to commonly used drugs, they have been transformed by complementation to prototrophy: *A. flavus* uracil auxotrophs by the heterologous *pyr*-4 gene of *N. crassa* [87], and *A. parasiticus* uridine and nitrate reductase auxotrophs by the homologous *pyr* G and *nia* D genes, respectively [85, 86]. *Aspergillus flavus* also has been transformed with the homologous benomyl resistance gene (*ben*) [88]. Transformation systems for both aflatoxigenic species generate stable transformants in the range of 20 transformants per microgram of DNA (*A. flavus* [87]) to 20–50 *pyr* G transformants/μg DNA (*A. parasiticus* [86]) and 110–120 *nia* D transformants/μg DNA (*A. parasiticus* [85]). *Aspergillus flavus* transformants resulted from single or multiple ectopic integrations of the transforming plasmids ([87]; G. Payne, personal communication). Analyses of *A. parasiticus* transformants showed both homologous and ectopic integration of single and multiple copies of plasmids; approximately 80% of *pyr* G and 20% of *nia* D transformants resulted from homologous integrations (J. Linz, personal communication). Homologous integration occurs when homologous DNA integrates into the region of homology in the chromosome, ectopic integration when DNA integrates into a nonhomologous region (Fig. 1). Both homologous and ectopic integration may result from incorporation of a single copy of the plasmid or of several copies of the plasmid. When multiple copies of the plasmid are inserted one after another at the same site, the process is referred to as tandem insertion.

B. Gene Identification and Function

Genes in the aflatoxin pathway can be identified by the same molecular techniques used to detect other fungal genes. Genetic complementation has been successful in detecting several genes [89, 96–98]. This method

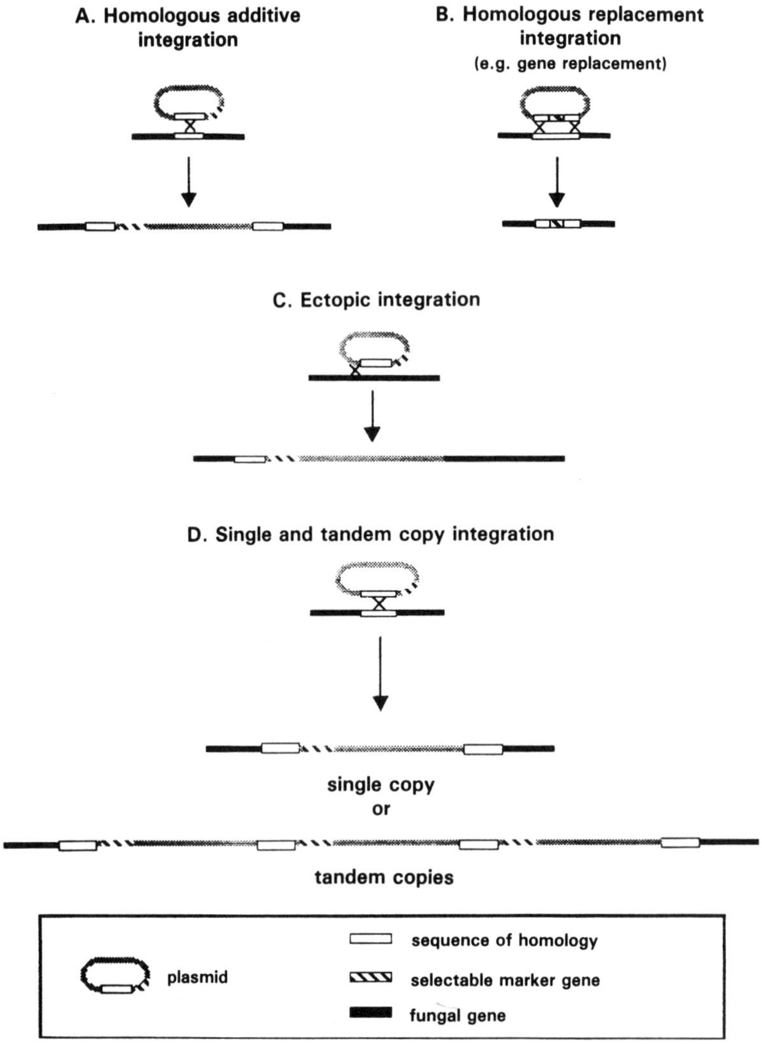

FIGURE 1. Plasmid integration into a fungal chromosome. (A) A plasmid integrates into the chromosome in a single crossover event at a region of sequence homology. (B) A plasmid exchanges a homologous DNA sequence with the chromosome in a double crossover event. (C) A plasmid integrates into the chromosome in a single crossover event at a region with no sequence homology. (D) Plasmids may integrate as single or tandem copies.

relies on the ability to identify a functioning gene by inserting the gene into an organism auxotrophic for the desired trait. For example, the *A. parasiticus trp* C gene was cloned by complementing an *Escherichia coli* tryptophan mutant with a DNA library of the fungal genome [99]. Alternatively, a characterized gene from one organism may be used to identify a similar gene in another organism if there is enough sequence similarity between the two genes. For instance, the *A. parasiticus nia* D, *pyr* G, *ras*, and *ben* genes were cloned after identification with heterologous genes from other fungi ([85, 86]; J. Linz, personal communication). Gene function also may be determined by disrupting a native DNA sequence in the fungus. A plasmid containing a mutated, nonfunctional form of a putative gene can be transformed into a fungus by homologous integration so that the wild-type gene will be replaced with the mutated gene (Fig. 1B). Any biological effects on the fungus following this gene replacement can then be elucidated. Targeted gene replacements in fungi include yeast genes [100, 101], the *N. crassa am* gene [102, 103], the *A. nidulans arg B* gene [104], and the *Ustilago maydis pyr*-6 gene [96].

A more direct approach in identification of genes is to first purify their gene products (proteins) and then use this purified protein to obtain antibodies and/or amino acid sequences. Antibodies raised to a purified protein can be used to screen an expressed cDNA library. The cDNA clone expressing the gene product should react with the antibodies. The primary N-terminus sequence and/or the sequence of internal fragments of a purified protein can be used to construct oligonucleotide sequences of the putative gene. These short DNA sequences should assist in identifying clones in cDNA or genomic libraries corresponding to the gene against which the oligonucleotide probes are prepared. These strategies have been used successfully to identify genes involved in tricothecene and HC toxin production in the plant pathogens *Fusarium sporotrichioides* and *Cochliobolus carbonum*, respectively [105–107]. Several enzymes in the aflatoxin biosynthetic pathway have been purified with the intent to identify pathway genes [108–111]. Oligonucleotide probes generated from the N-terminus of two separate methyltransferases [109, 111] and one reductase [108] are currently being used to screen *A. flavus* and *A. parasiticus* DNA libraries (Keller et al., unpublished data).

Once a gene is cloned, experimentally manipulable regulatory elements can be fused to the coding region of the cloned gene in order to study the consequences of altered gene expression [112]. Conversely, the influence of the promoter and other cis-acting regulatory sequences of the cloned gene can be investigated by fusing them to the coding region of reporter genes with easily detected activity. Promoters associated with cell differentiation in *Streptomyces coelicolor* [113], sporulation in *Bacillus subtilus* [114], and conidiophore development in *A. nidulans* [115] were visually detected with the β-galactosidase or luciferase encoding reporter genes. This latter approach has been especially useful in characterizing the temporal and spatial expression of genes and gene families [116].

C. Pulsed-Field Gel Electrophoresis

Unlike conventional electrophoresis, which uses a unidirectional electric field, pulsed-field gel electrophoresis uses two alternating electric fields. This recent advancement in electrophoretic technology has extended the size of DNA molecule separation to approximately 10 megabases (Mb) [117] and has been used successfully to resolve electrophoretic karyotypes of several fungi [118–122]. The ability to separate fungal chromosomes has application in gene mapping studies, especially for imperfect fungi, and has resulted in the unexpected discoveries that many fungal species have chromosomal length polymorphisms and may carry small supernumerary chromosomes [119, 121, 123, 124]. These observations suggest that researchers must reassess their view of fungal genetics and evolution.

Pulsed-field gel electrophoresis offers some advantages over conventional methods used to study fungal chromosomes. Fungal chromosomes, in comparison to those of other eukaryotes, are small and have been difficult to visualize using cytological techniques. Furthermore, creating genetic maps based on sexual or parasexual crosses is time-consuming. Nevertheless, cytological and genetic analyses have established linkage maps for *N. crassa* and *A. nidulans* [125–128]. The linkage maps of wild-type and translocation strains of these fungi were recently confirmed by pulsed-field gel electrophoresis [118, 122]. Most fungi lack the cytological and genetic analyses of *A. nidulans* and *N. crassa* with which to establish linkage groups. Instead, karyotypes are now being defined by electrophoresis and genetic maps by probing with cloned genes and RFLP analysis [118, 119, 120, 122, 124]. Using these techniques, a virulence gene has been assigned to a supernumerary chromosome in the pea pathogen *Nectria haematococca* [123], and the locus for toxin production has been associated with chromosome translocations in the corn pathogen *Cochliobolus heterostrophus* [124]. Studies of the aflatoxigenic aspergilli indicate that these species have six to eight chromosomes ranging in size from 3 to $\geq$7 Mb [129]. Furthermore, several *A. flavus* strains exhibit chromosome length polymorphisms and carry small (1–1.5 Mb) chromosomes [130].

IV. SECONDARY METABOLITES

A. Gene Clusters in Secondary Metabolism

Secondary metabolites, compounds not essential for growth, are produced most commonly by bacteria, actinomycetes, filamentous fungi, and plants. They are products of multistep enzymatic reactions that start from intermediates of primary metabolism [15, 50, 131, 132]. Common secondary metabolites include antibiotics, pigments, alkaloids, and toxins including aflatoxins. Perhaps the best characterized secondary metabolites are the antibiotics elaborated by actinomycetes [15]. The organization of genes in pathways of secondary metabolism are often in a clustered format in one region of a chromosome [133]. Examples of clustered genes have been found in bacteria [134–138], *Streptomyces* [139–142], and in yeast [143, 144], filamentous fungi [145–149], and other eukaryotes [150–152].

The significance of gene clustering is still speculative. Is this architecture a reflection of a local regulatory mechanism based on structural conformations of the chromosome location or a convenient "evolutionary package" conserved within and between species [133, 153]? The structural and sequence homology of the penicillin pathway among prokaryotes and eukaryotes provides evidence for the transfer of genetic material between species [148, 154]. Regardless, the clustering of genes provides a fortuitous format for DNA manipulations. Once the chromosome domain containing the cluster is identified, information on individual genes, their expression and regulation, can be achieved at a faster rate than if genes were spread throughout the genome. The clustering of genes also allows for convenient block transfer to heterologous expression systems via cosmids [155–158] or yeast artificial chromosomes (YACs) [152]. Furthermore, it is now feasible to identify clusters, as well as single genes, on specific chromosomes for mapping studies using pulsed-field gel electrophoresis [118, 159]. Alternatively, if a single gene has been cloned and is known to be linked to other genes in the pathway, it could be used to detect the appropriate chromosome, and a chromosome specific library could be constructed for analysis of the cluster.

Earlier results suggested that functionally related genes in eukaryotes were not linked. For example, the IPNS and DAOCS genes of *Cephalosporium acremonium* are on different chromosomes [160]. It is now known that many eukaryotic genes are clustered. In *N. crassa* and *A. nidulans*, pathway clusters have been cloned [146, 147, 161–163], assigned to chromosomes [147, 163, 164], and used in locating the corresponding cluster in another organism [154, 165]. As yet there is no direct proof that the aflatoxin pathway genes are clustered. However, the location of several aflatoxin mutants of *A. flavus* on linkage group VII (Section III.B) suggests a putative cluster, although it is not known if the mutants are allelic or not. One strategy in locating this potential cluster would be to identify the linkage group VII chromosome, through use of either an aflatoxin gene or some other marker. The identification of linkage group VII and/or identification of one structural gene in the pathway should provide a more rapid means of characterizing neighboring genes via chromosome specific libraries and/or chromosome walking techniques. To illustrate the feasibility of this type of strategy, Smith et al. [154] used probes of known genes in the penicillin pathway to isolate overlapping clones from a cosmid library on the assumption that the genes that are linked in prokaryotes would also be linked in *Penicillium*. The cloning of the *A. nidulans* proline catabolism gene cluster also involved identifying sets of overlapping cosmid clones [147].

Genes in the aflatoxin pathway might be isolated in fungal species other than *A. flavus* and *A. parasiticus*. The production of aflatoxin pathway intermediates in additional fungi (Section II) argues for some genetic conservation of the pathway. One of these other fungi may ultimately provide a more convenient model system than *A. flavus* or *A. parasiticus* to pursue the molecular dissection of the aflatoxin pathway. For example, *A. nidulans* is a low producer of penicillin but was used as a model system to study this pathway because of its established DNA transformation system and formal genetics [148]. Similarly, *A. nidulans* produces sterigmatocystin [35], a late intermediate in aflatoxin biosynthesis, and might prove useful in studying the molecular genetics of the aflatoxin pathway.

B. Coordinate Regulation of Gene Clusters

Most of the information on the expression of microbial secondary metabolites is based on the knowledge gleaned from *Streptomyces*. In general, the proposed analogies between prokaryotic and eukaryotic regulatory systems remain to be tested. Since the terminal product(s) in secondary metabolism are the result of many steps, the question arises: How is the timing of the pathway controlled such that each enzymatic step proceeds sequentially? Is sequential timing necessary? Evidence suggests that the steps do proceed sequentially both in *Streptomyces* [166] and in fungi [132, 167], including *A. parasiticus* [168]. In some cases, subclustering in "packages" of early, middle, and late genes occurs in a manner analogous to phage regulation [169]. Examples of subclustering include genes in the polyketide-derived compounds tylosin [170], tetracenomycin C [155], and actinorhodin [171, 172]. The polyketide origin of aflatoxin will be discussed below.

One convenient way to understand secondary metabolism is to consider it a form of biochemical differentiation [15, 76]. The characteristic appearance of secondary metabolites after active growth has ceased or in response to certain environmental signals is reminiscent of cellular and morphological differentiation. Several morphological changes in sporulation, sclerotia production, and colony morphology in *A. flavus* and *A. parasiticus* have been correlated with changes in aflatoxin production (Section III.D); this correlation may be coincidental or reflect a genetic relationship between these proesses. Genetic evidence has been found for global regulatory elements that control biochemical and cellular differentiation in actinomycetes. Three genes, *afs* A, *afs* B, and *saf*, cloned from *Streptomyces bikiniensis*, *S. coelicolor*, and *S. griseus*, respectively, encode positive regulatory proteins that control such functions as antibiotic biosynthesis and resistance, sporulation, pigment formation, production of extracellular enzymes, and differentiation [173–177]. The gene products of *afs* B and *saf* are DNA-binding proteins [178] that act at the transcriptional level. Whether the proteins interact directly with regulatory sequences of gene clusters or indirectly with one or more other controlling factors (e.g., *afs* A) is not known. Similarly, cell differentiation in yeast is controlled by a regulatory hierarchy of trans-acting DNA-binding proteins [179].

The implication that ancillary regulatory proteins could be involved in secondary metabolism makes the puzzle of control more intricate. One interesting type of global modification is exemplified by the change in life style of *Bacillus* wrought by sigma factors. Sigma factors are proteins that alter the specificity of RNA polymerase to recognize transcription start points, thus allowing for discretion in gene transcription. One sigma factor allows RNA polymerase to recognize and transcribe from distinct promoters of early sporulation genes, and another allows recognition and transcription of promoters heading middle sporulation genes [180]. Although no analogous sigma factor has been found in filamentous fungi, this does not rule out their existence. Sporulation-specific transcripts have been found and located on a gene cluster in *A. nidulans*; however, the regulatory elements controlling transcription have not been identified [161]. Regardless of route, secondary metabolite gene promoters are likely to have unique motifs that play a role in expression [113].

Besides genetic regulation of gene expression, a considerable amount of literature concerns the stimulation and/or repression of secondary metabolites by the availability of phosphate, carbon, and nitrogen [15, 50, 132, 181, 182]. The effects of these substances on aflatoxin production have been reviewed [50, 181–183].

C. Aflatoxin as a Secondary Metabolite

The aflatoxins are synthesized via a polyketide pathway, that is, from condensation of CoA esters (acetate) and β-keto acids (malonate) [15]. A presentation of the aflatoxin pathway by chemical structure results in this simplified scheme: acetate/polymalonate polyketide → polyhydroxyanthraquinones → xanthones → coumarins [16]. This simplified representation of the pathway suggests alternative strategies for molecular identification of the structural genes in the pathway, for example, the use of heterologous genes encoding enzymes that catalyze similar reactions in other organisms. One theory of evolution gaining supportive molecular evidence proposes that enzymes (or their functional domains) performing similar jobs in different organisms may arise from the same ancestor [184]. Conserved domains range from a few key amino acids, which may or may not have the same codons, to long stretches of nearly identical DNA sequences [154, 184–186]. For example, the *acv* A gene, which is the first gene in the penicillin biosynthetic pathway in *P. chrysogenum*, shares homology with genes for several bacterial peptide antibiotic synthetases [186]. This similarity suggests strategies for the elucidation of the aflatoxin pathway: One might look for a gene encoding for a polyketide synthase, or a fatty acid synthase from another organism, and use it to probe for the analogous gene in the aflatoxin pathway. Using genes cloned from the pathway of the polyketide actinorhodin, Malpartida et al. [187] successfully located related genes in other polyketide-producing *Streptomyces*. Recently two polyketide synthase genes from two different *Penicillium* species have been isolated [188, 189]. A region of DNA flanking the polyketide synthase gene cloned from *P. urticae* has been shown to strongly hybridize with DNA of *A. parasiticus* and other polyketide-producing fungi (M. Gaucher, personal communication); whether this DNA is involved in the aflatoxin pathway has not been established.

Many fungi produce intermediates in the aflatoxin pathway; *A. flavus* and *A. parasiticus* are distinct only in converting sterigmatocystin to aflatoxin. The enzymes catalyzing these two final steps are a methyltransferase and an oxidoreductase [109–111]. The aflatoxin-producing fungi may have picked up the pertinent enzymatic activities by piecing together active domains from various genetic sources. Some methyltransferases retain conserved motifs throughout the length of the protein [185, 190–192]. While consensus sequences are most notable among the DNA methyltransferases [191, 192], a putative *S*-adenosylmethionine binding site of three specific amino acids appears to be highly conserved in almost all methyltransferases [185, 190]. Although this particular amino acid sequence is not an encouraging one on which to construct an oligonucleotide probe for identification of the methyltransferases in the aflatoxin pathway, the identification of conserved motifs may serve in reconstructing the molecular origin of the genes involved in aflatoxin production.

V. CONCLUDING REMARKS

The aflatoxins, carcinogenic secondary metabolites of *Aspergillus flavus* and *A. parasiticus*, are synthesized in a series of enzymatic steps starting with a polyketide precursor. Detailed studies in the past three decades have established the structure and metabolic pathway of these compounds. Concomitant genetic studies, while hindered by the lack of a sexual cycle in the aflatoxigenic fungi, succeeded in producing mutants necessary for the elucidation of the aflatoxin pathway and in laying the foundation for linkage maps for both aspergilli.

Molecular techniques are now available for precise genetic dissection of the aflatoxin pathway. Transformation systems have been established for both toxin-producing fungi. Genes have been cloned and are presently being used as probes to create physical maps of the genome in chromosome separation studies. The molecular cloning of genes involved in secondary metabolite synthesis in other systems as described in this chapter provides the rationale for applying similar strategies to aid in the identification of genes involved in aflatoxin metabolism. The challenge is to take advantage of the tools available in a creative and thoughtful manner.

REFERENCES

1. Lancaster, M. D., Jenkins, F. P., and Phillip, J. M. Toxicity associated with certain samples of groundnuts. *Nature*, *192*:1095–1096, 1961.
2. Diener, U. F., Cole, R. J., Sanders, T. H., Payne, G. A., Lee, L. S., and Klich, M. A. Epidemiology of aflatoxin formation by *Aspergillus flavus*. *Ann. Rev. Phytopathol.*, *25*:249–270, 1987.
3. Corcos, L., Rousset, J.-P., Kiefer, F., Wiebel, F. J., and Weiss, M. C. Genetic analysis of aflatoxin B_1 activation in rat hepatoma cells. *Mol. Gen. Genet.*, *222*:291–296, 1990.
4. Guengerich, F. P. Roles of cytochrome P450 enzymes in chemical carcinogenesis and cancer chemotherapy. *Cancer Res.*, *48*:2946–2954, 1988.
5. Koser, P. L., Faletto, M. B., Maccubbin, A. E., and Gurtoo, H. L. The genetics of aflatoxin B_1 metabolism. *J. Biol. Chem.*, *263*:12584–12595, 1988.
6. McMahon, G., Davis, E. F., Huber, L.-J., Kim, W., and Wogan, G. N. Characterization of c-Ki-*ras* and N-*ras* in aflatoxin B_1-induced rat liver tumors. *PNAS*, *83*:9418–9422, 1986.
7. Rascati, R. J., and McNeely, M. Induction of retrovirus gene expression by aflatoxin B_1 and 2-acetylaminofluorene. *Mutat. Res.*, *122*:235–241, 3983.
8. Groopman, J. D., Busby, W. F., Jr., and Wogan, G. N. Nuclear distribution of aflatoxin B_1 and its interaction with histones in rat liver in vivo. *Cancer Res.*, *40*:4343–4351, 1980.
9. Martin, C. N., and Garner, R. C. Aflatoxin B_1-oxide generated by chemical or enzymatic oxidation of aflatoxin B_1 causes guanine substitution in nucleic acids. *Nature*, *267*:863–865, 1977.

10. Bonnett, M., and Taylor, E. R. The structure of aflatoxin B_1-DNA adduct at N^{-7} of guanine. Theoretical intercalation and covalent adduct models. *J. Biomol. Struct. Dyn.*, *7*:127–149, 1989.
11. Yu, F.-L., Bender, W., and Geronimo, I. H. Base and sequence specificities of aflatoxin B_1 binding to single- and double-stranded DNAs. *Carcinogenesis*, *11*:475–478, 1990.
12. Yu, F.-L., Geronimo, I. H., Bender, W., and Permthamsin, J. Correlation studies between the binding of aflatoxin B_1 to chromatin components and the inhibition of RNA synthesis. *Carcinogenesis*, *9*: 527–532, 1988.
13. Ong, T.-M., and de Serres, F. J. Mutagenicity of chemical carcinogens in *Neurospora crassa*. *Cancer Res.*, *32*:1890–1893, 1972.
14. Cundliffe, E. How antibiotic-producing organisms avoid suicide. *Annu. Rev. Microbiol.*, *43*:207–233, 1989.
15. Martin, J. F., and Liras, P. Organization and expression of genes involved in the biosynthesis of antibiotics and other secondary metabolites. *Annu. Rev. Microbiol.*, *43*:173–206, 1989.
16. Steyn, P. S., Vleggar, R., and Wessels, P. L. The biosynthesis of aflatoxin and its congeners. In *The Biosynthesis of Mycotoxins: A Study in Secondary Metabolism* (P. S. Steyn, Ed.), Academic, New York, 1980, pp. 105–155.
17. Hsieh, D. P. H., Lin, M. T., Yao, R. C., and Singh, R. Biogenesis of aflatoxins: conversion of norsolorinic acid and other hypothetical intermediates into aflatoxin B_1 by *Aspergillus parasiticus*. *J. Agric. Food Chem.*, *24*:1167–1170, 1976.
18. Bennett, J. W., Lee, L. S., Shoss, S. M., and Boudreaux, G. H. Identification of averantin as an aflatoxin B_1 precursor. *Appl. Environ. Microbiol.*, *39*:835–839, 1980.
19. McCormick, S. P., Bhatnagar, D., and Lee, L. S. Averufanin is an aflatoxin precursor between averantin and averufin in the biosynthetic pathway. *Appl. Environ. Microbiol.*, *53*:14–16, 1987.
20. Lin, M. T., Hsieh, D. P. H., Yao, R. C., and Donkersloot, J. A. Conversion of averufin into aflatoxins by *Aspergillus parasiticus*. *Biochemistry*, *12*:5167–5171, 1973.
21. Shroeder, H. W., Cole, R. J., Grigsby, R. D., and Hein, H., Jr. Inhibition of aflatoxin production and tentative identification of an aflatoxin intermediate "versiconal acetate" from treatment with dichlorvus. *Appl. Microbiol.*, *27*:394–399, 1974.
22. Lee, L. S., Bennett, J. W., Cucullu, A. F., and Ory, R. L. Biosynthesis of aflatoxin B_1—conversion of versicolorin A to aflatoxin B_1 by *Aspergillus parasiticus*. *J. Agric. Food Chem.*, *24*:1167–1170, 1976.
23. Hsieh, D. P. H., Lin, M. T., and Yao, R. C. Conversion of sterigmatocystin to aflatoxin B_1 by *Aspergillus parasiticus*. *Biochem. Biophys. Res. Commun.*, *52*:992–997, 1973.
24. Bhatnagar, D., McCormick, S. P., Lee, L. S., and Hill, R. A. Identification of *O*-methylsterigmatocystin as an aflatoxin B_1 and G_1 precursor in *Aspergillus parasiticus*. *Appl. Environ. Microbiol.*, *53*: 1028–1033, 1987.

25. Kulik, M. M., and Holaday, C. E. Aflatoxin: A metabolic product of several fungi. *Mycopathol. Mycol. Appl.*, *30*:137–140, 1967.
26. Kumar, G., and Sampath, S. Aflatoxins—their nature and biological effects. *Ind. Nutrit. Diet*, *4*:85–101, 1971.
27. Leitao, J., Le Bars, J., and Bailly, J.-R. Production of aflatoxin B_1 by *Aspergillus ruber* Thom and Church. *Mycopathologia*, *108*: 135–138, 1989.
28. Wilson, B . J., Campbell, T. C., Hayes, A. W., and Hanlin, R. T. Investigation of reported aflatoxin production by fungi outside the *Aspergillus flavus* group. *Appl. Microbiol.*, *16*:819–821, 1968.
29. Udagawa, S., Muroi, T., Kurata, H., Sekita, S., Yoshihira, K., Natori, S., and Umeda, M. The production of chaetoglobosins, sterigmatocystin, *O*-methylsterigmatocystin, and chaetocin by *Chaetomium* spp. and related fungi. *Can. J. Microbiol.*, *25*:170–177, 1978.
30. Aucamp, P. J., and Holzapfel, C. W. Polyhydroxyanthraquinones from *Aspergillus versicolor, Aspergillus nidulans* and *Bipolaris* sp.; their significance in relation to biogenetic theories on aflatoxin B_1. *J. S. Afr. Chem. Inst.*, *23*:40–56, 1970.
31. Holzapfel, C. W., Purchase, I. F. H., Steyn, P. S., and Gouws, L. The toxicity and chemical assay of sterigmatocystin, a carcinogenic mycotoxin and its isolation from two new fungal sources. *S. Medical J.*, *40*:1100–1101, 1966.
32. Rabie, C. J., Lubben, A., and Steyn, M. Production of sterigmatocystin by *Aspergillus versicolor* and *Bipolaris sorokiniana* on semisynthetic liquid and solid media. *Appl. Environ. Microbiol.*, *32*:206–208, 1976.
33. Danks, A. V., and Hodges, R. Polyhydroxyanthraquinones from *Dothistroma pini*. *Aust. J. Chem.*, *27*:1603–1606, 1974.
34. Assante, G., Camarda, L., Merlini, L., and Nasini, G. Dothistromin and 2-epidothistromin from *Cercospora smilacis*. *Phytochemistry*, *16*: 125–126, 1977.
35. Hajjar, J. D., Bennett, J. W., Bhatnagar, D., and Bahu, R. Sterigmatocystin production by laboratory strains of *Aspergillus nidulans*. *Mycol. Res.*, *94*:548–551, 989.
36. Frisvad, J. C. Secondary metabolites as an aid to *Emericella* classification. In *Advances in Penicillium and Aspergillus Systematics* (R. A. Samson and J. Pitt, Eds.), Plenum, New York, 1985, pp. 430–437.
37. Steyn, P. S., and Velggaar, R. Dihydrofuro[3,2':4,5]furo[3,2*b*]-xanthenones: the structures of austocystins G, H and I. *J. South Afr. Chem. Inst.*, *28*:375–377, 1975.
38. Rabie, C. J., Steyn, M., and VanSchalkwyk, G. C. New species of *Aspergillus* producing sterigmatocystin. *Appl. Environ. Microbiol.*, *33*:1023–1025, 1977.
39. Pontecorvo, G., Roper, J. A., Hemmons, L. M., MacDonald, K. D., and Bufton, A. W. J. The genetics of *Aspergillus nidulans*. *Adv. Genet.*, *5*:141–238, 1953.
40. Buxton, E. W. Heterokaryosis and parasexual recombination in pathogenic strains of *Fusarium oxysporum*. *J. Gen. Microbiol.*, *15*:133–139, 1956.

41. Anagnostakis, S. L. Vegetative incompatibility in *Endothia parasitica*. *Exp. Mycol.*, *1*:306–316, 1977.
42. Bradley, S. G. Parasexual phenomena in microorganisms. *Annu. Rev. Microbiol.*, *16*:35–52, 1962.
43. Casselton, L. A. Somatic recombination in fungi. *Sci. Prob. (London)*, *53*:107–115, 1965.
44. Roper, J. A. The parasexual cycle. In *The Fungi*, Vol. 2, *Fungal Organisms* (G. C. Ainsworth and A. S. Sussman, Eds.), Academic, New York, 1966, pp. 589–617.
45. Bennett, J. W. Aflatoxins and anthraquinones from diploids of *Aspergillus parasiticus*. *J. Gen. Microbiol.*, *113*:127–136, 1979.
46. Bennett, J. W., Vinnett, C. H., and Goynes, W. R., Jr. Aspects of parasexual analysis in *Aspergillus parasiticus*. *Can. J. Microbiol.*, *26*:706–713, 1980.
47. Papa, K. E. The parasexual cycle of *Aspergillus parasiticus*. *Mycologia*, *70*:766–773, 1978.
48. Bayman, P., and Cotty, P. J. Vegetative compatibility and genetic diversity in the *Aspergillus flavus* population of a single field. *Can. J. Bot.* (submitted).
49. Papa, K. E. Heterokaryon incompatibility in *Aspergillus flavus*. *Mycologia*, *78*:98–101, 1986.
50. Bennett, J. W., and Christensen, S. B. New perspectives on aflatoxin biosynthesis. *Adv. Appl. Microbiol.*, *29*:53–92, 1983.
51. Papa, K. E. Genetics of aflatoxin production in *Aspergillus flavus*: linkage between a gene for a high B_2-B_1 ratio and the histidine locus on linkage group VIII. *Mycologia*, *69*:1185–1190, 1977.
52. Lee, L. S. Metabolic precursor regulation of aflatoxin formation in toxigenic and non-toxigenic strains of *Aspergillus flavus*. *Mycopathologia*, *107*:127–130, 1989.
53. Bradshaw, R. E., Bennett, J. W., and Peberdy, J. F. Parasexual analysis of *Aspergillus parasiticus*. *J. Gen. Microbiol.*, *129*:2117–2123, 1983.
54. Lennox, J. E., and Davis, C. K. Selection of and complementation analysis among aflatoxin-deficient mutants of *Aspergillus parasiticus*. *Exp. Mycol.*, *7*:192–195, 1983.
55. Bennett, J. W., and Papa, K. E. The aflatoxigenic *Aspergillus* species. *Adv. Plant Pathol.*, *6*:263–280, 1988.
56. Papa, K. E. Linkage groups in *Aspergillus flavus*. *Mycologia*, *68*:159–165, 1976.
57. Papa, K. E. Genetics of *Aspergillus flavus*: complementation and mapping of aflatoxin mutants. *Genet. Res.*, *34*:1–9, 1979.
58. Papa, K. E. Dominant aflatoxin mutant of *Aspergillus flavus*. *J. Gen. Microbiol.*, *118*:279–282, 1980.
59. Gussack, G., Bennett, J. W., Cavalier, S., and Yatsu, L. Evidence for the parasexual cycle in a strain of *Aspergillus flavus* containing virus-like particles. *Mycopathologia*, *61*:159–165, 1977.
60. Leong, P. M., Bennett, J. W., and Ciegler, A. Protoplasts from auxotrophs of *Aspergillus flavus* and *A. parasiticus*. *Proc. 37th Gen. Meet. Soc. Ind. Microbiol.*, *22*:661–668, 1981.

61. Croft, J. H. Protoplast fusion and incompatibility in *Aspergillus*. In *Fungal Protoplasts: Applications in Biochemistry and Genetics* (J. F. Peberdy and L. Ferenczy, Eds.), Marcel Dekker, New York, 1985. pp. 225–240.
62. Kevei, F., and Peberdy, J. F. Interspecies hybridization after protoplast fusion in *Aspergillus*. In *Fungal Protoplasts: Applications in Biochemistry and Genetics* (J. F. Peberdy and L. Ferenczy, Eds.), Marcel Dekker, New York, 1985, pp. 241–257.
63. Ushijima, S., Nakadai, T., and Uchida, K. Breeding of new *Koji*-molds through interspecific hybridization between *Aspergillus oryzae* and *Aspergillus sojae* by protoplast fusion. *Agric. Biol. Chem.*, *54*: 1667–1676, 1990.
64. Croft, J. H. Genetic variation and evolution in *Aspergillus*. In *Evolutionary Biology of the Fungi* (A. D. M. Rayner, C. M. Brasier, and D. Moore, Eds.), Cambridge Univ. Press, New York, 1987, pp. 311–323.
65. Grindle, M. Heterokaryon compatibility of unrelated strains in the *Aspergillus nidulans* group. *Heredity, 18*:191–204, 1963.
66. Grindle, M. Heterokaryon compatibility of closely related wild isolates of *Aspergillus nidulans*. *Heredity, 18*:397–405, 1963.
67. Glass, N. L., Grotelueschen, J., and Metzenberg, R. L. *Neurospora crassa* A mating-type region. *PNAS, 87*:4912–4916, 1990.
68. Turcq, B., Denayrolles, M., and Begueret, J. Isolation of the two allelic incompatibility genes *s* and *S* of the fungus *Podospora anserina*. *Curr. Gen., 17*:297–303, 1990.
69. Bennett, J. W., Leong, P. M., Kruger, S., and Keyes, D. Sclerotial and low aflatoxigenic morphological variants from haploid and diploid *Aspergillus parasiticus*. *Experientia, 42*:848–851, 1986.
70. Bilgrami, K. S., Sinha, S. P., and Jeswal, P. Loss of toxigenicity of *Aspergillus flavus* strains during subculturing—a genetic interpretation. *Curr. Sci.*, *57*:551–552, 1988.
71. Lee, L. S., Lax, A. R., Mellon, J. E., and Klich, M. A. Stability for the character for aflatoxin production by *Aspergillus parasiticus*. *J. Agric. Food Chem.*, *24*:1167–1170, 1986.
72. Mayne, R. Y., Bennett, J. W., and Tallant, J. Instability of an aflatoxin-producing strain of *Aspergillus parasiticus*. *Mycologia, 53*: 644–648, 1971.
73. Torres, J., Guarro, J., Suarez, G., Sune, N., Calvo, M. A., and Ramirez, C. Morphological changes in strains of *Aspergillus flavus* Link ex Fries and *Aspergillus parasiticus* Speare related with aflatoxin production. *Mycopathologia, 72*:171–174, 1980.
74. De Vogel, P., Van Rhee, R., and Koelensmid, W. A. A. B. A rapid screening test for aflatoxin synthesizing aspergilli of the *flavus-oryzae* group. *J. Appl. Bacteriol.*, *28*:213–220, 1965.
75. Diener, U. L., and Davis, N. D. Aflatoxin production by isolates of *Aspergillus flavus*. *Phytopathology, 56*:1390–1393, 1966.
76. Bennett, J. W. Genetic perspective on polyketides, productivity, parasexuality, protoplasts and plasmids. In *Advances in Biotechnology*, Vol. III, Fermentation Products (M. Moo-Young, Ed.), Pergamon Press Canada, Toronto, 1981, pp. 409–415.

77. Bennett, J. W. Genetics of mycotoxin production with emphasis on aflatoxins. In *Overproduction of Microbial Products* (V. Krumphanzl, B. Sikyta, and Z. Vanek, Eds.), Academic, New York, 1982, pp. 551–561.
78. Mishra, N. C., and Tatum, E. L. Non-Mendelian inheritance of DNA-induced inositol independence in *Neurospora*. *PNAS, 70*:3875–3879, 1973.
79. Case, M. E., Schweizer, M., Kushner, S. R., and Giles, N. H. Efficient transformation of *Neurospora crassa* by utilizing hybrid plasmid DNA. *PNAS, 76*:5259–5363, 1979.
80. Hinnen, A., Hicks, J. B., and Fink, G. R. Transformation of yeast chimaeric ColE1 plasmid carrying LEU2. *PNAS, 75*:1929–1933, 1978.
81. Tilburn, J., Scazzocchio, C., Taylor, G. G., Zabicky-Zissman, J. H., Lockington, R. A., and Davies, R. W. Transformation by integration in *Aspergillus nidulans*. *Gene, 26*:205–221, 1983.
82. Fincham, J. R. S. Transformation in fungi. *Microbiol. Rev., 53*: 148–170, 1989.
83. Hynes, M. J. Transformation of filamentous fungi. *Exp. Mycol., 10*: 1–8, 1986.
84. Timberlake, W. E., and Marshall, M. A. Genetic engineering of filamentous fungi. *Science, 244*:1313–1317, 1989.
85. Horng, J. S., Chang, P.-K., Pestka, J. J., and Linz, J. E. Development of a homologous transformation system for *Aspergillus parasiticus* with the gene encoding nitrate reductase. *Mol. Gen. Genet., 224*:294–296, 1990.
86. Skory, C. D., Horng, J. S., Pestka, J. J., and Linz, J. E. Transformation of *Aspergillus parasiticus* with a homologous gene (*pyr*G) involved in pyrimidine biosynthesis. *Appl. Environ. Microbiol., 56*: 3315–3320, 1990.
87. Woloshuk, C. P., Seip, E. R., Payne, G. A., and Adkins, C. R. Genetic transformation system for the aflatoxin-producing fungus *Aspergillus flavus*. *Appl. Environ. Microbiol., 55*:86–90, 1989.
88. Seip, E. R., Woloshuk, C. P., Payne, G. A., and Curtis, S. E. Isolation and sequence analysis of a β-tubulin gene from *Aspergillus flavus* and its use as a selectable marker. *Appl. Environ. Microbiol., 56*:3686–3692, 1990.
89. Akins, R. A., and Lambowitz, A. M. A general method for cloning *Neurospora crassa* nuclear genes by complementation of mutants. *Mol. Cell. Biol., 5*:2272–2278, 1985.
90. Faugeron, G., Goyon, C., and Gregoire, A. Stable allele replacement and unstable non-homologous integration events during transformation of *Ascobolus immersus*. *Gene, 76*:109–119, 1989.
92. Parsons, K. A., Chumley, F. G., and Valent, B. Genetic transformation of the fungal pathogen responsible for rice blast disease. *PNAS, 84*:4161–4165, 1987.
92. Cooley, R. N., Shaw, R. K., Franklin, R. C. H., and Caten, C. E. Transformation for the phytopathogenic fungus *Septoria nodorum* to hygromycin B resistance. *Curr. Genet., 13*:383–389, 1988.
93. Dickman, M. B., and Partridge, J. E. Use of molecular markers for monitoring fungi involved in stalk rot of corn. *Theor. Appl. Genet., 77*:535–539, 1989.

94. Rodriguez, R. J., and Yoder, O. C. Selectable genes for transformation of the fungal plant pathogen *Glomerella cingulata* f. sp. *phaseoli* (*Colletotrichum lindemuthianum*). *Gene*, *54*:73–81, 1987.
95. Turgeon, B. G., Garber, R. C., and Yoder, O. C. Transformation of the fungal maize pathogen *Cochliobolus heterostrophus* using the *Aspergillus nidulans amd S* gene. *Mol. Gen. Genet.*, *201*:450–453, 1985.
96. Kronstad, J. W., Wang, J., Covert, S. F., Holden, D. W., McKnight, G. L., and Leong, S. A. Isolation of metabolic genes and demonstration of gene disruption in the phytopathogenic fungus *Ustilago maydis*. *Gene*, *79*:97–106, 1989.
97. Turgeon, B. G., MacRae, W. D., Garber, R. C., Fink, G. R., and Yoder, O. C. A cloned tryptophan-synthesis gene from the ascomycete *Cochliobolus heterostrophus* functions in *Escherichia coli*, yeast and *Aspergillus nidulans*. *Gene*, *42*:79–88, 1986.
98. Yelton, M. M., Timberlake, W. E., and van den Hondel, C. A. M. J. J. A cosmid for selecting genes by complementation in *Aspergillus nidulans*: selection of the developmentally regulated *yA* locus. *PNAS*, *82*:834–838, 1985.
99. Horng, J. S., Linz, J. E., and Pestka, J. J. Cloning and characterization of the *trp*C gene from an aflatoxigenic strain of *Aspergillus parasiticus*. *Appl. Environ. Microbiol.*, *55*:2561–2568, 1989.
100. Heinisch, J. Construction and physiological characterization of mutants disrupted in the phosphofructokinase genes of *Saccharomyces cerevisiae*. *Curr. Genet.*, *11*:227–234, 1986.
101. Parker, R., Simmons, T., Shuster, E. O., Siliciano, P. G., and Guthrie, C. Genetic analysis of small nuclear RNAs in *Saccharomyces cerevisiae*: viable sextuple mutant. *Mol. Cell. Biol.*, *8*:3150–3159, 1988.
102. Frederick, G. D., Asch, D. K., and Kinsey, J. A. Use of transformation to make targeted sequence alterations at the *am* (GDH) locus of *Neurospora*. *Mol. Gen. Genet.*, *217*:294–300, 1989.
103. Paietta, J. V., and Marzluf, G. A. Gene disruption by transformation in *Neurospora crassa*. *Mol. Cell. Biol.*, *5*:1554–1557, 1985.
104. Miller, B. L., Miller, K. Y., and Timberlake, W. E. Direct and indirect gene replacements in *Aspergillus nidulans*. *Mol. Cell. Biol.*, 5:1714–1721, 1985.
105. Hohn, T. M., and Beremand, P. D. Isolation and nucleotide sequence of a sesquiterpene cyclase gene from the trichothecene-producing fungus *Fusarium sporotrichioides*. *Gene*, *79*:131–138, 1989.
106. Hohn, T. M., and Van Middlesworth. Purification and characterization of the sesquiterpene cyclase trichodiene synthetase from *Fusarium sporotrichioides*. *Arch. Biochem. Biophys.*, *251*:756–761, 1986.
107. Panaccione. D. C., Scott-Craig, J. S., Pocard, J.-A., and Walton, J. D. Characterization of the *TOX* locus of *Cochliobolus carbonum*. Abstract A414, Am. Phytopath. Soc., Grand Rapids, Mich., 1990.
108. Bhatnagar, D., and Cleveland, T. E. Purification and characterization of a reductase from *Aspergillus parasiticus* SRRC 2043 involved in aflatoxin biosynthesis. *FASEB J.*, Abstract A2727, 1990.

109. Bhatnagar, D., Ullah, A. H. J., and Cleveland, T. E. Purification and characterization of a methyltransferase from *Aspergillus parasiticus* SRRC 163 involved in aflatoxin biosynthetic pathway. *Prep. Biochem.*, *18*:321–349, 1988.
110. Bhatnagar, D., Cleveland, T. E., and Kingston, D. G. Enzymological evidence for separate pathways for aflatoxin B_1 and B_2 biosynthesis. *Biochemistry*, *30*:4343–4350, 1991.
111. Keller, N. P., Dischinger, H. C., Jr., Bhatnagar, D., Cleveland, T. E., and Ullah, A. H. J. Purification of a second methyltransferase active in the aflatoxin biosynthetic pathway. *FASEB J.*, Abstract 2623, 1991.
112. Waring, R. B., May, G. S., and Morris, N. R. Characterization of an inducible expression system in *Aspergillus nidulans* using *alc*A and tubulin-coding genes. *Gene*, *79*:119–130, 1989.
113. Schauer, A., Ranes, M., Santamaria, R., Guijarro, J., Lawlor, E., Mendez, C., Chater, K., and Losick, R. Visualizing gene expression in time and space in the filamentous bacterium *Streptomyces coelicolor*. *Science*, *240*:768–772, 1988.
114. Bai, U., Lewandoski, M., Dubnau, E., and Smith, I. Temporal regulation of the *Bacillus subtilis* early sporulation gene *spoof*. *J. Bacteriol.*, *172*:5432–5439, 1990.
115. Aguirre, J., Adams, T. H., and Timberlake, W. E. Spatial control of developmental regulatory genes in *Aspergillus nidulans*. *Exp. Mycol.*, *14*:290–293, 1990.
116. Lamb, H. K., Hawkins, A. R., Smith, M., Harvey, I. J., Brown, J., Turner, G., and Roberts, C. F. Spatial and biological characterization of the complete quinic acid utilisation gene cluster in *Aspergillus nidulans*. *Mol. Gen. Genet.*, *223*:17–23, 1990.
117. Chu, G., Yollrath, D., and Davis, R. Separation of large DNA molecules by contour clamped homogeneous electric fields. *Science*, *234*:1582–1585, 1986.
118. Brody, H., and Carbon, J. Electrophoretic karyotype of *Aspergillus nidulans*. *PNAS*, *86*:6260–6263, 1989.
119. Kinscherf, T. G., and Leong, S. A. Molecular analysis of the karyotype of *Ustilago maydis*. *Chromosoma*, *96*:427–433, 1988.
120. Masel, A., Braithwaite, K., Irwin, J., and Manners, J. Highly variable molecular karyotypes in the plant pathogen *Colletotrichum gloeosporioides*. *Curr. Genet.*, *18*:81–86, 1990.
121. Rustchenko-Bulgac, E. P., Sherman, F., and Hicks, J. B. Chromosomal rearrangements associated with morphological mutants provide a means for genetic variation of *Candida albicans*. *J. Bacteriol.*, *172*:1276–1283, 1990.
122. Orbach, M. J., Vollrath, D., Davis, R. W., and Yanofsky, C. An electrophoretic karyotype of *Neurospora crassa*. *Mol. Cell. Biol.*, *8*:1469–1473, 1988.
123. Miao, V. P. W., Matthews, D. E., and Van Etten, H. D. Identification and chromosomal locations of a family of cytochrome P-450 genes for pisatin detoxification in the fungus *Nectria haematococca*. *Mol. Gen. Genet.*, *226*:214–223, 1991.
124. Tzeng, T. A restriction fragment length polymorphism map of *Cochliobolus heterostrophus*. Ph.D. Dissertation, Iowa State Univ., Ames, Iowa, 1990.

125. Elliot, C. G. The cytology of *Aspergillus nidulans*. *Genet. Res. (Camb.)*, *1*:462–476, 1960.
126. Kafer, E. An 8-chromosome map of *Aspergillus nidulans*. *Adv. Genet.*, *9*:105–145, 1958.
127. Krumlauf, R., and Marzluf, G. A. Characterization of the sequence complexity and organization of the *Neurospora crassa* genome. *Biochemistry*, *18*:3705–3713, 1979.
128. Perkins, D. D., Radford, A., Newmeyer, D., and Bjorkman, M. Chromosomal loci of *Neurospora crassa*. *Microbiol. Rev.*, *46*:426–570, 1982.
129. Keller, N. P. Electrophoretic karyotypes of aflatoxin producing and nonproducing *Aspergillus* spp. *Phytopathology*, *80*:979, 1990 (abstract).
130. Keller, N. P., Cleveland, T. E., and Bhatnagar, D. Variable electrophoretic karyotypes in *Aspergillus flavus* and *Aspergillus parasiticus*. *Curr. Gen.* (submitted).
131. Bennett, J. W., and Bentley, R. What's in a name?—microbial secondary metabolism. *Adv. Appl. Microbiol.*, *34*:1–28, 1989.
132. Martin, J. F., and Demain, A. L. Control of antibiotic biosynthesis. *Microbiol. Rev.*, *44*:230–251, 1980.
133. Malik, V. S. Genetics of secondary metabolism. In *Biotechnology*, Vol. 4 (H. J. Rehm and G. Reed, Eds.), VCH, Weinheim, 1986, pp. 39–68.
134. Brahmbhatt, H. N., Wyk, P., Quigley, N. B., and Reeves, P. R. Complete physical map of the *rfb* gene cluster encoding biosynthetic enzymes for the O antigen of *Salmonella typhimurium* LT2. *J. Bacteriol.*, *170*:98–102, 1988.
135. Dolph, P. J., Majerczak, D. R., and Coplin, D. L. Characterization of a gene cluster for exopolysaccharide biosynthesis and virulence in *Erwinia stewartii*. *J. Bacteriol.*, *170*:865–871, 1988.
136. Jacobson, M. R., Cash, V. L., Weiss, M. C., Laird, N. F., Newton, W. E., and Dean, D. R. Biochemical and genetic analysis of the *nif* USVWZM cluster from *Azotobacter vinelandii*. *Mol. Gen. Genet.*, *219*:49–57, 1989.
137. Marugg, J. D., Nielander, H. B., Horrevoets, A. J. G., van Megen, I., van Genderen, I., and Weiseek, P. J. Genetic organization and transcriptional analysis of a major gene cluster involved in siderophore biosynthesis in *Pseudomonas putida* WCS358. *J. Bacteriol.*, *170*:1812–1819, 1988.
138. Young, D. A., Bauer, C. E., Williams, J. C., and Marrs, B. L. Genetic evidence for superoperonal organization of genes for photosynthetic pigments and pigment-binding proteins in *Rhodobacter capsulatus*. *Mol. Gen. Genet.*, *218*:1–12, 1989.
139. Bibb, M. J., Biro, S., Motamedi, H., Collins, J. F., and Hutchinson, C. R. Analysis of the nucleotide sequence of the *Streptomyces glaucescens tcml* genes provides key information about the enzymology of polyketide antibiotic biosynthesis. *EMBO J.*, *8*:2727–2736, 1989.
140. Butler, M. J., Friend, E. J., Hunter, I. S., Kaczmarek, F. S., Sudgen, D. A., and Warren, M. Molecular cloning of resistance

genes and architecture of a linked gene cluster involved in biosynthesis of oxytetracycline by *Streptomyces rimosus*. *Mol. Gen. Genet.*, *215*:231–238, 1989.

141. Perkins, J. B., Guterman, S. K., Howitt, C. L., Williams, J. E., II, and Pero, J. *Streptomyces* genes involved in biosynthesis of the peptide antibiotic valinomycin. *J. Bacteriol.*, *172*:3108–3116, 1990.
142. Weber, J. M., Leung, J. O., Maine, G. T., Potenz, R. H. B., Paulus, T. J., and DeWitt, J. P. Organization of a cluster of erythromycin genes in *Saccharopolyspora erythraea*. *J. Bacteriol.*, *172*:2372–2383, 1990.
143. Denis-Duphil, M. Pyrimidine biosynthesis in *Saccharomyces cerevisiae*: the *ura2* cluster gene, its multifunctional enzyme product, and other structural or regulatory genes involved in de novo UMP synthesis. *Biochem. Cell. Biol.*, *67*:612–631, 1989.
144. Melnick, L., and Sherman, F. Nucleotide sequence of the COR region: a cluster of six genes in the yeast *Saccharomyces cerevisiae*. *Gene*, *87*:157–166, 1990.
145. Aramayo, R., Adams, T. H., and Timberlake, W. E. A large cluster of highly expressed genes is dispensable for growth and development in *Aspergillus nidulans*. *Genetics*, *122*:65–71, 1989.
146. Geever, R. F., Huiet, L., Baum, J. A., Tyler, B. M., Patel, V. B., Rutledge, B. J., Case, M. E., and Giles, N. H. DNA sequence, organization and regulation of the *qa* gene cluster of *Neurospora crassa*. *J. Mol. Biol.*, *207*:15–34, 1989.
147. Hull, E. P., Green, P. M., Arst, H. N., Jr., and Scazzocchio, C. Cloning and physical characterization of the *L*-proline catabolism gene cluster of *Aspergillus nidulans*. *Mol. Microbiol.*, *3*:553–559, 1989.
148. MacCabe, A. P., Riach, M. B., Unkels, S. E., and Kinghorn, J. R. The *Aspergillus nidulans npeA* locus consists of three contiguous genes required for penicillin biosynthesis. *EMBO J.*, *9*:279–287, 1990.
149. Raeder, U., Thompson, W., and Broda, P. RFLP-based genetic map of *Phanerochaete chrysosporium* ME446: lignin peroxidase genes occur in clusters. *Mol. Microbiol.*, *3*:911–918, 1989.
150. Mol, J. N. M., Koes, R., van den Berg, E. A., Reif, H. J., Kreuzaler, F., and Veltkamp, E. The genetics of secondary metabolite production in higher plants: the flavonoid genes of Petunia as a model system. In *Advances in Agricultural Biotechnology: Genetic Engineering of Plants and Microorganisms Important for Agriculture* (E. Magnien and D. deNettancourt, Eds.), Martinus Nijhoff/Dr. W. Junk, 1985, pp. 122–123.
151. Popodi, E., Minoo, P., Burke, T., and Waring, G. L. Organization and expression of a second chromosome follicle cell gene cluster in *Drosophila*. *Dev. Biol.*, *127*:248–256, 1988.
152. Rubock, M. J., Larin, Z., Cook, M., Papalopulu, N., Krumlauf, R., and Lehrach, H. A yeast artificial chromosome containing the mouse homeobox cluster *Hox-2*. *PNAS*, *87*:4751–4755, 1990.
153. Fincham, J. R. S. From auxotropic mutants to DNA sequences. In *Gene Manipulations in Fungi* (J. W. Bennett and L. L. Lasure, Eds.), Academic, New York, 1985, pp. 3–34.

154. Smith, D. J., Burnham, M. K. R., Bull, J. H., Hodgson, J. E., Ward, M. J., Browne, P., Brown, J., Barton, B., Earl A. J., and Turner, G. β-Lactam antibiotic biosynthetic genes have been conserved in clusters in prokaryotes and eukaryotes. *EMBO J.*, *9*: 741–747, 1990.
155. Burne, R. A., Parsons, D. T., and Marquis, R. E. Cloning and expression in *Escherichia coli* of the genes of the arginine deiminase system of *Streptococcus sanguis* NCTC 10904. *Infect. Immunol.*, *57*:3540–3548, 1989.
156. Motamedi, H., and Hutchinson, C. R. Cloning and heterologous expression of a gene cluster for the biosynthesis of tetracenomycin C, the anthracycline antitumor antibiotic of *Streptomyces glaucescens*. *PNAS*, *84*:4445–4449, 1987.
157. Smith, D. J., Burnham, M. K. R., Edwards, J., Earl, A. J., and Turner, G. Cloning and heterologous expression of the penicillin biosynthetic gene cluster from *Penicillium chrysogenum*. *Bio/technology*, *8*:39–41, 1990.
158. Weigel, B. J., Burgett, S. G., Chen, V. J., Skatrud, P. L., Frolik, C. A., Queener, S. W., and Ingolia, T. D. Cloning and expression in *Escherichia coli* of isopenicillin N synthetase genes from *Streptomyces lipmanii* and *Aspergillus nidulans*. *J. Bacteriol.*, *170*:3817–3826, 1988.
159. Monia, B. P., Haskell, K. M., Ecker, J. R., Ecker, D. J., and Crooke, S. T. Chromosomal mapping of the ubiquitin gene family in *Saccharomyces cerevisiae* by pulsed field gel electrophoresis. *Nucleic Acids Res.*, *17*:3611–3612, 1989.
160. Skatrud, P. L., and Queener, S. W. An electrophoretic molecular karyotype for an industrial strain of *Cephalosporium acremonium*. *Gene*, *78*:331–338, 1989.
161. Gwynne, D. I., Miller, B. L., Miller, K. Y., and Timberlake, W. E. Structure and regulated expression of the SpoC 1 gene cluster from *Aspergillus nidulans*. *J. Mol. Biol.*, *180*:91–109, 1984.
162. Johnstone, I. L., McCabe, P. C., Greaves, P., Gurr, S. J., Cole, G. E., Brown, M. A. D., Unkles, S. E., Clutterbuck, A. J., Kinghorn, J. R., and Innis, M. A. Isolation and characterization of the *crn*A-*nii*A-*nia*D gene cluster for nitrate assimilation in *Aspergillus nidulans*. *Gene*, *90*:181–192, 1990.
163. Schweizer, M., Case, M. E., Dykstra, C. C., Giles, N. H., and Kushner, S. R. Identification and characterization of recombinant plasmids carrying the complete *qa* gene cluster from *Neurospora crassa*, including the *qa*-1 + regulatory gene. *PNAS*, *78*:5086–5090, 1981.
164. Thomsett, A. B., and Cove, D. J. Deletion mapping of the *nii*A *nia*D gene region of *Aspergillus nidulans*. *Genet. Res.*, *34*:19–32. 1979.
165. Hawkins, A. R., Da Silva, A. J. F., and Roberts, C. F. Cloning and characterization of the three enzyme structural genes *qut*B, *qut*C, and *qut*E from the quininic acid utilization cluster in *Aspergillus nidulans*. *Curr. Genet.*, *9*:305–312, 1984.

166. Castro, J. M., Liras, P., Cortes, J., and Martin, J. F. Regulation of aminoadipyl-cysteinyl-valine, isopenicillin N synthetase, isopenicillin N isomerase and deacetoxycephalosporin C synthetase by nitrogen sources in *Streptomyces lactamdurans*. *Appl. Microbiol. Biotechnol.*, *22*:32–40, 1985.
167. Ramos, F. R., Lopez-Nieto, M. J., and Martin, J. F. Coordinate increase of isopenicillin N synthetase, isopenicillin N epimerase and deacetoxycephalosporin C synthetase in a high cephalosporin-producing mutant of *Acremonium chrysogenum* and simultaneous loss of the three enzymes in a non-producing mutant. *FEMS Microbiol. Lett.*, *35*:123–127, 1986.
168. Cleveland, T. E., Lax, A. R., Lee, L. S., and Bhatnagar, D. Appearance of enzyme activities catalyzing conversion of sterigmatocystin to aflatoxin B_1 in late-growth phase *Aspergillus parasiticus* cultures. *Appl. Environ. Microbiol.*, *53*:1711–1713, 1987.
169. Hendrix, R., Roberts, J., Stahl, F., and Weisberg, R., Eds. *Lambda II*, Cold Spring Harbor, N.Y., 1983.
170. Fishman, S. E., Cox, K., Larson, J. L., Reynolds, P. A., Seno, E. T., Yeh, W. K., Van Frank, R., and Hershberger, C. L. Cloning genes for the biosynthesis of a macrolide antibiotic. *PNAS*, *84*:8248–8252, 1987.
171. Cole, S. P., Rudd, B. A. M., Hopwood, D. A., Chang, C. J., and Floss, H. G. Biosynthesis of the antibiotic actinorhodin. Analysis of blocked mutants of *Streptomyces coelicolor*. *J. Antibiot.*, *40*:340–347, 1987.
172. Hopwood, D. A., Bibb, M. J., Chater, K. F., Janssen, G. R., Malpartida, F., and Smith, C. P. Regulation of gene expression in antibiotic-producing *Streptomyces*. In *Regulation of Gene Expression, 25 Years On* (I. R. Booth and C. F. Higgins, Eds.), Cambridge Univ. Press, New York, 1986, pp. 251–276.
173. Hara, O., and Beppu, T. Mutants blocked in streptomycin production in *Streptomyces griseus* – the role of A-factors. *J. Antibiot.*, *35*:349–358, 1982.
174. Horinouchi, S., Hara, O., and Beppu, T. Cloning of a pleiotropic gene that positively controls biosynthesis of A-factor, actinorhodin, and prodigiosin in *Streptomyces coelicolor* A3(2) and *Streptomyces lividans*. *J. Bacteriol.*, *155*:1238–1248, 1983.
175. Horinouchi, S., Nishiyama, M., Suzuki, H., Kumada, Y., and Beppu, T. The cloned *Streptomyces bikiniensis* A-factor determinant. *J. Antibiot.*, *38*:636–641, 1985.
176. Horinouchi, S., and Beppu, T. Regulation of secondary metabolism in *Streptomyces*. In *Biology of Actinomycetes, 88* (Y. Okami, T. Beppu, and H. Ogawara, Eds.), Jpn. Sci. Soc., Tokyo, 1988, pp. 71–75.
177. Martin, J. F., Daza, A., Asturias, J. A., Gil, J. A., and Liras, P. Transcriptional control of antibiotic biosynthesis at phosphate-regulated promoters and cloning of a gene involved in the control of the expression of multiple pathways in *Streptomyces*. In *Biology of Actinomycetes, 88* (Y. Okami, T. Beppu, and H. Ogawara, Eds.), Jpn. Sci. Soc., Tokyo, 1988, pp. 424–430.

178. Horinouchi, S., and Beppu, T. Production in large quantities of actinorhodin and undecylprodigiosin induced by *afs*B in *Streptomyces lividans*. *Agric. Biol. Chem.*, *48*:2131–2133, 1984.
179. Herskowitz, I. A regulatory hierarchy for cell specialization in yeast. *Nature*, *342*:749–757, 1989.
180. Losick, R., and Pero, J. Cascades of sigma factors. *Cell*, *25*: 582–584, 1981.
181. Abdollahi, A., and Buchanan, R. L. Regulation of aflatoxin biosynthesis: characterization of glucose as an apparent inducer of aflatoxin production. *J. Food Sci.*, *47*:143–146, 1981.
182. Maggon, K. K., Gupta, S. K., and Venkitasubramanian, T. A. Biosynthesis of aflatoxins. *Bacteriol. Rev.*, *41*:822–855, 1977.
183. Tice, G., and Buchanan, R. L. Regulation of aflatoxin biosynthesis: effect of exogenously supplied cyclic nucleotides. *J. Food Sci.*, *47*:153–157, 1981.
184. Jensen, R. A. Enzyme recruitment in evolution of new function. *Annu. Rev. Microbiol.*, *30*:409–425, 1976.
185. Smith, H. O., Annua, T., and Chandrasegaran, S. Finding sequence motifs in groups of functionally related proteins. *PNAS*, *87*:826–830, 1990.
186. Smith, D. J., Earl, A. J., and Turner, G. The multifunctional peptide synthetase performing the first step of penicillin biosynthesis in *Penicillium chrysogenum* is a 421073 dalton protein similar to *Bacillus brevis* peptide antibiotic synthetases. *EMBO J.*, *9*: 2743–2750, 1990.
187. Malpartida, F., Hallam, S. E., Kieser, H. M., Motamedi, H., Hutchinson, C. R., Butler, M. J., Sugden, D. A., Warren, M., McKillop, C., Bailey, C. R., Humphreys, G. O., and Hopwood, D. A. Homology between *Streptomyces* genes coding for synthesis of different polyketides used to clone antibiotic biosynthetic genes. *Nature*, *325*: 818–820, 1987.
188. Beck, J., Ripka, S., Siegner, A., Schlitz, E., and Schweizer, E. The multifunctional 6-methylsalicylic acid synthase gene of *Penicillium patulum*. *Eur. J. Biochem.*, *192*:487–498, 1990.
189. Wang, I.-K., Reeves, C., and Gaucher, G. M. Isolation and sequencing of a genomic DNA clone containing the 3'-terminus of the 6-methylsalicylic acid polyketide synthetase gene of *Penicillium urticae*. *Can. J. Microbiol.* (submitted).
190. Ingrosso, D., Fowler, A. V., Bleibaum, J., and Clarke, S. Sequence of the *D*-aspartyl/*L*-isoaspartyl protein methyltransferase from human erythrocytes. *J. Biol. Chem.*, *264*:20131–20139, 1989.
191. Lauster, R. Evolution of type II DNA methyltransferases, a gene duplication model. *J. Mol. Biol.*, *206*:313–321, 1989.
192. Posfae, J., Bhagwat, A. S., Posfai, G., and Roberts, R. Predictive motifs derived from cytosine methyltransferases. *Nucleic Acids Res.*, *17*:2421–2435, 1989.

12

STRAIN INSTABILITY IN FILAMENTOUS FUNGI

SHUBHA KALE and J. W. BENNETT *Tulane University, New Orleans, Louisiana*

I. MOLDS, MYCOTOXINS, AND SECONDARY METABOLITES

Mycotoxins are poisonous compounds produced by fungi that cause disease in animals. All mycotoxins are secondary metabolites produced by the group of fungi called "molds" in the vernacular. Mycotoxins are distinguished from mushroom poisons in that their toxic effects are caused by inadvertently ingesting, or otherwise coming in contact with, a substrate that contains the toxin-producing fungus. With mushroom poisoning, there is a deliberate intent to eat the basidiocarp [1].

Although the term *mold* does not have a precise scientific definition, it is commonly used by scientists to describe a group of nonmotile lower eukaryotes with chitinous cell walls and an absorptive, heterotrophic mode of nutrition. Mold colonies are visible to the naked eye, but individual cells and spores are microscopic. The individual filaments characteristic of mold growth are called *hyphae*; a group of hyphae constitutes a *mycelium*. Adjacent hyphae in a mold colony may fuse (anastomose), with subsequent intermingling of cellular contents. The filamentous growth form allows molds to grow rapidly over surfaces and to penetrate substrates, whereupon extracellular enzymes are secreted. Nutrients required for growth and development are reabsorbed and transported to the growing apical tips. The filamentous fungal cells are joined by pores and can be envisioned as a coenocytic system of branching tubes. Within the tubes, nuclei, mitochondria, and other subcellular organelles can move from place to place. A high surface/cell ratio is characteristic of the threadlike cell shape. Not only does this facilitate the secretion of extracellular enzymes, but it also allows other cellular products to cross the cell wall and enter the substrate. Secondary metabolites, including mycotoxins, are among the extracellular metabolic products associated with mold growth and secreted into the substrate.

Secondary metabolites are differentiation products of extremely diverse chemical structure produced by restricted taxonomic groups, are not essential to the growth of the producing organism, and are biosynthesized from one or more general (primary) metabolites by a wide variety of pathways [2]. Since mycotoxins are merely a subclass of secondary metabolites defined by their toxic effects on animals, the general genetic and biochemical questions raised by the problem of strain degeneration in mycotoxigenic fungi are applicable to all fungi that produce secondary metabolites.

Secondary metabolism is characteristic of certain differentiated cells of green plants, fungi, and bacteria; there are few secondary metabolites of animal origin [3]. They are low molecular weight compounds ("small molecules") that may accumulate in large quantities in families of related compounds (congeners). For example, the *Handbook of Toxic Fungal Metabolites* [4] lists 16 aflatoxins, 8 members of the sterigmatocystin group, and 49 trichothecenes. Enormous intrageneric, intraspecific, and intrastrain variation in secondary metabolism is found in nature. Most of the research in natural products chemistry consists of the isolation and characterization of new secondary metabolites, often from obscure and exotic species of plants and microbes. The best-known secondary metabolites are pharmacologically active [5–8].

Mycotoxins are economically detrimental. They can kill or sicken human beings and other animals. Antibiotics are economically beneficial. Antibiotics kill bacteria and are valued in medicine for killing the bacteria that cause disease. The division between mycotoxin-producing and antibiotic-producing species is an arbitrary one, based entirely on human expediency. If the fungus produces a secondary metabolite that is poisonous to us or one of our favored domesticated animal brethren, we deem it a toxin and do everything we can to eliminate the fungus and its metabolite from our environment. If the fungus produces a secondary metabolite that we can use as a drug, we do everything we can to pamper the mold so that it will produce maximum quantities of the desired compound.

The best-known fungal secondary metabolites are those with beneficial pharmacological activity: the penicillins, cephalosporins, ergot alkaloids, cyclosporin, and so forth. However, a number of compounds now classified as mycotoxins were discovered during the search for new antibiotics. Both citrinin and patulin are compounds with antibiotic activity that were rejected because of their toxicity to mammals. There has been more research on metabolites beneficial to human health than on those that are detrimental, so much of the relevant literature on loss of secondary metabolic capacity in filamentous fungi concerns antibiotic-producing industrial species.

II. CLASSIFICATION OF FUNGAL SECONDARY METABOLITES

The major compendium of known fungal secondary metabolites is the two-volume series by Turner [9] and Turner and Aldridge [10]. In their system, organic compounds produced by fungi are categorized not by biological activity but by biosynthetic origin. This classification emphasizes the relationship of compounds of different chemical types and highlights the fact that fungal metabolites, despite the enormous diversity of structure and activity, are derived from a few biosynthetic pathways.

Acetyl coenzyme A (acetyl CoA) is the most important single intermediate in fungal secondary metabolism. Condensation of acetyl CoA with malonyl CoA leads either to the polyketides or to the fatty acids, which can in turn give rise to secondary metabolites. Alternatively, condensation of three molecules of acetyl CoA yields mevalonic acid, the building block of terpene biosynthesis [9, 10]. In Turner's chemical taxonomy, secondary metabolites are placed into seven groups, three of which directly involve acetyl CoA: those derived from fatty acids, the terpenes and steroids, and the polyketides. Turner's other groups are the metabolites derived without the intervention of acetate, those derived from intermediates of the tricarboxylic acid cycle, and metabolites derived from amino acids. A final, miscellaneous category of "unclassifiable" compounds encompasses metabolites of mixed origin, or those about which too little is known to make an accurate assignment. The major mycotoxins are grouped according to their biosynthetic origin in Table 1; the major genera of producing species are also listed.

It is quickly apparent that most mycotoxins are derived from acetate (polyketides and terpenes) or amino acids. Four genera stand out as the major producers of mycotoxins: *Aspergillus, Claviceps, Fusarium,* and *Penicillium.* The majority of the mycotoxin-producing species are saprophytic. A few are facultative plant pathogens. The major exceptions to this generalization are the *Claviceps* species, which produce ergot alkaloids. These species are aggressive plant pathogens.

Several mycotoxins are produced by taxonomically disparate species—for example, sterigmatocystin by species of *Aspergillus, Bipolaris,* and *Chaetomium* and the cytochalasins by *Aspergillus, Helminthosporium, Metarrhizium, Phoma,* and *Zygosporium.*

It is safe to say that we are seeing only the tip of the iceberg. Most fungi have not been assayed for their mycotoxin-producing ability. If the definition of mycotoxin were extended to include fungal metabolites toxic to arthropods, nematodes, and other invertebrates, the number of mycotoxigenic compounds in nature would be difficult to estimate but is almost certainly large.

Pharmaceutical companies have conducted massive surveys of microorganisms in search of new antibiotics and other drugs. Unfortunately, most of these data have never been published. Since the emphasis in these screening programs is on finding compounds with useful, not harmful, physiological attributes, fungal extracts with mycotoxigenic properties "discovered" by drug companies are discarded rather than studied [11].

III. STRAIN INSTABILITY

A. Background

Over 200 years ago, Albert Von Haller described fungi as "a mutable and treacherous tribe" [12, 13]. His aphorism has been amply confirmed by mycologists ever since. Filamentous fungi grown in a constant environment display a tendency toward spontaneous change that is truly remarkable. This change occurs with higher frequency than is normal for mutation. It involves morphological characteristics such as sporulation (both asexual and sexual), formation of aerial mycelia, and pigmentation, as well

TABLE 1. Major Mycotoxins Grouped According to Biosynthetic Origin

Biosynthetic origin	Mycotoxin	Producing genus
Polyketide	Aflatoxin	*Aspergillus*
	Alternariol	*Alternaria*
	Citrinin	*Aspergillus, Penicillium*
	Citreovirdin	*Penicillium*
	Luteroskyrin	*Penicillium*
	Ochratoxin	*Aspergillus, Penicillium*
	Patulin	*Aspergillus, Penicillium*
	Penicillic acid	*Aspergillus, Penicillium*
	Rugulosin	*Penicillium*
	Secalonic acid	*Claviceps, Penicillium*
	Sterigmatocystin	*Aspergillus, Chaetomium, Bipolaris*
	Viomellein	*Aspergillus, Penicillium*
	Xanthomegnin	*Penicillium*
	Zearalenone	*Fusarium*
Amino acids	Brevianamide	*Penicillium*
	Cyclopiazonic acid	*Penicillium*
	Cytochalasins	*Aspergillus, Helminthosporium, Metarrhizum, Phoma, Zygospora*
	Ergot alkaloids	*Claviceps*
	Malformins	*Aspergillus*
	Roquefortine	*Penicillium*
	Slaframine	*Rhizoctonia*
Terpenes	Pasapaline	*Claviceps*
	Penitrem	*Penicillium*
	Trichothecenes	*Fusarium, Myrothecium, Stachybotris*
Tricarboxylic acid	Rubratoxin	*Penicillium*
	Tenuazonic acid	*Alternaria, Aspergillus*

Source: After Cole and Cox [4], Turner [9], and Turner and Aldridge [10].

as physiological characteristics such as virulence, mating type, and secondary metabolite production. Morphological variants are often observed as sectors on radially growing fungal colonies on laboratory media. Plant pathologists often classify members of plant pathogenic species into different biotypes based on variable pathogenicity. Terms such as *variation, saltation, mutation, strain degeneration,* and *pleomorphism* are used, often interchangeably and without carefully stipulated definitions. The enormous plasticity of many fungi when grown in culture poses problems for the taxonomist, the plant pathologist, and the industrial microbiologist. Culture collections have coped with strain instability in a largely pragmatic fashion, developing empiric techniques that maximize both longevity and stability in stock cultures. (See Smith [14] for a review.)

An early review by Hansen [15] cited references dating back to the early 1900s that dealt with the variability of imperfect fungi that could not be explained on the basis of Mendelian behavior, implying an exeptionally high degree of genetic instability in filamentous fungi, especially when grown on synthetic media. Hansen presented evidence that much of the variability was due, not to mutation, but rather to the presence of what he called the "dual phenomenon." This phenomenon was defined as the ability of numerous fungi to exist in nature as combinations of distinct genetic entities. Indeed, in many fungi, genetically dissimilar nuclei share the same mycelium. Such a mycelium is termed *heterokaryotic*, while *homokaryotic* colonies contain genetically identical nuclei. The dual phenomenon was later identified as heterokaryosis. Heterokaryosis is one cause of variation in filamentous fungi.

B. Culture Conditions and Systematics

Molds are amenable to many of the same laboratory procedures as bacteria: growth in pure culture, production of colonies on agar media, the use of simple environmental triggers to induce sporulation, and so forth. Unlike bacteria, colony-forming units are not single cells but spores or mycelial fragments that give rise to hyphae that extend by apical growth. This means that a mycelial mass contains both young and old cells and cells at various stages of development. The classical bacteriologists' growth curve (lag phase, exponential or logarithmic phase, and stationary phase) was developed for a population of single cells and does not apply to the fungal population. For this reason, Bu'Lock [16] introduced the terms *trophophase* (feeding phase) and *idiophase* (peculiar phase) to describe the growth kinetics of filamentous organisms in which there is a heterogeneous mixture of growing and nongrowing cells. Trophophase is roughly equivalent of the logarithmic phase of bacteria and describes the time of active nutrient uptake and increase in dry weight. Idiophase roughly corresponds with the stationary phase of bacteria and describes the phase associated with the depletion of nutrients, a leveling off of dry weight, and the time during which most secondary metabolites, including mycotoxins, are synthesized. The terms *trophophase* and *idiophase* are widely used by scientists in the antibiotics industry to describe submerged batch fermentation of not only filamentous fungi but also filamentous bacteria (actinomycetes).

The "bible" of *Aspergillus* taxonomy is Raper and Fennell's 1965 monograph, *The Genus Aspergillus* [17], which is a revision of Thom and Raper's 1945 monograph, *A Manual of the Aspergilli* [18]. Both works devote an entire chapter to the topic of variation and are to be commended for the careful definitions:

> The term *variant* applies to strains arising through gradual change from normal members of identifiable species. The characters of a

> variant are generally not stable but subject to continued change and further variation. Variants frequently appear as colony sectors, overgrowths, or other localized areas of changed appearance; when isolated in pure culture, they may or may not retain their distinguishing characteristics... . The term *mutant* is applied to strains that show abrupt, marked, and persistent differences from known parent cultures [18, p. 68].

Neither variants nor mutants are afforded taxonomic status, but rather are accepted as evidence for the range of genetic variation within a given species or group.

Filamentous fungi are classified largely on their morphological characteristics in laboratory culture. Since fungi are so variable, especially when grown in the laboratory over the course of several years, systematicists have had to contend with morphological plasticity in devising their taxonomies. Spore type, size, and shape have proved to be among the most reliable taxonomic criteria.

As stated earlier, the major mycotoxin-producing species fall into four genera: *Aspergillus, Claviceps, Fusarium,* and *Penicillium* (Table 1). The literature on all four genera is filled with references to strain variability and pleomorphism, particularly in the case of *Fusarium*. One *Fusarium* expert has written, "*Fusarium* has always been considered a highly variable organism. One might even say that *Fusarium* and variation are synonymous" [19].

In certain cases, strain instability has been attributed to extrachromosomal inheritance. Although none of the well-studied models for cytoplasmic inheritance involve a mycotoxin-producing species, these examples provide useful contrasts to the phenomena observed with respect to loss and attenuation of secondary metabolic capabilities.

C. The Heterokaryon Test

Heterokaryosis "is the coexistence of genetically different nuclei in cytoplasmic continuity with one another" [20, p. 256]. In an actively growing mycelium, cells are connected one to another along the hyphae through septal pores; cytoplasmic streaming facilitates movement of nutrients, metabolites, and organelles. Many nuclei may participate in the determination of the phenotype; thus, "any mycelium must be viewed simultaneously as an integrated organism, and as a population of nuclei in a morphologically patterned cytoplasmic environment" [20, p. 567].

Laboratory studies of natural and artificial heterokaryons by Jinks in the 1950s [21–23] demonstrated a difference between variation of nuclear and cytoplasmic origin. This work led to the elucidation of the "heterokaryon test" to distinguish whether a given variant phenotype is due to a nuclear or a cytoplasmic determinant. Two stocks that differ in a variant character are combined in a heterokaryon. One stock carries an easily identifiable nuclear marker such as a spore color mutant. The other stock carries the phenotype of unknown origin. The fused heterokaryotic mycelium is then resolved into its component homokaryons. If both homokaryons carry the unknown phenotype, then it can be concluded that the trait is under cytoplasmic control. If the traits segregate in the monokaryon, nuclear control is likely. The heterokaryon test allows analysis of cytoplasmic inheritance in anamorphic (imperfect) species that lack a sexual phase and thus cannot be subjected to Mendelian analysis [24].

Cytoplasmic inheritance provides a useful model for explaining differences that occur in sister colonies derived from individual homokaryotic clones. Variant phenotypes reflect changes in the population of cytoplasmic elements, complete loss of cytoplasmic elements, or mutation of cytoplasmic elements.

Temperature is an important factor associated with the production of cleistothecia in *A. nidulans* [25]. When propagated by hyphae at 25°C, all the lines of the fungus remain stable. At 37°C, variation is frequent. The variation has been attributed to an increased probability of mutation in the cytoplasmic elements at 37°C, while at 25°C the reproduction of the cytoplasmic elements is hypothesized to be more or less in step with cell division.

A red variant of *A. nidulans*, isolated by ultraviolet irradiation of conidiospores, was found to be a cytoplasmic element by its behavior in a heterokaryon test [26]. Such cytoplasmic variants have been dubbed "dauer modifications" [27]. Dauer modification has been defined as "an environmentally induced phenotypic change in a cell that survives in the generative or vegetative descendants of the cell in the absence of the original stimulus. However, with time, the trait weakens and eventually disappears" [28]. Such environmentally induced changes in cytoplasmic elements could persist for many generations or could decrease in expression in succeeding generations. The classical literature on fungal cytoplasmic inheritance including the origins of extrachromosomal variants and mechanisms of their inheritance was reviewed by Jinks [24].

D. Cytoplasmic Inheritance

The study of genetic traits transmitted separately from nuclear genes is variously called cytoplasmic, extranuclear, or extrachromosomal inheritance. Extrachromosomal genes are located in cell organelles or in infectious particles such as plasmids and viruses. These genes display non-Mendelian ratios in genetic crosses and yield different phenotypic ratios in reciprocal crosses. Cytoplasmic inheritance was first described in green plants and associated with variegated leaf color. These variegated phenotypes were eventually correlated with changes in chloroplasts, specifically alterations in chloroplast DNA.

The status of filamentous fungi as coenocytes, and that of yeasts as facultative anaerobes, have combined to make fungi the ideal biological material for studying cytoplasmic inheritance. While some of the classical examples of cytoplasmic inheritance in fungi have not had their molecular basis elucidated, a number of other, well-known phenotypes have been associated with either aberrant mitochondrial genomes or viruslike particles or plasmids (see Table 2).

The mitochondrial genomes of yeasts and molds are larger and more complex than animal mitochondrial DNA (mtDNA). Moreover, major alterations of fungal mtDNA are fairly common. The best studied example of

TABLE 2. Cytoplasmic Inheritance in Fungi

System	Example	Species
Classical "variant," plasmagene not identified	Red, purple, mycelial[a]	*Aspergillus nidulans*
	Vegetative death,[b] compact	*Aspergillus glaucus*
	Mycelial, low sporulation	*Penicillium* spp.
Mitochondrial DNA or mitochondrial plasmid	Petite (*rho*)	*Saccharomyces cerevisiae*
	poky (mi-1); stopper	*Neurospora crassa*
	ragged	*Aspergillus amstelodami*
	Senescence	*Podospora anserina*
	Senescence	*Podospora curvicolla*
	Senescence (kalilo)	*Neurospora intermedia*
Virus, viruslike particle	Killer	*Saccharomyces cerevisiae*
	Killer	*Kluyveromyces lactis*
	Hypovirulence	*Ophiostoma (Endothia) parasiticus*
	D-Factor	*Cryphonectria (Ceratocystis) ulmi*

[a]These variants are no longer available and cannot be reexamined [31, p. 412].
[b]Phenotype is similar to *ragged* of *A. amstelodami*.
Source: After Jinks [24], Dujon and Belcour [29], and Buck [30].

mitochondrial inheritance subjected to genetic analysis comes from yeast. *Saccharomyces cerevisiae* is a facultative anaerobe. Cells with altered mtDNA are found as respiratory-deficient mutants. These slow-growing forms were named "petites" and were first studied by Ephrussi and colleagues [32]. They found that crosses between petites and wild-type gave non-Mendelian segregation in progeny, and they postulated that a cytoplasmic factor controlling the petite phenotype was localized in mitochondria [33]. The literature of classical molecular biology contains an enormous output of research on petites [34].

The respiratory-deficient cytoplasmic petites (or *rho* mutants) are the consequence of deletions of wild-type mtDNA and amplification of the remaining fragment. The retained mtDNA exists as populations of linear and/or circular multimers. Different petites have different amplified mtDNA fragments each of characteristic size and sequence. (See Bockelmann et al. [35] and Dujon and Belcour [29] for reviews.) A number of phenotypes affecting growth in filamentous fungi, including *poky* in *Neurospora crassa* and *senescence* in *Podospora anserina*, are also caused by structural changes in the mitochondrial genome (Table 2).

Many fungi contain viruses, especially dsRNA viruses. These were first discovered by their association with interferon-inducing properties of *Penicillium* species. Most of these viruses confer few, if any, detectable phenotypes on the fungal host. Some viruses are the genetic determinants of killer proteins, and several others cause fungal disease (Table 2). Shortly after their discovery, they were invoked as the possible cause of certain degenerative phenomena in both antibiotic- and mycotoxin-producing fungi, but closer scrutiny has not supported these hypotheses (see below). For comprehensive reviews of fungal virology, see Lemke [36], Buck [30], and Koltin and Leibowitz [37].

IV. SECONDARY METABOLISM AND STRAIN INSTABILITY

Loss of secondary metabolite production is one example of strain instability in filamentous fungi. It is frequently correlated with changes in sporulation or formation of macroscopic structures such as cleistothecia and sclerotia and other cultural characteristics, but the strain may grow as vigorously as the parental forms that produce a full complement of morphological and biochemical differentiation products. Sermonti [38] has suggested the term *yield decay* to describe decreasing product formation since degeneration connotes a loss in the strain's vigor or growth potential.

Review articles on secondary metabolism frequently allude to strain degeneration. For example, Weinberg writes, "Degeneration can be of considerable nuisance in the industrial use of microorganisms" [39, p. 5]. Demain cautions,

> An additional characteristic common to all secondary metabolites is the ease with which the microbial producers of these compounds lose their ability to form the product, i.e., "strain degeneration." Thus, mutation is of importance for secondary product formation in the negative, as well as in the positive, direction [40, p. 178].

Genetic instability in antibiotic-producing *Streptomyces* is common, and during the 1970s the loss of plasmids or other cytoplasmic factors was a popular hypothesis to explain this instability. Subsequently, it was discovered that gene amplification sometimes accompanies mutation of unstable genes, whereas loss of plasmids is only rarely the etiological factor. (See Malik [41] for a review.)

In fungi, few molecular mechanisms for strain decay have been elucidated. Some researchers have been able to demonstrate that the decay phenomenon is not due to cytoplasmic factors; in most cases, the data are largely descriptive.

A. *Penicillium* and Penicillin

Simultaneous loss of penicillin and the yellow pigment chrysogenin was reported in *Penicillium notatum* in 1932 [42]. During World War II, as research on penicillin intensified, the problem of strain decay received increased attention. Active strains were most likely to degenerate after continued serial transfers on synthetic media. By holding the number of transfers to an absolute minimum and using spore inocula stored in dry sterile sand at low temperatures, the degeneration was minimized [43]. The degree of variability was influenced by the strain used, the conditions of growth, and composition of the medium [44]. Other workers simply made a large number of agar tubes of a high-yielding strain, stored them at 1–3°C, and used these as inocula for fermentation broths [45].

A single spore analysis of *P. notatum* revealed that some strains contained two distinct genetic elements, that is, they exhibited the "dual phenomenon" (heterokaryosis) [46, 47]. Use of conidia while avoiding mycelial fragments was recommended as a preventive step against degeneration.

From a comprehensive study of *penicillin rundown*, a term coined for degeneration occurring in penicillin production, it was found that mass spore transfers of a high-yielding strain of *P. notatum* caused the fungus to undergo rapid natural variation [48]. No explanation was offered for the process of variation, but the use of vegetative mycelial transfers over spore transfers was recommended for preventing the rundown. This recommendation was opposite to the one given by Foster et al. [43] and Hansen and Snyder [46], who advocated the use of conidial transfers as the best way to prevent degeneration.

During the 1960s, parasexual methods were applied to *P. chrysogenum* to attempt to improve penicillin yield. Certain cases of decrease in penicillin titers were attributed to mutation in the parental diploid before haploidization [49]. The problem of penicillin rundown was solved empirically. Modern industrial strains are mutants isolated by classical methods of "brute force" mutagenesis and selection. Along with high penicillin titer on inexpensive substrates and ease of growing in submerged culture, strain stability is one of the attributes selected for in creating the high-yielding penicillia used in the pharmaceutical industry today.

When dsRNA viruses were discovered, scientists at Lilly Corporation examined a production strain of *P. chrysogenum*. It was infected with viruses. After treatment with mutagenic agents, 30 strains of varying virus titers were assayed for penicillin. All retained the ability to produce penicillin; that is, no correlation between titer of virus and titer of penicillin was detected [50].

In effect, the issue of strain decay in the penicillia has become moot. Industrial strains are generally stable, dsRNA viruses can coexist with a high penicillin yielding genome, and instability is not considered a problem in contemporary penicillin fermentations.

B. *Aspergillus* and Aflatoxin

In *A. nidulans* a number of morphological changes, including changes in pigmentation, have been associated with duplications and transpositions of chromosome segments that are unstable at mitosis [51–53]. However, most of the data concerning instability of secondary metabolite production in the genus *Aspergillus* concern the aflatoxigenic species.

Aflatoxins are decaketides with toxigenic, carcinogenic, mutagenic, and teratogenic potential. The aflatoxin-producing species, *A. flavus* and *A. parasiticus*, are common storage contaminants and falcultative pathogens. Aflatoxin contamination of peanuts, corn, cottonseed, rice, millet, and many other important crops constitutes an international health problem.

There is wide variation in aflatoxigenic potential of *A. flavus* and *A. parasiticus* strains isolated from nature. Some wild-type strains are nontoxigenic; producing strains range from low to high in toxigenic potential. Several early reports also noted loss of aflatoxin-producing ability after prolonged maintenance of cultures in the laboratory [54, 55].

Mutant strains that did not produce aflatoxin were induced experimentally in *A. flavus* by exposure of an aflatoxigenic isolate to barium ions for six successive "generations." The entire population of resultant colonies was nontoxigenic [56]. In another study, using an aflatoxigenic strain of *A. parasiticus*, single-spore and polysporous ("mass spore") cultures were transferred on defined and complex media. After six successive transfers, polysporous transfers were attenuated; however, when single spores were isolated from these attenuated cultures, there was a return to high aflatoxigenicity. The authors hypothesized that a cytoplasmic factor in the mycelium, but not the spores, was associated with the degenerative process [57]. Similar loss of toxigenicity upon serial transfer was described by Torres et al. [58] on shredded wheat and by Clevstrom and Ljungren [59] on complex media in the presence of formic acid and ammonium formate.

A red mycelial mutant of *A. parasiticus* (ATTC 24690) partially blocked in aflatoxin biosynthesis has been useful in studying strain decay in this species. The red pigment (norsolorinic acid) and aflatoxin are biosynthetically related through the polyketide pathway. Loss of norsolorinic acid production is easily monitored as loss of red color. In an attempt to repeat the Lee and Townsley experiment [56], this strain was transferred successively for five "generations" in a defined medium containing barium ions. The majority of resultant colonies were unpigmented. Controls in which the serial transfers were conducted in medium lacking barium were also unpigmented, indicating that the serial transfer rather than the barium ion was the important variable. Unpigmented forms were stable and remained unpigmented and nonaflatoxigenic even after passage through a single-spore phase and after many years in laboratory culture. Conditions in which the inocula for serial transfer contained mycelia rather than conidia favored the loss of pigment production [60].

In a similar study, two classes of nonaflatoxigenic morphological variants were isolated from *A. parasiticus*, using the procedure of mycelial maceration. The variants were dubbed *fan* and *fluff* based on their morphology on agar plates. Neither variant type produced detectable aflatoxin. A somatic (parasexual) diploid synthesized using the *fan* phenotype produced high levels of aflatoxin, while another diploid containing the *fluff* phenotype produced low levels of aflatoxin [61]. When mutant strains carrying spore color and auxotrophic markers were tested with the *fan* and *fluff* phenotypes in a heterokaryon test, no segregation of the variant phenotypes was noted; that is, the *fan* and *fluff* phenotypes did not appear to be under cytoplasmic control [62]. In a later study, serial transfer of mycelial macerates was conducted to isolate *fan* and *fluff* variants in a diploid carrying anthraquinone markers for norsolorinic acid and versicolorin A. No detectable anthraquinones or aflatoxins were detected in the *fan* or *fluff* diploids, but unstable pigmented segregants were isolated from both after treatment with a haploidization agent. Thus, these segregants retained the structural genes for polyketide biosynthesis, suggesting that the production of the *fan* and *fluff* phenotypes involved changes at a regulatory locus [63].

More recently, Lee [64] conducted precursor feeding experiments with two nontoxigenic strains of *A. flavus* isolated by repeated transfer on laboratory media and four nontoxigenic strains isolated from nature. None of the strains isolated from nature was capable of biotransforming sterigmatocystin or *O*-methylsterigmatocystin to aflatoxin, but the nontoxigenic strains isolated by laboratory transfer did biotransform these precursors into aflatoxin [64]. Similar biotransformation experiments in our laboratory with *fan* and *fluff* gave no evidence of biotransformation (unpublished data), indicating a difference between the *A. parasiticus* degenerate forms and the *A. flavus* forms studied by Lee [64].

During the 1970s, yet another form of nonaflatoxigenicity was correlated with proteinaceous viruslike particles lacking detectable nucleic acids [65]. Later research indicated that cycloheximide induced aflatoxin biosynthesis in this strain. It was hypothesized that cycloheximide had cured the strain of a genetic determinant for the control of aflatoxin biosynthesis and that this element might provide a biological control mechanism against aflatoxin if it were transmitted to aflatoxigenic strains [66]. This hypothesis has not been confirmed in other laboratories.

C. *Claviceps* and Ergot Alkaloids

The ergot alkaloids constitute a biogenetic family of pharmacologically active fungal metabolites, most of which contain an ergoline ring system. Many of the best-known members of the family are amide derivatives of lysergic acid. They are produced by various *Claviceps* species that parasitize grasses and cereals. The overwintering sclerotial stage of the life cycle is called *ergot*. Ergot acts to constrict blood vessels and may cause convulsions and hallucinations. When ergot-infected grains are eaten in sufficient quantity, a toxic syndrome called gangrenous ergotism or St. Anthony's fire can result. Reports of St. Anthony's fire date back to medieval times. The line between drug and toxin is blurred with respect to these secondary metabolites. Several ergot alkaloids are used in medical practice to treat migraine and to stimulate uterine contractions.

When *Claviceps purpurea* is cultured from high-alkaloid-containing sclerotia from nature, a high proportion of the resultant progeny are nonproducing. Strain degeneration is also common in *C. paspali*. Degenerate colonies frequently are distinguished by alterations in colony morphology. Degenerate forms can usually be returned to high production by passage through the host plant. Although some theorists have attributed instability to simple heterokaryosis, there is considerable evidence that this hypothesis does not suffice to explain most examples of strain decay in *Claviceps* [67].

D. *Fusarium*, Trichothecenes, and Zearalenone

The genus *Fusarium* is a large and taxonomically difficult group of saprophytic and parasitic molds. The major mycotoxins associated with these fungi are zearalenone and the trichothecenes. Several human and veterinarian fusariotoxicoses have been elucidated, including alimentary toxic aleukia, fescue foot, "inebriant bread syndrome," moldy sweet potato toxicosis, and scabby grain intoxication (akakabi-byo) [68].

Trichothecenes are sesquiterpenes characterized by a 12,13-epoxytrichothecene ring system. T-2, nivalenol, deoxynivalenol (vomitoxin), and diacetoxyscirpenol are among the most commonly encountered *Fusarium* derivatives. Other genera synthesizing trichothecenes are *Myrothecium* (verrucarins and roridins), *Stachybotrys* (satratoxins), and *Trichoderma* (trichodermin). Trichothecenes function as potent inhibitors of protein synthesis [69].

The trichothecenes gained brief notoriety when they were implicated as alleged components in the chemical warfare agent Yellow Rain. Subsequently, this connection was discredited [68, 71, 72].

Fusarium has long been cursed by mycologists for its variability [73]. It is quite common for wild-type strains isolated from nature to change radically in appearance when subcultured in the laboratory. These changes usually involve loss of aerial mycelium and an increase in macroconidia, and sometimes an increase in pigmentation. Most studies on spontaneous variability are descriptive. Quantitative data of any sort, much less of a genetic or molecular nature, are woefully lacking [74]. To further complicate the matter, different taxonomic systems have been used in different countries. Nomenclature varies from laboratory to laboratory, and many species have been misidentified.

A consortium of scientists interested in the toxigenic fusaria have attempted to bring order to this chaos. They collected approximately 200 *Fusarium* strains that had been identified in the literature as toxin producers. About half bore names that did not correspond with current taxonomic concepts, and numerous strains had lost the ability to make the reported toxin. They reduced the 25 trichochecene-producing fusaria as named in the published literature to 11 accepted species. An International Toxic *Fusarium* Reference Collection (ITFRC) was established, and reference stocks are maintained at the South African Medical Research Council in Pretoria, South Africa, and the *Fusarium* Research Center at the Pennsylvania State University, University Park, Pennsylvania.

The efforts of the ITFRC should alert readers of the literature on *Fusarium* mycotoxins to possible nomenclatural problems with reported mycotoxin-producing species. Moreover, researchers new to the field must recognize the high likelihood of spontaneous strain degeneration and make arrangements to deposit reference strains with ITFRC [75, 76].

Almost the entire emphasis of research on vegetative instability in the trichothecene-producing fusaria has been taxonomic in intent. In contrast, there have been some experimental studies on the zearalenone-producing species *F. graminearum* (*Gibberella zeae*). Zearalenone is a polyketide with estrogenic potential. It causes an estrogenic syndrome in swine, resulting in enlarged vulvae and mammary glands, but with low toxicity. A closely related analog of zearalenone is marketed as a feed supplement for cattle, and a nontoxigenic strain of *F. graminearum* is grown commercially in Britain as a meat substitute for vegetarians ("mycoprotein").

Duncan and Bu'Lock [77] isolated two classes of mutants from a high-zearalenone strain of *F. graminearum* by serial subculture on rich nutrient media. In addition to loss of zearalenone-producing ability, the strains exhibited altered colony morphology and pigmentation. RUP (red under-pigment) mutants produced low levels of zearalenone and a red pigment; from these a further set of non-zearalenone-producing morphological mutants, designated PR (pale RUP) were isolated by multiple passages. RUP and PR did not reassort in a heterokaryon test. Their morphology appeared similar to spontaneous degenerate forms of *Fusarium* described earlier by Booth [78] and Burnett [79]; in their mode of isolation and behavior in heterokaryon tests they were similar to the nonaflatoxigenic *fan* and *fluff* mutants of *A. parasiticus* isolated by Bennett et al. [63]. The Bu'Lock group introduced the designation *lazy* to describe this general type of pleiotropic mutation leading to loss of zearalenone production accompanied by morphological change [80].

V. SUMMARY, CONCLUSIONS, AND HYPOTHESES

Stability implies continuance without change, permanence, while instability is the opposite: change, impermanence. Degeneration describes a decline or deterioration in qualities, or reversion to a less organized, simpler type. Given the high frequency of phenotypic change encountered in clones of filamentous fungi, the term "strain instability" is preferable to "strain degeneration." In some species and some strains, spontaneous phenotypic change occurs at such high frequency that it could be viewed as a normal part of the developmental repertoire rather than as a degeneration to an inferior form.

The biochemical and morphological changes displayed by filamentous fungi in culture have, in some cases, been attributed to heterokaryosis, chromosome instability, or cytoplasmic inheritance. The coenocytic growth habit and high frequency of anastomoses between adjacent hyphae no doubt contribute to pleomorphism in molds. The formation of heterokaryons following fusion of genetically dissimilar hyphae creates a cellular environment in which cytoplasmic components intermingle while nuclei remain intact. The heterokaryon test is an important tool for distinguishing phenotypes based on cytoplasmic or nuclear factors, especially for anamorphic (imperfect) species.

The strain decay described for loss of penicillin, aflatoxin, and zearalenone production resembles in some ways the phenotypic changes associated with the loss of certain cytoplasmic elements. In *Streptomyces*, "curing" of plasmids leads to changes in colony morphology and loss of metabolite production [81], while alterations of the mitochondrial genome, such as senescence in *Podospora* [33] create a mitochondrial syndrome that at least superficially resembles the alteration from wild-type growth form correlated with loss of secondary metabolite production.

The phenomenon of fungal strain instability involving loss of secondary metabolite production suffers from lack of a common, distinguishing name. In the antibiotics industry, the terms *penicillin rundown* and *yield decay* emphasize the absence of a desirable metabolite. In the mycotoxin literature, the nomenclatural convention has been to identify nontoxigenic variants individually, for example, *fan* and *fluff* for the nonaflatoxigenic and morphologically altered forms of *A. parasiticus* [63] and *lazy* for a similar family of mutants in *Fusarium* [80]. In culture, *fans, fluffs*, and *lazys* share some things in common with the mitochondrial syndrome of filamentous fungi. Morphology is altered from wild-type to a less organized form, and the mutants are not isolated as rare, random, and diverse types, but as populations of mutant colonies with identical phenotypes.

However, they also exhibit important differences from the mitochondrial syndromes. The *fan, fluff*, and *lazy* phenotypes are not infectious nor do they reassort in a heterokaryon test. In general, data indicate that these cases of lost toxigenicity, as well as strain decay in antibiotic-producing species, are controlled by nuclear factors. The underlying genetic control mechanisms affecting the stability of secondary metabolite production in filamentous fungi have not been elucidated.

Some reasonable speculation seems in order. Every student of secondary metabolism has been struck by the contrast between the enormous number and extraordinary diversity of the metabolites and the relative economy and simplicity of the biochemical pathways that generate them. The aflatoxins produced by *Aspergillus*, the patulins produced by *Penicillium*, and the zearalenones produced by *Fusarium* are all polyketide-derived, originating through condensation of acetyl CoA and malonyl CoA. It seems unlikely that nature has evolved a separate polyketide synthase for each pathway and genus. More plausibly, closely related multienzyme complexes interact with other, species-specific enzymes to generate minor modification in the number and arrangement of carbon atoms in the basic skeleton. Similarly, secondary metabolites derived from amino acids or terpenes are likely to be synthesized from evolutionarily related enzymes.

The lack of molecular studies on fungal secondary metabolism is a major impediment to further understanding of this process. To date, only a few structural genes for fungal secondary metabolite pathways have been cloned. Virtually nothing is known about the regulatory sequences that coordinate the expression of these structural genes. There is a large body of empirical data that correlate nutritional deprivation with the trophophase—idiophase transition; hence, it is not unreasonable to hypothesize common triggering mechanisms for secondary metabolite production. The "global" effects associated with the *fan, fluff*, and *lazy* phenotypes would support the hypothesis that strain instability may involve changes in regulatory loci.

An attractive model that would explain the high frequency with which secondary metabolic capability is lost in the absence of obvious environmental mutagens invokes intragenomic mobile DNA and karyotype plasticity. Modern geneticists no longer envision the genome as a highly stable entity in which genetic maps reflect immovable genes only rarely altered by mutation or gross chromosome rearrangements. Since McClintock's discovery of transposable controlling elements in corn, numerous mobile DNAs have been found in both prokaryotes and eukaryotes. Well-known examples include mammalian DNA arrangements involved in the generation of antibody diversity, invertible DNA segments causing phase variation in *Salmonella* and host range variation in phage Mu, antigen variation in trypanosomes, and the high frequency of yeast mating type switches. It is well documented that the deletion, insertion, rearrangement, and amplification of transposable elements have a wide range of important biological consequences in a wide number of species. Mobile DNAs may inactivate genes by inserting themselves within either the structural gene or its controlling site. See Berg and Howe [82] for a comprehensive review.

Fungal strain instability associated with laboratory subculture is a common theme in the mycological and industrial microbiological literature. The high frequency with which such instability is encountered in the absence of external mutagens, the subsequent stability of many of these variant forms, and the pleiotropic nature of the mutational event could all be modeled as the inactivation of a regulatory element resulting from a subtle DNA rearrangement. The nutritionally rich environment of the laboratory may create a metabolic milieu in which genetic transposition is favored.

At the time of this writing (1990) we have virtually no molecular understanding of either mycotoxin biosynthesis or the loss of mycotoxin biosynthesis in clonal derivatives of toxigenic strains. However, the recent development of transformation systems in both *A. flavus* [83, 84] and *A. parasiticus* [85] and other approaches for identifying genes in aflatoxin biosynthesis [86] should facilitate molecular genetic manipulations within these species. In addition, the recent application of electrophoretic karyotyping, using orthagonal field gel electrophoresis, has revealed widespread intraspecific chromosomal polymorphism among filamentous fungi, including aflatoxigenic species ([87] and N. Keller, pers. comm.).

In summary, filamentous fungi exhibit an extremely high rate of variation when grown and observed in the laboratory. The mechanisms by which toxigenic potential is lost or suppressed in mycotoxin producers is not understood. It is likely to involve changes in global regulatory signals, and these regulatory signals are likely to be similar among secondary metabolic pathways sharing common precursors of primary metabolism. Subtle DNA rearrangement within the somatic genome provide a tantalizing and attractive model for explaining the phenomenon.

Besides the intrinsic biological interest of studying strain instability in filamentous fungi, this research may have practical consequences. An understanding of the regulatory mechanism controlling secondary metabolism could lead to rational methods for the isolation of stable industrial strains and may eventually allow us to control natural outbreaks of mycotoxin contamination.

ACKNOWLEDGMENTS

We thank L. S. Lee and D. Bhatnagar for the insights into aflatoxin biosynthesis and fungal strain degeneration, J. D. Bu'Lock and Morris Grindle for reviewing the text, and P. Crickenberger for manuscript preparation. Research in this laboratory has been supported by a grant from the Newcomb Foundation.

REFERENCES

1. Bennett, J. W. Mycotoxins, mycotoxicosis, mycotoxicology and mycopathologia. *Mycopathologia, 100*:3, 1987.
2. Bennett, J. W., and Bentley, R. What's in a name? —microbial secondary metabolism. *Adv. Appl. Microbiol., 34*:1, 1989.
3. Luckner, M. *Secondary Metabolism in Microorganisms, Plants and Animals*, VEB, Gustav Fischer Verlag, Jena, 1984.
4. Cole, R. J., and Cox, R. H. *Handbook of Toxic Fungal Metabolites*, Academic, New York, 1981.
5. Weinberg, E. D. Secondary metabolism: raison d'etre. *Perspect. Biol. Med., 14*:565, 1971.
6. Demain, A. L. How do antibiotic producing microorganisms avoid suicide? *Ann. N.Y. Acad. Sci., 235*:601, 1974.
7. Bennett, J. W., and Ciegler, A. *Secondary Metabolism and Differentiation in Fungi*, Marcel Dekker, New York, 1983.
8. Campbell, I. M. Secondary metabolism and microbial physiology. *Adv. Microb. Physiol., 25*:1, 1984.
9. Turner, W. B. *Fungal Metabolites*, Academic, London, 1971.
10. Turner, W. B., and Aldridge, D. C. *Fungal Metabolites II*, Academic, London, 1983.
11. Bu'Lock, J. D., Nisbet, L. J., and Winstanley, D. J. *Bioactive Microbial Products: Search and Discovery* (for Society for General Microbiology), Academic, London, 1982.
12. Ingold, C. D. Genetics of the microfungi. *Nature, 157*:614, 1948.
13. Foster, J. W. Some introspections on mold metabolism. *Bacteriol. Rev., 11*:167, 1947.
14. Smith, D. Culture and preservation. In *Living Resources for Biotechnology* (D. L. Hawksworth and B. E. Kirsop, Eds.), Cambridge Univ. Press, New York, 1988, p. 75.
15. Hansen, H. N. The dual phenomenon in imperfect fungi. *Mycologia, 30*:442, 1938.
16. Bu'Lock, J. D. *The Biosynthesis of Natural Products: An Introduction to Secondary Metabolism*, McGraw-Hill, London, 1965.
17. Raper, K. B., and Fennell, D. I. *The Genus Aspergillus*, Williams and Wilkins, Baltimore, 1965.
18. Thom, C., and Raper, K. B. *A Manual of the Aspergilli*, Williams and Wilkins, Baltimore, 1945.
19. Toussoun, T. A. Prologue. In *Fusarium: Diseases, Biology, and Taxonomy* (P. E. Nelson, T. A. Toussoun, and R. J. Cook, Eds.), Pennsylvania State Univ. Press, University Park, Penna., 1981, p. 289.
20. Davis, R. H. Mechanisms of inheritance, heterokaryosis. In *The Fungi*, Vol. II, *The Fungal Organism* (G. C. Ainsworth and A. S. Sussman, Eds.), Academic, New York, 1966, p. 567.

21. Jinks, J. L. Somatic selection in fungi. *Nature (London), 174*:40, 1954.
22. Jinks, J. L. Naturally occurring cytoplasmic changes in fungi. *C.R. Lab. Carlberg Ser. Physiol., 26*:183, 1956.
23. Jinks, J. L. Selection of cytoplasmic differences. *Proc. Roy. Soc. London, B, 146*:537, 1957.
24. Jinks, J. L. Mechanisms of inheritance, extranuclear inheritance. In *The Fungi*, Vol. II, *The Fungal Organism* (G. C. Ainsworth and A. S. Sussman, Eds.), Academic, New York, 1966, p. 619.
25. Arlett, C. F. A system of cytoplasmic variation in *Aspergillus nidulans. Heredity, 15*:377, 1960.
26. Arlett, C. F., Grindle, M., and Jinks, J. L. The red cytoplasmic variant of *Aspergillus nidulans. Heredity, 17*:197, 1962.
27. Jinks, J. L. Cytoplasmic inheritance in fungi. In *Methodology in Basic Genetics* (W. J. Burdette, Ed.), Holden-Day, San Francisco, 1963, p. 325.
28. King, R. C. *A Dictionary of Genetics*, Oxford Univ. Press, New York, 1972, p. 78.
29. Dujon, B., and Belcour, L. Mitochondrial DNA instabilities and rearrangements in yeasts and fungi. In *Mobile DNA* (D. E. Berg and M. M. Howe, Eds.), ASM, Washington, D.C., 1989, p. 861.
30. Buck, K. W., Ed. *Fungal Virology*, CRC Press, Boca Raton, Fla., 1986.
31. Turner, G. Cytoplasmic inheritance and senescence. In *The Filamentous Fungi*, Vol. III, *Developmental Mycology* (J. E. Smith and D. R. Berry, Eds.), Wiley, New York, 1978, p. 406.
32. Ephrussi, B., Hottinguer, H., and Chimenes, A. M. Action de l'acriflavine sur les levures. I. La mutation "petite colonie." *Ann. Inst. Pasteur Paris, 76*:351, 1949.
33. Slonimiski, P. P., and Ephrussi, B. Action de l'acriflavine sur les levureres. V. Le systeme des cytochromes des putants "petite colonie." *Ann. Inst. Pasteur Paris, 77*:47, 1949.
34. Wilkier, B. Genetic and functional aspects of yeast mitochondria. In *Yeast Genetics, Fundamental and Applied Aspects* (J. F. T. Spencer and A. R. W Smith, Eds.), Springer, New York, 1983, p. 255.
35. Bockelmann, B., Ostewaczh, H. D., Schmidt, F. R., and Schulte, E. Extrachromosomal DNA in fungi — organization and function. In *Fungal Virology* (K. W. Buck, Ed.), CRC Press, Boca Raton, Fla., 1986, p. 237.
36. Lemke, P. A., Ed. *Viruses and Plasmids in Fungi*, Marcel Dekker, New York, 1979.
37. Koltin, Y., and Leibowitz, M. J. *Viruses of Fungi and Simple Eukaryotes*, Marcel Dekker, New York, 1988.
38. Sermonti, G. *Genetics of Antibiotic Producing Microorganisms*, Wiley-Interscience, New York, 1969.
39. Weinberg, E. D. Biosynthesis of secondary metabolites: roles of trace metals. *Adv. Microbiol. Physiol., 4*:1, 1970.
40. Demain, A. L. Mutation and the production of secondary metabolites. *Adv. Appl. Microbiol., 16*:177, 1973.

41. Malik, V. S. Genetics of secondary metabolism. *Biotechnology, 4*:39, 1986.
42. Clutterbuck, D. W., Lowell, R., and Raistrick, H. The formation from glucose by members of the *Penicillium chrysogenum* series of a pigment, an alkali-soluble protein and penicillin — the antibacterial substance of Fleming. *Biochem. J., 26*:1907, 1932.
43. Foster, J. W., Woodruff, H. B., and McDaniel, L. E. Microbial aspects of penicillin III. Production of penicillin in surface cultures of *Penicillium notatum. J. Bacteriol., 46*:421, 1943.
44. Walksman, S. A., and Reilly, H. C. Strain specificity and production of antibiotic substances. III. *Penicillium notatum-chrysogenum* group. *Proc. Natl. Acad. Sci. USA, 30*:99, 1944.
45. Raper, K. B., Alexander, D. F., and Coghill, R. D. Penicillin II. Natural variation and penicillin production in *Penicillium notatum* and allied species. *J. Bacteriol., 48*:639, 1945.
46. Hansen, H. N., and Snyder, W. C. Relation of dual phenomenon in *Penicillium notatum* to penicillin production. *Science, 49*:264, 1944.
47. Baker, G. E. Heterokaryosis in *Penicillium notatum. Bull. Torr. Bot. Club, 71*:367, 1944.
48. Whiffen, A. J., and Savage, G. M. The relation of natural variation in *Penicillium notatum* to the yield of penicillin in surface culture. *J. Bacteriol.*, 53:231, 1947.
49. Ball, C. Improvement of penicillin productivity in *Penicillium chrysogenum* by recombination. In *Genetics of Industrial Microorganisms* (Z.-Z. Vanek and J. Cudlin Host'alek, Eds.), Academic, Prague, 1973.
50. Lemke, P. A., Nash, H., and Pieper, S. W. Lytic plague formation and variation in virus titre among strains of *Penicillium chrysogenum. J. Gen. Microbiol., 76*:285, 1973.
51. Nga, B. H., and Roper, J. A. A system generating spontaneous intrachromosomal changes at mitosis in *Aspergillus nidulans. Gen. Res., 14*:63, 1969.
52. Azevedo, J. L., and Roper, J. A. Mitotic non-conformity in *Aspergillus*: successive and transposable genetic changes. *Genet. Res. Camb., 16*:79, 1970.
53. Upshall, A., and Roberts, M. D. Somatic instabilities in wild isolates of *Aspergillus nidulans. FEMS. Microbiol. Lett., 9*:311, 1980.
54. Diener, V. L., and Davis, N. D. Aflatoxin production by isolates of *Aspergillus flavus. Phytopathology, 56*:1393, 1966.
55. Wilson, B. J., Campbell, T. C., Hayes, A. W., and Hamlin, R. T. Investigation of reported aflatoxin production by fungi outside the *Aspergillus flavus* group. *Appl. Microbiol., 16*:819, 1968.
56. Lee, E. G. H., and Townsley, P. M. Chemical induction of mutation or variation in aflatoxin-producing cultures of *Aspergillus flavus. J. Food Sci., 33*:420, 1968.
57. Mayne, R. Y., Bennett, J. W., and Tallant, J. Instability of an aflatoxin producing strain of *Aspergillus parasiticus. Mycologia, 63*: 644, 1971.
58. Torres, J., Guarro, J., Suarez, G., Sune, N., Calvo, M. A., and Ramiez, C. Morphological changes in strains of *Aspergillus flavus*

Link ex Fries and *Aspergillus parasiticus* Speare related with aflatoxin production. *Mycopathology, 72*:171, 1980.
59. Clevstrom, G., and Ljungren, H. Aflatoxin formation and the dual phenomenon in *Aspergillus flavus* Link. *Mycopathologia, 92*:129, 1985.
60. Bennett, J. W. Loss of norsolorinic acid and aflatoxin production by a mutant of *Aspergillus parasiticus. J. Gen. Microbiol., 124*:429, 1981.
61. Bennett, J. W., Silverstein, R. B., and Kruger, S. J. Isolation and characterization of two nonaflatoxigenic classes of morphological variants of *Aspergillus parasiticus. J. Am. Oil Chem. Soc., 58*:952, 1981.
62. Bennett, J. W., Wheeler, D. G., and Dunn, J. J. Genetic analysis of aflatoxin production by *Aspergillus parasiticus*. In *Advances in Biotechnology*, Vol. 3, *Fermentation Products* (M. Moo-Young, C. Vezina, and K. Singh, Eds.), Pergamon, Toronto, 1981.
63. Bennett, J. W., Leong, P.-M., Kruger, S., and Keyes, D. Sclerotial and low aflatoxigenic morphological variants from haploid and diploid *Aspergillus parasiticus. Experientia, 42*:848, 1985.
64. Lee, L. S. Metabolic precursor regulation of aflatoxin formation in toxigenic and non-toxigenic strains of *Aspergillus flavus. Mycopathologia, 107*:127, 1989.
65. Wood, H. A., Bozarth, R. F., Adler, J. P., and Mackenzie, D. W. Proteinaceous virus-like particles from an isolate of *Aspergillus flavus. J. Virol., 37*:215, 1974.
66. Schmidt, F. R., Davis, N. D., Diener, U. L., and Lemke, P. A. Cycloheximide induction of aflatoxin synthesis in a nontoxigenic strain of *Aspergillus flavus. Biotechnology, 1*:794, 1983.
67. Vining, L. C., and Taber, W. A. Ergot alkaloids. In *Secondary Products of Metabolism* (A. H. Rose, Ed.), Academic, London, 1979, p. 389.
68. Joffe, A. Z. *Fusarium Species: Their Biology and Toxicology*, Wiley-Interscience, New York, 1986.
69. Mirocha, C. J., Pathre, S. V., and Christensen, C. M. Mycotoxins. In *Secondary Products of Metabolism* (A. H. Rose, Ed.), Academic, London, 1979, p. 467.
70. Seagrave, S. *Yellow Rain: A Journey through the Terror of Chemical Warfare*, M. Evans, New York, 1981.
71. Seeley, T. D., Nowicke, J. W., Meselson, M., Guillemin, J., and Akratanakul, P. Yellow rain. *Sci. Am.*, September 1985, p. 128.
72. Marshall, E. Yellow rain evidence slowly whittled away. *Science, 233*:18, 1986.
73. Snyder, W. C., and Hansen, H. N. The species concept in *Fusarium. Am. J. Bot., 27*:64, 1940.
74. Puhalla, J. E. Genetic considerations of the genus *Fusarium*. In *Fusarium: Diseases, Biology, and Taxonomy* (F. E. Nelson, T. A. Toussoun and R. J. Cook, Eds.), Penn. State Univ. Press, University Park, 1981, p. 291.
75. Marasas, W. F. O., Nelson, P. E., and Toussoun, A. *Toxigenic Fusarium Species: Identification and Mycotoxicology*, Penn. State Univ. Press, University Park, 1984.

76. Lacey, J., Ed. *Trichothecenes and Other Mycotoxins*, Wiley, Chichester, 1985.
77. Duncan, J. S., and Bu'Lock, J. D. Degeneration of zearalenone production in *Fusarium graminearum*. *Exp. Mycol.*, *9*:133, 1985.
78. Booth, C. *The Genus Fusarium*. Commonwealth Agricultural Bureau, Kew, Surrey, U.K., 1971.
79. Burnett, J. H. Aspects of *Fusarium* genetics. In *The Applied Mycology of Fusarium* (M. O. Moss and J. E. Smith, Eds.), Cambridge Univ. Press, Cambridge, 1984, p. 39.
80. Bu'Lock, J. D., Mooney, J. P., and Wright, C. E. Regulation of mycotoxin production by *Fusarium graminearum*: complementation effects between two mutant types. *Biotech. Lett.*, *8*:323, 1986.
81. Kirsop, B. E., Ed. *The Stability of Industrial Organisms*, Commonwealth Mycological Institute, Kew, Surrey, U.K., 1981.
82. Berg, D. C., and Howe, M. M. *Mobile DNA*, Am. Soc. Microbiology, Washington, D.C., 1989.
83. Woloshuk, C. P., Seip, E. R., Payne, G. A., and Adkins, C. R. Genetic transformation system for the aflatoxin-producing fungus *Aspergillus flavus*. *Appl. Environ. Microbiol.*, *55*:86, 1989.
84. Payne, G. A., and Woloshuk, C. P. Transformation of *Aspergillus flavus* to study aflatoxin biosynthesis. *Mycopathologia*, *107*:139, 1989.
85. Horng, J. S., Linz, J. E., and Pestka, J. J. Cloning and characterization of the *trpC* gene from an aflatoxigenic strain of *Aspergillus parasiticus*. *Appl. Environ. Microbiol.*, *55*:256, 1989.
86. Bhatnagar, D., Cleveland, T. E., and Lillehoj, E. B. Enzymes in aflatoxin B_1 biosynthesis: strategies for identifying pertinent genes. *Mycopathologia*, *107*:75, 1989.
87. Skinner, D. Z., Budde, A. D., and Leong, S. A. In *More Gene Manipulations in Fungi* (J. W. Bennett and L. L. Lasure, Eds.), Academic, San Diego, CA., 1991, p. 86.

13

GENETIC APPROACHES TO THE CHEMICAL ECOLOGY OF PHYTOPATHOGENIC *FUSARIUM* SPECIES

ANNE E. DESJARDINS *Agricultural Research Service, National Center for Agricultural Utilization Research, U.S. Department of Agriculture, Peoria, Illinois*

I. INTRODUCTION

Fungi of the genus *Fusarium* are cosmopolitan plant pathogens and soil saprophytes. Many common species, and indeed individual strains, of *Fusarium* can infect a wide range of plant hosts [1–3]. These adaptable species are among some of the most serious pathogens of crop plants worldwide. The deleterious effects of *Fusarium* infection on crop yields have been well documented and are exacerbated by the fact that many *Fusarium* species contaminate plant tissues with a wide variety of toxins that are injurious to human and animal health [4]. The control of *Fusarium* diseases and mycotoxicoses will depend upon prevention of fungal infection and development in the field and in storage. One long-term approach to this problem is the elucidation of the factors involved in the pathogenicity of *Fusarium* species and in plant resistance to them. Considerable progress has been made in understanding the roles of many putative pathogenicity and resistance factors in fungus–plant interactions. Critical reviews of different aspects of this field can be found in several books, including those by Durbin [5] and Bailey and Mansfield [6], and in the *Annual Review of Phytopathology*.

Both fungi and higher plants produce a great diversity of low molecular weight secondary metabolites. While the chemistry of many of these compounds is now well understood, more recently attention has focused on their biosynthesis, their biological activities in vitro, and their roles in plant–pathogen interactions. An increasing body of evidence indicates that the chemical interactions between fungi and higher plants are both complex and highly integrated. For example, as plants have evolved toxins (e.g., phytoalexins) that limit fungal growth in their tissues, fungi have adapted, in many cases, by evolving means to tolerate their hosts' toxins [7]. Due primarily to the complexities and limitations of the experimental systems, however, the actual importance of many fungal and plant toxins in pathogenicity and resistance has often been controversial, a subject that has been well reviewed [7, 8]. Certainly the most powerful technique for the analysis of these complex systems has been that of genetic manipulation. In this regard, fungi of the genus *Fusarium* have proved to be particularly amenable to analysis by a variety of genetic methods, including classical meiotic recombination and induced mutation as well as molecular genetics.

Under the appropriate culture conditions, some species of *Fusarium* can produce a heterothallic ascomycete sexual stage. In this review, I will use the binomial of the sexual stage where it is known, as is the recommended practice for plant pathogenic fungi [9]. Heterothallic *Fusarium* species have a simple mating system, with two mating types determined by alternative alleles at a single locus. For a fertile cross to occur, one of the strains also must be able to function as a female and the other as a male. A successful cross between strains of the appropriate mating type and sex results in production of perithecia from which haploid ascospores can be isolated for random ascospore or tetrad analysis. To date, these conventional genetic techniques have been applied most extensively and successfully to the analysis of pathogenicity of natural variants and mutants of *F. solani* (sexual stage: *Nectria haematococca*, mating population VI) and more recently to natural variants of *F. sambucinum* (sexual stage: *Gibberella pulicaris*). Mutational analyses have also been conducted in *F. sporotrichioides*, *F. solani* f. sp. *phaseoli*, *F. oxysporum* f. sp. *vasinfectum*, and *F. oxysporum* f. sp. *lycopersici*. DNA-mediated transformation technologies have recently been developed and are being applied to analysis of pathogenicity factors in *N. haematococca*, *G. pulicaris*, and *F. sporotrichioides*, for example (see Chapter 14).

Although there have been countless correlative studies of putative pathogenicity factors of *Fusarium* species, this review will deal only with those studies where genetic techniques and analyses have been applied. It has been well documented that pathogenicity and resistance factors include high molecular weight fungal and plant polymers and the enzymes involved in their degradation. This review, however, will deal only with low molecular weight toxic metabolites produced by *Fusarium* or by its plant hosts, and with their detoxification. Genetic analysis of the role of various *Fusarium* toxins in pathogenicity is discussed first. Next, isoflavonoid, sesquiterpene, and glycoalkaloid toxins from plants and genetic evidence for a role for their detoxification in pathogenicity are discussed. Finally, information and speculation on the regulation and integration of plant and fungal chemical responses in their common environment in diseased tissue are presented.

II. GENETIC ANALYSES OF THE ROLE OF TOXINS IN PATHOGENICITY

A. Fungal Toxins

Trichothecenes

The trichothecenes (Fig. 1) are a large family of sesquiterpene toxins produced by many *Fusarium* species and related fungi. Trichothecenes are

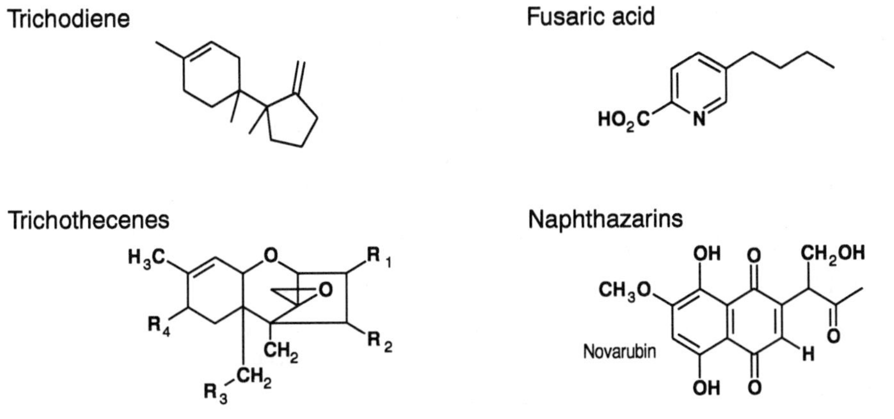

	R_1	R_2	R_3	R_4
T-2 Toxin	OH	OAc	OAc	$OCOCH_2CH(CH_2)_3$
Neosolaniol	OH	OAc	OAc	OH
Diacetoxyscirpenol	OH	OAc	OAc	H
Deacetylated Calonectrin	OAc	H	OH	H
Dideacetylated Calonectrin	OH	H	OH	H

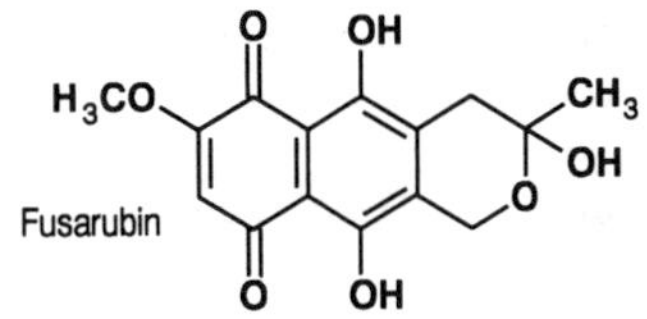

FIGURE 1. Fungal toxins.

potent eukaryotic protein synthesis inhibitors and host-nonspecific phytotoxins, and they have also been implicated in mycotoxicoses of humans and animals [4]. There is considerable circumstantial evidence for a role for trichothecenes in plant pathogenesis, including correlations of virulence with ability to produce toxins in vitro [10] and isolation of toxins from diseased plant tissues [4, 11]. Chlorosis, necrosis, and other symptoms are produced in a wide variety of plants treated with very low concentrations of trichothecenes (10^{-5}–10^{-6} M) [12, 13].

Mutants blocked or altered in trichothecene production have recently been generated in a T-2 toxin-producing strain of *F. sporotrichioides* by ultraviolet light (UV) treatment [14]. Three of these mutants were shown by complementation tests and chemical analysis to each be blocked in a different step in the T-2 biosynthetic pathway and to accumulate different end products (see Chapter 14 and Refs. 14–17) (Table 1). These three mutants and the wild-type parent were used to test the causal relationship between trichothecene production and virulence on parsnip (*Pastinaca sativa*) root disks [18]. Only the T-2 toxin-producing wild-type strain and a 4,15-diacetoxyscirpenol-producing mutant were of high virulence. A calonectrin analog-producing mutant and a trichodiene-producing mutant were of no or very low virulence but complemented each other to restore T-2 toxin production in vitro and to restore virulence on parsnip root (Tables 1 and 2). These mutants have been carefully characterized, and all are prototrophs and identical to wild-type in growth rate [15] and metabolism of parsnip phytoalexins [18]. Furthermore, it seems unlikely that independently isolated, complementary mutants, selected solely for their

TABLE 1. Major Trichothecenes Produced by Wild-Type and Mutant Strains of *Fusarium sporotrichioides*

Strain	Major trichothecenes produced in liquid culture (No. of hydroxyl groups)
NRRL 3299 (wild-type)	T-2 toxin (4)
MB1716	4,5-Diacetoxyscirpenol (3)
MB2972	15-Deacetylcalonectrin (2) 3,15-Deacetylcalonectrin (2)
MB5493	Trichodiene (0)
MB2972 and MB5493 (coculture)	T-2 toxin (4)

Source: Data from Refs. 15–17.

inability to produce T-2 toxin and blocked at different steps within the T-2 biosynthetic pathway, would also carry second-site mutations that not only render them nonpathogenic but are also complementary. These results therefore strongly suggest that production of certain highly oxygenated trichothecenes is required for pathogenicity of *Fusarium* on parsnip root. Genetic crosses with these mutants have not been possible because the sexual stage of *F. sporotrichioides* is not known.

Naphthazarins

Naphthazarins (Fig. 1) are a group of red-pigmented polyketide toxins produced by several *Fusarium* species, including *F. solani* f. sp. *pisi*, *F. javanicum*, *F. moniliforme*, *F. graminearum*, and *F. oxysporum* [19, 20]. Naphthazarins have been shown to cause a variety of symptoms, including wilting and chlorosis (at 10^{-3}–10^{-4} M) in treated plants and have been detected in *Fusarium*-infected plant tissues [19, 21]. Kern [22] selected and analyzed natural variants and laboratory mutants of *F. solani* var. *martii* that did not produce naphthazarins and found them to be weakly pathogenic on pea. The causal relationship between naphthazarin production and virulence was tested further by a genetic cross between one of these mutants and a naphthazarin-producing, highly pathogenic strain of *N. haematococca*, M.P. VI. Among ascospore progeny in the 50 tetrads analyzed, there was no significant correlation between naphthazarin (fusarubin, novarubin) production and pathogenicity. Furthermore, some recombinant progeny showed high virulence and no naphthazarin production [23]. Similar results were obtained in an analysis of 435 progeny from crosses between five virulent, naphthazarin-producing field strains and one avirulent nonproducing strain

TABLE 2. Virulence of Wild-Type and Mutant Strains of *Fusarium sporotrichioides* on Parsnip Root Disks

Exp. no.	Percentage of tissue rotted (mean + SD) NRRL 3299	MB1716	MB2972	MB5493	MB5493 and MB2972
1	53 ± 29 (4)	49 ± 31 (4)	1 ± 1 (3)	3 ± 3 (3)	nd
2	50 ± 27 (10)	62 ± 38 (10)	0 (8)	0 (7)	nd
3	87 ± 19 (10)	83 ± 17 (9)	4 ± 5 (8)	0 (8)	nd
4	53 ± 15 (5)	44 ± 15 (6)	2 ± 2 (8)	1 ± 3 (7)	nd
5	62 ± 17 (5)	85 ± 21 (5)	0 (10)	0 (10)	20 ± 12 (7)
6	57 ± 13 (5)	nd	1 ± 1 (5)	0 (5)	25 ± 11 (20)
7	41 ± 15 (5)	nd	1 ± 1 (5)	0 (5)	24 ± 15 (20)

Virulence was measured as grams rotted of parsnip root disks. Numbers in parentheses indicate number of disks measured for each treatment. nd, not done.
Source: Data from Ref. 18.

of *N. haematococca* M.P. VI [24]. These genetic analyses do not support a major role for naphthazarins in virulence of *Fusarium* on pea.

Fusaric Acid

Several *Fusarium* species, including *F. moniliforme* and form species of *F. oxysporum*, produce fusaric acid (Fig. 1) and related phytotoxic pyridine derivatives in vitro and in infected plant tissues [19, 20]. Fusaric acid can cause wilting, necrosis, and other symptoms in treated plants but, in general, only at relatively high (10^{-3}–10^{-4} M) concentrations [25]. Surveys of natural variants have not revealed clear correlations between fusaric acid production and pathogenicity of *F. oxysporum* [19]. Mutants blocked or altered in fusaric acid production were generated in a high-toxin-producing strain of *F. oxysporum* f. sp. *lycopersici* by UV treatment. Thirty-one of these mutants were used to test the causal relationship between fusaric acid production and virulence on tomato (*Lycopersicon esculentum*) seedlings. These results showed no correlation between pathogenicity and in vitro fusaric acid production of mutant strains; some low toxin producers were highly pathogenic [25]. Venkata Ram [26] also found that in vitro fusaric acid production of UV-generated mutants of *F. vasinfectum* did not correlate with their virulence on cotton (*Gossypium arboreum*) plants. These genetic analyses therefore do not support a major role for fusaric acid in virulence of *Fusarium* on tomato or cotton.

B. Plant Toxins

Isoflavonoid Phytoalexins

Isoflavonoid phytoalexins (Fig. 2a) are produced by many leguminous plants in response to fungal infection [27] and have demonstrated in vitro fungitoxicity to a number of *Fusarium* species, including *F. oxysporum* f. spp. *phaseoli* and *pisi*, *F. solani* f. sp. *phaseoli*, *N. haematococca* M.P. I. and M.P. VI, and *F. roseum* [7, 28]. Extensive analyses of natural variants of *N. haematococca* M.P. VI from a wide range of habitats have demonstrated strong correlations between the ability to detoxify pisatin and virulence on pea (*Pisum sativum*) [29] and between the ability to detoxify both medicarpin and maackiain and virulence on chickpea (*Acer arietinum*) [30]. Similarly, among natural variants of *F. solani* (sp. *phaseoli*), there is a strong correlation between kievitone detoxification and virulence on bean (*Phaseolus vulgaris*) [31].

Smith and co-workers [32] have also selected and analyzed laboratory mutants of *F. solani* f. sp. *phaseoli* with reduced ability to detoxify kievitone and corresponding reduced virulence on bean. Interpretation of these results is not unambiguous, however, because the mutants differed from wild-type in a number of traits in addition to kievitone detoxification [33]. Genetic crosses with these mutants have not been possible because the sexual stage of *F. solani* f. sp. *phaseoli* is not known.

The availability of a heterothallic genetic system in *N. haematococca* M.P. VI has allowed genetic tests of whether there is a causal relationship between detoxification of isoflavonoid phytoalexins and pathogenicity. VanEtten and co-workers [34–36] have made dozens of genetic crosses between field strains and their progeny and analyzed thousands of progeny for pisatin detoxification and virulence. No progeny has yet been found that is unable to detoxify pisatin but is still highly virulent on pea, which provides the strongest classical genetic evidence to date that phytoalexin detoxification is required for virulence and, indirectly, that isoflavonoid phytoalexins can be a disease-resistance factor. This extensive body of work has recently been reviewed [37, 38].

Pisatin is detoxified via demethylation catalyzed by a cytochrome P-450 monooxygenase [39]. The gene for this enzyme has been isolated from *N. haematococca* by expression in *Aspergillus nidulans* [40] and has recently been used in transformation of two pisatin-sensitive fungi that are not pathogenic on pea, a strain of *N. haematococca* and a strain of the maize pathogen *Cochliobolus heterostrophus*. In both cases, selected transformants were able to detoxify pisatin and were pathogenic on pea [41, 42].

Sesquiterpene Phytoalexins

Sesquiterpene phytoalexins (Fig. 2b) are produced by several plant families, but most research has focused on their identification and characterization in potato (*Solanum tuberosum*) and related food crops [43]. In vitro fungitoxicity of selected sesquiterpenes has been demonstrated against a limited number of *Fusarium* species, including *F. oxysporum* f. sp. *vasinfectum* and *F. sulphureum* [44, 45]. Analyses of natural variants of *Gibberella pulicaris* from a wide range of habitats have demonstrated strong correlations between the ability to detoxify both rishitin and lubimin and virulence on potato tubers [46, 47].

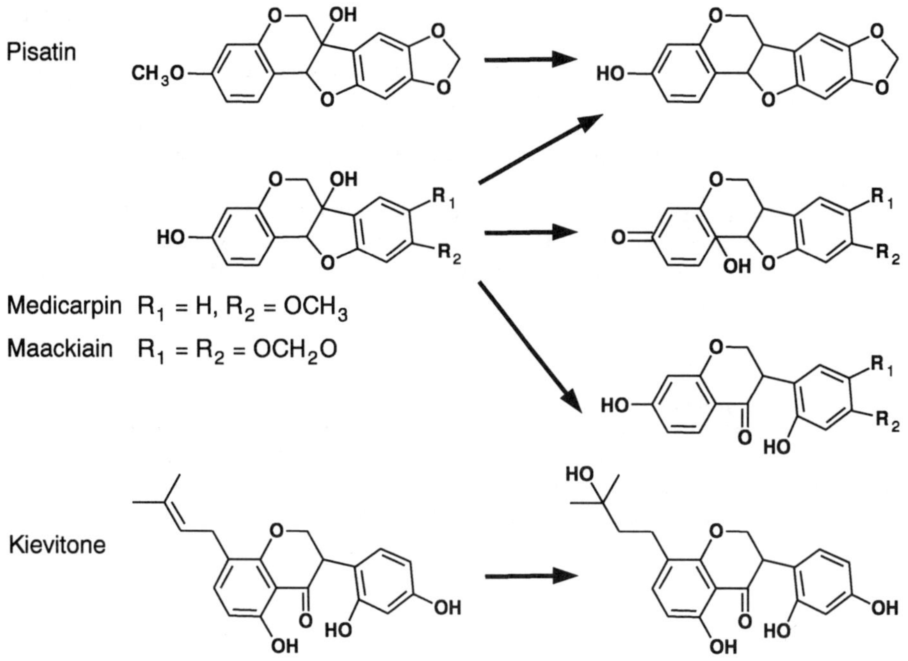

(a)

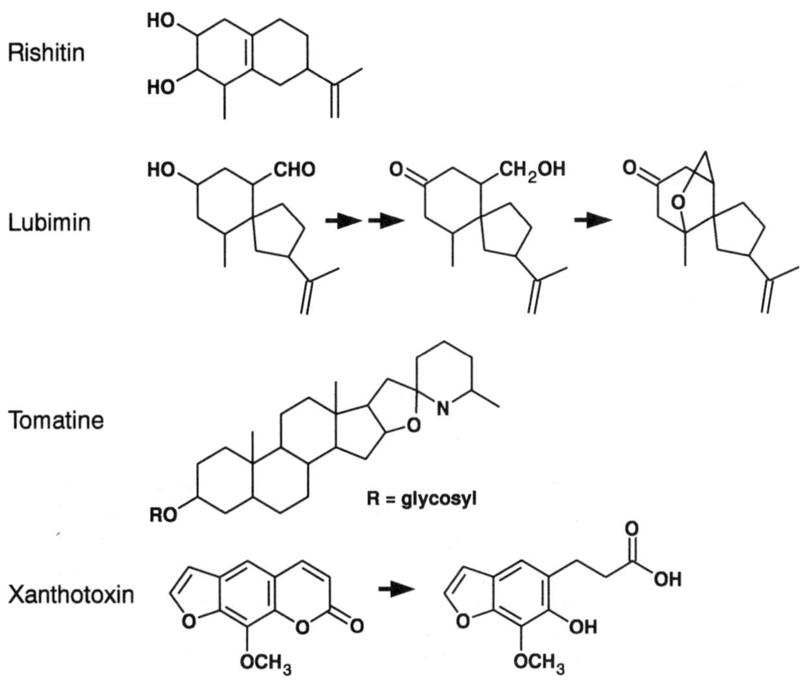

(b)

FIGURE 2. (a) Plant isoflavonoid toxins and their fungal metabolites; (b) more plant toxins and their fungal metabolites.

The recent availability of a heterothallic genetic system in *G. pulicaris* [48] has allowed genetic tests of whether there is a causal relationship between detoxification of sesquiterpene phytoalexins and pathogenicity. Tetrad analysis of more than 200 progeny from six crosses between field strains and progeny have shown that rishitin detoxification is necessary but not sufficient for virulence on potato tubers (Fig. 3) ([47]; Desjardins and Gardner, unpublished). A similar genetic analysis of lubimin metabolism has not yet been possible because genetically fertile lubimin-sensitive strains have not been identified in this species.

Steroid Glycoalkaloids

Green tomatoes, potatoes, and other solanaceous plants contain constitutive steroid glycoalkaloids (Fig. 2b) that are toxic to a variety of fungi in vitro. Défago and co-workers [49, 50] selected and analyzed laboratory mutants of *N. haematococca* with increased tolerance of tomatine and increased virulence on green tomatoes. The causal relationship between

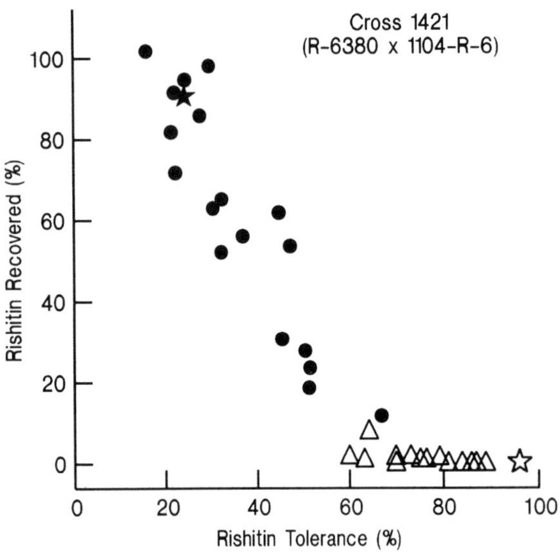

FIGURE 3. The relationship between tolerance of rishitin, metabolism of rishitin, and virulence on potato tubers in tetrad progeny from cross 1421 of *Gibberella pulicaris*. Virulence was measured as grams rotted of three potato tuber slices. Tolerance was measured as the effect of 1 mM rishitin on radial growth of cultures on an agar medium, which was then extracted and assayed by gas chromatography for rishitin recovery. Filled symbols indicate avirulent strains, and open symbols indicate virulent strains. Circles and triangles indicate progeny, and stars indicate the two parental strains. (Figure from Ref. 47.)

tomatine tolerance and virulence was tested further by genetic crosses between two allelic or closely linked mutant strains and wild-type strains. Among 60 random ascospore progeny tested, pathogenicity always co-segregated with tomatine tolerance. It should be noted that increased tolerance of tomatine in these mutants is apparently not due to increased metabolism of tomatine but rather correlates with a reduction in sterol content of the fungal mycelium.

III. REGULATION AND INTEGRATION OF FUNGAL AND PLANT CHEMICAL RESPONSES

A. Trichothecenes

Because many *Fusarium* secondary metabolites are phytotoxic, it has been tempting to associate this activity with a role in plant pathogenesis. Correlation studies have identified several classes of *Fusarium* host-nonspecific toxins, including trichothecenes, naphthazarins, and fusaric acid, as potential pathogenicity factors. Mutational and/or recombinational genetic analyses have supported a role for trichothecenes in pathogenesis but have not supported a role for naphthazarins or fusaric acid, at least in the systems studied to date (see also Ref. 51). Further studies of *F. sporotrichioides*-infected parsnip root have provided additional evidence for a role for trichothecenes in this disease [18]. T-2 toxin was found in infected root tissues, although not in all experiments and not at the infection boundary as determined by gas chromatography/mass spectrometry analysis of extracts of parsnip root cross sections. Although toxicity of T-2 and other trichothecenes to parsnip root cells is not known, levels of T-2 toxin in infected parsnip root strips were within the range (10^{-5}–10^{-6} M) known to be toxic to other plants [12, 13]. Several trichothecenes, including 4,15-diacetoxyscirpenol, 15-monoacetoxyscirpenol, neosolaniol, and T-2 toxin, were also detected at levels greater than 10^{-5} M in potato tubers experimentally infected with *G. pulicaris* [52], and even higher levels have been found in naturally infected potato tubers [53]. Attempts of others to detect trichothecenes in infected plant tissues have not always been successful; for example, Bean and co-workers [54] found no trichothecenes in muskmelon seedlings infected with toxin-producing strains of *Myrothecium roridum*. As discussed by Yoder [8], Mitchell [55], and others, the inability to detect toxins in infected plant tissues does not necessarily indicate that toxins are not involved in pathogenesis. Toxins may be inactivated by plant enzymes or otherwise unstable, or they may not be detectable owing to interference by plant constituents. Some evidence to support these hypotheses has been obtained in *Fusarium* infections of grains. Levels of the trichothecene deoxynivalenol increased and then declined in wheat heads naturally and experimentally infected with *F. graminearum* [56, 57], and suspension cultures of wheat transformed added deoxynivalenol to a variety of products among which 3-acetoxydeoxynivalenol has been identified [58].

Due largely to the involvement of trichothecenes in human and animal mycotoxicoses, their mechanism of action has been extensively investigated.

Trichothecenes inhibit eukaryotic protein synthesis by binding to the 60S ribosomal subunit [59]. The trichothecenes are known to vary in toxicity and in effectiveness as protein-synthesis inhibitors. The most toxic trichothecenes are those, like T-2 toxin and 4,15-diacetoxyscirpenol, that inhibit the initiation of protein synthesis [59]. Studies by Wei and McLaughlin [60] showed that less-oxygenated trichothecenes such as 15-deacetylcalonectrin are less toxic and inhibit the elongation and termination of protein synthesis rather than its initiation. Among the *F. sporotrichioides* mutants assayed for pathogenicity [18], the mutant that accumulated 4,15-diacetoxyscirpenol was as pathogenic on parsnip root as the T-2 toxin-producing parent, while the mutant that produced only the calonectrin analogs 15-deacetylcalonectrin and 3,15-dideacetylcalonectrin was only slightly pathogenic. One possible interpretation of these results is that not only are trichothecenes important for pathogenicity, but also particular, highly oxygenated trichothecenes may be required.

The trichothecene biosynthetic pathway is discussed in detail elsewhere in this volume (Chapter 14), and only certain aspects will be briefly presented here. The trichothecenes are complex, highly oxygenated sesquiterpenes biosynthesized from farnesyl pyrophosphate via the hydrocarbon trichodiene (Fig. 1). Incorporation studies with $^{18}O_2$ and $H_2{}^{18}O$ in whole-cell cultures of *F. sporotrichioides* have shown that the oxygenations of trichodiene to T-2 toxin are catalyzed by molecular oxygen-dependent monooxygenases or dioxygenases [61]. Inhibition of these oxygenations by a variety of cytochrome P-450 inhibitors provides supporting evidence for a role for microsomal cytochrome P-450 monooxygenases in trichothecene biosynthesis [62]. The production of *F. sporotrichioides* mutants that accumulate trichothecenes with different degrees of oxygenation [14–16] supports the hypothesis that oxygenations of at least some of the carbons of the trichothecene nucleus are catalyzed by separate enzymes. Genetic analysis in *G. pulicaris* have identified a locus that controls hydroxylation of 4,15-diacetoxyscirpenol to neosolaniol [63], but further study is needed to determine whether this gene codes for the oxygenase itself. Attempts to obtain trichodiene or trichothecene oxygenation by *Fusarium* microsome preparations have so far been unsuccessful. *Fusarium* and other filamentous fungi have in general proven to be difficult experimental organisms for isolation of membrane-bound oxygenase enzymes.

Production of trichothecenes by *Fusarium* in pure culture is highly dependent on nutrient composition, temperature, aeration, and other conditions of fermentation as well as on the particular species and strains being grown [64]. There is little known, however, about how these environmental and genetic factors specifically regulate trichothecene biosynthesis. In this regard, the trichothecenes are similar to many other complex secondary metabolites produced by filamentous fungi under various conditions of nutrient limitation. Even less is known about the regulation of trichothecene production in *Fusarium*-infected plants. Miller and Greenhalgh [11] recently summarized evidence that fungal growth in plant tissues is oscillatory in nature and may be subject to nutrient limitations analogous to those that induce secondary metabolite formation during fermentation. It may therefore be informative to extrapolate from simpler liquid culture conditions to the more complex environment of the plant–

fungus interaction. For example, as mentioned above, some plant tissue cultures are able to degrade or otherwise modify trichothecenes [58] although it is not known which, if any, of these biotransformations are detoxifications. There is no evidence to date for a role of any specific plant constituent in inducing or increasing trichothecene biosynthesis. Biosynthesis can, however, be blocked in liquid cultures of *F. sporotrichioides* and *G. pulicaris* by the addition of certain flavonoids, furanocoumarins, and other naturally occurring plant metabolites at concentrations noninhibitory to fungal growth ([65]; Desjardins, Spencer, and Plattner, unpublished). Most of these inhibitors appear to block the pathway after formation of the nontoxic precursor trichodiene and before formation of highly oxygenated trichothecenes, suggesting an effect on enzyme(s) catalyzing trichodiene oxygenation. There is also some indirect evidence that furanocoumarins may partially block trichothecene biosynthesis in *Fusarium*-infected parsnip roots [18]. In infected, viable root tissues, furanocoumarins accumulated to high levels, whereas T-2 toxin was present at very low levels and trichodiene accumulated. In contrast, in microwave-cooked parsnip root infected with *F. sporotrichioides*, very low levels of furanocoumarins and trichodiene were detected and the level of T-2 toxin was comparable to that observed in liquid cultures.

B. Phytoalexin Detoxification Enzymes

Fungal genetic analyses have provided convincing evidence that phytoalexin detoxification can be important for pathogenicity of at least some *Fusarium* species. It has also been consistently observed that phytoalexin-tolerant fungal strains, whether natural variants or ascospore progeny from genetic crosses, can be of low pathogenicity. These data suggest that phytoalexin detoxification alone is not sufficient for high pathogenicity. This is not surprising because pathogenicity is undoubtedly a complex process depending on many genes, and plant resistance to fungal pathogens can involve many defense responses. It is also possible that fungal phytoalexin detoxification systems that show in vitro bioassay activity may not be active, or active enough, in infected plants. In these cases, the outcome of the plant-pathogen interaction could depend not only upon the presence of detoxification genes but also upon their regulation. It is therefore important to identify the products of phytoalexin detoxification and the enzyme reactions involved, as well as the mechanisms achieved in the initiation and the quantitative and qualitative control of these enzymes and their genes. The following discussion will focus on regulation of phytoalexin detoxification in the genetic systems already presented.

Fusarium species detoxify phytoalexins by a wide variety of enzymatic mechanisms (Figs. 2a and b) [17]. *Nectria haematococca* M.P. VI, as previously mentioned, detoxifies the isoflavonoid pisatin by a demethylation that is catalyzed by a cytochrome P-450 monooxygenase [39]. *Nectria haematococca* M.P. VI also detoxifies medicarpin and maackiain by one of three alternative oxygenations of the isoflavonoid ring system [30], two of which are known to be catalyzed by monooxygenases [66]. The isoflavonoid kievitone, in contrast, is detoxified by *F. solani* f. sp. *phaseoli* via hydration of the isopentenyl side chain by an extracellular enzyme

[67]. Preliminary mass spectral data suggest that *G. pulicaris* metabolizes the sesquiterpene rishitin by oxygenation, but the detailed structure is not yet known (Gardner and Desjardins, unpublished). The related compound lubimin, however, can be detoxified by 3,4-dehydrogenation of the intermediate isolubimin followed by cyclization to the nontoxic cyclodehydroisolubimin, which can be further reduced and/or oxygenated [68]. Somewhat unexpectedly, *G. pulicaris* and *F. sporotrichioides* do not demethylate xanthotoxin, a methoxylated furanocoumarin phytoalexin, to the nontoxic xanthotoxol but instead hydrogenate the lactone ring, which leads to decyclization to a nontoxic carboxylic acid [69].

Most phytoalexin detoxification studies have been done by feeding pure compounds to fungal cultures, but in some cases the fungal metabolites have also been found in infected plant tissues. For example, kievitone hydrate has been detected in *Fusarium*-infected beans [28], and cyclodehydroisolubimin in *Fusarium*-infected potatoes [46]. These observations provide more direct evidence that phytoalexin detoxification occurs in diseased tissue.

Some phytoalexin detoxification enzymes of *Fusarium* have been shown to be substrate-inducible rather than constitutive. Pisatin demethylase of *N. haematococca* M.P. VI has been most extensively studied in this regard. Demethylase activity of a highly virulent strain was found to be most strongly induced by pisatin and, to a lesser extent, by certain structurally similar compounds [38]. Kievitone hydratase as well, although present at low levels constitutively, can be induced severalfold by treatment with the substrate and related flavonoids [38]. Substrate inducibility of enzymes for medicarpin, maackiain, rishitin, and lubimin metabolism has not yet been studied in detail. However, there is a puzzling lack of metabolism of exogenously supplied 15-dihydrolubimin by a *G. pulicaris* strain that can make this compound from lubimin and can rapidly metabolize it further. One possible explanation of this observation is that the enzymes responsible for lubimin detoxification are not constitutive and are not induced by the metabolite 15-dihydrolubimin [46].

The tolerance of *G. pulicaris* to the potato phytoalexins rishitin and lubimin, and of *N. haematococca* M.P. VI to the chickpea phytoalexins medicarpin and maackiain, illustrates several aspects of specificity in phytoalexin detoxification. Rishitin-tolerant natural variants of *G. pulicaris* can be sensitive to lubimin and vice versa, which suggests, not unexpectedly, that different enzymes are responsible for detoxification of these two biogenetically related but structurally somewhat dissimilar compounds [47]. In contrast, in 53 natural variants of *N. haematococca* M.P. VI, tolerance of medicarpin was strongly associated with tolerance of maackiain. As might be predicted from their structural similarity, maackiain and medicarpin were metabolized to analogous products oxygenated at carbons 6a, 1a, or 11a [30]. Tetrad analyses of crosses among field strains have identified three loci for maackiain metabolism. One locus is specific for 1a-hydroxylation of maackiain but closely linked or identical to a locus for pisatin demethylation, while a second locus is specific for the 1a position, and a third locus is specific for the 6a position of maackiain [38, 70, 71]. These metabolism studies have all been done by feeding substrate to whole-cell cultures. Studies of the specificity of phytoalexin detoxification in cell-free preparations have been very limited. VanEtten et al. [38] examined the ability of microsomes from one strain of *N. haematococca* M.P. VI to demethylate a series of isoflavonoids and related compounds. Pisatin was the preferred substrate for demethylation, which suggests that cytochrome P-450 monooxygenases that catalyze pisatin demethylation have relatively narrow substrate specificities.

Genetic analyses of phytoalexin detoxification in *Fusarium* reported to date support the hypotheses that field strains have multiple detoxification genes and that a high level of detoxification is essential for a high level of virulence. By extensive genetic crosses in *N. haematococca* M.P. VI, VanEtten and co-workers identified and isolated a family of genes for pisatin detoxification. They crossed many different pisatin-detoxifying natural variants and analyzed ascospore progeny for their ability to detoxify pisatin. Detection of pisatin-sensitive, nondetoxifying progeny was interpreted to mean that the parental demethylase genes were at different loci. In this system, six different loci for pisatin demethylation were identified: no field strain was found with more than two loci; and only two field strains were found with genes for pisatin demethylation at the same locus [38]. These conclusions, however, may need to be revised owing to the recent observation that pisatin demethylase genes can be located on small, meiotically unstable chromosomes [72, 73].

Preliminary genetic analyses of rishitin detoxification in *G. pulicaris* identified two loci for rishitin detoxification in one selected field strain. Rishitin nonmetabolizers were not recovered among the progeny from eight crosses among six rishitin-metabolizing field strains. Although the number of progeny tested was relatively small (50 in one cross and 20 in each of seven crosses), these results could be interpreted as indicating multiple loci for rishitin metabolism in field strains such that all recombinant progeny inherit at least one gene and consequently a rishitin-metabolizing phenotype. Alternatively, all field strains may have rishitin-metabolism genes at the same locus or loci or at closely linked loci, and for that reason rishitin-nonmetabolizing progeny were not recovered. To test these hypotheses, crosses between rishitin-detoxifying field strains and nondetoxifying progeny strains are in progress ([47]; Desjardins and Gardner, unpublished).

Genetic and biochemical analyses in *N. haematococca* M.P. VI identified several phenotypes of pisatin demethylase that were characterized in detail by their lag time and rate of demethylation in liquid cultures. Only those phenotypes associated with shorter lag times and higher rates of demethylation appear to be associated with high levels of virulence [38]. And, again, preliminary tetrad analyses of rishitin detoxification in one field strain of *G. pulicaris* indicate the presence of one gene for a high detoxification phenotype that confers virulence and one nonallelic gene for a lower detoxification phenotype that does not ([47]; Desjardins and Gardner, unpublished).

With the recent developments in the molecular biology of filamentous fungi, regulation of phytoalexin detoxification at the gene level can now be investigated. As mentioned in Section II.B.1, a gene for pisatin demethylation from a highly virulent strain of *N. haematococca* M.P. VI has been cloned and partially sequenced. Preliminary results indicate that induction of demethylation genes by pisatin occurs in this strain and in others at the level of transcription. The cloned gene has also been used for hybridization experiments to determine that there is a high degree of DNA homology among pisatin demethylase genes that confer the short lag time, higher-rate phenotypes that confer virulence, and a low degree of homology with genes that confer low activity and low virulence [38].

C. Chemical Ecology and Evolution of *Fusarium*-Plant Systems

Evolution via natural selection argues that production of toxic chemicals provides some selective advantage for the producing organisms, whether fungi or plants. Genetic change in plant or fungal chemistry in response to natural selection is a difficult hypothesis to test by controlled experimentation, and the limited experimental research done to date has been predominantly in the related field of insect–plant interactions. Berenbaum and co-workers, for example, compared field-grown, webworm-infested populations of wild parsnip (*Pastinaca sativa*) to those grown insect-free in a greenhouse. They found significant genetic variation in furanocoumarin levels in these wild populations and significant selection for an increase in levels of certain toxic furanocoumarins in the insect-damaged field-grown plants [74]. For cultivated plants, however, there is little known about either the extent of genetic variation in phytoalexin levels or the relationship of phytoalexin accumulation to resistance to fungal diseases. For example, eight potato cultivars varied more than 20-fold in levels of sesquiterpene phytoalexins induced in tubers by fungal infection but were very similar in disease resistance [75]. Interpretation of these results, however, is not straightforward because of the different genetic backgrounds of the potato cultivars tested. Pea mutants that produce pisatin constitutively have been obtained but have apparently not been tested for disease resistance [76].

As Gould has argued [77], hypotheses about mechanisms of natural selection can be critically tested by historical as well as experimental approaches. As related to *Fusarium*, this would mean the reconstruction of past events based on the detailed observation and description of present populations. Some *Fusarium* species are particularly amenable to historical analysis because they are broadly adapted both as saprophytes and as plant pathogens, and because they provide large populations from diverse habitats and widely separated geographic locations (see also Ref. 78). In the collections of *N. haematococca* M.P. VI and *G. pulicaris*, whose chemical ecology has received the most study so far, there is a wide range of selectable, natural variation for a number of traits that may be relevant to plant pathogenesis. Furthermore, many of these ecologically and geographically diverse natural variants have maintained their sexual interfertility, which allows confirmation of their identity as members of a reproductively unique biological species. Analyses of strains of *G. pulicaris* isolated either from soils or from diseased plants from six continents have revealed clear differences between strains from these two habitats in a number of traits relevant to plant pathogenesis. For example, 19 of 22 (86%) plant-derived strains, but only 2 of 37 (5%) soil-derived strains were highly tolerant of the phytoalexin xanthotoxin (Fig. 4) [79]. Twenty-

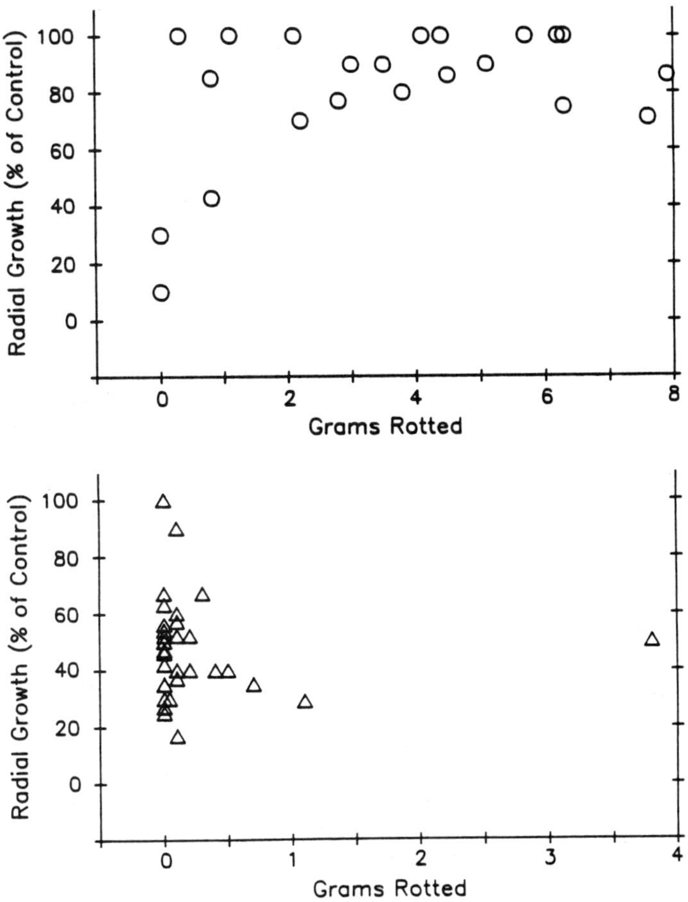

FIGURE 4. The relationship between strain habitat, tolerance of xanthotoxin, and virulence on parsnip root of 61 field strains of *Gibberella pulicaris*. Virulence was measured as grams rotted of three parsnip root disks. Tolerance was measured as the effect of 1 mM xanthotoxin on radial growth of cultures on an agar medium. Circles represent plant-derived strains. Triangles represent soil-derived strains. (Figure from Ref. 3.)

three of these strains were also tested for tolerance of sesquiterpene phytoalexins, and 10 of 16 (62%) plant-derived strains but only one of seven (14%) soil-derived strains were highly tolerant of rishitin and lubimin [47]. Tolerance of *G. pulicaris* to all three phytoalexins has been shown to correlate with their detoxification (see Sections II.B.1 and II.B.2). These relationships between phytoalexin detoxification and plant pathogenicity support the hypothesis that selection for the ability to detoxify plant chemicals has directed evolution of *G. pulicaris* from a soil saprophyte to a plant pathogen. Strains of *G. pulicaris* isolated from plants that do not produce xanthotoxin, lubimin, or rishitin can be tolerant of these phytoalexins and virulent on the plants that make them. VanEtten and co-workers [37] also found that many strains of *N. haematococca* M.P. VI from host plants other than pea—that is, habitats that apparently do not contain pisatin—were tolerant of pisatin and that some of these strains were also virulent on pea. They speculated that these strains might have had prior contact with pea or that *N. haematococca* M.P. VI might be broadly adapted to a variety of plant hosts. The virulence on parsnip of *G. pulicaris* strains from potato, corn, and other plants might be similarly explained. It should be noted that xanthotoxin and other furanocoumarins, unlike pisatin, rishitin, or lubimin, are quite widespread in higher plants and have been reported to occur in the species, genera, or families of plants from which many of the furanocoumarin-tolerant strains in this study were isolated [79].

Preliminary analysis of 82 strains of *G. pulicaris* for their ability to produce trichothecene toxins in liquid culture also revealed significant differences between plant-derived and soil-derived populations. Twenty of 37 (54%) plant-derived but only 6 of 41 (15%) soil-derived strains produced trichothecenes in greater than trace (>1 μg/mL) amounts after 7 days incubation in liquid shake cultures. All but one of the highest (>10 μg/mL) trichothecene producers were from plants (Desjardins and Beremand, unpublished). These correlations support the hypothesis that selection for the ability to produce high levels of trichothecenes has directed the evolution of *G. pulicaris* toward plant pathogenicity. It should be noted, however, that strains isolated from soil and from plants are not always necessarily saprophytes and pathogens, respectively. In *G. pulicaris*, for example, we identified one soil-derived strain that is highly virulent on field corn (Desjardins, Caldwell, and Bennett, unpublished). We have also isolated from potato tubers strains that are not highly virulent on potato tubers, at least under our experimental conditions [52]. It is probable that all pathogenic strains can survive as soil saprophytes and that some saprophytic strains can survive in diseased plant tissues due to the concurrent or prior presence of other, pathogenic strains. In *G. pulicaris*, however, there seem to be relatively rare exceptions to the general pattern that strains from plants are pathogenic and strains from soil are not.

Most species of *Fusarium* are very common in soil, where they can persist for years on soil debris as hyphae or as various resistant spore forms. It is probable that ancestral *Fusarium* species were soil saprophytes that escaped from the intense microbial competition of the soil environment by evolving mechanisms for plant pathogenesis. An unresolved question in the history of *Fusarium* and its chemical ecology is the role of coevolution, that is, whether *Fusarium* and its host plants have undergone stepwise genetic changes of adaptation and counteradaptation over time. Figure 5 presents a simple diagram for such a hypothetical pairwise fungal—plant system based on a model for herbivore—plant coevolution published by Gould [80]. This diagram is a speculative composite of results from our experiments on *G. pulicaris* and *F. sporotrichioides* infection of parsnip

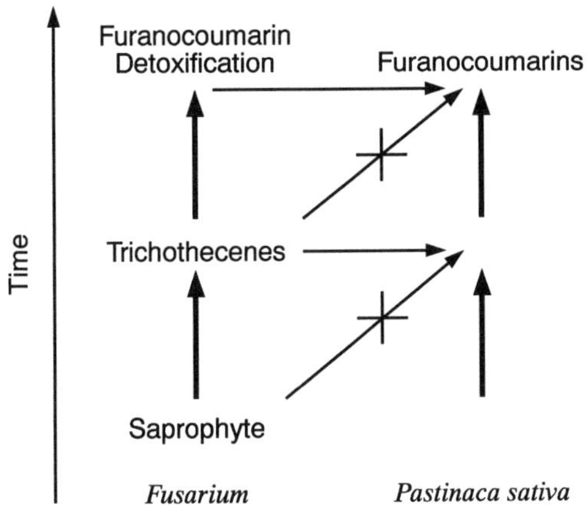

FIGURE 5. Model of *Fusarium–P. sativa* coevolution. Vertical arrows indicate evolution through time. Diagonal arrows with crosses indicate avirulence. Horizontal arrows indicate virulence. (Diagram adapted from Ref. 80, Figure 2.)

root [4, 18, 65]. In Fig. 5, a soil saprophytic *Fusarium* becomes pathogenic by evolving the ability to produce trichothecenes (which may inhibit synthesis of proteins required for disease resistance). The plant is eventually able to prevent pathogenicity by evolving the ability to synthesize furanocoumarins that block trichothecene biosynthesis. The fungus then counteradapts by detoxifying the plant furanocoumarins, thereby recovering pathogenicity. As coevolution proceeds, additional adaptations may occur in the fungus (e.g., new toxins or suppressors of furanocoumarin biosynthesis) or in the plant (e.g., detoxification of trichothecenes or trichothecene-resistant ribosomal proteins). The complexity of the final system would depend in large part on the physiological cost to plant and pathogen of maintaining each adaptation and on its usefulness in the general chemical ecology of the organism. For example, although *Fusarium* can detoxify phytoalexins, many other plant pathogenic fungi cannot; *Phytophthora infestans* is unable to detoxify sesquiterpene phytoalexins [43]. Although trichothecene biosynthesis was blocked by a wide variety of furanocoumarin phytoalexins produced by umbelliferous and rutaceous plants, it was not blocked by isoflavonoid phytoalexins typically produced by leguminous plants [65]. The fact that a large proportion of *G. pulicaris* strains isolated from soil have demonstrated limited ability to synthesize trichothecenes and to detoxify furanocoumarin and sesquiterpene phytoalexins argues that there may be some selection against these traits in certain environments as has been observed in other systems [80]. Since fusaria are not obligate pathogens, they can successfully respond to a plant defense by adapting, by finding a new host that has not evolved the defense, or by surviving as saprophytes. On the basis of multiplicity of genes for pisatin demethylation and their DNA homology, VanEtten and co-workers [73] speculated that pisatin detoxification genes evolved from a cytochrome P-450 present in an ancestral saprophyte. The multiple copies of less active, less specific forms of this gene still present in field strains of *N. haematococca* may therefore be tracks of fungal history.

IV. PROSPECTS FOR OTHER SELECTED *FUSARIUM*-PLANT SYSTEMS

Nectria haematococca and *G. pulicaris* have demonstrated their usefulness for genetic analysis of putative pathogenicity and resistance factors in fungal interactions with legumes and potatoes. Extensive collections of both of these *Fusarium* species include strains that are pathogenic to many different species of plants [1, 3]. It should therefore be possible to use these strain collections to study pathogenicity in a variety of plant systems. For example, 62 strains of *G. pulicaris* were tested for pathogenicity on parsnip (*Pastinaca sativa*) and for in vitro tolerance of the furanocoumarin xanthotoxin, a major parsnip phytoalexin. Twenty-three of the strains were highly tolerant of xanthotoxin, and there was a strong correlation between xanthotoxin tolerance and virulence on parsnip. Furthermore, tolerance of xanthotoxin and of other naturally occurring furanocoumarins such as angelicin, pimpinellin, and isopimpinellin correlated in large part with their detoxification [3]. As mentioned in Section III.B, *G. pulicaris* detoxifies xanthotoxin by hydrogenation of the lactone ring, leading to ring opening [69]. Genetic analysis in this system has been limited to date by low fertility of xanthotoxin-sensitive field strains. A search for xanthotoxin-sensitive recombinants among progeny from crosses between xanthotoxin-tolerant field strains is in progress.

Gibberella fujikuroi (*Fusarium moniliforme*) can infect a wide range of plant hosts and is a serious pathogen of maize (*Zea mays*) worldwide. *Gibberella fujikuroi* has been implicated in a variety of human and animal mycotoxicoses, and its causative role in equine leukoencephalomalacia has been conclusively established [4]. Mycotoxins and/or phytotoxins produced by *G. fujikuroi* include fusarins, a family of polyketides; moniliformin, a cyclobutenedione, and fumonisins, a family of esterified aminoheptadodecanols [4, 81]. The structural similarity between fumonisins and host-specific phytotoxins produced by *Alternaria alternata* f. sp. *lycopersici* [82] suggests a possible role for fumonisins in plant pathogenicity of *G. fujikuroi*. Genetically fertile heterothallic strains that produce fumonisins in vitro have recently been identified in this species (Plattner, Leslie, and Desjardins, unpublished), which indicates that this may be a useful system for genetic analysis of the role of fumonisins in plant pathogenicity. Many strains of *G. fujikuroi* produce high levels of gibberellic acids in vitro. Although gibberellins are plant growth regulators rather than phytotoxins per se, their role in pathogenesis of *G. fujikuroi* has been the subject of considerable speculation. In the bakanae disease of rice (*Oryza satica*) in particular, abnormal elongation of diseased plant stems is a common symptom, but the role of gibberellins produced by the pathogen has not been clearly established [83]. Although a number of mutants blocked or altered in gibberellin production have been generated and characterized in *G. fujikuroi*, their pathogenicity has apparently never been tested [84].

Very little is known about the role of plant toxins and their detoxification in *G. fujikuroi* and other *Fusarium* diseases of Gramineaceae. Constitutive and/or induced antifungal compounds produced by graminaceous plants include avenacins, triterpenoids from oats (*Avena sativa*); hordatines, coumaroyl agmatines from barley (*Hordeum vulgare*); cyclic hydroxamates from maize and wheat (*Triticum aestivum*); and diterpene momilactones from rice [85]. Recent advances in the development of *G. fujikuroi* as a genetic system, by Leslie and co-workers in particular (see Ref. 86, for example), and the concurrent increased awareness of the prevalence of *G. fujikuroi* mycotoxicoses foreshadow a renewal of interest in mechanisms of pathogenicity of this major plant pathogen.

Gibberella zeae (*Fusarium graminearum*) is a wide host range species of major economic importance as a pathogen of wheat, maize, and other grains. Many strains of *G. zeae* produce trichothecene toxins as well as zearalenones, a family of phytotoxic lactones that can cause an estrogenic syndrome in animals that eat contaminated feed [4]. A role for zearalenone in plant pathogenesis and in sexual reproduction of *G. zeae* has been proposed but has not yet been critically tested [87]. Any genetic analyses in *G. zeae* would be complicated by the fact that all known genetically fertile strains of this species are apparently self-fertile.

V. CONCLUSION

Genetic analysis has been invaluable in confirming the chemical basis of pathogenicity, host selectivity, and disease resistance in *Fusarium*—plant interactions. Our knowledge, although far from complete, is sufficient to suggest potentially useful experimental approaches to chemically mediated prevention of *Fusarium* infection and to the genetic engineering of *Fusarium*-resistant plants. For example, naturally occurring or synthetic inhibitors of fungal toxin biosynthesis or phytoalexin detoxification enzymes might prevent pathogenesis. Another obvious approach is to use trichothecenes or other toxins for in vitro selection of toxin-resistant and, hopefully, *Fusarium*-resistant plant material. A further strategy, recently described in detail [38], is to genetically alter phytoalexin biosynthetic pathways to yield novel phytoalexins that might be less susceptible to fungal detoxification. In addition, it is clear that *Fusarium* species also provide useful experimental systems for the study of mechanisms of natural selection in the coevolution of plants and their fungal pathogens.

ACKNOWLEDGMENTS

I thank Paul Nelson for identifying and supplying most of the strains used for the *G. pulicaris* studies discussed in this review, Hans VanEtten for supplying articles prior to their publication, and both of them for their advice and encouragement.

REFERENCES

1. VanEtten, H. D. Identification of additional habitats of *Nectria haematococca* mating population VI. *Phytopathology, 68*:1552–1556, 1978.
2. Kuhlman, E. G. Varieties of *Gibberella fujikuroi* with anamorphs in *Fusarium* section Liseola. *Mycologia, 74*:759–768, 1982.
3. Desjardins, A. R., Spencer, G. F., and Plattner, R. D. Tolerance and metabolism of furanocoumarins by the phytopathogenic fungus *Gibberella pulicaris* (*Fusarium sambucinum*). *Phytochemistry, 28*: 2963–2969, 1989.
4. Marasas, W. F. O., Nelson, P. E. and Toussoun, T. A. *Toxigenic Fusarium Species*, Penn. State Univ. Press, University Park, 1984.
5. Durbin, R. D. *Toxins in Plant Disease*, Academic, New York, 1981.
6. Bailey, J. A., and Mansfield, J. W. *Phytoalexins*, Wiley, New York, 1982.
7. VanEtten, H. D., Matthews, D. E., and Smith, D. A. Metabolism of phytoalexins. In *Phytoalexins* (J. A. Bailey and J. W. Mansfield, Eds.), Wiley, New York, 1982, pp. 181–217.
8. Yoder, O. C. Toxins in pathogenesis. *Ann. Rev. Phytopathol., 18*: 103–129, 1980.
9. Yoder, O. C., Valent, B., and Chumley, F. Genetic nomenclature and practice for plant pathogenic fungi. *Phytopathology, 76*:383–385, 1986.
10. Manka, M., Visconti, A., Chelkowski, J., and Bottalico, A. Pathogenicity of *Fusarium* isolates from wheat, rye and triticale towards seedlings and their ability to produce trichothecenes and zearalenone. *Phytopathol. Z., 13*:24–29, 1985.
11. Miller, J. D., and Greenhalgh, R. Metabolites of fungal pathogens and plant resistance. In *Biotechnology for Crop Protection* (P. A. Hedin, J. J. Menn, and R. M. Hollingsworth, Eds.), American Chemical Society, Washington, D.C., 1988, pp. 118–129.
12. Cutler, H. G., and Jarvis, B. B. Preliminary observations on the effects of macrocyclic trichothecenes on plant growth. *Environ. Exp. Bot., 25*:115–128, 1985.
13. Kuti, J. O., Ng, T. J., and Bean, G. A. Possible involvement of a pathogen-produced trichothecene metabolite in *Myrothecium* leaf spot of muskmelon. *Physiol. Mol. Plant. Pathol., 34*:41–54, 1989.
14. Beremand, M. N. Isolation and characterization of mutants blocked in T-2 toxin biosynthesis. *Appl. Environ. Microbiol., 53*:1855–1859, 1987.
15. Beremand, M. N., Black, P. J., and Plattner, R. D. Isolation and characterization of two new mutants blocked in T-2 toxin biosynthesis. *J. Cell. Biochem. Suppl.*, 12c, p. 261 (Abstract), 1988.
16. Plattner, R. D., Tjarks, L. W., and Beremand, M. N. Trichothecenes accumulated in liquid culture of a mutant of *Fusarium sporotrichioides* NRRL 3229. *Appl. Environ. Microbiol, 55*:2190–2194, 1989.

17. McCormick, S. P., Taylor, S. L., Plattner, R. D., and Beremand, M. N. New modified trichothecenes accumulated in solid culture by mutant strains of *Fusarium sporotrichioides*. *Appl. Environ. Microbiol.*, *55*:2195–2199, 1989.
18. Desjardins, A. E., Spencer, G. F., Plattner, R. D., and Beremand, M. N. Furanocoumarin phytoalexins, trichothecene toxins, and infection of *Pastinaca sativa* by *Fusarium sporotrichioides*. *Phytopathology*, *79*:170–175, 1989.
19. Kern, H. Phytotoxins produced by fusaria. In *Phytotoxins in Plant Diseases* (R. K. S. Wood, A. Ballio, and A. Graniti, Eds.), Academic, New York, 1972, pp. 35–48.
20. Stöessl, A. Structure and biogenetic relations: fungal nonhost-specific. In *Toxins in Plant Disease* (R. D. Durbin, Ed.), Academic, New York, 1981, pp. 110–220.
21. Nemec, S., Phelps, D., and Baker, R. Effects of dihydrofusarubin and isomarticin from *Fusarium solani* on carbohydrate status and metabolism of rough lemon seedlings. *Phytopathology*, *79*:700–705, 1989.
22. Kern, H. Les naphthazarines des *Fusarium*. *Ann. Phytopathol.*, *10*: 327–345, 1978.
23. Holenstein, J., and Défago, G. Inheritance of naphthazarin production and pathogenicity to pea in *Nectria haematococca*. *J. Exp. Bot.*, *34*:927–935, 1983.
24. Tegtmeier, K. The role of naphthazarin production by *Nectria haematococca* M.P. VI in pathogenicity on pea (*Pisum sativum*). *J. Cell. Biochem. Suppl.*, *8C*:255 (Abstract) (1984).
25. Kuo, M. S., and Scheffer, R. P. Evaluation of fusaric acid as a factor in development of fusarium wilt. *Phytopathology*, *54*:1041–1044, 1964.
26. Venkata Ram, C. S. Production of ultraviolet induced mutations in *Fusarium vasinfectum* with special reference to fusaric acid synthesis. *Proc. Natl. Inst. Sci. India Biol. Sci.*, *23*:117–122, 1957.
27. Ingram, J. L. Phytoalexins from the Leguminosae. In *Phytoalexins* (J. A. Bailey and J. W. Mansfield, Eds.), Wiley, New York, 1982, pp. 21–80.
28. VanEtten, H. D., and Pueppke, S. G. Isoflavonoid phytoalexins. *Ann. Proc. Phytochem. Soc.*, *13*:239–289, 1976.
29. VanEtten, H. D., Matthews, P. S., Tegtmeier, K. J., Dietert, M. F., and Stein, J. I. The association of pisatin tolerance and demethylation with virulence on pea in *Nectria haematococca*. *Physiol. Plant Pathol.*, *16*:257–268, 1980.
30. Lucy, M. C., Matthews, P. S., and VanEtten, H. D. Metabolic detoxification of the phytoalexins maackiain and medicarpin by *Nectria haematococca* field isolates: relationship to virulence on chickpea. *Physiol. Mol. Plant Pathol.*, *33*:187–199, 1988.
31. Smith, D. A., Harrer, J. M., and Cleveland, T. E. Relation between prodution of extracellular kievitone hydratase by isolates of *Fusarium* and their pathogenicity on *Phaseolus vulgaris*. *Phytopathology*, *72*: 1319–1323, 1982.
32. Smith, D. A., Wheeler, H. E., Banks, S. W., and Cleveland, T. E. Association between lowered kievitone hydratase activity and reduced

virulence to bean in variants of *Fusarium solani* f. sp. *phaseoli*. *Physiol. Plant Pathol.*, *25*:135–147, 1984.
33. Choi, G. H., Richey, M. G., Turbek, C. S. M., and Smith, D. A. Comparative characteristics of a kievitone hydratase-producing isolate of *Fusarium solani* f. sp. *phaseoli* and of two strains deficient in this enzyme. *Physiol. Mol. Plant Pathol.*, *30*:215–224, 1987.
34. Tegtmeier, K. J., and VanEtten, H. D. The role of pisatin tolerance and degradation in the virulence of *Nectria haematococca* on peas: a genetic analysis. *Phytopathology*, *72*:608–612, 1982.
35. Kistler, H. C., and VanEtten, H. D. Regulation of pisatin demethylation in the fungus *Nectria haematococca*. *J. Gen. Microbiol.*, *130*: 2605–2613, 1984.
36. Mackintosh, S. F., Matthews, D. E., and VanEtten, H. D. Two additional genes for pisatin demethylation and their relationship to the pathogenicity of *Nectria haematococca* on pea. *Mol. Plant-Microbe Interactions*, *2*:354–362, 1989.
37. VanEtten, H. D., Matthews, D. E., and Mackintosh, S. F. Adaptation of pathogenic fungi to toxic chemical barriers in plants: the pisatin demethylase of *Nectria haematococca* as an example. In *Molecular Strategies for Crop Protection* (C. Arntzen and C. Ryan, Eds.), Alan Liss, New York, 1987, pp. 59–70.
38. VanEtten, H. D., Matthews, D. E., and Matthews, P. S. Phytoalexin detoxification: importance for pathogenicity and practical implications. *Ann. Rev. Phytopathol.*, *27*:143–164, 1989.
39. Matthews, D. E., and VanEtten, H. D. Detoxification of the phytoalexin pisatin by a fungal cytochrome P-450. *Arch. Biochem. Biophys.*, *224*:494–505, 1983.
40. Weltring, K.-M., Turgeon, B. G., Yoder, O. C., and VanEtten, H. D. Isolation of a phytoalexin-detoxification gene from the plant pathogenic fungus *Nectria haematococca* by detecting its expression in *Aspergillus nidulans*. *Gene*, *68*:335–344, 1988.
41. Ciufetti, L. M., Weltring, K.-M., Turgeon, B. G., Yoder, O. C., and VanEtten, H. D. Transformation of *Nectria haematococca* with a gene for pisatin demethylating ability, and the role of pisatin detoxification in virulence. *J. Cell. Biochem. Suppl.*, *12C*:278 (abstract), 1988.
42. Schäfer, W., Straney, D., Ciufetti, L., VanEtten, H. D., and Yoder, O. C. One enzyme makes a fungal pathogen, but not a saprophyte, virulent on a new host plant. *Science*, *246*:247–249, 1989.
43. Kuć, J. Phytoalexins from the solanaceae. In *Phytoalexins* (J. A. Bailey and J. W. Mansfield, Eds.), Wiley, New York, 1982, pp. 81–105.
44. Ward, E. W. B., Unwin, C. H., and Stöessel, A. Postinfectional inhibitors from plants. *Can. J. Bot.*, *52*:2481–2488, 1974.
45. Ward, E. W. B., and Stoessl, A. Phytoalexins from potatoes: evidence for the conversion of lubimin to 15-dihydrolubimin by fungi. *Phytopathology*, *67*:468–471, 1977.
46. Desjardins, A. E., Gardner, H. W., and Plattner, R. D. Detoxification of the potato phytoalexin luminin by *Gibberella pulicaris*. *Phytochemistry*, *28*:431–437, 1989.
47. Desjardins, A. E., and Gardner, H. W. Genetic analysis in *Gibberella pulicaris*: rishitin tolerance, rishitin metabolism, and virulence on potato tubers. *Mol. Plant-Microbe Interactions*, *2*:26–34, 1989.

48. Desjardins, A. E., and Beremand, M. N. A genetic system for trichothecene toxin production in *Gibberella pulicaris (Fusarium sambucinum)*. *Phytopathology, 77*:678–683, 1987.
49. Défago, G., and Kern, H. Induction of *Fusarium solani* mutants insensitive to tomatine, their pathogenicity and aggressiveness to tomato fruits and pea plants. *Physiol. Plant Pathol., 22*:29–37, 1983.
50 Défago, G., Hern, H., and Sedlar, L. Genetic analysis of tomatine insensitivity, sterol content and pathogenicity for green tomato fruits in mutants of *Fusarium solani*. *Physiol. Plant Pathol., 22*:39–43, 1983.
51. VanEtten, H. D., and Kistler, H. C. *Nectria haematococca*, mating populations I and VI. *Adv. Plant Pathol., 6*:189–206, 1988.
52. Desjardins, A. E., and Plattner, R. D. Trichothecene toxin production by strains of *Gibberella pulicaris (Fusarium sambucinum)* in liquid culture and in potato tubers. *J. Agric. Food Chem., 37*:388–392, 1989.
53. Lafont, P., Girard, T., Payen, J., Sarfati, J., and Gaillardin, M. Contamination de pommes de terres de consommation par des fusariotrichothecenes. *Microbiol. Alimont. Nutr., 1*:147–152, 1983.
54. Bean, G. A., Fernando, T., Jarvis, T. B., and Bruton, B. The isolation and identification of trichothecene metabolites from a plant pathogenic strain of *Myrothecium roridum*. *J. Nat. Prod., 47*:727–728, 1984.
55. Mitchell, R. E. The relevance of non-host-specific toxins in the expression of virulence by pathogens. *Ann. Rev. Phytopathol., 22*: 215–245, 1984.
56. Miller, J. D., and Young, J. C. Deoxynivalenol in an experimental *Fusarium graminearum* infection of wheat. *Can. J. Plant Pathol., 7*: 132–134, 1985.
57. Scott, P. M., Nelson, K., Kanhere, S. R., Karpinski, K. F., Hayward, S., Weish, G. A., and Tecih, A. Decline in deoxynivalenol concentrations in 1983 Ontario winter wheat before harvest. *Appl. Environ. Microbiol., 48*:884–886, 1984.
58. Miller, J. D., and Arnison, P. G. Degradation of deoxynivalenol by suspension cultures of the fusarium bead blight resistant wheat cultivar Frontana. *Can. J. Plant Pathol., 8*:147–150, 1986.
59. Bamburg, J. R. Chemical and biochemical studies of the trichothecene mycotoxins. In *Mycotoxins and Other Fungal Related Food Problems* (J. V. Rodrichs, Ed.), American Chemical Society, Washington, D.C., 1976, pp. 144–162.
60. Wei, C., and McLaughlin, C. S. Structure-function relationship in the 12,13-epoxytrichothecenes. *Biochem. Biophys. Res. Commun., 57*:838–844, 1974.
61. Desjardins, A. E., Plattner, R. D., and VanMiddlesworth, F. Trichothecene biosynthesis in *Fusarium sporotrichioides*: origin of the oxygenations of T-2 toxin. *Appl. Environ. Microbiol., 51*:493–497, 1986.
62. Desjardins, A. E., Plattner, R. D., and Beremand, M. N. Ancymidol blocks trichothecene biosynthesis and leads to accumulation of trichodiene in *Fusarium sporotrichioides* and *Gibberella pulicaris*. *Appl. Environ. Microbiol., 63*:1860–1865, 1987.

63. Beremand, M. N., and Desjardins, A. E. Trichothecene biosynthesis in *Gibberella pulicaris*: inheritance of C-8 hydroxylation. *J. Ind. Microbiol.*, *3*:167–174, 1988.
64. Committee on Protection Against Mycotoxins, *Protection Against Trichothecene Mycotoxins*, National Academy Press, Washington, D.C., 1983.
65. Desjardins, A. E., Plattner, R. D., and Spencer, G. F. Inhibition of trichothecene toxin biosynthesis by naturally occurring shikimate aromatics. *Phytochemistry*, *27*:767–771, 1988.
66. Matthews, D. E., Weiner, E. J., Matthews, P. S., and VanEtten, H. D. Role of oxygenases in pisatin biosynthesis and in the fungal degradation of maackiain. *Plant Physiol.*, *83*:365–370, 1987.
67. Kuhn, P. J., and Smith, D. A. Isolation from *Fusarium solani* f. sp. *phaseoli* of an enzymic system responsible for kievitone and phaseollidin detoxification. *Physiol. Plant Pathol.*, *14*:179–190, 1979.
68. Gardner, H. W., Desjardins, A. E., Weisleder, D., and Plattner, R. D. Biotransformation of the potato phytoalexin lubimin by *Gibberella pulicaris*: identification of major products. *Biochim. Biophys. Acta*, *966*:347–356, 1988.
69. Spencer, G. F., Desjardins, A. E., and Plattner, R. D. 5-(2-Carboxyethyl)-6-hydroxy-7-methoxybenzofuran, a fungal metabolite of xanthothoxin. *Phytochemistry*, *29*:2495–2497, 1990.
70. Miao, V., Lucy, M., and VanEtten, H. D. Detoxification of the phytoalexin maackiain is important for virulence of *Nectria haematococca* on chickpea. *J. Cell. Biochem. Suppl.*, *10C*:25 (Abstract), 1986.
71. Miao, V., and VanEtten, H. D. Inheritance of maackiain metabolizing ability in *Nectria haematococca*. *Can. J. Plant Pathol.*, *9*:283 (Abstract), 1987.
72. Miao, V., and VanEtten, H. D. Meiotic loss of a gene for pisatin demethylase from *Nectria haematococca* M.P. VI. *J. Cell Biochem. Suppl.*, *12C*:285 (Abstract), 1988.
73. VanEtten, H. D., Matthews, D., Matthews, P., Miao, V., Maloney, A., and Straney, D. A family of genes for phytoalexin detoxification in the plant pathogen *Nectria haematococca*. In *Signal Molecules in Plants and Plant–Microbe Interactions* (B. Ludenburg, Ed.), Springer-Verlag, Berlin, 1989, pp. 219–228.
74. Berenbaum, M. R., and Zangerl, A. R. Stalemates in the coevolutionary arms race. In *Chemical Mediation of Coevolution* (K. C. Spencer, Ed.), Academic, New York, 1989, pp. 113–132.
75. Price, K. R., Howard, B., and Coxon, D. T. Stress metabolite production in potato tubers infected by *Phytophthora infestans*, *Fusarium avenaceum* and *Phoma exigua*. *Physiol. Plant Pathol.*, *9*:189–197, 1976.
76. Hadwiger, L. A., Sander, C., Eddyvean, J., and Ralston, J. Sodium azide induced mutants of peas that accumulate pisatin constitutively. *Phytopathology*, *66*:629–630, 1976.
77. Gould, S. D. *Wonderful Life*: *The Burgess Shale and the Nature of History*, Norton, New York, 1989.
78. Burnett, J. H. Aspects of *Fusarium* genetics. In *The Applied Mycology of Fusarium* (M. O. Moss and J. E. Smith, Eds.), Univ. Cambridge Press, Cambridge, 1984, pp. 40–69.

79. Murray, R. P. H., Mendez, J., and Brown, S. A. *The Natural Coumarins*, Wiley, New York, 1982.
80. Gould, F. Genetics of pairwise and multi-species plant-herbivore co-evolution. In *Chemical Mediation of Coevolution* (K. C. Spencer, Ed.), Academic, New York, 1989, pp. 13–56.
81. Bezuidenhout, S. C., Gelderblom, C. A., Gorst-Allman, C. P., Horak, R. M., Marasas, W. F. O., Spiteller, G., and Vleggar, R. STructure elucidation of the fumonisins, mycotoxins from *Fusarium moniliforme*. *J. Chem. Soc. Chem. Commun.*, *1988*:743–745, 1988.
82. Bottini, A. T., Bowen, J. R., and Gilchrist, D. G. Phytotoxins. II. Characterization of a phytotoxic fraction from *Alternaria alternata* f. sp. *lycopersici*. *Tetrahedron Lett.*, *22*:2723–2726, 1981.
83. Sun, S. K., and Snyder, W. C. The bakanae disease of the rice plant. In *Fusarium*: Diseases, Biology and Taxonomy (P. E. Nelson, T. A. Toussoun, and R. J. Cook, Eds.), Penn. State Univ. Press, University Park, 1981, pp. 105–113.
84. Panopoulous, N. J., and Staskawicz, B. J. Genetics of production. In *Toxins in Plant Disease* (R. D. Durbin, Ed.), Academic, New York, 1981, pp. 79–107.
85. Mansfield, J. W. Antimicrobial compounds. In *Biochemical Plant Pathology* (J. A. Callow, Ed.), Wiley, New York, 1983, pp. 237–265.
86. Klittich, C. J. R., and Leslie, J. F. Chlorate-resistant, nitrate-utilizing (crn) mutants of *Fusarium moniliforme* (*Gibberella fukikuroi*). *J. Gen. Microbiol.*, *135*:721–727, 1989.
87. Cullen, D., Caldwell, R. W., and Smalley, E. B. Cultural characteristics, pathogenicity and zearalenone production by strains of *Gibberella zeae* isolated from corn. *Phytopathology*, *72*:1415–1418, 1982.

14

BIOSYNTHESIS AND REGULATION OF TRICHOTHECENE PRODUCTION BY *FUSARIUM* SPECIES

MARIAN N. BEREMAND* and SUSAN P. McCORMICK *Agricultural Research Service, National Center for Agricultural Utilization Research, U.S. Department of Agriculture, Peoria, Illinois*

I. INTRODUCTION

Trichothecenes are tricyclic sesquiterpene metabolites produced by several genera of fungi: *Trichothecium*, *Cephalosporium*, *Myrothecium*, *Trichoderma*, *Stachybotrys*, *Cylindrocarpon*, *Verticimonosporium*, and *Fusarium* [1, 2]. Interest in these fungal secondary metabolites stems from their involvement in a number of mycotoxicoses of humans and animals [3–7]. These compounds, which can enter the food supply via fungally infected grains and produce [8–10], are potent inhibitors of eukaryotic protein synthesis, and it is this inhibitory activity that is the primary basis of their toxicity [11–13].

As a group, the trichothecenes are characterized by the presence of an olefinic bond at the 9,10 position and an epoxy group at the 12,13 position (Fig. 1); they differ from one another in the degree and type of oxidations and esterifications at positions 3, 4, 7, 8, and 15. These variations are not only responsible for the diversity and large number of trichothecenes that have been identified (over 90 to date) [14] but are also directly associated with the level of toxicity displayed by individual trichothecenes toward susceptible organisms, including plants, animals, fungi, insects, and humans [11–15].

The sesquiterpene nature of trichothecenes, as well as their biosynthetic origin, was established by early studies that demonstrated that the trichothecenes are derived from mevalonic acid via trichodiene, an alicyclic hydrocarbon produced by the cyclization of farnesyl pyrophosphate [16]. Since that time, elucidation of the details of how and why these compounds are elaborated by fungi has constituted an ongoing active area of research. The purpose of this review is to discuss new information pertaining to the biosynthetic pathway and the regulation of trichothecene production.

Present affiliation: Texas A&M University, College Station, Texas.

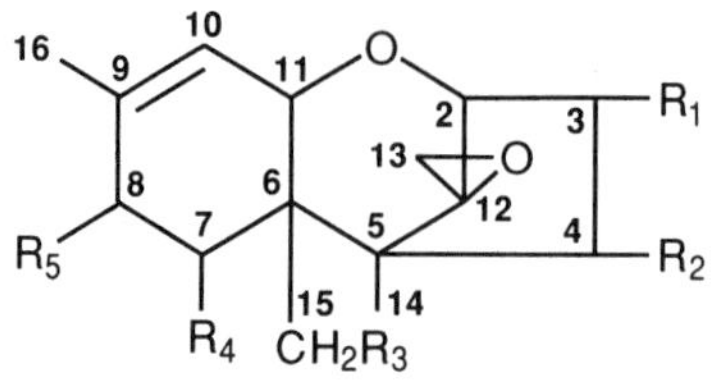

Trichothecene	R_1	R_2	R_3	R_4	R_5
Diacetoxyscirpenol	OH	OAc	OAc	H	H
Neosolaniol	OH	OAc	OAc	H	OH
8-acetyl neosolaniol	OH	OAc	OAc	H	OAc
T-2 Toxin	OH	OAc	OAc	H	OX[a]

[a] X = isovalerate

FIGURE 1. Basic trichothecene ring structure and several common trichothecenes.

In recent years, significant progress has been made in many areas regarding trichothecene biosynthesis, largely owing to advances made in research approaches and techniques. The increasing technological sophistication of analytical instrumentation and methodology (gas chromatography, mass spectrometry, tandem mass spectrometry, and nuclear magnetic resonance) has allowed both the detection and evaluation of minor trichothecenes and related metabolites as possible biosynthetic intermediates or shunt products. These approaches have been presented in several recent articles [17–19]. During this same time, complementary approaches utilizing the tools of genetics, biochemistry, and molecular biology have also been developed. These last three areas will be the primary focus of this review. Most of this latter research has been done with two major trichothecene-producing *Fusarium* species, *F. sambucinum* and *F. sporotrichioides*.

II. GENETIC SYSTEMS

A. Possible Systems

Of the dozen or so genetically fertile *Fusarium* species, five include well-documented trichothecene-producing strains (Table 1). However, the perfect stage has been routinely obtained in the laboratory for only two of these, *Gibberella zeae* (*F. graminearum*) and *G. pulicaris* (*F. sambucinum*). While *G. zeae* is one of the best-known and most studied trichothecene-producing organism [3, 4, 30–33], it is predominantly a homothallic (self-fertile) fungus, and it has yet to be developed as a useful genetic system [4, 22, 23, 33]. On the other hand, *G. pulicaris* is a heterothallic (self-sterile) fungus [20] and is rapidly becoming an important genetic organism for studying many specialized fungal processes including trichothecene production. Throughout this chapter, the binomial of the teleomorph will be used for all fertile *Fusarium* species (Table 1).

TABLE 1. Trichothecene Production Among Genetically Fertile *Fusarium* Species

Asexual stage (anamorph)	Sexual stage (teleomorph)	Sexuality hetero/	homothallic	Trichothecene production[a] [ref.]
F. sambucinum[b]	*Gibberella pulicaris*	+	—	+ [4, 9, 20, 21]
F. graminearum	*Gibberella zeae*	+[c]	+	+ [4, 9, 22, 23]
F. equisiti	*Gibberella intricans*	+	—	+ [4, 24]
F. acuminatum	*Gibberella acuminata*	+	—	+ [4, 9, 21]
F. heterosporum	*Gibberella gordonii*	+	—	+ [4]
F. tricinctum	*Gibberella tricincta*	+	—	– [4, 24]
F. moniliforme	*Gibberella fujikuroi*	+	—	– [4, 25]
F. lateritium	*Gibberella baccata*	+	+	– [4]
F. stilboides	*Gibberella stilboides*	+	—	– [4]
F. decemcellulare	*Calonectria rigidiuscula*	+	+	– [4, 26, 27]
F. nivale	*Calonectria nivalis*	+	—	– [4, 28]
F. solani	*Nectria haematocca*	+	+[c]	– [4]

[a]According to Marasas et al. [4].
[b]*F. sambucinum* (*G. pulicaris*) = *F. sulphureum* (*G. cyanogena*) [4, 29].
[c]Rarer form.

In 1987, the first genetic analysis of trichothecene production by sexual crosses was reported for studies with *G. pulicaris* [29]. Prior to this time, little attention had been given to the use of *G. pulicaris* as a genetic tool. Only a few laboratories reported the production of the sexual stage [20, 25, 34], and crosses were limited to determining the inheritance of three genes [34, 35]. However, *G. pulicaris* has proven to be an exceptionally amenable organism, and, as described below, it has the best developed genetic system known for a trichothecene-producing organism.

B. *Gibberella pulicaris* as a Model System

Several features make *G. pulicaris* an excellent organism for genetic and molecular studies aimed at investigating both the biosynthesis and function of trichothecenes and the regulation of their production. In particular, *G. pulicaris* is a heterothallic ascomycete [20, 21]; this means that it is self-sterile and that all the products of a single meiosis are retained together in a sac called an ascus. This first property permits controlled sexual crosses between two strains of opposite mating type, and the second allows tetrad analysis of the resulting progeny.

In practice, crosses are made by pairing two strains of opposite mating type and appropriate sex: individual strains can be either mating type 1 or mating type 2, and either male, female, hermaphroditic (both male and female), or neuter (neither male nor female); no female-only strains have been found in *G. pulicaris*, although they do exist in other fertile bisexual *Fusarium* species [36, 37]. The female (or recipient) strain is grown alone first (on mulberry twigs/water agar at 20–25°C) to allow formation of the female reproductive structures called protoperithecia. The male (or donor) strain is then added by covering the surface of the protoperithecia with a freshly prepared suspension of conidia. After the cross is incubated 2 or more weeks at 15°C, mature perithecia may begin to appear [29, 38]. When mature, the perithecia are filled with clusters or rosettes of asci. Each ascus contains eight haploid spores called ascospores. These spores represent the four products of a single meiosis, each of which has undergone an immediate mitotic cell division such that every ascus has four sets of spores that are twins.

The entire mating process generally requires 4–12 weeks: 2–4 weeks for a culture to produce protoperithecia and another 2–8 weeks after fertilization for the perithecia to mature. The mature ascospores can then be dissected, under magnification, and collected as unordered tetrads [29, 38]. The details of these procedures and those of the *G. pulicaris* life cycle have been published [20, 21, 25, 29, 34, 35, 38–41]. Finally it should be noted that although several earlier workers reported the rapid, and apparently abundant, formation of the sexual stage of this fungus on wheat straw [21, 25], others have not been able to repeat these conditions [29].

Except for the transient diploid stage at the beginning of meiosis, *Fusarium* cells contain haploid (ln) nuclei. This provides another advantage in that all of the alleles are expressed in the ascospore progeny. Consequently, expression of each parental allele can be determined in one generation (rather than 2) and recessive mutant alleles can be readily recognized and recovered. One slight disadvantage is the inability to determine by standard crosses whether alleles are dominant or recessive to one another.

The above features are similar to those of other filamentous, heterothallic fungi. Further details regarding information about *Fusarium* genetic crosses at the cellular level and general methods of genetic analysis in fungi can be found in Refs. 42–45.

One of the factors that defines the usefulness of a genetic system is strain stability. Segregation data from a large number of crosses indicate that strains of this species are relatively stable. Genes for mating type, femaleness, pigmentation, auxotrophy, trichothecene production, and other biochemical processes are inherited in a normal 1:1 Mendelian fashion [21, 29, 34, 38, 46, 47].

Gibberella pulicaris strains meet three other criteria for a well-developed genetic system. Strains of *G. pulicaris* are readily amenable to genetic manipulation, mutagenesis, and transformation. These properties have facilitated the development and production of many useful strains, including a highly fertile isogenic line for strain R-6380 [46]. The availability of this isogenic line will alleviate some of the problems encountered in crosses between "field" strains. Such crosses do not always yield complete tetrads, and the diverse genetic backgrounds of the parents place restrictions on the interpretation of the data and the types of experiments that can be undertaken. Additional studies have also shown that strain R-6380 is readily mutable [46] and transformable [48] (see Section V). Both biochemical mutants (auxotrophs) and morphological mutants have been obtained by UV treatment of this strain; some of these mutants should be useful for mapping and cloning studies.

Gibberella pulicaris is also particularly well suited to investigate the possible role or function of trichothecenes as virulence or pathogenicity factors. It is a serious plant pathogen and a major cause of dry rot on potato tubers [4, 49]. A rapid potato disk virulence assay is available and has already been used to identify and genetically characterize several pathogenicity genes in this fungus [47]. Once *G. pulicaris* mutants deficient in trichothecene production are obtained, this assay can also be used to critically test whether trichothecenes are involved in the pathogenicity of this fungus.

Finally, fertile field strains of *G. pulicaris* not only make trichothecenes but also demonstrate considerable variation in their ability to do so [29, 38, 50, 51]. With proper storage, minimal subculturing, and uniform culture conditions, individual strains of *G. pulicaris* display stable toxin phenotypes with respect to both the kinds and amounts of trichothecenes they produce ([29, 38, 46, 50]; Beremand and Desjardins, unpublished results). As discussed below, this natural variability has been used to gain new information about the trichothecene toxin (*Tox*) genes.

C. Genes Involved in Trichothecene Biosynthesis in *Gibberella pulicaris*

Genes controlling the production and formation of trichothecenes have been investigated by genetic analysis [29, 38]. To date these studies have involved crosses between natural variants, as laboratory-induced *Tox* mutants have not yet been isolated in *G. pulicaris*. As noted above, these natural variants differ in both the type and amount of trichothecenes they produce. Diacetoxyscirpenol (DAS) and 8-acetyl neosolaniol (8-AcNeo) are the two most common trichothecenes made by strains of this species in liquid culture. Since 8-AcNeo differs from DAS by having an extra oxygen group at C-8 (Fig. 1), strains that make only DAS are designated $C8^-$, while those that make some 8-AcNeo are designated $C8^+$ [38]. Both "high" and "low" toxin-producing strains also exist; under the same growth conditions, some strains make as little as 1–2 μg/mL DAS, while others make >100 μg/mL DAS [29, 38, 50]. By making appropriate crosses between these various strains it has been possible to both identify genes responsible for different "trichothecene" phenotypes and to generate new "trichothecene" phenotypes [38]. The results from these crosses have yielded information about the location, arrangement, type, and number of *Tox* genes.

In all crosses to date, Mendelian segregation ratios have been observed for differences in trichothecene production [29, 38]. These results demonstrated that genes for trichothecene biosynthesis are located on nuclear chromosomes. Thus far, there is no evidence for extranuclear *Tox* genes, although their occurrence cannot be ruled out.

Crosses between different $C8^+$ and $C8^-$ strains have yielded either a 1:1 or 1:3 ratio of $C8^+$:$C8^-$ progeny [38]. These results suggest that there are at least two unlinked *Tox* loci, each of which can confer the $C8^-$ phenotype. These data also provide some information about the possible regulatory/metabolic relationships between these loci. Even though these two genes are not linked physically, they are linked metabolically. Both genes appear to affect the same biosynthetic step (i.e., hydroxylation at C-8) and, based on the segregation data, the "$C8^-$" allele at one locus, if not both, is epistatic to the "$C8^+$" allele at the other locus. The biochemical basis of this interaction remains to be determined, but once known it should provide additional information about the mechanisms that regulate trichothecene biosynthesis.

Altogether, a minimum of five *Tox* loci have been identified by genetic recombination [29, 38]. Three determine the type of trichothecene produced: two effect the hydroxylation at C-8, and one determines the acetylation state of the oxygen at C-8. Another two effect the observed trichothecene levels. One of these appears to be a regulatory gene that may suppress the expression of one or more other *Tox* genes.

Limited information is available concerning linkage relationships for the *Tox* genes. Thus far, all of the *Tox* genes that have been identified appear to be unlinked to either a locus for femaleness, one for red pigmentation, or the one for mating type. Likewise, they do not appear to be linked to each other except in one case [38].

These studies provided some of the first information pertaining to the genetic control of trichothecene biosynthesis. They showed that trichothecene production is determined, at least in part, by multiple, unlinked nuclear genes.

III. INDUCED MUTANTS

A. Mutants Blocked in Trichothecene Biosynthesis

Mutants blocked or altered in T-2 toxin biosynthesis are listed in Table 2. All of these mutants (designated Tox^-) were produced by UV irradiation of *F. sporotrichioides* NRRL 3299 [52, 53], a widely studied T-2 toxin-producing strain. They are the first mutants specifically blocked in a trichothecene biosynthetic pathway.

Isolation of these mutants was made possible by the availability of a monoclonal antibody, 15H6, to T-2 toxin [55]. The specificity and sensitivity of this antibody permitted the development of a rapid screen for T-2 toxin minus mutants. 15H6 is highly specific for T-2: it cannot bind to trichothecenes unless an isovalerate (or a closely related) moiety is present at C-8. Consequently, any mutation that blocks the pathway prior to the addition of the isovalerate moiety can be detected; even though other trichothecenes may accumulate, they will not be detected by the antibody. In addition, 15H6 requires only 50 μL of medium from a liquid culture to test the ability of a strain to make T-2 toxin. Thus a large number of potential mutants can be simultaneously grown in small individual cultures (150 μL each) in 96-well microtiter plates. After 3–5 days growth, several hundred cultures can be assayed in a single day for T-2

TABLE 2. Laboratory-Derived Mutants of *F. sporotrichioides* NRRL 3229

Mutant	Mutagen	Genotype	Major compound accumulated[a]
Tox$^-$ mutants			
MB1370	UV	*Tox1-1*$^-$	Diacetoxyscirpenol
MB1716	UV	*Tox1-2*$^-$	Diacetoxyscirpenol
MB2972	UV	*Tox3-1*$^-$	Dideacetyl calonectrin, deacetyl calonectrin
MB5493	UV	*Tox4-1*$^-$	Trichodiene
Auxotrophs			
MB1026	UV	*Gln1-1*$^-$	T-2 toxin
MB1083	UV	*Met1-1*$^-$	T-2 toxin
MB1572	UV	*Arg1-1*$^-$	T-2 toxin
MB2402	UV	*Lys1-1*$^-$	T-2 toxin
MB2915	UV	*Leu1-1*$^-$	T-2 toxin, neosolaniol, 8-propionyl neosolaniol, 8-isobutyryl neosolaniol

[a]Data from liquid cultures [52–54, 59].

toxin. With this procedure, over 20,000 UV-mutagenized survivors have been screened for the production of T-2 toxin (Beremand, unpublished results).

The *Tox* mutants listed in Table 2 fall into three complementation groups [53]. When grown together, all pairwise combinations, except MB1370 + MB1716, yield large amounts of T-2. Chemical characterization of these four mutants (in liquid cultures) has shown that MB5493 is blocked early in the T-2 pathway and accumulates large amounts of trichodiene [53]. MB2972 is blocked at a later step and accumulates dideacetyl calonectrin and deacetyl calonectrin (Fig. 1) [53, 54]. Both MB1370 and MB1716 are blocked near the end of the pathway and accumulate DAS [52]. These results, along with the data from the complementation and feeding studies ([56]; Beremand and Plattner, unpublished results), indicate that each of these mutants is blocked at a single step in the trichothecene biosynthetic pathway. Growth of these mutants on solid substrate has allowed the production and isolation of additional compounds [57].

These mutants have a wide range of applications in genetic, biochemical, physiological, and molecular studies pertaining to trichothecene production. Some of these will be discussed in the following portions of this chapter. One application, which is discussed in detail in the next chapter, has been the use of these mutants to demonstrate that trichothecenes are pathogenicity factors for the infection of parsnips by *F. sporotrichioides* [58].

B. Auxotrophs

Auxotrophic mutants have also been obtained from both *F. sporotrichioides* NRRL 3299 [52] and *G. pulicaris* R-6380 [46] by UV mutagenesis. Some of those obtained for *F. sporotrichioides* are listed in Table 2. Again, these mutants have a range of potential uses in cloning studies, complementation tests, or as markers for gene mapping or cell fusions. In addition, one mutant, the leucine auxotroph MB2915, has had a direct bearing on the analysis of the T-2 toxin biosynthetic pathway. This mutant displays an altered pattern of trichothecene production (Table 2), and, as will be discussed in greater detail in Section VI, has been useful in demonstrating that leucine is a direct precursor to T-2 toxin [59].

IV. ENZYMES IN TRICHOTHECENE BIOSYNTHESIS

A. Trichodiene Synthase

In 1973, Evans et al. [60, 61] developed a cell-free system from *Trichothecium roseum* that catalyzed the conversion of trans–trans farnesyl pyrophosphate to trichodiene. This conversion, which marks the first step leading from the isoprenoid biosynthetic pathway to the trichothecene biosynthetic pathway, is catalyzed by the enzyme trichodiene synthase. The product of this reaction, trichodiene, is the first cyclic intermediate in the trichothecene biosynthetic pathway [62].

Trichodiene synthase from *F. sporotrichioides* NRRL 3299 has been purified and characterized [63]. This enzyme appears to be a dimer composed of identical subunits (M_r 45,000) [63]. A detailed stereochemical model for the trichodiene synthase–catalyzed reaction has been derived from mechanistic studies [64, 65]. The reaction proceeds by conversion of trans–trans farnesyl pyrophosphate to trans–trans nerolidyl pyrophosphate, which is then isomerized to cis–trans nerolidyl pyrophosphate. Cyclization occurs next, with a 1,4 hydride shift and two methyl shifts, to yield trichodiene.

B. Other Pathway Enzymes

Considerably less is known about the remaining pathway enzymes that are responsible for the subsequent oxygenations and acetylations that result in the production of various trichothecenes. Labeling studies with ^{18}O have shown that six of the oxygenation steps required for T-2 toxin formation by *F. sporotrichioides* use molecular oxygen [66]. Such reactions are often catalyzed by cytochrome P-450 (cyt P-450) monooxygenases. The identification of cyt P-450 inhibitors, including ancymidol, that effectively block T-2 production further suggests that at least some of these enzymes are cyt P-450s [67, 68]. Finally, both the isolation of mutants from *F. sporotrichioides* that are blocked at different steps in the T-2 biosynthetic pathway and the segregation of *Tox* genes in *G. pulicaris* indicate that specific enzymes may exist for some, if not all, of these steps.

V. MOLECULAR APPROACHES

A. Gene Cloning

The trichodiene synthase gene from *F. sporotrichioides* NRRL 3299 has recently been cloned and sequenced by Hohn and Beremand [69]. Trichodiene synthase gene fragments were obtained by using antiserum against the trichodiene synthase protein to screen an expression (λgt11) vector library prepared with *F. sporotrichioides* DNA. These fragments were then used as hybridization probes to isolate an intact copy of the gene. This is the first trichothecene gene to be cloned.

Hybridization studies have shown that the trichodiene synthase gene is present as a single copy in the *F. sporotrichioides* genome; this locus has been designated *Tox5* [69]. The linkage relationships between this locus and the other four identified *Tox* loci (see Section III and Ref. 52) in *F. sporotrichioides* NRRL 3299 are currently unknown.

Sequence data have shown that the *Tox5* gene contains a 1218-nucleotide open reading frame [69]. The coding sequence, which specifies an M_r 43,999 protein, begins at nucleotide 37 and is 1122 nucleotides long; it is interrupted at nucleotide 469 by a single 60-nucleotide-long intron that does not contain any termination codons. Removal of this intron allowed production of both an active enzyme and trichodiene in *E coli*; copies of the trichodiene synthase gene containing the intron were also transcribed and translated in *E. coli*, but neither enzyme activity nor trichodiene was detected [70].

Sequences similar to those known to be important in the transcription of higher eukaryotic genes are present in the 5' noncoding region of the *Tox5* gene [69]. These include two sequences resembling the TATA box and three resembling the CCAAT box sequences. However, it remains to be determined whether or not these sequences are involved in regulating the transcription of the *Tox5* gene.

B. Transformation Systems

The transformation of filamentous fungi has seen tremendous success within the past five years. Several excellent reviews have been written that provide a complete, in-depth overview of the techniques, strategies, and applications fundamental to these gene transfer systems [71–75].

Transformation systems have recently been reported for a number of *Fusarium* species, including *F. oxysporum* [76, 77], *N. haematococca* [78, 79], *G. fujikuroi* [80], *G. zeae* [80], *F. sporotrichioides* [81], and *G. pulicaris* [48]. The latter two systems were developed specifically to be used to further investigate the genes and cellular regulatory systems that control trichothecene biosynthesis. *Fusarium sporotrichioides* was chosen as one system primarily due to the availability of the *Tox*⁻ mutants, the ultimate goal being to clone the *Tox* genes by complementation of these mutants. *Gibberella pulicaris* was chosen as a second system because of the ability to apply both classical genetic approaches and molecular approaches to future studies.

Transformation of both of these organisms was accomplished by the introduction of plasmid or cosmid DNA vectors into protoplasts followed by selection for the vector-encoded, dominant drug resistance gene hygromycin B phosphotransferase (hph) [48, 81]. This gene allows cells to grow in the presence of hygromycin B (HygB). Wild-type cells from both *F. sporotrichioides* and *G. pulicaris* are very sensitive to this drug; consequently, only those protoplasts that acquire and express the vector-encoded hph sequences are able to grow in the presence of HygB.

The advantages of a selectable dominant drug resistance marker are severalfold [72], and two of these will be briefly discussed. One is that any cell can be transformed because it is not necessary to construct strains that carry specific genetic markers. Such genetic markers are usually lethal mutations that impose a growth requirement for a particular amino acid; during transformation these mutations are complemented by wild-type alleles carried on the vectors. The second feature involves the intended use of these transformation systems. One of the main future goals of this research is to reintroduce cloned *Tox* gene sequences into the *Fusarium* cells for the specific purpose of investigating gene structure—function relationships. Some of these studies will rely on a two-step "gene replacement," which involves excision of the vector sequences. When the selectable marker on the vector is for drug resistance, excision of the vector restores the cell to the background wild-type genotype. On the other hand, excision of a vector with a primary metabolism gene as the selectable marker returns the cell to auxotrophy, an often undesirable state in an investigation of the regulation of secondary metabolism.

The same group of vectors was used to transform both *F. sporotrichioides* and *G. pulicaris*. These vectors contain the hph gene linked to either promoter 1 from *Cochliobolus heterostrophus* (vectors pUCH1 and CosHyg1) [82] or the trpC promoter from *Aspergillus nidulans* (vector pDH25) [83]. In general, the *F. sporotrichioides* and *G. pulicaris* transformants had the same overall features, which are commonly seen in transformants obtained from other filamentous fungi [71, 72]. In all cases, transformation occurred by integration of the vector DNA sequences into the host genome. Integration occurred at random, at one or more sites, either as a single-copy sequence or as multiple tandem-copy sequences (Salch and Beremand, unpublished results).

Currently the transformation frequencies are low in both systems. This poses a problem in attempting to select cloned genes by complementation of the *Tox*$^-$ mutants because such an approach requires a higher transformation frequency. This low frequency, however, does not limit introduction of cloned genes into the organism. Preliminary studies with vectors carrying a copy of the cloned *F. sporotrichioides* trichodiene synthase gene indicate that this sequence integrates into the chromosomal *Tox5* locus during transformation (Beremand and Hohn, unpublished results). Homologous integration is the foundation for many key experiments and critical areas of research. The ability to transform both *F. sporotrichioides* and *G. pulicaris* will undoubtedly provide many new avenues of research.

VI. BIOSYNTHETIC PATHWAYS

As indicated in the introduction, significant advances have been made recently in determining the order of the reactions in the trichothecene biosynthetic pathways. This work has involved several *Fusarium* species and a variety of experimental approaches, including the use of blocked mutants. Previous work has been extensively reviewed [16, 64].

As described in Section IV, it is generally accepted that the biosynthesis of trichothecenes proceeds from trichodiene, which is a cyclization product of farnesyl pyrophosphate. In early studies with *Trichothecium roseum*, Machida and Nozoe [62] isolated several trichothecene compounds of possible biogenetic significance. Feeding studies with these compounds indicated that trichodiene, trichodiol and its cyclization product 12,13-epoxytrichothecene, and trichothecolone were precursors of trichothecin. This information resulted in a scheme in which the pyran ring oxygen originates from the C-2 hydroxyl group of trichodiol, and following this ring formation, the primary oxidation site is at C-4, followed by secondary oxidations and esterifications.

Studies on trichothecene biosynthesis in *Fusarium* species have been more recent. Experiments in which liquid cultures of *F. sporotrichioides* were incubated with $^{18}O_2$ or $H_2{}^{18}O$ demonstrated that the pyran ring oxygen, the 12,13-epoxide, and the hydroxyl groups at C-3, C-4, C-8, and C-15 of T-2 toxin are all derived from molecular oxygen and that the oxygens in the three acyl groups are derived from water [66]. As discussed in Section IV, addition of oxygenase inhibitors such as ancymidol [67, 68] to *F. sporotrichioides* cultures blocked T-2 toxin production and resulted in the accumulation of trichodiene. This was the first evidence for the role of trichodiene in *Fusarium* trichothecene synthesis.

Clues to what some of the oxygenated precursors of T-2 toxin might be began to emerge with the isolation of minor constituents, including some uncyclized trichothecenes (see Fig. 2), from large-scale solid fermentations of *F. sporotrichioides* and *G. pulicaris* [84–88]. One of these compounds, trichotriol, is of particular interest. Trichotriol cyclized to isotrichodermol in solution during NMR studies, leading to speculation that primary oxidation at C-3 may occur prior to the second cyclization [85, 89].

Liquid and solid fermentations of the *F. sporotrichioides Tox*$^-$ mutants have supported this hypothesis and have also provided additional information on the biosynthesis of T-2 toxin. These mutants, as described in Section III, are *F. sporotrichioides* MB5493, which accumulates large quantities of trichodiene [53]; *F. sporotrichioides* MB1716, which accumulates diacetoxyscirpenol [52]; and *F. sporotrichioides* MB2972, which accumulates two calonectrin analogs [53, 54] and, under certain growth conditions, an interesting group of new uncyclized trichothecenes [57] (Fig. 2).

The range of compounds accumulated by these blocked mutant strains suggested an order to the oxygenations that result in the formation of T-2 toxin, which undergoes oxygenation at C-3, C-4, C-8, and C-15. MB2972 accumulates compounds with two oxygens at C-3 and C-15; MB1716 accumulates a compound with three oxygens at C-3, C-15, and C-4. This

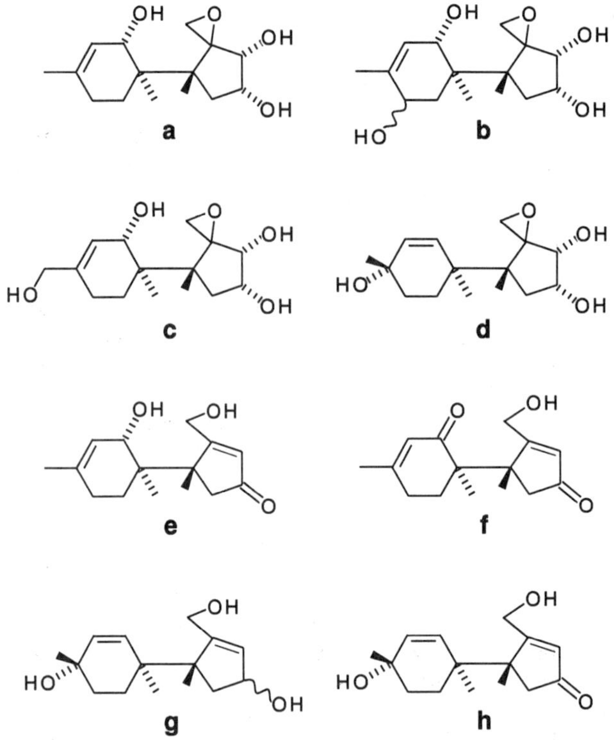

FIGURE 2. Modified trichothecenes isolated from *F. sporotrichioides* and *F. sambucinum*. (a) Tricho-9-ene-2,3,11-triol; (b) tricho-9-ene-2,3,8,11-tetraol; (c) tricho-9-ene-2,3,11,16-tetraol; (d) trichotriol; (e) FS-1; (f) FS-3; (g) FS-2; (h) FS-4. (a)–(d) from Ref. 57; (e)–(h) from Refs. 85, 87, 88.

pattern of accumulation indicated that the oxygenations occur in the following order: (1) C-3 or C-15; (2) C-4; and (3) C-8.

Experiments in which possible oxygenated precursors were fed to liquid cultures of *F. sporotrichioides* MB5493 have clearly demonstrated that the oxidation at C-3 precedes that at C-4 and C-15 [56]. Thus the order of oxygenations in the biosynthesis of T-2 toxin is C-3, C-15, C-4, C-8. In addition, these feeding experiments demonstrated (1) that oxygenation at C-3 occurs prior to the second cyclization and (2) that this second cyclization involves two steps (conversion of tricho-9-ene-2,3,11-triol to trichotriol and conversion of trichotriol to isotrichodermol) (Fig. 3, Scheme c); these two steps may be nonenzymatic. Subsequently, isotrichodermol is converted to isotrichodermin, an intermediate leading to T-2 toxin. These results are consistent with time-course experiments with MB2972, which indicate that the uncyclized compounds trichotriol and tricho-9-ene-2,3,11-triol appear transiently, suggesting that they are

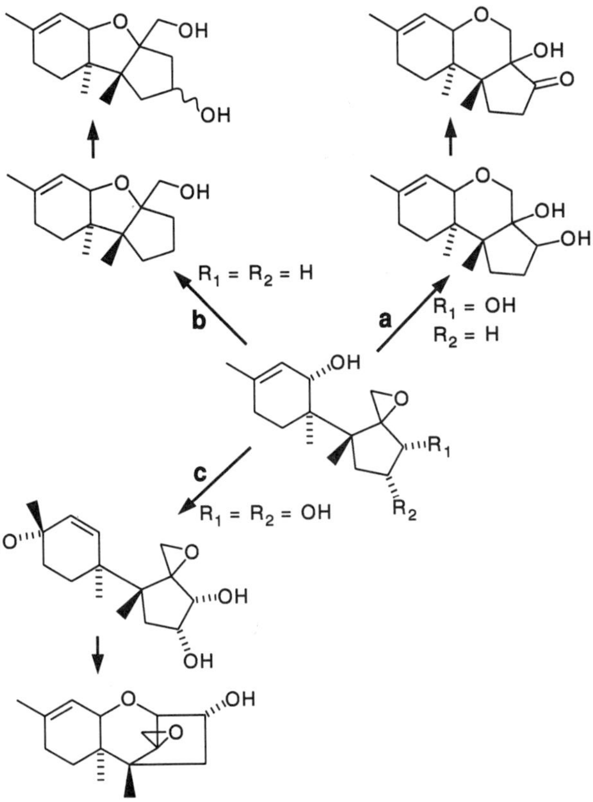

FIGURE 3. Proposed schemes for biosynthesis of (a) sambucoin, (b) apotrichodiol, and (c) isotrichodermol [56, 99].

intermediates in the biosynthesis of the calonectrin analogs [57]. Two intermediates in trichothecin biosynthesis in *T. roseum*, 12,13-epoxytrichothecene and trichodiol, were not bioconverted to T-2 toxin by MB5493.

Another aspect of trichothecene biosynthesis that has been investigated in *F. sporotrichioides* is the origin of the ester side chains. Labeling studies with ^{13}C and analysis of a leucine auxotroph have shown that leucine is the direct biosynthetic precursor of the isovalerate side chain in T-2 toxin and that propionyl and isobutyryl neosolaniol are branch pathway cometabolites [59, 90]. Consequently, T-2 toxin is the product of a mixed biogenesis that involves two primary metabolic pathways, one for isoprenoids and the other for an amino acid. A proposed pathway for the biosynthesis of T-2 toxin is presented in Fig. 4.

Recent information about trichothecene biosynthesis is also available from a series of studies involving *F. culmorum*. Zamir et al. [91] confirmed the role of trichodiene in *F. culmorum* biosynthesis of 3-acetyldeoxynivalenol (3-ADON) with an NMR study of the incorporation of

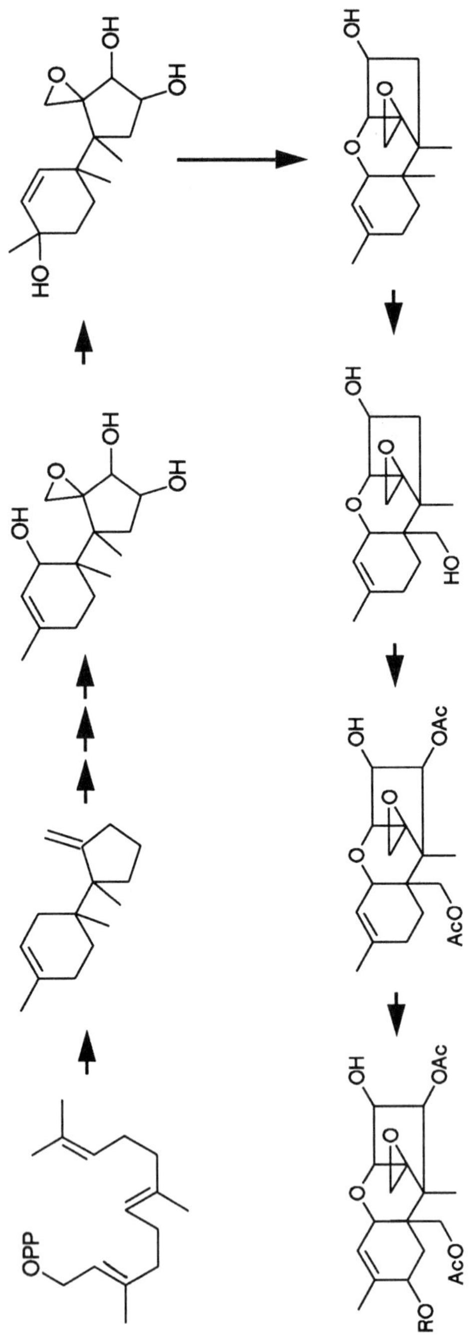

FIGURE 4. Proposed pathway for biosynthesis of T-2 toxin by *F. sporotrichioides* [56]. R = isovalerate.

specifically labeled deuterated trichodiene. Biosynthetic schemes for the conversion of trichodiene to 3-ADON have evolved with the isolation of 15 trichothecenes, including isotrichodermin, calonectrins, and 12,13-epoxytrichothecene, from *F. culmorum* cultures [92–94]. Based on the trichothecenes isolated and time-course experiments [95], Baldwin et al. [93] and Greenhalgh et al. [95] proposed a biosynthetic scheme that begins with a cyclization to 12,13-epoxytrichothecene, followed by a primary oxidation at C-3 and then secondary oxidations at C-15 and C-8. The isolation of both 3,8- and 3,15-oxygenated trichothecenes suggested that secondary oxidations may occur simultaneously although at different rates.

In more recent studies, two of the suspected precursors of 3-ADON in *F. culmorum,* isotrichodermin and 12,13-epoxytrichothecene, were investigated by a kinetic-pulse label (KPL) technique [17, 89, 90, 92–98]. Incorporation of label from isotrichodermin into 3-ADON confirmed that isotrichodermin was a precursor of 3-ADON. In contrast, the other suspected precursor, 12,13-epoxytrichothecene, although initially reported to also be converted to 3-ADON [17], was later found to be converted to sambucinol, a modified trichothecene, rather than 3-ADON [92].

The fact that 12,13-epoxytrichothecene is a precursor to neither T-2 nor 3-ADON whereas isotrichodermin is a precursor to both suggests that the biosynthesis of trichothecenes in *F. culmorum, G. pulicaris,* and *F. sporotrichioides* probably involves the same initial sequence of events, which is different from that for the biosynthesis of trichothecin by *T. roseum.* In addition, the conversion of 12,13-epoxytrichothecene to sambucinol suggests that although 12,13-epoxytrichothecene is not involved in *Fusarium* T-2 toxin or 3-ADON biosynthesis, it may be involved in other, closely related biosynthetic routes in *Fusarium* species, as described below.

The isolation of modified trichothecenes, which are derived from trichodiene but lack the 12,13-epoxide group, has led to a series of proposed schemes with different cyclizations. Mohr et al. [99] proposed schemes for the biosynthesis of sambucinol and sambucoin (Fig. 3, Scheme a), two modified trichothecenes initially isolated from *G. pulicaris.* The scheme for sambucinol involves oxidation of 12,13-epoxytrichothecene at both C-11 and C-3, followed by opening of the epoxide and acetalization. The scheme for sambucoin requires two allylic hydroxylations, at C-11 and C-2 of trichodiene, followed by an epoxidation. Nucleophilic attack by the C-11 hydroxyl at C-13 to form an oxygen bridge would result in a diol intermediate, which upon oxidation at C-2 would lead to sambucoin.

A third group of modified trichothecenes, the apotrichothecenes, have also been isolated from *F. sporotrichioides, F. culmorum, G. zeae,* and *F. crookwellense* [89, 100, 101]. The characterization of one apotrichothecene, apotrichodiol, has led to a proposed scheme for apotrichothecene biosynthesis (Fig. 3, Scheme b) in which 11-hydroxy-12,13-epoxytricho-10-ene is an intermediate [95]. The scheme involves formation of an oxygen bridge between C-11 and C-12 and opening of the epoxide. Zamir et al. [101] showed with KPL experiments that apotrichidiol is a dead-end metabolite in *F. culmorum* [101]. However, Greenhalgh et al. [95] detected oxidation products of apotrichodiol in crude fungal extracts of various *Fusarium* species, which suggests that apotrichodiol may not be a dead-end metabolite. In fact, they proposed an alternative scheme for the biosynthesis of sambucinol in which apotrichodiol is a precursor to sambucinol via one of these oxidation products, 11-epiapotrichothecene [95].

Finally, uncyclized 11-hydroxy trichothecenes, similar to those required by the schemes proposed for biosynthesis of some of the modified trichothecenes (sambucoins and apotrichothecenes), have been isolated from *F. sporotrichioides* and *G. pulicaris* [57, 85, 87, 88]. Feeding studies with the *Tox*$^-$ mutant strains of *F. sporotrichioides* suggest that these compounds also play a role in trichothecene biosynthesis [56]. Thus, the biosynthesis of true trichothecenes and these modified trichothecenes, probably share two additional steps in common after trichodiene before the two pathways branch.

VII. REGULATION OF TRICHOTHECENE PRODUCTION

Understanding the regulation of trichothecene biosynthesis is of fundamental importance. Although numerous studies have examined some of the factors that influence trichothecene production, none of the underlying regulatory mechanisms have been determined. The factors and possible mechanisms that can determine the final composition and concentration of trichothecenes in a given culture or sample involve both genetic and environmental components and are outlined in Table 3.

TABLE 3. Factors and Possible Mechanisms Involved in Regulating Trichothecene Production

Type of factor	Mechanisms
Genetic	
gene expression	Induction
natural variants	
mutants	Precursor control
Nutritional	
substrates	Feedback inhibition
Physical	
temperature, pH, light/dark	Bioregulators
H_2O, CO_2, O_2, etc.	
Biological	
plants, microbes	Catabolite repression
Chemical	
enzymes, compounds	Phosphate repression

Very little is known about the genetic factors that regulate trichothecene biosynthesis. Preliminary studies with crosses in *G. pulicaris* have indicated that one of the genetically defined *Tox* genes may repress the expression of at least one other *Tox* gene [38]. This repression appears to be gene-specific rather than allele-specific. Consequently, these crosses may have allowed the identification of a trans-acting regulatory gene; further analysis of these gene interactions are in progress. In another set of experiments, an examination of the expression of the trichodiene synthase enzyme in *F. sporotrichioides* and *G. pulicaris* showed that large increases in enzyme activity (50-fold and 10-fold, respectively), occurred just prior to the detectable onset of trichothecene biosynthesis [102]. These data indicate that regulation of trichodiene synthase activity is accomplished through increases in its cellular concentration and are consistent with trichodiene synthase being an inducible enzyme. Now that the trichodiene synthase gene is cloned, it will be possible both to determine whether this regulation occurs at the transcriptional or translational level and to investigate the necessary regulatory signals at the molecular level.

The relationship between the onset of trichothecene synthesis and glucose levels in liquid shake cultures has also been examined. Two groups working with strains of *F. sporotrichioides* observed that the initiation of trichothecene biosynthesis occurred with a high concentration of glucose remaining in the media [102, 103]. This implied that trichothecene biosynthesis is not subject to glucose repression. However, similar experiments with *G. zeae* indicated that the initiation of trichothecene synthesis was dependent upon the reduction of glucose to low levels [104]. These results may reflect strain or species differences.

Although it is well known that trichothecene profiles and levels change in response to different temperature and light conditions as well as to differences in the substrate or media composition [3, 7, 105, 106], a coherent picture of how these factors influence trichothecene production has not yet emerged. Again, the same factors are often reported to have opposite effects with respect to increasing or decreasing toxin levels.

One of the most interesting and challenging aspects of understanding how trichothecene production is regulated involves investigating how other organisms that coexist with trichothecene-producing strains in nature affect trichothecene production. These factors will be mentioned only briefly here as they are discussed in detail in Chapter 13. Trichothecenes can be chemically modified by other organisms. Microbes are known to produce enzymes that can transform or degrade trichothecenes [107–109]. Maize leaf tissue can also convert 15-acetyldeoxynivalenol to deoxynivalenol [110]. These organisms can also have a more direct effect on toxin biosynthesis. Phytoalexins produced by plants in response to fungal infections are known to specifically block trichothecene biosynthesis at the beginning of the pathway such that trichodiene accumulates [58, 111]. Clearly, the effects of both the host organism and the microbial community must be evaluated in studies concerning the regulation of trichothecene biosynthesis.

VIII. FUTURE STUDIES

It has been the intent of this review to highlight new areas of research aimed at addressing a broad range of questions concerning trichothecene production. The experimental tools necessary to provide critical information and hence rapid advancements in this field are now at hand. Thus, the investigation of many key aspects of trichothecene biosynthesis are at a pivotal point and in some cases, for the first time, at an attainable point. These include further elucidation of the trichothecene biosynthetic pathway, a precise identification and analysis of the remaining genes and enzymes involved in trichothecene production, a critical examination of the mechanisms that regulate trichothecene production, and, perhaps most intriguing, the determination of the function(s) of these interesting secondary metabolites.

A key element to these possible advancements is the fact that many of the genetic, molecular, biochemical, and chemical approaches, as well as much of the information derived from them, complement one another. For example, identification of the intermediate metabolites in the trichothecene biosynthetic pathway provides not only the order of the pathway but also the possible identification of substrates for the pathway enzymes. This information is critical to the development of appropriate in vitro assays for the identification, characterization, and isolation of these enzymes. Identification of the enzymes could also proceed via molecular approaches by using complementation of the Tox^- mutants to clone the corresponding genes.

Once a gene is cloned, it can be manipulated in a wide variety of ways. Many of these manipulations are based on the ability to introduce or remove genes from the fungal genome at will. The occurrence of homologous integration in *F. sporotrichioides* makes it amenable to such studies. Undoubtedly, the molecular techniques that delete, replace, or otherwise modify gene loci with respect to their regulatory and structural regions will be heavily exploited to determine the critical structure–function relationships with respect to the *Tox* gene products and the regulatory control mechanisms that determine their expression. In conjunction with these studies, the techniques of chromosome separation and RFPL (restriction fragment length polymorphism) analyses can also be applied to map these cloned genes.

In addition to utilizing the Tox^- mutants to clone *Tox* genes, these mutants have already proven useful both in defining steps in the biosynthetic pathway and in providing a critical test to demonstrate that trichothecenes can be pathogenicity factors. Most interestingly, these same mutants are also altered in conidial development, thus providing the first evidence that trichothecenes may in fact be morphogens that play an important role in normal fungal development and differentiation (Beremand, unpublished results).

Clearly, future studies involving trichothecene biosynthesis will proceed in several fascinating directions. Many of these studies will be focused on areas that have tremendous potential to provide new insight into the biological functions of the trichothecenes and the underlying mechanisms that trigger and regulate their production.

REFERENCES

1. Busby, W. F., Jr., and Wogan, G. N. Trichothecenes. In *Mycotoxins and N-Nitroso Compounds: Environmental Risks*, Vol. II (R. C. Shank, Ed.), CRC Press, Boca Raton, Fla., 1981, pp. 29–45.
2. Cole, R. J., and Cox, R. H. *Handbook of Toxic Fungal Metabolites*, Academic, New York, 1981, pp. 152–263.
3. Joffee, A. Z. *Fusarium Species: Their Biology and Toxicology*, Wiley, New York, 1986.
4. Marasas, W. F. O., Nelson, P. E., and Toussoun, T. A. *Toxigenic Fusarium Species: Identity and Mycotoxicology*, Penn. State Univ. Press, University Park, 1984.
5. Mirocha, C. J. Mycotoxins associated with *Fusarium*. In *Applied Mycology of Fusarium* (M. O. Moss and J. E. Smith, Eds.), Cambridge Univ. Press, Cambridge, 1984, pp. 141–155.
6. Ueno, Y. *Trichothecenes: Chemical, Biological and Toxicological Aspects*, Elsevier, New York, 1983.
7. Smalley, E. B., Marasas, W. F. O., Strong, F. M., Bamburg, J. R., Nichols, R. E., and Kosuri, N. R. Mycotoxicoses associated with moldy corn. In *Toxic Microorganisms* (M. Herzberg, Ed.), U.S. Dept. of Interior, Washington, D. C., 1970, pp. 163–173.
8. Betina, V. *Bioactive Molecules*, Vol. 9, *Mycotoxins: Chemical, Biological and Environmental Aspects*, Elsevier, Amsterdam, 1989.
9. Booth, C. The *Fusarium* problem: historical, economic and taxonomic aspects. In *The Applied Mycology of Fusarium* (M. O. Moss and J. E. Smith, Eds.), Cambridge Univ. Press, Cambridge, 1984, pp. 1–13.
10. Ishii, K., Kobayashi, J., Ueno, Y., and Ichinoe, M. Occurrence of trichothecin in wheat. *Appl. Environ. Microbiol.*, *52*:331–333, 1986.
11. Cundliffe, E., Cannon, M., and Davies, J. Mechanisms of inhibition of eukaryotic protein synthesis by trichothecene fungal toxins. *Proc. Natl. Acad. Sci. USA*, *71*:30–34, 1974.
12. McLaughlin, C. S., Vaughan, M. H., Campbell, I. M., Wei, C. M., Stafford, M. E., and Hansen, B. S. Inhibition of protein synthesis by trichothecenes. In *Mycotoxins in Human and Animal Health* (J. V. Rodricks, C. W. Hesseltine, and M. A. Mehlman, Eds.), Pathotox, Park Forest South, Ill., 1977, pp. 263–273.
13. Wei, C., and McLaughlin, C. S. Structure–function relationship in the 12,13-epoxytrichothecenes, novel inhibitors of protein synthesis. *Biochem. Biopys. Res. Commun.*, *57*:838–844, 1974.
14. Anderson, D. W., Black, R. M., Lee, C. G., Pottage, C., Rickard, R. L., Sandford, M. S., Webber, T. D., and Williams, N. E. Structure–activity studies of trichothecenes: cytotoxicity of analogs and reaction products derived from T-2 toxin and neosolaniol. *J. Med. Chem.*, *32*:555–562, 1989.
15. Bamburg, J. R., and Strong, F. M. 12,13-Epoxytrichothecenes. In *Microbial Toxins*, Vol. 7 (S. Kadis, A. Ciegler, and S. J. Ajl, Eds.), Academic, New York, 1971, pp. 207–292.

16. Tamm, C. L., and Breitenstein, W. The biosynthesis of trichothecene mycotoxins. In *The Biosynthesis of Mycotoxins* (P. S. Steyn, Ed.), Academic, New York, 1980, pp. 69–104.
17. Zamir, L. O. Biosynthesis of 3-acetyldeoxynivalenol and sambucinol. *Tetrahedron, 45*:2277–2305, 1989.
18. Greenhalgh, R., Blackwell, B. A., and Savard, M. The NMR spectra of trichothecenes and related fungal metabolites. *Tetrahedron, 45*: 2373–2383, 1989.
19. Plattner, R. D., Beremand, M. N., and Powell, R. G. Analysis of trichothecene mycotoxins by mass spectrometry and tandem mass spectrometry. *Tetrahedron, 45*:2251–2262, 1989.
20. Gordon, W. L. The occurrence of *Fusarium* species in Canada. II. *Can. J. Bot., 30*:209–251, 1952.
21. Gordon, W. L. Sex and mating types in relation to the production of perithecia by certain species of *Fusarium*. *Proc. Can. Phytopathol. Soc., 28*:11, 1961.
22. Francis, R. G., and Burgess, L. W. Characteristics of two populations of *Fusarium roseum* 'Graminearum' in eastern Australia. *Trans. Br. Mycol. Soc., 68*:421–427, 1977.
23. Messiaen, C. M., and Cassini, R. Taxonomy of *Fusarium*. In *Fusarium, Diseases, Biology, and Taxonomy* (P. E. Nelson, T. A. Toussoun, and R. J. Cole, Eds.), Penn. State Univ, University Park, 1981, pp. 426–455.
24. El-Groll, N. E., McRitchie, J. J., Schoulties, C. L., and Ridings, W. H. The identification, induction of perithecia, and pathogenicity of *Gibberella (Fusarium) tricincta* n. sp. *Can. J. Bot., 56*:2203–2205, 1978.
25. Booth, C. *The Genus Fusarium*, Commonwealth Mycological Institute, Kew, Surrey, England, 1971.
26. Reichler, R. E., and Snyder, W. C. Heterothallism and ascospore number in *Calonectria rigidiuscula*. *Phytopathology, 54*:1297–1299, 1964.
27. Ford, E. J., Bourret, J. A., and Snyder, W. C. Biologic specialization in *Calonectria (Fusarium) rigidiuscula* in relation to green point gall of cocoa. *Phytopathology, 57*:710–712, 1967.
28. Cook, R. J., and Bruehl, G. W. *Calonectria nivalis*, perfect stage of *Fusarium nivale*, occurs in the field in North America. *Phytopathology, 56*:1100–1101, 1966.
29. Desjardins, A. E., and Beremand, M. N. A genetic system for trichothecene toxin production in *Gibberella pulicaris (Fusarium sambucinum)*. *Phytopathology, 77*:678–683, 1987.
30. Ichinoe, M., Kurata, H., Suguira, Y., and Ueno, Y. Chemotaxonomy of *Gibberella zeae* with special reference to production of trichothecenes and zearalenone. *Appl. Environ. Microbiol., 46*:1364–1369, 1983.
31. Mirocha, C. J., Abbas, H. K., Windels, C. E., and Xie, W. Variation in deoxynivalenol, 15-acetyldeoxynivalenol, 3-acetyldeoxynivalenol, and zearalenone production by *Fusarium graminearum* isolates. *Appl. Environ. Microbiol., 55*:1315–1316, 1989.

32. Blackwell, B. A., Miller, J. D., and Greenhalgh, R. ^{13}C NMR study of the biosynthesis of toxins by *Fusarium graminearum*. *J. Biol. Chem.*, *260*:4243–4247, 1985.
33. Windels, C. E., Mirocha, C. J., Abbas, H. K., and Xie, W. Perithecium production in *Fusarium graminearum* populations and lack of correlation with zearalenone production. *Mycologia, 81*:272–277, 1989.
34. El-Ani, A. S. Cytogenetics of sex in *Gibberella cyanogena* (Desm.) Sacc. *Science, 123*:850, 1956.
35. Gordon, W. L. Geographical distribution of mating types in *Gibberella cyanogena* (Desm.) Sacc. *Nature, 173*:505, 1954.
36. Matuo, T., and Snyder, W. C. Use of morphology and mating populations in the identification of formal species in *Fusarium solani*. *Phytopathology, 63*:562–565, 1973.
37. Burnett, J. H. Aspects of *Fusarium* genetics. In *The Applied Mycology of Fusarium* (M. O. Moss and J. E. Smith, Eds.), Cambridge Univ. Press, Cambridge, 1984, pp. 39–69.
38. Beremand, M. N., and Desjardins, A. E. Trichothecene biosynthesis in *Gibberella pulicaris*: inheritance of C-8 hydroxylation. *J. Ind. Microbiol.*, *3*:167–174, 1988.
39. Booth, C. Perfect states (teleomorphs) of *Fusarium* species. In *Fusarium: Diseases, Biology, and Taxonomy* (P. E. Nelson, T. A. Toussoun, and R. J. Cook, Eds.), Penn. State Univ. Press, University Park, 1981, pp. 446–452.
40. Nelson, P. E., Toussoun, T. A., and Marasas, W. F. O. *Fusarium Species: An Illustrated Manual for Identification*, Penn. State Univ. Press, University Park, 1983.
41. Howson, W. T., McGinnis, R. C., and Gordon, W. L. Cytological studies on the perfect stages of some species of *Fusarium*. *Can. J. Genet. Cytol.*, *5*:60–64, 1963.
42. Puhalla, J. E. Genetic considerations of the genus *Fusarium*. In *Fusarium Diseases, Biology and Taxonomy* (P. E. Nelson, T. A. Toussoun, and R. J. Cook, Eds.), Penn. State Univ. Press, University Park, 1981, pp. 291–305.
43. Snyder, W. C., Georgopoulous, S. G., Webster, R. K., and Smith, S. N. Sexuality and genetic behavior in the fungus *Hypomyces* (*Fusarium*) *solani* f. sp. *cucurbitae*. *Hilgardia, 43*:161–185, 1975.
44. Fincham, J. R. S., Day, P. R., and Radford, A. *Fungal Genetics*, 4th ed., Blackwell, Oxford, 1979.
45. Bistis, G. N., and Georgopoulous, G. S. Some aspects of sexual reproduction in *Nectria haematococca* var. *cucurbitae*. *Mycologia, 71*: 127–143, 1979.
46. Beremand, M. N. Genetic and mutational tools for investigating the genetics and molecular biology of trichothecene production in *Gibberella pulicaris* (*Fusarium sambucinum*). *Mycopathology, 107*:67–74, 1989.
47. Desjardins, A. D., and Gardner, H. W. Genetic analysis in *Gibberella pulicaris*: rishitin tolerance, rishitin metabolism, and virulence on potato tubers. *Mol. Plant–Microbe Interactions, 2*:26–34, 1989.

48. Salch, Y. P., and Beremand, M. N. Development of a transformation system for *Fusarium sambucinum*. *J. Cell. Biochem. Suppl. 12C*:290, 1988.
49. Boyd, A. E. W. Potato storage diseases. *Rev. Plant Pathol., 51*: 297–321, 1972.
50. Desjardins, A. E., and Plattner, R. D. Trichothecene toxin production by strains of *Gibberella pulicaris* (*Fusarium sambucinum*) in liquid culture and in potato tubers. *J. Agric. Food Chem., 37*:388–392, 1989.
51. Ishii, K., Pathre, S. V., and Mirocha, C. J. Two new trichothecenes produced by *Fusarium roseum*. *J. Agric. Food Chem., 26*:649–653, 1978.
52. Beremand, M. N. Isolation and characterization of mutants blocked in T-2 toxin biosynthesis. *Appl. Environ. Microbiol., 53*:1855–1859, 1987.
53. Beremand, M. N., Black, P. J., and Plattner, R. D. Isolation and characterization of two new mutants blocked in T-2 toxin biosynthesis. *J. Cell. Biochem., 12C*:261, 1988.
54. Plattner, R. D., Tjarks, L. W., and Beremand, M. N. Trichothecenes accumulated in liquid culture of a mutant *Fusarium sporotrichioides* NRRL 3299. *Appl. Environ. Microbiol., 55*:2190–2194, 1989.
55. Hunter, K. W., Brimfield, A. A., Miller, M., Finkelman, F. D., and Chu, F. S. Preparation and characterization of monoclonal antibodies to the trichothecene mycotoxin T-2. *Appl. Environ. Microbiol., 49*: 168–172, 1985.
56. McCormick, S. P., Taylor, S. L., Plattner, R. D., and Beremand, M. N. Bioconversion of possible T-2 toxin precursors by a mutant strain of *Fusarium sporotrichioides* NRRL 3299. *Appl. Environ. Microbiol., 56*:702–706.
57. McCormick, S., Taylor, S. L., Plattner, R. D., and Beremand, M. N. New modified trichothecenes accumulated in solid culture by mutant strains of *Fusarium sporotrichioides*. *Appl. Environ. Microbiol., 55*: 2195–2199, 1989.
58. Desjardins, A. E., Spenser, G. F., Plattner, R. D., and Beremand, M. N. Furanocoumarin phytoalexins, trichothecene toxins, infection of *Pastinaca sativa* by *Fusarium sporotrichioides*. *Phytopathology, 79*: 170–175, 1989.
59. Beremand, M. N., VanMiddlesworth, F., Taylor, S., Plattner, R. D., and Weisleder, D. Leucine auxotrophy specifically alters the pattern of trichothecene production in a T-2 toxin-producing strain of *Fusarium sporotrichioides*. *Appl. Environ. Microbiol., 54*:2759–2766, 1988.
60. Evans, R., Hanson, J. R., and Marten, T. Biosynthesis of 2-*cis*-farnesol. *J. Chem. Soc. Chem. Commun., 14*:465, 1973.
61. Evans, R., and Hanson, J. R. Studies in terpenoid biosynthesis. Part XIV. Formation of the sesquiterpene trichodiene. *J. Chem. Soc., Perkin Trans., 1*:326–329, 1976.
62. Machida, Y., and Nozoe, S. Biosynthesis of trichothecin and related compounds. *Tetrahedron, 28*:5113–5117, 1972.

63. Hohn, T. M., and VanMiddlesworth, F. Purification and characterization of the sesquiterpene cyclase trichodiene synthetase from *Fusarium sporotrichioides*. *Arch. Biochem. Biophys.*, *251*:756–761, 1986.
64. Cane, D. E., Ha, H., Paegellis, C., Waldmeir, F., Swanson, S., and Murthy, P. P. N. Trichodiene biosynthesis and the stereochemistry of the enzymatic cyclization of farnesyl pyrophosphate. *Bioorg. Chem.*, *13*:246–265, 1985.
65. Cane, D. E., and Ha, H. Trichodiene biosynthesis and the enzymatic cyclization of nerolidyl pyrophosphate. *J. Am. Chem. Soc.*, *108*: 3097–3099, 1986.
66. Desjardins, A. E., Plattner, R. D., and VanMiddlesworth, F. L. Trichothecene biosynthesis in *Fusarium sporotrichioides*: origin of the oxygen atoms of T-2 toxin. *Appl. Environ. Microbiol.*, *51*:493–497, 1986.
67. VanMiddlesworth, F., Desjardins, A. E., Taylor, S. L., and Plattner, R. D. Trichodiene accumulation by ancymidol treatment of *Gibberella pulicaris*. *J. Chem. Soc., Chem. Commun.*, *1986*:1156–1157.
68. Desjardins, A. E., Plattner, R. D., and Beremand, M. N. Ancymidol blocks trichothecene biosynthesis and leads to accumulation of trichodiene in *Fusarium sporotrichioides* and *Gibberella pulicaris*. *Appl. Environ. Microbiol.*, *53*:1860–1865, 1987.
69. Hohn, T. M., and Beremand, P. D. Isolation and nucleotide sequence of a sesquiterpene cyclase gene form the trichothecene-producing fungus *Fusarium sporotrichioides*. *Gene*, *79*:131–138, 1989.
70. Hohn, T. M., and Plattner, R. D. Expression of the trichodiene synthase gene of *Fusarium sporotrichioides* in *Escherichia coli* results in sesquiterpene production. *Arch. Biochem. Biophys.*, *275*:92–97, 1989.
71. Hynes, M. J. Transformation of filamentous fungi. *Exp. Mycol.*, *10*: 1–8, 1986.
72. Rambosek, J., and Leach, L. Recombinant DNA in filamentous fungi: progress and prospects. *CRC Crit. Rev. Biotechnol.*, *6*:357–393, 1987.
73. Rine, J., and Carlson, M. *Saccharomyces cerevisiae* as a paradigm for modern molecular genetics of fungi. In *Gene Manipulations in Fungi* (J. W. Bennett and L. L. Lasure, Eds.), Academic, Orlando, Fla., 1985, pp. 125–160.
74. Ballance, D. J., and Turner, G. Development of high-frequency transforming vector for *Aspergillus nidulans*. *Gene*, *36*:321–331, 1985.
75. Timberlake, W. E. *Molecular Genetics of Filamentous Fungi*, UCLA Symposia on Molecular and Cellular Biology, New Series, Vol. 34, Alan R. Liss, Inc., New York, 1985.
76. Maladier, L., Daboussi, M. J., Julien, J., Roussel, F., Scazzocchio, C., and Brygoo, Y. Cloning of the nitrate reductase gene (*nig* D) of *Aspergillus nidulans* and its use for transformation of *Fusarium oxysporum*. *Gene*, *78*:147–156, 1989.

77. Kistler, H. C., and Benny, U. K. Genetic transformation of the fungal plant wilt pathogen *Fusarium oxysporum*. *Curr. Genet.*, *13*:145–149, 1988.
78. Soliday, C. L., Dickman, M. B., and Kolattukudy, P. E. Structure of the cutinase gene and detection of promoter activity in the 5'-flanking region of fungal transformation. *J. Bacteriol.*, *171*:1942–1951, 1989.
79. Marek, E. T., Schardl, C. L., and Smith, D. A. Molecular transformation of *Fusarium solani* with antibiotic resistance marker having no fungal DNA homology. *Curr. Genet.*, *15*:421–428, 1989.
80. Dickman, M., and Partridge, J. E. Use of molecular marker for monitoring fungi involved in stalk rot of corn. *Theor. Appl. Genet.*, *77*: 535–539, 1989.
81. Beremand, M. N., Turgeon, B. G., and Yoder, O. C. Transformation of *Fusarium sporotrichioides*. 14th Fungal Biology Conference (Abstr. 27), 1987.
82. Turgeon, B. G., Garber, R. C., and Yoder, O. C. Development of a fungal transformation system based on selection of sequences with promoter activity. *Mol. Cell Biol.*, *7*:3297–3305, 1987.
83. Cullen, D., Smalley, E. B., and Dimond, R. L. Heterokaryosis in *Fusarium tricinctum* and *F. sporotrichioides*. *J. Gen. Micro.*, *129*: 3035–3041, 1983.
84. Corley, D. G., Rottinghaus, G. E., and Tempesta, M. S. Secondary metabolites from *Fusarium sporotrichioides* MC-72083. *J. Nat. Prod.*, *50*:897–902, 1987.
85. Corley, D. G., Rottinghaus, G. E., and Tempesta, M. S. Toxic trichothecenes from *Fusarium sporotrichioides* (MC-72083). *J. Org. Chem.*, *52*:4405–4408, 1987.
86. Corley, D. G., Rottinghaus, G. E., and Tempesta, M. S. Novel trichothecenes from *Fusarium sporotrichioides*. *Tetrahedron Lett.*, *27*: 427–430, 1986.
87. Corley, D. G., Rottinghaus, G. E., Tracy, J. K., and Tempesta, M. S. New trichothecene mycotoxins of *Fusarium sporotrichioides* (MC 72083). *Tetrahedron Lett.*, *27*:4133–4136, 1986.
88. Sanson, D. R., Corley, D. G., Barnes, C. L., Searles, S., Schlemper, E. O., and Tempesta, M. S. New mycotoxins from *Fusarium sambucinum*. *J. Org. Chem.*, *54*:4313–4318, 1989.
89. Greenhalgh, R., Fielder, D. A., Morrison, L. A., Charland, J.-P., Blackwell, B. A., Savard, M. E., and ApSimon, J. W. Secondary metabolites of *Fusarium* species: apotrichothecene derivatives. *J. Agric. Food Chem.*, *37*:699–705, 1989.
90. VanMiddlesworth, F., Beremand, M. N., Isbell, T. A., and Weisleder, D. T-2 toxin biosynthesis: origin of isovalerate side chain. *J. Org. Chem.*, 1237–1240.
91. Zamir, L. O., Gauthier, M. J., Devor, K. A., Nadeau, Y., and Sauriol, F. Trichodiene is a precursor to trichothecenes. *J. Chem. Soc., Chem. Commun.*, *1989*:598–600.
92. Baldwin, N. C. P., Bycroft, B. W., Dewick, P. M., Gilbert, J., and Holden, I. Biosynthesis of trichothecene mycotoxins in *Fusarium culmorum* cultures. *Z. Naturforsch.* *40c*:514–518, 1985.

93. Baldwin, N. C. P., Bycroft, B. W., Dewick, P. M., and Gilbert, J. Metabolic conversions of trichothecene mycotoxins: biotransformation of 3-acetyldeoxynivalenol into fusarenon-X. *Z. Naturforsch., 41c*: 845–850, 1986.
94. Baldwin, N. C. P., Bycroft, B. W., Dewick, P. M., Marsh, D. C., and Gilbert, J. Trichothecene mycotoxins from *Fusarium culmorum* cultures. *Z. Naturforsch., 42c*:1043–1049, 1987.
95. Greenhalgh, R., Blackwell, B. A. Pare, J. R. J., Miller, J. D., Levandier, D., Meier, R. M., Taylor, A., and ApSimon, J. W. Isolation and characterization by mass spectrometry and NMR spectroscopy of secondary metabolites of some *Fusarium* species. In *Mycotoxins and Phycotoxins* (P. P. Steyn and R. Vleggaar, Eds.), Elsevier, Amsterdam, 1986, pp. 137–152.
96. Zamir, L. O. Biosynthesis of 3-acetyldeoxynivalenol. In *Mycotoxins and Phycotoxins* (P. S. Steyn and R. Vleggaar, Eds.), Elsevier, Amsterdam, 1986, pp. 109–124.
97. Zamir, L. O., and Devor, K. A. Kinetic pulse-labeling study of *Fusarium culmorum. J. Biol. Chem., 262*:15348–15353, 1987.
98. Zamir, L. O., Devor, K. A., Nikolakakis, A., and Sauriol, F. Biosynthesis of *Fusarium culmorum* trichothecenes. *J. Biol. Chem., 265*:6713–6725, 1990.
99. Mohr, P., Tamm, C., Zurcher, W., and Zehnder, M. Sambucinol and sambucoin, two metabolites of *Fusarium sambucinum* possessing modified trichothecene structures. *Helv. Chim. Acta, 67*:406–412, 1984.
100. ApSimon, J. W., Blackwell, B., Greenhalgh, R., Meier, R. M., Miller, D., Pare, J. R. J., and Taylor, A. Secondary metabolites produced by some *Fusarium* species. In *Mycotoxins and Phycotoxins* (P. S. Steyn and R. Vleggaar, Eds.), Elsevier, Amsterdam, 1986, pp. 125–136.
101. Zamir, L. O., Devor, K. A., and Sauriol, F. Structure determination and biosynthesis of a novel metabolite of *Fusarium culmorum*, apotrichodiol. *J. Biol. Chem., 262*:15354–15358, 1987.
102. Hohn, T., and Beremand, M. N. Regulation of trichodiene synthase in *Fusarium sporotrichioides* and *Gibberella pulicaris* (*Fusarium sambucinum*). *Appl. Environ. Microbiol., 55*:1500–1503, 1989.
103. Ueno, Y., Sawano, M., and Ishii, K. Production of trichothecene mycotoxins by *Fusarium* species in shake culture. *Appl. Microbiol., 30*:4–9, 1975.
104. Miller, J. D., Taylor, A., and Greenhalgh, R. Production of deoxynivalenol and related compounds in liquid culture by *Fusarium graminearum. Can. J. Microbiol., 29*:1171–1178, 1983.
105. Cuero, R. G., Murray, A., and Smith, J. E. Control of toxigenic molds in cereal seeds. In *Biocontrol of Plant Diseases*, Vol. 2 (K. G. Mukerji and K. L. Garg, Eds.), CRC Press, Boca Raton, Fla., 1988, pp. 67–84.
106. El-Banna, A. A., Scott, P. M., Lau, P.-Y., Sakuma, T., Platt, H. W., and Campbell, V. Formation of trichothecenes by *Fusarium solani* var. *coreuleum* and *Fusarium sambucinum* in potatoes. *Appl. Environ. Microbiol., 47*:1169–1171, 1984.

107. Claridge, C. A., and Schmitz, H. Microbial and chemical transformations of some 12,13-epoxytrichothec-9,10-enes. *Appl. Environ. Microbiol.*, *36*:63–67, 1978.
108. Yoshizawa, T., and Morooka, N. Biological modification of trichothecene mycotoxins: acetylation and deacetylation of deoxynivalenol by *Fusarium* sp. *Appl. Microbiol.*, *29*:54–58, 1975.
109. Yoshizawa, T., and Morooka, N. Comparative studies on microbial and chemical modifications of trichothecene mycotoxins. *Appl. Environ. Microbiol.*, *30*:38–43, 1975.
110. Miller, J. D., Young, J. C., and Trenholm, H. L. *Fusarium* toxins in field corn. 1. Time course of fungal growth and production of deoxynivalenol and other mycotoxins. *Can. J. Bot.*, *61*:3080–3087, 1983.
111. Desjardins, A. E., Plattner, R. D., and Spencer, G. F. Inhibition of trichothecene toxin biosynthesis by naturally occurring shikimate aromatics. *Phytochemistry*, *27*:767–771, 1988.

15

CYCLOPIAZONIC ACID: SPECULATION ON ITS FUNCTION IN FUNGI

RONALD T. RILEY *Richard B. Russell Agricultural Research Center, Agricultural Research Service, U.S. Department of Agriculture, Athens, Georgia*

DOUGLAS E. GOEGER *Division of Human Nutrition, University of Texas Medical Branch, Galveston, Texas*

I. INTRODUCTION

Bu'Lock [1] concluded his introductory chapter in the book *The Biosynthesis of Mycotoxins* with the statement

> If we are to find specific functions—from the microbes' point of view—for mycotoxins, we must try to establish their role in the microorganism's own interactions with its habitat, its survival and proliferation parameters, and its interaction with predators, host, and commensal species. So far as I am aware, this has hardly been attempted to date—a comment which would perhaps be better made in a postscript rather than in an introduction!

Cyclopiazonic acid is an indole tetramic acid (Fig. 1) that is produced by many species of *Aspergillus* and *Penicillium*. The current understanding of the role that cyclopiazonic acid plays in the ecology of the fungi that produce it has changed little since Bu'Lock's comments a decade ago. With the exception of a study in which Wicklow and Cole [2] hypothesized a role for cyclopiazonic acid in the chemical defense system of *Aspergillus flavus*, there has been little effort to expand the state of knowledge concerning the role of cyclopiazonic acid in the ecology of fungi. The objective of the present work is not to report evidence on the function of cyclopiazonic acid in fungal ecology, but to summarize discoveries of the biological activity of cyclopiazonic acid in animal systems and extend these observations to consideration of the biological activity of cyclopiazonic acid in the ecological context. For example, understanding the interaction of cyclopiazonic acid in animal systems may help to identify areas that may prove profitable for developing an understanding of the functions of cyclopiazonic acid in fungal systems. If it turns out that there are similar biological responses to cyclopiazonic acid in both fungal and animal cells,

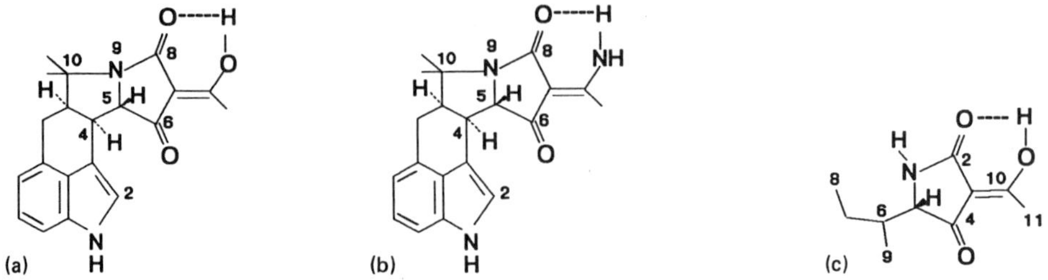

FIGURE 1. The chemical structure of (a) cyclopiazonic acid, (b) cyclopiazonic acid imine, and (c) tenuazonic acid.

then a link will have been made between the biochemical language of cells that might help to explain why fungal metabolites in foods and feeds can ultimately cause toxicity in animals.

Three areas of the biological activity of cyclopiazonic acid in animal systems will be considered. All three areas have been studied extensively since 1983. The first is the ability of cyclopiazonic acid to alter calcium homeostasis [3–7]. The second is the ability of cyclopiazonic acid to induce charge alterations in plasma membranes and mitochondria [7–10]. The third area is the ability of cyclopiazonic acid to behave as an antioxidant [6]. Table 1 summarizes the effects of tetramic acids on these three areas.

II. CYCLOPIAZONIC ACID–INDUCED ALTERATIONS IN CALCIUM HOMEOSTASIS

In animal systems, cyclopiazonic acid appears to alter calcium homeostasis via two distinct processes. First, it is now well established that cyclopiazonic acid is an extremely potent and specific inhibitor of calcium-dependent ATPase of cardiac and skeletal muscle sarcoplasmic reticulum [3, 4, 11] and possibly the endoplasmic reticulum [6] of animal cells. In both rat [3] and rabbit [11] sarcoplasmic reticulum vesicles, significant inhibition occurs at concentrations as low as 5 nmol of cyclopiazonic acid per milligram of sacroplasmic reticulum protein, indicating a possible stoichiometric reaction between cyclopiazonic acid and the calcium-dependent ATPase. Second, cyclopiazonic acid alters the calcium permeability of vesicles via some mechanism that not only involves inhibition of the ATPase [4] but also alters the kinetics of $^{45}Ca^{2+}$ uptake by cells [5, 6].

A. Specific Inhibition of Calcium-Dependent ATPases

The work of Seidler et al. [11] indicates that cyclopiazonic acid inhibits the calcium transport ATPase of sarcoplasmic reticulum by interfering with the conformational transition between the E_1 and E_2 states of the enzyme, an essential step for calcium transport. The findings are in agreement with earlier findings that cyclopiazonic acid prevents phosphorylation of the

TABLE 1. A Summary of the Actions of Cyclopiazonic Acid, Cyclopiazonic Acid Imine, and Tenuazonic Acid on Vesicle and Cell Preparations

Action	Concentration		
	CPA	CPAI	TA
Effects in SRV[a]			
calcium uptake by SRV	30	150	>1000
calcium-dependent SRV ATPase activity	30	150	>1000
phosphorylation of SRV	5	n.d.	n.d.
Calcium uptake by lysates of renal cells[b]	0.5	n.d.	n.d.
Altered calcium permeability[c]			
calcium efflux from SRV	10, 33	n.d.	n.d.
^{45}Ca uptake by intact cells	50, 100	n.d.	n.d.
Fixed charge alterations[d]			
surface charge alterations in cells	10	50	>750
Antioxidant activity[e]			
prevention of lipid peroxidation in cells	30	65	>500

Abbreviations: CPA, cyclopiazonic acid; CPAI, cyclopiazonic acid imine; TA, tenuazonic acid; SRV, sarcoplasmic reticulum vesicles; n.d., not done.

[a]The concentration, in nmoles/mg SR protein, at which 50% inhibition was measured [3, 4, 7].

[b]The micromolar concentration at which 50% inhibition was measured [6].

[c]The micromolar concentration at which the experiments were conducted [4–6].

[d]The micromolar concentration at which a 100% increase in tetraphenylphosphonium was measured in muscle myoblasts, L6 cells [7–10].

[e]The micromolar concentration that inhibited malondialdehyde formation by 50% in renal cells, LLC-PK_1 cells [6].

ATPase and inhibits calcium binding to a high-affinity binding site [4] that precedes the phosphorylation of the ATPase [12]. Seidler et al. [11] also found that at high ATP concentrations cyclopiazonic acid inhibition of the ATPase was reduced. However, inhibition of the ATPase could not be attributed to competition for the nucleotide triphosphate binding site since cyclopiazonic acid also inhibits the hydrolysis of acetyl phosphate, *p*-nitrophenylphosphate, and carbamylphosphate. The fact that high concentrations of ATP reduce the extent of inhibition may offer some interesting insights into the function of cyclopiazonic acid in fungi.

Cyclopiazonic acid inhibition of the calcium transport ATPase of sarcoplasmic and endoplasmic reticulum is important, but more exciting is the specificity of the inhibition. Seidler et al. [11] found that cyclopiazonic acid had no effect on hydrolysis of ATP by the following ATPases: kidney Na^+,K^+-ATPase, gastric H^+,K^+-ATPase, erythrocyte Ca^{2+} transport ATPase, the beef heart mitochondrial F_1-ATPase, brain microsome Na^+,K^+-ATPase activity, and T-tubule and sarcolemma Mg^{2+}-stimulated ATPase. Other plasma membrane ATPases (unpublished results), have been tested, and no ATPase in animal cells other than the sarcoplasmic and endoplasmic reticulum calcium-dependent ATPase is affected by cyclopiazonic acid.

Ion-motive ATPases are grouped into three classes: the electrogenic F-type ATPases, the ion-transporting P-type ATPases [13], and the V-type (vacuolar), which transport only protons [14]. Amino acid sequencing studies indicate that all P-type ATPases have a common ancestral origin and that the F-type ATPases have evolved independently [13]. The conserved regions of P-type ATPases are the kinase domain, which binds ATP (E_1 conformation), and the phosphatase domain, which catalyzes the hydrolysis of the phosphorylated intermediate (E_2 conformation). Transition between the conformations (E_1 to E_2) is required for translocation of the cation. Seidler et al. [11] hypothesized that cyclopiazonic acid stabilizes the calcium ATPase in an enzymatically inactive conformation; they concluded that inhibition by cyclopiazonic acid involves a feature of the calcium pump that makes it distinct from other P- and F-type ATPases. The fact that cyclopiazonic acid inhibits calcium binding to a high-affinity calcium binding site [4] and that the only P-type calcium-dependent ATPases affected are those that do not require calmodulin suggests that the specificity of cyclopiazonic acid is due to its interaction with that part of the ATPase which confers ion specificity.

B. Altered Calcium Permeability

In addition to its action as a specific, high-affinity inhibitor of the calcium transport ATPase, cyclopiazonic acid alters calcium permeability in vesicles and cells in a manner that appears to involve a different mechanism of action. Because the sarcoplasmic reticulum is enriched in calcium transport ATPase, sarcoplasmic reticulum vesicles are able to accumulate high concentrations of calcium. When the ATPase is inhibited, the accumulated calcium rapidly exits the vesicles; this should occur whether the ATPase is inhibited because of ATP depletion (hexokinase treatment) or from direct inhibition by nonspecific inhibitors such as sulfhydryl reactant mersalyl acid [4] or cyclopiazonic acid. Interestingly, cyclopiazonic acid-treated vesicles exhibit a significantly reduced rate of calcium efflux relative to untreated vesicles or vesicles treated with either hexokinase or mersalyl acid [4]. In addition to cyclopiazonic acid, there are currently three other chemicals (thapsigargin [15], 2,5-di(*tert*-butyl)-1,4-benzohydroquinone [16], and lasalocid [17]) that are potent inhibitors of microsomal calcium transport and calcium-dependent ATPase activity. The specificity of lasalocid has not yet been determined. However, like cyclopiazonic acid, thapsigargin and 2,5-di(*tert*-butyl)-1,4-benzohydroquinine inhibit only the microsomal calcium-dependent ATPase, have no effect on the plasma membrane calcium-dependent ATPase, and in the case of 2,5-di(*tert*-butyl)-1,4-benzohydroquinone, have no effect on the mitochondrial $F_1 \cdot F_0$ ATPase. However, in vesicles treated with either thapsigargin [15] or 2,5-di(*tert*-butyl)-1,4-benzohydroquinone [18], calcium efflux was as rapid as efflux induced by ATP depletion. Interestingly, thapsigargin has no effect on skeletal muscle sarcoplasmic reticulum calcium transport, and the effect of 2,5-

di(*tert*-butyl)-1,4-benzohydroquinone has not been reported. Cyclopiazonic acid is currently the only compound identified as a specific high-affinity inhibitor of both sacroplasmic and endoplasmic reticulum calcium transport.

In addition, calcium permeability is altered by cyclopiazonic acid in both passively and actively calcium-loaded sarcoplasmic reticulum vesicles [4], and we have found it to alter $^{45}Ca^{2+}$ uptake by cultured renal and muscle cells [5, 6]. The effects of cyclopiazonic acid on $^{45}Ca^{2+}$ uptake by cultured cells appears to be produced by a nonspecific stabilization of the plasma membrane resulting in reduced calcium permeability [6]. There is evidence that cyclopiazonic acid can allosterically alter the binding of the calcium channel blocker nitrendipine to calcium channels in rat cortical tissue [19]. However, since it had no effect on nifedipine-displacable ^{3}H-nitrendipine binding [19], these results support the hypothesis that cyclopiazonic acid acted in a nonspecific manner to alter the binding.

C. Implications for Fungal Systems

Specific interactions between plant/fungal chemicals and animal systems often have pointed the way to the discovery of biologically active endogenous compounds and/or important biochemical systems in animals. Such is the case with the plant opiates, which were discovered long before the receptor in animals or the endogenous opiates, the enkephalins and endorphins. Similarly, the foxglove plant was used to treat congestive heart failure 200 years before the cardiac steroid digitalis was identified as the active component of the foxglove or its target, the Na^{+},K^{+} ATPase, was discovered. Examples for fungal chemicals also exist. There is often little or no idea of the function of biologically active chemicals in the plant and fungal species that produce them because the focus of most scientific inquiry is to improve the human condition. Since there is a great quantity of data on the biological activity of fungal metabolites in animal systems relative to their activity in fungal systems, the discovery of specific biological activity of fungal compounds in animals may prove profitable in characterization of their biological function in the fungi. Bu'Lock [1] favored the position that it was the process of secondary metabolism that was advantageous to the fungus and not the products. However, as Bu'Lock recognized, the conditions under which secondary metabolism are studied are generally designed to optimize metabolite production and not to better understand secondary metabolism as it occurs in the natural environment. Thus, the research emphasis has been the study of secondary metabolism in the fungi, whereas the products of this process have been studied primarily in animals. It appears desirable to engage in speculations based on the premise that the biological activity of biologically active products in animals mimics their activity in the fungus under natural environmental conditions.

It has been suggested that cyclopiazonic acid functions as a defense mechanism in *Aspergillus flavus* [2]. It seems unlikely that defense against animals is its primary function. If cyclopiazonic acid was involved in defense against animal predation, then it seems logical that the outer surface of animal plasma membranes (rather than the sarcoplasmic reticulum) should be the site of action. Animal plasma membrane functions do not appear to be *specifically* affected by cyclopiazonic acid. Also, cyclopiazonic acid's activity as an inhibitor of sarcoplasmic reticulum ATPases would not make it a very impressive toxin against vertebrate and invertebrate predators for two reasons: (1) the time required for cyclopiazonic acid to migrate to the sarcoplasmic reticulum would be long (minutes to hours), and (2) the amount of cyclopiazonic acid required to cause an observable effect would be high because the calcium ATPase constitutes 50–80% of the total protein content [20] of the sarcoplasmic reticulum. Cyclopiazonic acid is not as toxic to cultured cells (renal, muscle, liver) [5] or animals as other chemicals that are produced for the specific purpose of chemical defense or offense (i.e., snake venoms, insect venoms, toxins produced by hydrozoans and shellfish).

A possible function for cyclopiazonic acid involves regulation of calcium homeostasis in the fungi that produce the compound. The importance of intracellular calcium in the transduction of extracellular signals into intracellular events in animal cells is undisputed [21, 22], and it has been hypothesized that calcium may play an equally important role in fungi [23]. Cyclopiazonic acid is produced (peaks) during the balance growth phase [24] in addition to the later period of secondary metabolite production, which supports the hypothesis that cyclopiazonic acids function is more than simply an end product of secondary metabolism. Perhaps the compound's specificity and potency as an inhibitor of the calcium dependent ATPase are more than coincidental. Intracellular calcium homeostasis plays an important role in regulating various functions of fungi. For example, conidiation in *Penicillium notatum* [25, 26], binding of proteases to plasma membranes in *Allomyces arbuscula* [27], hyphal extension in *Saprolegia ferax* [28], and the protein secretory [29] and nuclear envelope translocation [30] pathways in yeast. Presumably, calcium has similar or equally important functions in fungi that produce cyclopiazonic acid.

Cyclopiazonic acid was the first-identified naturally occurring specific, high-affinity inhibitor of the sarcoplasmic reticulum calcium-dependent ATPase, an enzyme that plays a pivotal role in the regulation of intracellular calcium in muscle cells. It is now well documented that the calcium transport ATPase of endoplasmic reticulum is exceedingly important in regulating the intracellular calcium concentration in animal cells and that the intracellular calcium concentration is a key regulatory point in the inositol phospholipid signal transduction pathway [31]. The potential for compounds that alter calcium transport to modify the metabolism of cellular systems is immense when one considers the number of enzymatic processes that are calcium-dependent, including most protein phosphorylations and dephosphorylations. It has become increasingly clear that intracellular calcium and inositol phospholipid hydrolysis play an important role in fungi. The cytoplasmic calcium levels in fungal cells are as low (10^{-7} M) as in animal cells [32]; the observations suggest that the pertinent regulatory mechanisms in animal cells also apply to fungal cells. Miller and Sanders [32] demonstrated that ATP-dependent calcium extrusion is required in order for fungal cells to maintain nanomolar cytosolic calcium concentrations. While the cytosolic concentration is nanomolar, the vacuolar concentration is millimolar [33]. Thus, the vacuole possesses a powerful calcium accumulation mechanism, and it has been speculated that vacuoles of plant and fungal cells are functionally analogous to the endoplasmic reticulum of animal cells with respect to calcium regulation [23]. However, calcium accumulation by plant and fungal vacuoles is either coupled to the vacuolar H^+-ATPase via an H^+/Ca^{2+} antiporter [34] or occurs by way of a H^+/Ca^{2+}-ATPase [32].

It is now well established that both plants [13] and fungi [29] possess a calmodulin-stimulated calcium ATPase. In plants calmodulin-insensitive calcium-transport ATPase exists in the endoplasmic reticulum [35, 36]. Unlike animal cells, which have a highly developed endoplasmic reticulum, the structure in fungi is not as well developed, and its function is poorly understood [37]. Under certain environmental conditions, and during ergot alkaloid production, the endoplasmic reticulum can be extensive and pronounced in the fungus *Claviceps purpura* [38, 39]. Limited studies suggest that the endoplasmic reticulum of fungi is biochemically similar to that of higher plant and animal cells [40, 41]. However, there is currently no evidence to suggest that fungi possess a P-type calcium transport ATPase functionally analogous to the animal calcium ATPase of the endoplasmic reticulum. In addition, there is no evidence that the endoplasmic reticulum in fungi functions in the inositol triphosphate (IP_3) signal transduction pathway; this does not mean that an endoplasmic reticulum, calcium-dependent ATPase does not exist, nor does it mean that if found in small amounts its role is insignificant. There is evidence for a calmodulin-insensitive, calcium transport ATPase associated with the plasma membrane and endocytic vesicles in *Dictyostelium discoideum* [42]. Thus, an ATPase that is similar in many respects to that of animal endoplasmic reticulum, calcium-dependent ATPase exists in the ameboid state of the fungus *D. discoideum*. However, the phylogenetic connection between fungi and the cellular slime molds is not universally agreed upon [43, 44]. The ATPase is associated with a non-IP_3-releasable calcium store and therefore is not like the endoplasmic reticular system in mammalian cells, where IP_3 triggers the release of stored calcium from the endoplasmic reticulum. It may be worthwhile to explore for a cyclopiazonic acid–inhibitable P-type ATPase in fungi. If a calcium-dependent membrane ATPase exists in fungal endoplasmic reticulum or some other organelle membrane, then cyclopiazonic acid could prove to be an invaluable biochemical probe for its presence.

In addition to its specific effects, cyclopiazonic acid causes several apparently nonspecific effects that result in altered calcium permeability in mammalian systems; these effects may also have some relevance in fungal systems. Reduced calcium permeability via a nonspecific membrane alteration could modify the release of calcium from the fungal vacuole or slow the entry of extracellular calcium across the plasma membrane. It is possible that the rate of change in intracellular calcium in response to extracellular first messengers (hormones) could be altered by cyclopiazonic acid–induced alterations in calcium permeability. If the effects of calcium as a second messenger are dependent on the rate of change, magnitude, and/or duration of intracellular calcium, then the effect might alter the ultimate functional effects (e.g., phosphorylation of specific proteins). Thus, cyclopiazonic acid could modulate functional effects in fungal cells by altering the kinetics of calcium entry or efflux from extracellular or intracellular calcium pools. For example, recently it has been demonstrated that calcium influx across the plasma membrane of yeast is triggered by the mating pheromone, α-factor [45]. If this calcium influx is prevented, the cells will die following exposure to α-factor, indicating that the transient, α-factor-induced increase in intracellular calcium is essential for the mating pheromone response.

III. CYCLOPIAZONIC ACID–INDUCED ELECTRIC CHARGE ALTERATIONS

A. Fixed Electric Charge Alterations in Animal Cell Membranes

Intact or permeabilized cells (renal, liver, muscle) coincubated with the lipophilic membrane potential probe tetraphenylphosphonium bromide and cyclopiazonic acid show an increased accumulation of tetraphenylphosphonium relative to cells exposed to only tetraphenylphosphonium [9, 10]. In order to understand the significance of this observation, the behavior of tetraphenylphosphonium in cellular membranes must be understood. Tetraphenylphosphonium is a lipophilic quaternary cation that has been used to measure transmembrane potentials in vesicles, mitochondria, and plant and animal cells and has been used as a probe of charge alterations in both lipid and protein domains of biomembranes [46–50].

Increased accumulation of tetraphenylphosphonium in cyclopiazonic acid–treated cells was mistakenly attributed to an increased transmembrane potential difference resulting from stimulation of the electrogenic plasma membrane Na^+,K^+ pump [8]. Alternatively, dipole effects or change in surface charge can dramatically alter the zero-potential partition coefficient for hydrophobic ions [46]. In cells treated with cyclopiazonic acid, the accumulated tetraphenylphosphonium was not in passive equilibrium with the transmembrane potential difference; tetraphenylphosphonium was recovered noncovalently bound in the plasma membrane and mitochondrial fractions [9]. The interaction between tetraphenylphosphonium, cyclopiazonic acid, and membranes was weak since tetraphenylphosphonium could be separated from the membranes by $CHCl_3$ extraction or by passage through a gel filtration column (Ref. 9 and unpublished results). Further investigations revealed that freeze–thaw lysed cells exhibited cyclopiazonic acid–stimulated tetraphenylphosphonium accumulation in the absence of a transmembrane potential difference; this result indicated that both tetraphenylphosphonium and cyclopiazonic acid must have access to the intracellular space for increased tetraphenylphosphonium accumulation. It appeared that the site of interaction between cyclopiazonic acid and tetraphenylphosphonium was intracellular and possibly due to a fixed-charge alteration on the cytoplasmic side of the plasma membrane.

Subsequent investigations revealed that different cell types exhibited age- and growth-related differences in the extent of tetraphenylphosphonium accumulation and in their sensitivity to cyclopiazonic acid–induced alterations in tetraphenylphosphonium accumulation [10]. Differences in sensitivity to cyclopiazonic acid by different cell types may be related to differences in the growth characteristics of the cells. Susceptibility of cells to cyclopiazonic acid–induced charge alterations in, on, or across cell membranes appeared related to age- and growth-dependent changes (unique to each cell type) and perhaps to basic biochemical properties of cell membranes that play an important role in the targeting of tissues by cyclopiazonic acid.

B. Significance for Membrane Function and Implications for Fungal Systems

The relationship between charge alterations on or within biomembranes and the function of membranes is not well established. Nonetheless, in vitro studies and theoretical considerations indicate that pertinent membrane alterations can or should have profound effects on transport systems and receptors associated with cell membranes [51–53]. For example, it has been shown that the charge carried by lipids in close proximity to voltage-gated ion channels can affect the gating of the channel [54]. This effect is produced by differences in the surface potential on either side of the channel protein and also in proximity to the proteins that comprise the voltage-sensing machinery. Consequently, an electric field generated by fixed charge differences across the membrane and in close proximity to critical membrane proteins can change the biological activity of channels, ATPases, transport proteins, and so on.

Electric fields also function as a means of transcellular signaling [55]. It is now fairly well established that membrane ATPases are able to use oscillating electric fields to perform chemical work. In fact, many cell functions can be altered by weak oscillating electromagnetic fields. It is believed that the membrane itself acts as an amplifier of weak electric fields and that certain membrane proteins have the ability to decipher the electric signal and initiate processes resulting in functional effects; this concept has been called *electroconformational coupling* [56] or *membrane transductive coupling* [57]. In electroconformational coupling, only oscillating local electric fields are important. Thus the transient electric field in close proximity to the target protein is much more important than the overall electrical properties of the cell. Tsong [55] hypothesizes that locally fluctuating transmembrane electric fields are a common feature of cellular signaling and that through an understanding of this cellular laguage it may be possible to regulate cell function.

Cyclopiazonic acid and probably many other fungal metabolites induce charge alterations in cellular membranes. It is quite likely that these charge alterations modify the electrical properties of closely associated proteins and thus could alter the electroconformational coupling. This could be a way for fungi to effect transcellular signaling and could explain the diversity of different fungal metabolites, with each perhaps targeting specific membrane components that regulate cell function.

IV. ANTIOXIDANT ACTIVITY OF CYCLOPIAZONIC ACID

A. Prevention of Patulin-Induced Lipid Peroxidation in Animal Cells

Cyclopiazonic acid can prevent the lipid peroxidation induced by patulin [6]. Patulin, like cyclopiazonic acid, is a fungal metabolite produced by

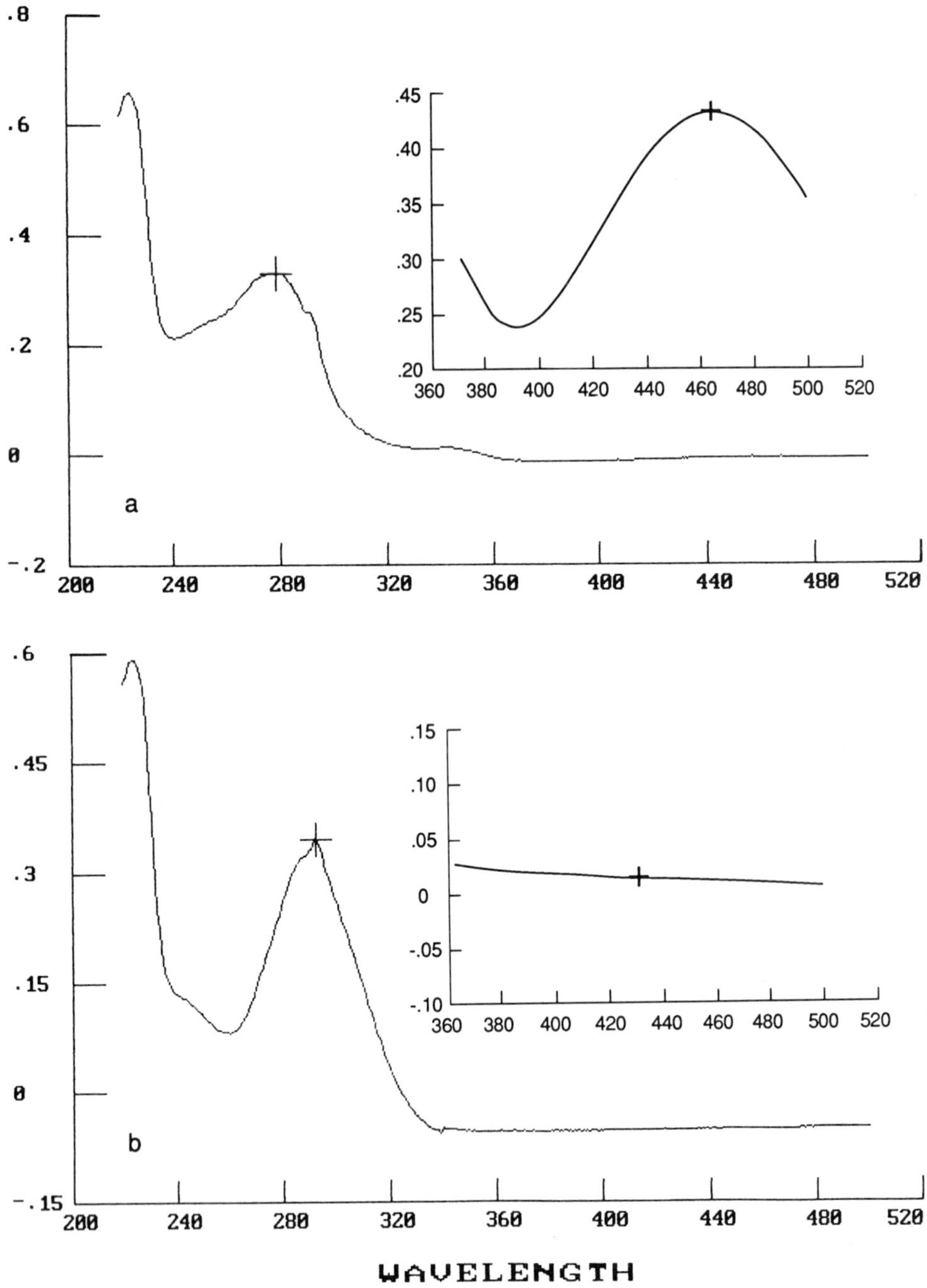

FIGURE 2. Chelation of iron by cyclopiazonic acid but not cyclopiazonic acid imine. Ultraviolet and visible spectra of materials extracted with chloroform after mixing 10 μmoles of ferrous iron ($FeSO_4$) in 1 mL of water with 1 μmole of cyclopiazonic acid (a) or cyclopiazonic acid imine (b) in 1

many *Penicillium* and *Aspergillus* species [58]. Several of these species can produce both cyclopiazonic acid and patulin [59]. Although simultaneous production of patulin and cyclopiazonic acid has not been demonstrated, aflatoxin and cyclopiazonic acid can be produced simultaneously by many isolates of *Aspergillus flavus* [60]. The ability of cyclopiazonic acid to prevent patulin-induced lipid peroxidation indicates that cyclopiazonic acid has antioxidant activity. An antioxidant is defined by Halliwell [61] as "any substance that, when present at low concentrations compared to those of an oxidizable substrate, significantly delays or prevents oxidation of that substrate." The oxidizable substrate in the case of patulin is membrane lipid. Other tetramic acids have been tested for antioxidant activity, and only the indole tetramic acids (cyclopiazonic acid and cyclopiazonic acid imine) prevented the patulin-induced lipid peroxidation in animal cells [6]. As with charge alteration, the ability of cyclopiazonic acid to prevent lipid peroxidation was different between cell types and may be dependent on properties of the plasma membrane that can change as cells age and differentiate [10].

Although the mechanism by which cyclopiazonic acid prevents lipid peroxidation has not been extablished, one well-studied mechanism of antioxidant activity is iron chelation [61]. Since the iron chelator deferoxamine also prevented patulin-induced lipid peroxidation, iron chelation might account for the antioxidant activity of cyclopiazonic acid [6]. Like many other tetramic acids [62], cyclopiazonic acid is an extremely effective chelator of iron. The interaction of cyclopiazonic acid with ferrous iron results in a chloroform-extractable chelate (Fig. 2). Tenuazonic acid also chelates iron in a chloroform-extractable form [62], but cyclopiazonic acid imine does not form an iron chelate [6]; this is not surprising because the 3-acetyl group of tenuazonic acid, which is equivalent to the 7-acetyl group in cyclopiazonic acid, is responsible for the metal-complexing properties of tenuazonic acid [63]. It is the acetyl group that is modified to form the imine of cyclopiazonic acid [64]. Since cyclopiazonic acid imine also prevented lipid peroxidation and tenuazonic acid did not, it appears that the protective effect of tetramic acids is restricted to compounds that contain an indole nucleus; thus, the ability of indole tetramic acids to prevent lipid peroxidation does not involve iron chelation.

FIGURE 2. (Continued)
mL of methanol. Mixtures were allowed to stand for 60 min, during which time the cyclopiazonic acid–iron mixture formed an insoluble red precipitate and the cyclopiazonic acid imine–iron mixture remained clear and colorless. After 60 min, 0.5 mL of $CHCl_3$ was added to each tube, and the tubes contents were mixed. Solvent blanks were prepared without the indole tetramic acids but with $FeSO_4$. The red precipitate in the cyclopiazonic acid–iron mixture dissolved readily into the $CHCl_3$ phase, producing an intense red color. The $CHCl_3$ phase from each treatment was removed and evaporated to dryness, and then the residue was dissolved in methanol. A 1:20 dilution of the methanol extract was scanned on an AMINCO DW2 UV/VIS spectrophotometer from 200 nm to 350 nm (UV spectra), and the undiluted extract was scanned from 360 nm to 500 nm (visible spectra) [6].

Interestingly, cyclopiazonic acid imine, but not tenuazonic acid, causes a fixed electric charge alteration in L6 myoblasts [7], and the indole tetramic acids, but not tenuazonic acid, inhibit ATP-dependent calcium accumulation in sarcoplasmic reticulum vesicles [7]. Thus, the fixed charge alteration on plasma membranes and/or the inhibition of a calcium-dependent ATPase may also be involved in the ability of indole tetramic acids to protect cells against lipid peroxidation.

B. Possible Interactions with Other Fungal Metabolites

Lipid peroxidation induces aflatoxin biosynthesis by *Aspergillus* species [65, 66]. Addition of antioxidants to the culture medium will prevent carbon tetrachloride–induced lipid peroxidation and aflatoxin congener biosynthesis in both *Aspergillus flavus* and *Aspergillus parasiticus* [65]. Patel et al. [66] showed that iron-catalyzed peroxidative damage to cell membranes induces aflatoxin production by *A. parasiticus*, and thus lipid peroxidation may be an important factor in initiating aflatoxin biosynthesis under field conditions. The concept is plausible because many breakdown products of cell membranes are now known to act as intracellular messengers [67]. Endogenous antioxidants, such as cyclopiazonic acid (in *A. flavus*), could serve to regulate aflatoxin biosynthesis by preventing the oxidation of membrane lipids leading to the production of precursors or regulators of aflatoxin biosynthesis. However, in culture there is no evidence to support the hypothesis that aflatoxin and cyclopiazonic acid production are linked [68].

Cyclopiazonic acid production in liquid culture by *Penicillium cyclopium* is associated with decreased metabolic activity and growth, reduced carbon flow through the Krebs cycle, increased protein catabolism, lipid mobilization, and mitochondrial degeneration [69]. Concurrent with these changes, carbon is being diverted into the production of cyclopiazonic acid. Addition of ATP to cultures depressed cyclopiazonic acid production [69]; this observation suggested to Holzapfel [24] that ATP might function as a catabolite repressor of cyclopiazonic acid synthesis. It is possible that in fungal cells the ability of cyclopiazonic acid to interact with some key enzyme in the biosynthetic pathway is prevented by ATP. However, as intracellular ATP pools become depleted, cyclopiazonic acid might be able to interact with the regulatory site, diverting carbon into secondary synthesis and resulting in the production of large amounts of cyclopiazonic acid. The advantage conferred by the metabolic shift remains ambiguous, but perhaps production of siderophores would act to stockpile iron for later use by the growing fungus. Neilands [70] defined siderophores as "relatively low-molecular weight (500–1000), virtually ferric specific ligands, the biosynthesis of which is tightly regulated by iron." However, Holzapfel [24] concluded that cyclopiazonic acid production by *Penicillium cyclopium* is not dependent on the presence of trace iron. Thus, cyclopiazonic acid is not a siderophore as defined by Neilands [70] but is nonetheless an excellent chelator of iron. Since cyclopiazonic acid production is greatest during the period just preceding dormancy, it would be important for subsequent fungal growth that an easily available supply of iron be present. Since concentrations of free iron in soil are usually low, stockpiles of chelated iron would be beneficial for rapid growth. The chelate would need to be highly specific, bind iron very tightly, and be recognized only by the fungi that secreted the chelator. In mammalian cells, chelated iron (iron transferrin) is internalized via the process of receptor-mediated endocytosis [71]. Acid conditions in the endosome promote disassociation of iron from transferrin [71]. A similar process may be involved in iron uptake by fungi. However, the limited studies of the iron chelate of cyclopiazonic acid indicate that the iron is not easily dissociated even with acid treatment (unpublished results).

V. CONCLUSIONS

What we have presented is highly speculative. We have tried to present some of our data and the data of others in a way that suggests how it might be extrapolated to fungal systems. Cyclopiazonic acid is a very specific inhibitor of a key enzyme in animal cells, is a lipophilic and highly charged molecule that can cause alterations in the electrical properties of animal cell membranes, has antioxidant properties, and is a very good iron chelator. We are beginning to understand how these properties of cyclopiazonic acid can alter animal cell function, but it is yet to be determined how this compound might alter fungal cell function. Studies using fungal cells could provide insight into the role of cyclopiazonic acid in the ecology of the fungi that produce it.

REFERENCES

1. Bu'Lock, J. D., Mycotoxins as secondary metabolites. In *The Biosynthesis of Mycotoxins* (J. D. Bu'Lock, Ed.), Academic, New York, 1980, pp. 1–16.
2. Wicklow, D. T., and Cole, R. J. Tremorgenic indole metabolites and aflatoxins in sclerotia of *Aspergillus flavus*: an evolutionary perspective. *Can. J. Bot.*, *60*:525–528, 1982.
3. Goeger, D. E., Riley, R. T., Dorner, J., and Cole, R. Cyclopiazonic acid inhibition of Ca^{2+}-transport ATPase in rat skeletal muscle sarcoplasmic reticulum vesicles. *Biochem. Pharmacol.*, *37*:978–981, 1988.
4. Goeger, D. E., and Riley, R. T. Interaction of cyclopiazonic acid with rat skeletal muscle sarcoplasmic reticulum vesicles: effect on Ca^{+2} binding and Ca^{+2} permeability. *Biochem. Pharmacol.*, *38*:3995–4003, 1989.
5. Riley, R. T., Goeger, D. E., and Hinton, D. M. Mycotoxin-induced alterations in ion transport across cell membranes. In *Cellular and Molecular Mode of Action of Selected Microbial Toxins in Foods and Feeds* (A. E. Pohland and J. Richard, Eds.), Plenum, New York, 1990, pp. 451–466.
6. Riley, R. T., and Showker, J. L. The mechanism of patulin's cytotoxicity and the antioxidant activity of indole tetramic acids. *Toxicol. Appl. Pharmacol.*, *109*:108-126, 1991.

7. Riley, R. T., and Goeger, D. E. The ability of three tetrazonic acids, cyclopiazonic acid, cyclopiazonic acid imine, and tenuazonic acid, to inhibit calcium uptake by rat sarcoplasmic reticulum vesicles and to potentiate tetraphenylphosphonium uptake by cultured cells. *FASEB J.* *3*:A235, 1989.
8. Riley, R. T., Norred, W. P., Dorner, J., and Cole, R. J. Increased accumulation of the lipophilic cation tetraphenylphosphonium by cyclopiazonic acid-treated renal epithelial cells. *J. Toxicol. Environ. Health*, *15*:779–788, 1985.
9. Riley, R. T., Showker, J. L., Cole, R. J., and Dorner, J. The mechanism by which cyclopiazonic acid potentiates accumulation of tetraphenylphosphonium in cultured renal epithelial cells. *J. Biochem. Toxicol.*, *1*:13–30, 1986.
10. Riley, R. T., Goeger, D. E., Showker, J. L., Cole, R. J., and Dorner, J. Age and growth related changes in cyclopiazonic acid-potentiated lipophilic cation accumulation by cultured cells and binding to freeze-thaw lysed cells. *J. Biochem. Toxicol.*, *2*:251–264, 1987.
11. Seidler, N. W., Jona, I., Vegh, M., and Martonosi, A. Cyclopiazonic acid is a specific inhibitor of the Ca^{2+}-ATPase of sarcoplasmic reticulum. *J. Biol. Chem.*, *264*:17816–17823, 1989.
12. de Meis, L., and Vianna, A. L. Energy interconversion by the Ca^{2+}-dependent ATPase of the sarcoplasmic reticulum. *Annu. Rev. Biochem.*, *48*:275–292, 1979.
13. Serrano, R. Plasma membrane ATPase. In *The Plant Plasma Membrane* (C. Larsson and I. M. Moller, Eds.), Springer-Verlag, Berlin, 1990, pp. 127–153.
14. Stryer, L. *Biochemistry*, W. H. Freeman, New York, 1988.
15. Thapstrup, O., Cullen, P. J., Drobak, B. K., Hanley, M. R., and Dawson, A. P. Thapsigargin, a tumor promoter, discharges intracellular Ca^{2+} stores by specific inhibition of the endoplasmic reticulum Ca^{2+}-ATPase. *Proc. Natl. Acad. Sci. USA*, *87*:2466–2470, 1990.
16. Moore, G. A., McConkey, D. J., Kass, E. N., O'Brien, P. J., and Orrenius, S. 2,5-Di-(*tert*-butyl)-1,4-benzohydroquinone—a novel inhibitor of liver microsomal Ca^{2+} sequestration. *FEBS Lett.*, *224*:331–336, 1987.
17. Kawashima, T., Hara, H., and Kanazawa, T. Selective inhibition by lasalocid of hydrolysis of ADP-insensitive phosphoenzyme in the catalytic cycle of sarcoplasmic reticulum Ca^{2+}-ATPase. *J. Biol. Chem.*, *265*:10993–10999, 1990.
18. Kass, G. E., Duddy, S. K., Moore, G. A., and Orrenius, S. 2,5-di-(*tert*-butyl)-1,4-benzohydroquinone rapidly elevates cytosolic Ca^{2+} concentration by mobilizing the inositol 1,4,5-triphosphate-sensitive Ca^{2+} pool. *J. Biol. Chem.*, *264*:15192–15198, 1989.
19. Valdes, J. J., Wolff, V. L., and Ross, D. H. Dihydropyridine receptors: possible allosteric regulation by tremorgenic toxins. Chem. Res. Development & Engineering Center Tech. Report CRDEC-TR-87008, National Technical Information Service, U.S. Dept. of Commerce, Springfield, Va., 1986.
20. Martonosi, A. N. Mechanisms of Ca^{2+} release from sarcoplasmic reticulum of skeletal muscle. *Physiol. Rev.*, *64*:1240–1320, 1984.

21. West, C. A., Bruce, R., and Ren, Y.-L. Second messengers in animals and their possible relevance for plants. In *Signal Molecules in Plants and Plant-Microbe Interactions* (B. J. J. Lugtenberg, Ed.), Springer-Verlag, Berlin, 1989, pp. 27–40.
22. Poovaiah, B. W., Reddy, A. S. N., and McFadden, J. J. Calcium messenger system: role of protein phosphorylation and inositol bisphospholipids. *Physiol. Plantarum, 69*:569–573, 1987.
23. Cornelius, G., Gebauer, G., and Techel, D. Inositol triphosphate induces calcium release from *Neurospora crassa* vacuoles. *Biochem. Biophys. Res. Commun., 162*:852–856, 1989.
24. Holzapfel, C. W. The biosynthesis of cyclopiazonic acid and related tetramic acids. In *The Biosynthesis of Mycotoxins* (J. D. Bu'Lock, Ed.), Academic, New York, 1980, pp. 327–354.
25. Pitt, D., and Mosley, M. J. Oxidation of carbon sources via tricarboxylic acid cycle during calcium-induced conidiation of *Penicillium notatum. Antonie van Leeuwenhoek, 52*:467–482, 1986.
26. Mosley, M. J., Pitt, D., and Barnes, J. C. Adenine and pyridine nucleotide levels during calcium-induced conidiation in *Penicillium notatum. Antonie van Leeuwenhoek, 56*:191–199, 1989.
27. Ojha, M. *Allomyces* Ca^{2+}-activated neutral protease: interaction with phospholipids and plasma membranes. *Plant Sci., 59*:151–158, 1989.
28. Jackson, S. L., and Heath, I. B. Effects of exogenous calcium ions on tip growth, intracellular Ca^{2+} concentration, and actin arrays in hyphae of the fungus *Saprolegnia ferax. Exp. Mycol., 13*:1–12, 1989.
29. Rudolph, H. K., Antebi, A., Fink, G. R., Buckley, C. M., Dorman, T. E., LeVitre, J., Davidow, L. S., Mao, J.-I., and Moir, D. T. The yeast secretory pathway is perturbed by mutations in PMR1, a member of Ca^{2+} ATPase family. *Cell, 58*:133–145, 1989.
30. Kalinich, J. F., and Douglas, M. G. In vitro translocation through the yeast nuclear envelope. *J. Biol. Chem., 264*:17979–17989, 1989.
31. Hannun, Y. A., and Bell, R. M. Mechanism of activation of protein kinase C: role of diacylglycerol and calcium secondary messengers. In *Cell Calcium and the Control of Membrane Transport* (L. J. Mandel and D. C. Eaton, Eds.), Rockefeller Univ. Press, New York, 1987, pp. 229–240.
32. Miller, A. J., and Sanders, D. The energetics of cytosolic calcium homeostasis in fungal cells. *Plant Physiol. Biochem., 27*:551–556, 1989.
33. Halachmi, D., and Eilam, Y. Cytosolic and vacuolar Ca^{2+} concentration in yeast cells measured with the Ca^{2+}-sensitive fluorescence dye indo-1. *FEBS Lett., 256*:55–61, 1989.
34. Blackford, S., Rea, P. A., and Sanders, D. Voltage sensitivity of H^+/Ca^{2+} antiport in higher plant tonoplast suggests a role in vacuolar calcium accumulation. *J. Biol. Chem., 265*:9617–9620, 1990.
35. Buckhout, T. J. Characterization of Ca^{2+} transport in purified endoplasmic reticulum vesicles from *Lepidium sativum* L. roots. *Plant Physiol., 76*:962–967, 1984.

36. Giannini, J. L., Gildensoph, L. H., Reynolds-Niesman, I., and Briskin, D. P. Calcium transport in sealed vesicles from red beet (*Beta vulgaris* L.) storage tissue. *Plant Physiol.*, *85*:1129–1136, 1987.
37. Burnee, J. H. *Fundamentals of Mycology*, Edward Arnold, London, 1976.
38. Rehacek, Z. Biosynthesis of ergot alkaloids. In *Biotechnology Applications and Research* (P. N. Cheremisinoff and R. P. Ouellette, Eds.), Technomic, Basel, 1985, pp. 565–585.
39. Vorisek, J., and Rehacek, Z. Fine structural localization of alkaloid synthesis in endoplasmic reticulum of submerged *Claviceps purpurea*. *Arch. Microbiol.*, *117*:297–302, 1978.
40. Borgeson, C. E., and Bowman, B. J. Isolation and characterization of the *Neurospora crassa* endoplasmic reticulum. *J. Bacteriol.*, *156*: 362–368, 1983.
41. Bowman, B. J., Borgeson, C. E., and Bowman, E. J. Composition of *Neurospora crassa* vacuolar membranes and comparison to endoplasmic reticulum, plasma membranes, and mitochondrial membranes. *Exp. Mycol.*, *11*:197–205, 1987.
42. Milne, J. L., and Coukell, M. B. Identification of a high-affinity Ca^{2+} pump associated with endocytic vesicles in *Dictyostelium discoideum*. *Exp. Cell Res.*, *185*:21–32, 1989.
43. Olive, L. S. *The Mycetozoan*, Academic, New York, 1975.
44. Raper, K. B. *The Dictyostelids*, Princeton Univ. Press, Princeton, N.J., 1984.
45. Iida, H., Yagawa, Y., and Anraku, Y. Essential role for induced Ca^{2+} influx followed by $[Ca^{2+}]_i$ rise in maintaining viability of yeast cells late in the mating pheromone response pathway. *J. Biol. Chem.*, *265*:13391–13399, 1990.
46. Cafiso, D. S., and Hubbell, W. L. Estimation of transmembrane potentials from phase equilibria of hydrophobic paramagnetic ions. *Biochemistry*, *17*:187–195, 1978.
47. Ritchie, R. J. A critical assessment of the use of lipophilic cations as membrane potential probes. *Prog. Biophys. Mol. Biol.*, *43*:1–32, 1984.
48. Levitsky, D. O., Loginov, V. A., and Lebedev, A. V. Charge changes in sarcoplasmic reticulum and Ca^{2+}-ATPase induced by calcium binding and release: A study using lipophilic ions. *Membrane Biochem.*, *6*:291–307, 1986.
49. Cafiso, D. S. Paramagnetic hydrophobic ions as probes for electrically active conformational transitions in ion channels. *Biophys. J.*, *45*:6–7, 1984.
50. Riley, R. T., Goeger, D. E., and Norred, W. P. Lipophilic cations as membrane probes in cultured renal ($LLC\text{-}PK_1$) and muscle cells (L6). In *Nephrotoxicity: In Vitro to In Vivo, Animals to Man* (P. H. Bach and E. A. Lock, Eds.), Plenum, London, 1989, pp. 743–748.
51. Edmonds, D. T. Electrostatic models for ion channels and pumps. *Biochem. Soc. Symp.*, *50*:1–10, 1985.
52. Krueger, B. K. Toward an understanding of structure and function of ion channels. *FASEB J.*, *3*:1906–1914, 1989.

53. Zimanyi, L., and Garab, G. Configuration of the electric field and distribution of ions in energy transducing biological membranes: model calculations in a vesicle containing discrete changes. *J. Theor. Biol.*, *138*:59–76, 1990.
54. Cukierman, S., Zinkand, W. C., French, R. J., and Krueger, B. K. Effects of membrane surface charge and calcium on the gating of rat brain sodium channels in planar bilayers. *J. Gen. Physiol.*, *92*:431–447, 1988.
55. Tsong, T. Y. Deciphering the language of cells. *TIBS*, *14*:89–92, 1989.
56. Tsong, T. Y., and Astumian, R. D. Electroconformational coupling: how membrane-bound ATPase transduces energy from dynamic electric fields. *Ann. Rev. Physiol.*, *50*:273–290, 1988.
57. Adey, W. R. The sequence and energetics of cell membrane transductive coupling to intracellular enzyme systems. *Bioelectrochem. Bioenerg.*, *15*:447–456, 1986.
58. Cole, R. J., and Cox, R. H. *Handbook of Toxic Fungal Metabolites*, Academic, New York, 1981.
59. Trucksess, M. W., Mislivec, P. B., Young, K., Bruce, V. R., and Page, S. W. Cyclopiazonic acid production by cultures of *Aspergillus* and *Penicillium* species isolated from dried beans, corn meal, macaroni, and pecans. *J. Assoc. Off. Anal. Chem.*, *70*:123–126, 1987.
60. Gallagher, R. T., Richard, J. L., Stahr, H. M., and Cole, R. J. Cyclopiazonic acid production by aflatoxigenic and non-aflatoxigenic strains of *Aspergillus flavus*. *Mycopathologia*, *66*:31–36, 1978.
61. Halliwell, B. How to characterize a biological antioxidant. *Free Radical Res. Commun.*, *9*:1–32, 1990.
62. Lebrun, M.-H., Duvert, P., Gaudemer, F., Gaudemer, A., Deballon, C., and Boucly, P. Complexation of the fungal metabolite tenuazonic acid with copper(II), iron(III), nickel(II), and magnesium(II) ions. *J. Inorg. Biochem.*, *24*:167–181, 1985.
63. Lebrun, M.-H., Nicolas, L., Boutar, M., Gaudemer, F., Ranomenjanahary, S., and Gaudemer, A. Relationship between the structure and the phytotoxicity of the fungal toxin tenuazonic acid. *Phytochemistry*, *27*:77–84, 1988.
64. Holzapfel, C. W., Hutchison, R. D., and Wilkins, D. C. The isolation and structure of two indole derivatives from *Penicillium cyclopium* Westling. *Tetrahedron*, *26*:5239–5246, 1970.
65. Fanelli, C., Fabbri, A. A., Panfili, G., Castoria, R., De Luca, C., and Passi, S. Aflatoxin congener biosynthesis induced by lipoperoxidation. *Exp. Mycol.*, *13*:61–68, 1989.
66. Patel, U. D., Bapat, S. R., and Dave, P. J. Induction of aflatoxin biosynthesis in *Aspergillus parasiticus* by ascorbic acid-mediated lipid peroxidation. *Curr. Microbiol.*, *20*:159–164, 1990.
67. Needleman, P., Turk, J., Jakschik, B. A., Morrison, A. R., and Lefkowith, J. B. Arachidonic acid metabolism. *Annu. Rev. Biochem.*, 55:69–102, 1986.
68. Blaney, B. J., Kelly, M. A., Tyler, A. L., and Connole, M. D. Aflatoxin and cyclopiazonic acid production by Queensland isolates

of *Aspergillus flavus* and *Aspergillus parasiticus*. *Aust. J. Agric. Res.*, *40*:395–400, 1989.

69. Neethling, D. C., and McGrath, R. M. Metabolic development and mitochondrial changes during cyclopiazonic acid production in *Penicillium cyclopium*. *Can. J. Microbiol.*, *23*:856–872, 1977.
70. Neilands, J. B. Iron absorption and transport in microorganisms. *Annu. Rev. Nutr.*, *1*:27–46, 1981.
71. Theil, E. C., and Aisen, P. The storage and transport of iron in animal cells. In *Iron Transport in Microbes, Plants and Animals* (G. Winkelmann, D. van der Helm, and J. B. Neilands, Eds.), VCH, Weinheim, 1987, pp. 491–520.

16

OCHRATOXIN A—A MYCOTOXIN OF CONCERN

FREDRIK CARL STØRMER *National Institute of Public Health, Oslo, Norway*

I. INTRODUCTION

Since ochratoxin A was discovered 25 years ago [1], it has received increased attention; the toxin is produced by certain species of *Aspergillus* and *Penicillium*. Ochratoxin occurrence in plant products, especially cereals, beans, nuts [2, 3], and in meats and dried fish [4] has been well documented [5]. In plant products, concentrations up to 27.5 ppm have been observed, whereas the highest concentration of residues in animal products (bacon from pigs) is 0.067 ppm [6].

Ochratoxin A-induced nephropathy produces in domestic animals morphological and functional changes that are similar to those seen in patients with Balkan endemic nephropathy [7, 8]. Ochratoxin B has been encountered in only a few circumstances as a natural contaminant; it has been reported in moldy bread at a concentration of 0.6 mg/kg (dry weight) together with ochratoxin A [9], and it is considerably less toxic than ochratoxin A. Ochratoxin B administered to trout at levels of 66.7 mg/kg [10, 11] caused no deaths, but pathological changes in the liver and kidneys similar to those produced by low levels of ochratoxin A were observed.

II. HISTORICAL BACKGROUND

A. Mortality Crisis and Population Growth in England

The last mortality crisis in England with 25% or more deaths above trend occurred in 1741–1742 [12, 13]. After 1750, infant and child mortality declined among commoners and even more rapidly among the well-to-do. Significant changes also took place in the seasonality of conceptions and of crisis mortality after 1750. Mortality crisis tended to follow severe winters; death rates varied from one region to another and from one parish to another.

Economic development after 1750 served to stimulate English population growth, but it could not have been responsible for the population growth that occurred elsewhere in Europe during the same time period. It seems clear that improvements in sanitation and medical care, decline in war casualties and deaths associated with famine, or even smallpox inoculation cannot account for increased population growth [14]. The answer probably lies in changes in nutrition [15–17]. An improvement in quantity of calories, balance of nutrients, and absence of toxic compounds in food appeared to reduce human mortality.

B. Changes in Diet

The introduction of the potato into the diet of northwestern Europe has been considered since 1843 as the answer to the puzzle of population growth [18, 19]. However, increased efficiency in caloric intake alone does not explain improvements in health. In terms of calories and nutrient balance, the English were better fed in the first half of the eighteenth century than the Irish. Nevertheless, this was a sickly period in England, one of the sickliest ever, and the English population grew more slowly than the Irish [20].

C. Toxins in Food

An important cause of the mortality crisis may have been the presence of toxins in the food. Cereals are seldom free of mold contamination, even before harvesting; field fungi include *Alternaria*, *Fusarium*, *Helminthosporium*, and *Cladosporium* spp. Storage fungi such as *Aspergillus* and *Penicillium* spp. dominate after harvest, and finally advanced decay fungi, *Papulospora* spp., *Sordaria* spp., *Fusarium graminearum*, and *Mucorales*, develop [21].

In current European agricultural practices, cereals at harvest may have water content of 16% and even 18% in wet seasons, whereas in the United States conditions routinely differ (14% is expected for soft wheats and 14.5% for hard wheats). It is advisable that cereal grains not exceed a water content of 13.5% for storage over several months, and it is necessary to reduce the mositure level to 12% for long-term storage to control fungal development [22].

Some fungi exhibit a wide temperature range for growth, but for others the minimum and maximum cardinal temperatures are relatively close (Fig. 1). *Penicillium viridicatum*, an undesirable pathogen in silos and an ochratoxin A producer, is able to grow at temperatures below 10°C.

There is considerable evidence that in the years of greatest human mortality and least fertility in England there were serious and widespread epidemics of mold poisoning. In the period 1541–1587 there were four crises. The worst was during the years 1557–1560. The weather in these years was favorable for development of ergot and *Fusarium* toxins [20].

The year 1557 was extraordinarily wet, and all the grain was reportedly "choked and blasted." Mortality was 60.5% above the normal trend in 1557–1560. The seasonality of death and low birth rates in 1557 and 1558 support a diagnosis of ergotism. The seasonality of death in 1559 and 1560 suggests a diagnosis of alimentary toxic aleukia caused by T-2 toxin; since T-2 toxin does not suppress fertility, the normal birth rate for 1560–1561 confirms the premise [12, 23, 24].

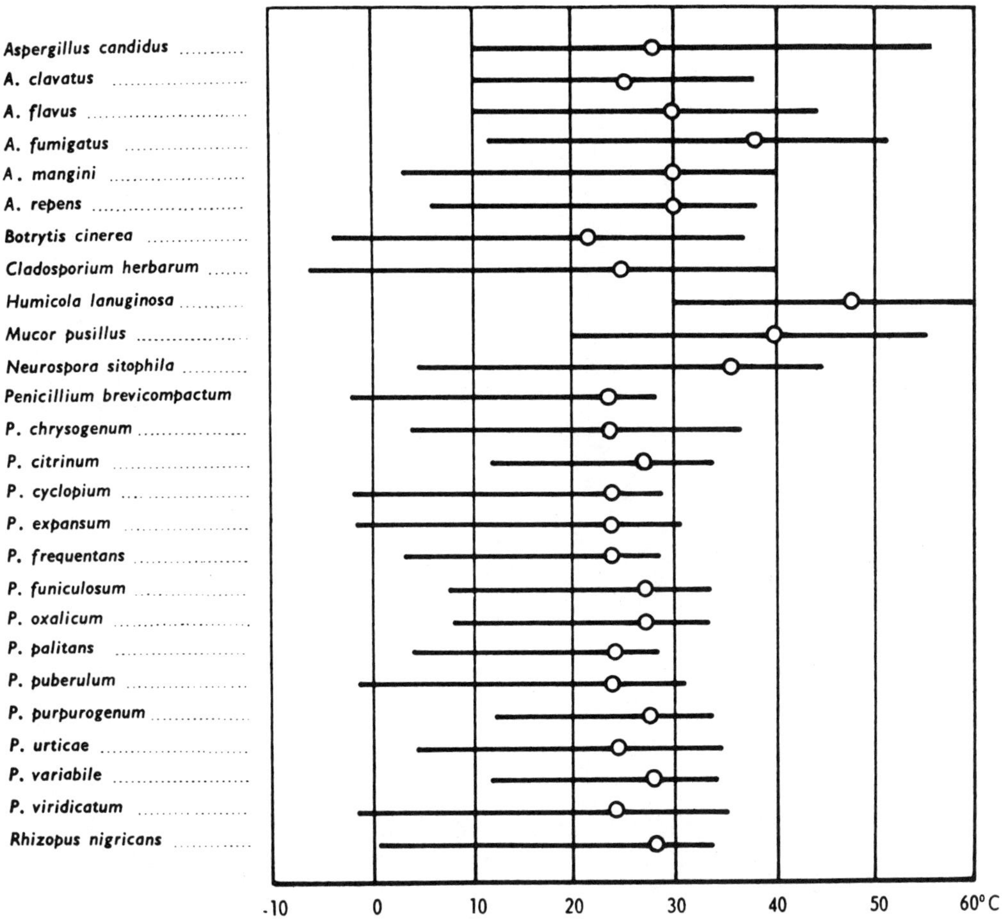

FIGURE 1. Minimum and maximum temperatures for the growth of a number of molds from food commodities [21].

D. Decline in Mortality—A Possible Explanation

A study of the mortality caused by lung tuberculosis was carried out in London [25] (Fig. 2). A sharp and unreversed decline in the disease appeared about 1830 and continued without interruption. The decline in relevant deaths cannot be attributed to medical knowledge or improved treatment, nor could it have been produced by better air quality or improving climatic conditions. However, a dramatic change in the diet of the poor did occur: a rapid increase in potato consumption. Potatoes do not generally harbor the strains of fungi that produce immunosuppressant mycotoxins that may be found in badly stored grain. The ingestion of T-2 toxin by mice increased the susceptibility to tuberculosis infection. A shift in diet from cereals to potatoes could have reduced the chances

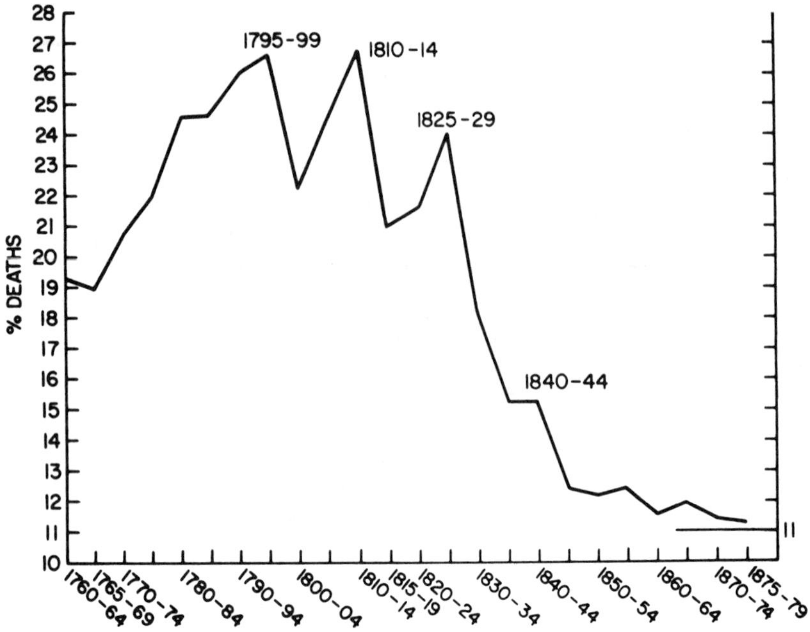

FIGURE 2. Lung tuberculosis in London, 1760–1879 (specific death ratios) [20].

of ingesting dangerous toxins [26]; these findings suggest that medical invention had less influence than dietary change on improved health conditions in London. A great deal of writing on the pertinent history reflects a naive awe of the medical profession, which the facts do not always justify [25].

A recently updated study of mycotoxins in food and feed [27] reviews the worldwide levels and occurrence of mycotoxins in various commodities since 1976. It is apparent that intensified efforts are needed to establish control of mycotoxins in the global food supply. Extensive deoxynivalenol contamination of grains, such as wheat, was demonstrated as well as co-contamination by other *Fusarium* toxins, especially nivalenol. When contamination of feeds by ochratoxin A was significant, the toxin occurred in swine kidney and smoked meat in high levels.

III. CHEMISTRY AND METABOLISM OF OCHRATOXINS

A. Chemistry

Ochratoxins A and B are dihydroisocoumarin derivatives linked through a carboxy group to *L*-phenylalanine by an amide group (Fig. 3). On acid hydrolysis, ochratoxins yield phenylalanine and isocoumarin parts, for

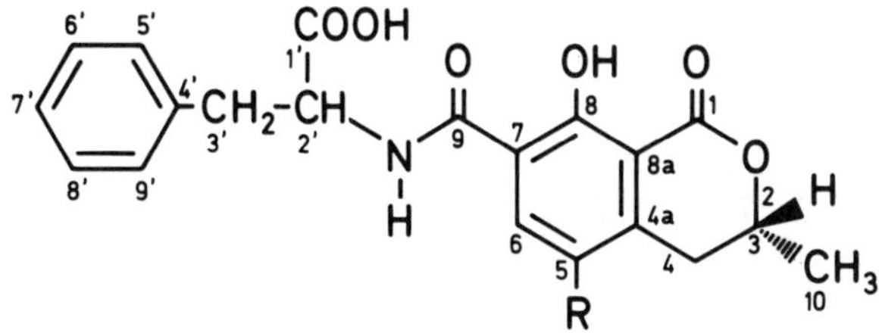

FIGURE 3. Ochratoxins A (R = Cl) and B (R = H).

example, ochratoxin α from ochratoxin A and ochratoxin β from ochratoxin B. Physical characteristics of the ochratoxins have been characterized [28].

B. Metabolism of Ochratoxin A

When ochratoxin A is incubated with liver microsomes from rat, pig, or human in the presence of NADPH, two hydroxylated metabolites are formed; they are (4*R*)-4-hydroxyochratoxin A and (4*S*)-4-hydroxyochratoxin A [29]. If ochratoxin A is incubated with liver microsomes from rabbit, an additional hydroxylated metabolite is formed, namely 10-hydroxyochratoxin A [30] (Fig. 4). Formation of the hydroxylated metabolites was inhibited by carbon monoxide and metyrapone, and their production was stimulated when the animals were pretreated with phenobarbital (Table 1) [31]. Ochratoxin A induces a type 1 difference spectrum when added to a microsomal suspension; the results indicate the cytochrome P-450 is catalyzing the hydroxylation reactions. The finding that (4*R*)-4-hydroxyochratoxin A was not converted to the 4*S* isomer during incubation and extraction shows that the metabolites are not laboratory artifacts. V_{max} values for (4*R*)- and (4*S*)-4-hydroxyochratoxin A increase in different ratios when rats are pretreated with phenobarbital; this indicates that the two alcohols are formed by different species of cytochrome P-450 [29] (Table 1).

When ochratoxin A is given intraperitoneally or per os to rats, ochratoxin α, ochratoxin A, and (4*R*)-4-hydroxyochratoxin A are excreted in the urine [32, 33]. Approximately 1–2% of the administered toxin was recovered as the hydroxylated metabolite, and total recovery of administered toxin was in the range of 30–40%. The presence of ochratoxin A and ochratoxin α in rat was also observed in other studies [34–37]. The 4*R* isomer may be identical to "compound IV" described by Galtier et al. [35] because thin layer chromatographic R_f values are similar and the 4*S* epimer was not detected; only ochratoxin A and α were found in feces. The fate of the remaining toxin is unknown; it may have been metabolized to conjugated metabolites.

Isolated hepatocytes from rats can form significant amounts of the 4*R* isomer from ochratoxin A but only traces of (4*S*)-4-hydroxyochratoxin A [38]. The fact that only (4*R*)-4-hydroxyochratoxin A is excreted in the

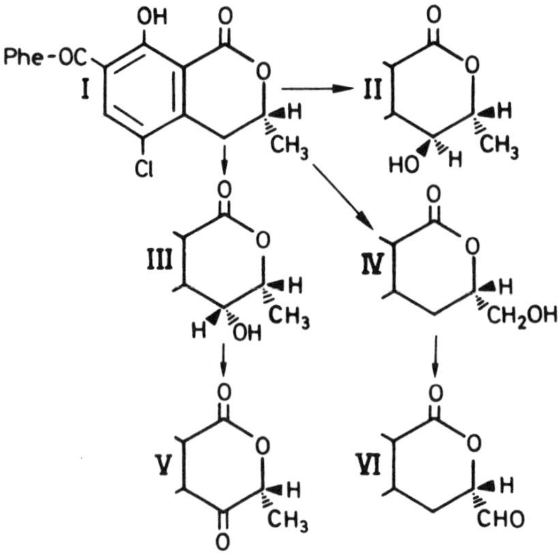

FIGURE 4. Metabolites (II—IV) formed from ochratoxin A in the presence of NADPH and microsomes from various species, and possible metabolites (V and VI) formed from III and IV in the presence of alcohol dehydrogenase and NAD. In II—VI, only the right-hand parts of the molecules are shown. II, (4*S*)-4-hydroxyochratoxin A; III, (4*R*)-4-hydroxyochratoxin A; IV, 10-hydroxyochratoxin A; V, 4-oxo-ochratoxin A; VI, 10-oxo-ochratoxin A [39].

urine and formed by hepatocytes could be due to differences in K_m and V_{max} values. There are significant differences in the rate constant values for the two epimers. Low V_{max} values for the formation of the three metabolites suggests that hydroxylation is probably of minor importance when cells are exposed to ochratoxin A. (4*R*)-4-Hydroxyochratoxin A and 10-hydroxyochratoxin A could be oxidized further in vivo. It has been reported [39] that the metabolites are oxidized by horse liver alcohol dehydrogenase in the presence of NAD (Fig. 3). However, the compounds were poor substrates in the reaction.

The Lewis strain of rats (extensive metabolizers of debrisoquine) and the dark agouti (DA) strain of rats (poor metabolizers of debrisoquine) show a genetic polymorphism for the gene regulating debrisoquine-4-hydroxylase (P-450 IID1) in a manner thought to be analogous to human metabolism [40, 41]. Metabolic ratios similar to those observed for debrisoquine:4-hydroxydebrisoquine were observed for the hydroxylation of ochratoxin A to (4*R*)-4-hydroxyochratoxin A [42]. In both cases, hydroxylase activity was one-third to one-fourth as high in DA rats as in Lewis strains. Slight differences between rat strains were found in the induction of several hepatic drug-metabolizing enzymes except for ochratoxin A 4-hydroxylase; the latter was inducible only by 3-methyl cholanthrene in the DA

TABLE 1. K_m and Maximum Velocity for the Formation of Ochratoxin A Metabolites by Liver Microsomal Fractions from Various Species in the Presence of NADPH

	K_m (μM)			Maximum velocity [nmol product/(mg protein/h)]		
Animal	4*R*	4*S*	10-OHOA	4*R*	4*S*	10-OHOA
Rat, albino	60.2	175.5	—	6.8	0.89	—
albino[a]	61.3	182.5	—	23.0	10.2	—
brown	48	75	—	0.77	1.02	—
brown[a]	56	71	—	3.7	8.7	—
Rabbit	133	341	210	0.16	0.34	0.24
Rabbit[a]	105	—	210	0.39	0.63	1.06
Pig	31.8	68.8	—	1.30	1.78	—
Human	77.5	136	—	0.57	0.39	—

Abbreviations: 4*R* = (4*R*)-4-hydroxyochratoxin A; 4*S* = (4*S*)-4-hydroxyochratoxin A; 10-OHOA = 10-hydroxyochratoxin A.
[a]Phenobarbital-induced [29, 30].

strain and appeared to be linked genetically with other Ah locus-associated monooxygenases [41]. Based on the different metabolic ratios for ochratoxin α:(4*R*)-4-hydroxyochratoxin A and debrisoquinone:4-hydroxydebrisoquine, the results suggested that the two hydroxylase activities have cosegregated in the two strains [42]. The observations also suggested a connection between polymorphic oxidation reaction(s) and ochratoxin A-induced tumorigenesis.

C. Metabolism of Ochratoxin B

Like ochratoxins A and α, ochratoxins B and β are also excreted together when ochratoxin B is administered [43]. Størmer et al. [44] observed an additional metabolite in the urine from rats given ochratoxin B, namely 4-hydroxyochratoxin B.

The ratios of excreted ochratoxin A or B metabolites were markedly different. For example, 10 times as much 4-hydroxyochratoxin B was excreted as (4*R*)-4-hydroxyochratoxin A [44]. The results are not surprising, since the large differences in relevant K_m values and lower affinity for plasma proteins enhance excretion [45]. Rats have also been given a mixture of ochratoxins A and B (i.p., 4 mg of each toxin/kg body weight), and the amounts of excreted ochratoxin A and B metabolites were determined; no differences in the amounts of excreted ochratoxin A metabolites were found. The results are in agreement with the kinetics, which indicate that ochratoxin B does not inhibit the formation of (4*R*)-4-hydroxyochratoxin A and does not interfere with the uptake of ochratoxin A by the hepatocytes or inhibit the cleavage of the peptide bond in the molecule [38].

D. Metabolism of Ochratoxin C

Ochratoxin C (ethyl ester of ochratoxin A) was found in feed containing high concentrations of ochratoxin A [46]. When ochratoxin A is incubated in vitro with rumen fluid, it is readily converted to ochratoxin C [47]. The conversion of ochratoxin C to ochratoxin A occurs in rats after oral and intravenous administration of toxin [48]. Time-based concentrations of ochratoxin A in blood were essentially the same after oral administration of equivalent amounts of either ochratoxin A or ochratoxin C. Intravenously, ochratoxin C was also converted to ochratoxin A. The conversion of ochratoxin C into ochratoxin A is a possible explanation of the similar toxicity of the two toxins [48].

Caco-2 cells isolated from human colon carcinoma [49] spontaneously differentiate in culture to morphology with characteristics of normal intestinal tissue. Experiments with this cell line [50] did not yield any metabolite after incubation with either ochratoxin A or ochratoxin B. However, ochratoxin C was hydrolyzed to some extent to ochratoxin A [51]. The results indicate that some ochratoxin C could be hydrolyzed to ochratoxin A during passage through the intestinal mucosa.

E. Tissue Distribution and Excretion

The toxicokinetic profile of ochratoxin A varies considerably among different species [34, 35, 37, 52]. Variation in persistence of ochratoxin A in test animals has been attributed to different affinities of the toxin to serum albumin of the species [52]. In animals with albumin deficiency, ochratoxin A disappears from plasma in a short time, and its excretion in urine and bile is 20–70 times higher than in normal rats; this indicates that albumin bound to ochratoxin A will delay toxin elimination by limiting its transfer from the bloodstream to the hepatic and renal cells [53]. The effects of intravenous injection of ochratoxin A on absorption of toxin from the jejunum in rats have been studied in vivo [54]. Ochratoxin A was absorbed from the jejunum even when its level was higher in the plasma than in the jejunal lumen; the results indicated that ochratoxin A was absorbed in its nonionized form.

Whole-body radioautography studies of rainbow trout [55] have shown that the highest concentrations of ochratoxin A or its metabolites are found in kidney, urine, and bile, and not in muscular tissue. Fish seem to be able to exclude the toxin from the body either through hydrolysis or through conjugation and subsequent elimination [45].

Recently, a study was carried out on the excretion patterns of ochratoxin A in fish (carp), bird (quail), mouse, rat, and monkey [45]. The toxin was given either orally or by intravenous administration of 0.05 mg of ochratoxin A/kg body weight; this is the amount of toxin occasionally found in domestic animals and humans [56, 57]. The elimination half-life varied from 0.68 h after oral administration in fish up to 840 h in monkeys.

Quail were similar to carp in eliminating ochratoxin A quickly, but the amount of ochratoxin A that enters the general systemic circulation after oral administration (bioavailability) was much greater in birds (62%) than in fish (1.6%). Ochratoxin A was found not only in plasma but also in other tissues. A considerable fraction of the toxin was found in the gastrointestinal system, kidney, and liver [58]. In quail plasma, two unique ochratoxin A-binding proteins were found with different affinities for the toxin. Mice had the same ochratoxin A levels in plasma regardless of administration route and the highest bioavailability of the toxin among the species investigated [45]. Two ochratoxin A-binding proteins were also found in mice; whole-body radioautography studies showed a slow clearance of the toxin from the bloodstream [59]. Rats had the highest maximal concentration of ochratoxin A in plasma of all the test species, and the toxin was easily detected 2 weeks after either oral or intravenous administration. The elimination half-life of the toxin was greater than 100 h. In binding experiments the fraction of free toxin was extremely low [45].

Administration of ochratoxin A intraduodenally to rats (300 μg/kg body weight) yielded some toxin recovery from intestinal lymph (3%) during the first hours, with essentially complete toxin elimination within 9 h [33]. Small amounts of ochratoxin α were detected during the first 2 h; no other metabolites were detected. Owing to its early appearance, the small amount of ochratoxin α suggests formation of the metabolite within the intestinal mucosa. Ochratoxin A (administered to rats orally) appeared in the bile, with maximum excretion taking place within the first 6 h [60]. Bile from treated animals was given to other rats by gavage into the stomach or duodenum; ochratoxin A in plasma of these rats (measured 24 h later) was higher when the bile was inserted into the duodenum than when administered via the stomach. The results suggested that the difference in ochratoxin A persistence in plasma of various animals may not be only a result of different binding to plasma constituents; it might also be due to enterohepatic circulation of ochratoxin A.

Among several animal species, the monkey had the longest elimination half-life, 35 days after intravenous administration, and toxin absorption from the gastrointestinal tract was slow; this toxin activity in monkeys seems to be a good model for humans [45]. In studies of Balkan endemic nephropathy, serum concentrations of ochratoxin A of 50–100 ng/mL have been detected [57]. In a few cases where repeated samples were drawn, a high persistence of the toxin in the blood was observed. The toxicokinetics and binding properties of ochratoxin B in rats have also been determined [45]; this compound was more readily eliminated and had a lower affinity for plasma proteins than ochratoxin A.

IV. BIOLOGICAL AND BIOCHEMICAL EFFECTS OF OCHRATOXINS A AND B

A. Nephropathy

The kidney receives 25% of the cardiac output, which exposes the organ to a variety of chemical agents. One event related to the nephrotoxic response to xenobiotics is alteration in active transport systems of the renal tubules. It has been shown that the proximal tubule in pigs is the target segment of the nephron in ochratoxin A-induced nephropathy [61]. The brush border was reduced in height and density in damaged areas, and some cells with light cytoplasm exhibited essentially no brush border in pigs [62] (Fig. 5). Mitochondria in slightly damaged epithelial cells were reduced in number and were occasionally swollen, but generally they appeared unaltered. An increase in peroxisomes with a fine granular material

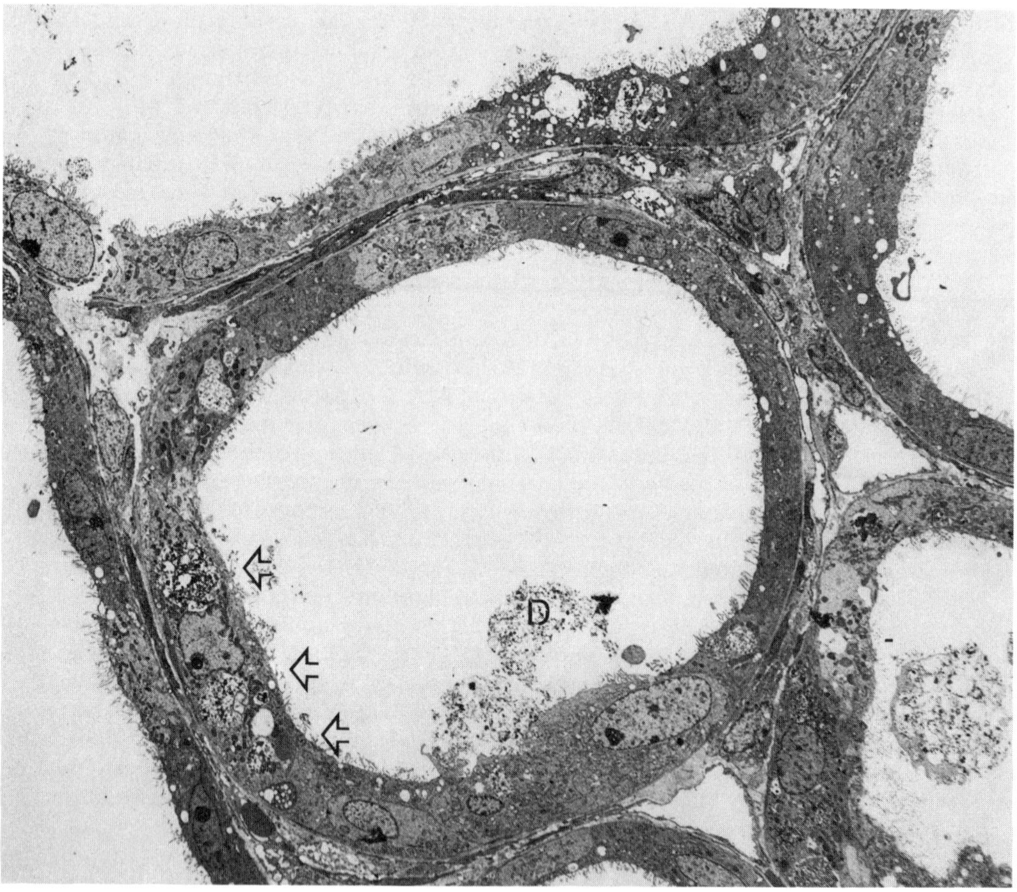

FIGURE 5. Electron micrograph of the kidney of a pig given 800 μg/kg ochratoxin A for 5 days. The epithelium of the lower part of the proximal convoluted tubule is flattened, and some cells are almost without brush border. Damaged cells with very condensed chromatin are located at different levels in the epithelial wall (arrows). Cellular debris is desquamated into the lumen (D). From Ref. 62.

and one or two paracrystalline plates in the membranes appeared in many of the damaged cells (Fig. 6). In target cells, peroxisomes appeared to have lost membrane integrity, and the organelles were leaking materials into the cytosol (Fig. 6). Reduction of structural integrity was associated with an increase in the presence of catalase and cyanide-insensitive fatty acid oxidase activity in the soluble pig kidney fractions.

In rats treated with ochratoxin A, food intake was reduced by 50%. Microscopic lesions were observed in some of the nephrons from rats given ochratoxin A. The cell damage ranged from vacuolization of the cytoplasm to necrosis and desquamation of epithelial cells and left a naked basement membrane (Fig. 7). When ochratoxin A and ochratoxin B were given in combination, the animals were clinically unaffected, and histologically there was only slight damage of proximal tubules [44] (Fig. 8).

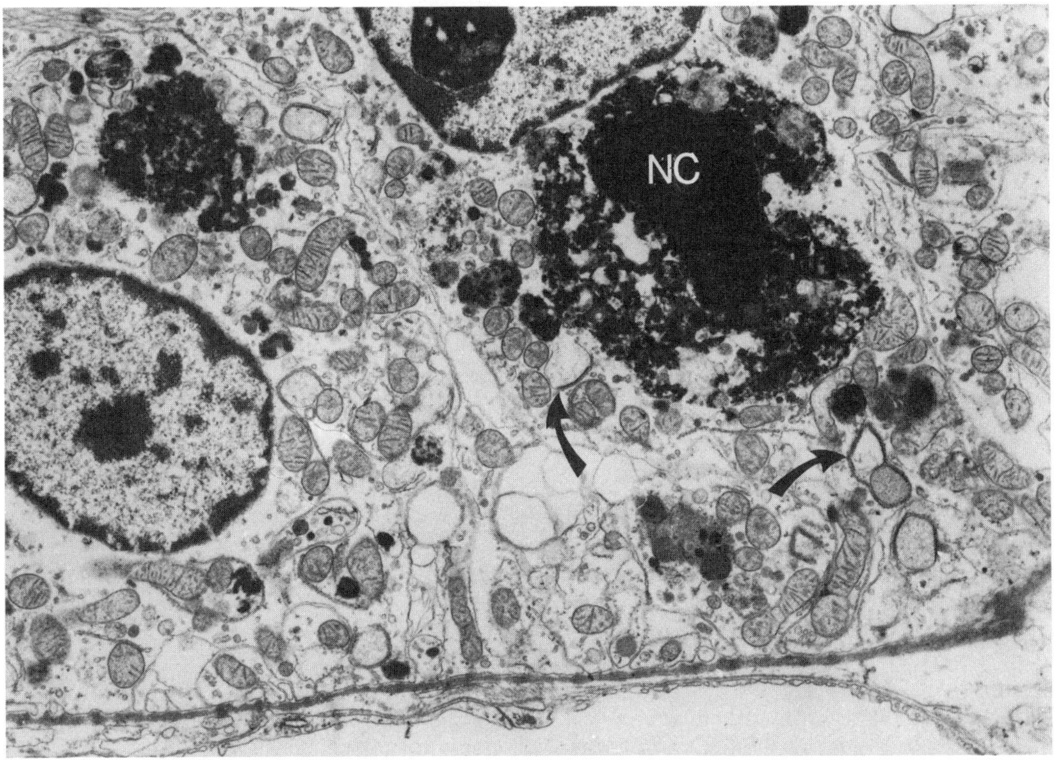

FIGURE 6. Same tissue as Fig. 5. Electron micrograph from a damaged proximal epithelial cell showing peroxisomes with paracrystalline plates. Some peroxisomes appear to be leaking into the cytoplasm (arrows). The necrotic cell (NC) exhibits a very condensed nucleus, and the cell organelles are almost indistinguishable. From Ref. 62.

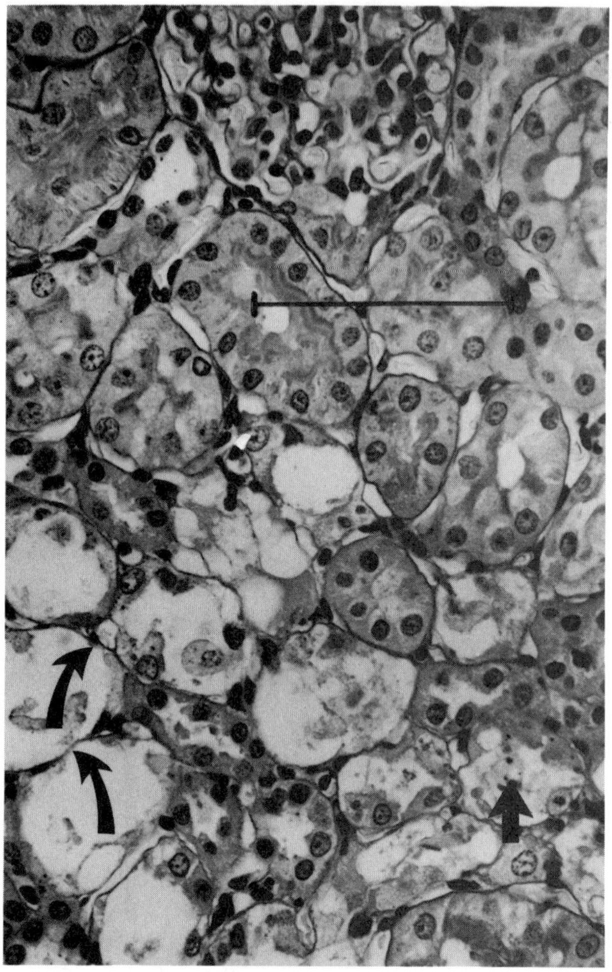

FIGURE 7. Micrograph from kidney of rat given 4 mg of ochratoxin A per kilogram body weight. Note the necrosis of epithelial cells (straight arrow) and the naked basement membranes (curved arrows). Bar 100 μm. From Ref. 44.

Ochratoxin A alters not only *p*-aminohippuric acid accumulation as an organic anion transport parameter but also the cation transport pathway [63]. Probenecid is a competitive inhibitor of the organic anion transport pathway in renal tubules [64, 65]. Elimination of ochratoxin A from the kidney may be altered when active tubular secretion is inhibited by probenecid where it blocks the entry of ochratoxin A into proximal tubule cells [66]. Probenecid also protected against ochratoxin A-induced decrease in cellular adenosine triphosphate in isolated nephron segments

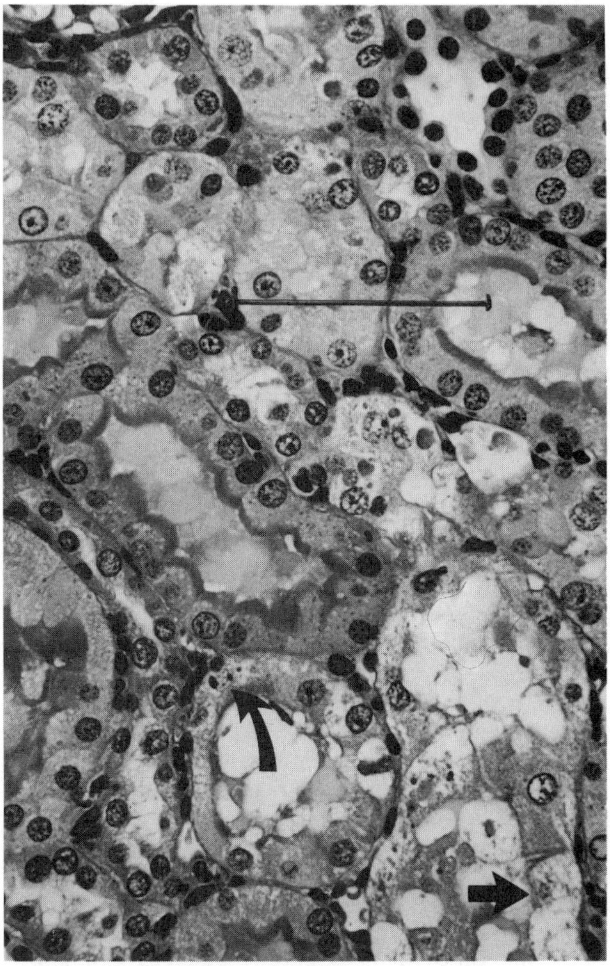

FIGURE 8. Micrograph from kidney of rat given 4 mg of ochratoxin A and 4 mg ochratoxin B per kilogram body weight. The cytoplasm is vacuolated in some epithelial cells (straight arrow), and a few epithelial cells appear necrotic (curved arrow). Bar, 100 μm. From Ref. 44.

[67]. The results suggested that ochratoxin A might enter the plasma membrane in the middle and terminal portions of the proximal tubule through the anion transport pathway and inhibit mitochondrial oxidative phosphorylation.

Although the biochemical basis for ochratoxin-mediated kidney damage remains unknown, both renal glyconeogenesis [68] and cytosolic phosphoenolpyruvate carboxykinase activity are reduced in rats fed ochratoxin A for several days. This effect appears to be specific since several other enzyme activities located in renal proximal convoluted tubules and hepatic cytosolic phosphoenolpyruvate carboxykinase activity in toxin-treated rats were unaffected [69]. Ochratoxin A produced a reduction in specific activities of phosphoenolpyruvate carboxykinase and γ-glutamyl transpeptidase after addition of toxin to primary cultures of rat kidney proximal convoluted tubule cells [70]. Phosphate-dependent glutaminase was not affected by the toxin. The investigations also found cyclic adenosine monophosphate (cAMP)-mediated increases in phosphoenolpyruvate carboxykinase and mRNA content, which were abolished when ochratoxin A was added simultaneously with cAMP.

B. Teratogenicity

Ochratoxin A is a potent teratogen in laboratory animals; it has been reported to elicit prenatal dysmorphogenesis in rats [71–75], mice [76], hamsters [77], and chick embryos [78]. The mechanism of ochratoxin A-induced teratogenesis has not been clearly defined and may involve an indirect effect through maternal action [79] and/or a direct effect on the developing embryo [72, 77]. In support of the contention of a direct fetal effect of ochratoxin, it has been demonstrated that maternal ochratoxin A can cross the placenta and accumulate in fetal tissues within 12 h following its administration to pregnant rats [80]. A more recent study [81] demonstrates that ochratoxin A can deleteriously affect a number of parameters associated with embryonic growth and differentiation in the rat in a concentration-dependent manner; the results provide evidence for direct action of this mycotoxin in the developing embryo.

A study was undertaken to determine the effects of ochratoxins A and B on protein synthesis and the development of chicken embryo limb bud micromass [82]; this process involves a cartilagenous stage that is characterized by the biosynthesis of collagen and cartilage proteoglycans. Cells were incubated with various amounts of either ochratoxin A or ochratoxin B in the presence of [^{3}H]proline or $^{35}SO_4$ for 6 days (Fig. 9). Incorporation of radiolabeled precursors decreased with increasing concentrations of the toxins. The average IC_{50} for ochratoxins A and B was 1.9 and 6.1 μM, respectively. The cells were also simultaneously incubated with several concentrations of ochratoxin A together with nontoxic concentrations of ochratoxin B (0.5 and 1.0 μM) in combination. The treatments led to concentration responses that were similar to exposing cultures to ochratoxin A alone. Preincubation with ochratoxin B for up to 2 h prior to the addition of ochratoxin A did not change the activity of the latter toxin.

C. Lipid Peroxidation

Addition of ochratoxin A to rat liver microsomes greatly enhanced the rate of NADPH- or ascorbate-dependent lipid peroxidation [83]. NADPH-dependent lipid peroxidation in kidney microsomes was similarly enhanced by ochratoxin A. The process required the presence of trace amounts of iron, but cytochrome P-450 and free active oxygen appeared not to be involved.

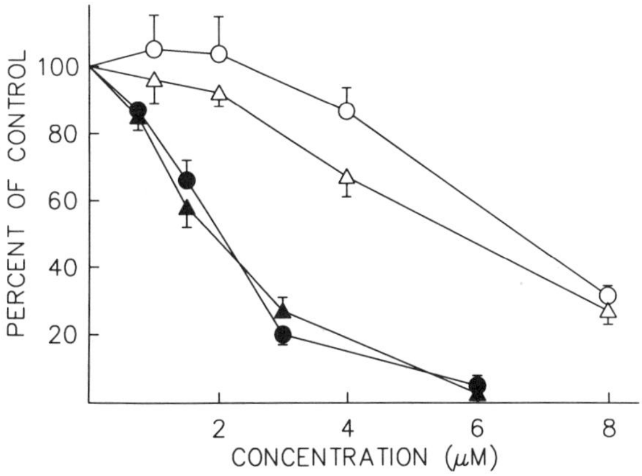

FIGURE 9. The ability of ochratoxin A (filled symbols) and ochratoxin B (open symbols) to affect differentiation ($^{3}SO_4^{2-}$) (O) and general protein synthesis ([^{3}H]proline) (Δ) in chicken limb bud biomass cultures. Incorporation control values = 1213 ± 131 cpm (sulfate) and 52,870 ± 2410 cpm (proline). From Ref. 82.

The efficiency of several ochratoxins (A, B, C, α, and *O*-methylochratoxin C; 125 μM of each) in enhancement of lipid peroxidation was related to the presence and reactivity of a phenolic hydroxyl group. Furthermore, the ability of the ochratoxins to enhance lipid peroxidation in microsomes correlated with their known toxicities in chicks. Administration of ochratoxin A to rats also resulted in enhanced lipid peroxidation in vivo. However, no effect was observed when a mixture of (4*R*)- and (4*S*)-4-hydroxyochratoxin was tested (2.5 μM). The results suggested that toxin-mediated lipid oxidation may play a role in the observed toxicity in animals.

D. Calcium Metabolism

Administration of ochratoxin A to rodents alters a variety of biochemical parameters in liver, kidney, and the urinary tract [83–88]. The effect of ochratoxin A on hepatic microsomal calcium sequestration has been studied both in vivo and in vitro [89]. The rate of ATP-dependent calcium uptake was inhibited by 42–45% in ochratoxin A-intoxicated rats. In the presence of NADPH, addition of ochratoxin A (2.5–100 μM) caused a concentration-dependent inhibition of calcium uptake (28–94%) by rat liver microsomes. Various agents that inhibited ochratoxin A-enhanced lipid peroxidation [83] were also able to block the destruction of calcium uptake activity. Lipid peroxidation enhanced by ochratoxin A was also accompanied by leakage of calcium from calcium-loaded microsomes. The authors suggested that ochratoxin A disrupts microsomal calcium homeostasis by an impairment of the endoplasmic reticulum membrane probably via enhanced lipid peroxidation.

E. Interaction with Other Mycotoxins

The toxicity and clinical signs observed in animals when more than one mycotoxin is presented in feed are complex and diverse. Some mycotoxins, such as the combination of aflatoxin with either ochratoxin A or T-2 toxin, interact to produce synergistic toxicity in broiler chicks. Multiple mycotoxin exposure can produce effects that differ greatly from the toxic response to a single mycotoxin. Aflatoxicosis causes fatty liver in poultry, but the toxic effect is prevented by the simultaneous presence of ochratoxin A. Of the mycotoxin combinations that have been investigated in poultry and swine, the aflatoxin + ochratoxin A and aflatoxin + T-2 toxin interactions appear to be the most toxic [90–92].

F. Genetic Toxicology

Ochratoxin A was tested in numerous laboratories for induction of gene mutation in a variety of *Salmonella typhimurium* strains with and without exogenous metabolic activation; results were uniformly negative [93–98]. In addition, no growth inhibition due to DNA damage was observed in *Bacillus subtilis* strain M15/H17 [99], and no induction of mitotic recombination was reported in *Saccharomyces cerevisiae* D3 after treatment with ochratoxin A [96].

Tests with mammalian cells have indicated some genotoxic activity by ochratoxin A. Unscheduled DNA synthesis was reported in ACI rat and C3H mouse hepatocyte cultures treated with ochratoxin A [100]. DNA damage in the form of single-strand breaks was observed in kidney, liver, and spleen of male BALB/c mice injected intraperitoneally with ochratoxin A. The DNA damage to splenic cells was confirmed by an in vitro procedure using phytohemagglutinin-stimulated cultured spleen cells incubated with ochratoxin A [101]. DNA single-strand breaks have also been reported in renal and hepatic cells of rats administered ochratoxin A by gavage [88]. Although DNA damage in mammalian cells provides limited evidence of mutagenic potential, no induction of gene mutation was detected in cultured mouse FM3A cells exposed to ochratoxin A in the absence of metabolic activation [102]. In cytogenic tests with cultured Chinese hamster ovary cells, a small but significant dose-related increase in sister chromatid exchange was observed after treatment of the cells with ochratoxin A. However, ochratoxin A did not induce chromosomal aberrations either with or without Aroclor 1254-induced male Sprague-Dawley rat liver microsomes [95].

G. Carcinogenicity

One-year feeding studies in male DDY mice showed that exposure of 40 ppm ochratoxin A for 50 weeks caused renal and hepatic tumors [103, 104]. All of nine mice fed 40 ppm ochratoxin A had cystic renal adenomas compared with none in 10 control mice. The carcinogenic effects of ochratoxin A in mice were confirmed in a 24-month feed study in male and female $B6C3F_1$ mice exposed to ochratoxin A [105].

In a recent 2-year gavage study [95], there was clear evidence of carcinogenic activity of ochratoxin A in male F344/N rats; exposure to the toxin substantially increased incidences of uncommon tubular cell adenomas and of tubular cell carcinomas of the kidney. These results provided clear evidence of carcinogenic activity in female F344/N rats by increased incidences of uncommon tubular cell adenomas and of tubular cell carcinomas of the kidney and by increased fibroadenomas of the mammary gland. Ochratoxin A also caused nonneoplastic renal changes including tubular cell hyperplasia, tubular cell proliferation, cytoplasmic alteration, karyomegali, and degeneration of the tubular epithelium.

There is increasing evidence that Balkan endemic nephropathy is associated with a dramatic increase in urinary tract tumors [42]. Balkan endemic nephropathy may be related to ochratoxin A; of 600 blood samples collected in an endemic region in Yugoslavia, about 7% were positive for ochratoxin A [57]. Ochratoxin A residues have been found in pork products and are associated with the presence of nephropathy in swine [106]. From the results of all the relevant studies and because ochratoxin A can be quite widespread in products for human consumption, it would seem that this compound presents a true potential hazard for humans [95].

H. Ochratoxin A Versus Ochratoxin B

Ochratoxin B can reverse some of the nephrotoxic effects of ochratoxin A. Also the effect on immunity by ochratoxin A is influenced by ochratoxin B (Section V.C). When human peripheral blood lymphocytes were pulsed with a mixture of ochratoxin A and ochratoxin B, or exposed to ochratoxin A before the addition of ochratoxin B, the inhibition of phytohemagglutinin-induced T-cell proliferation was about the same as was seen with ochratoxin A alone. If the cells were pulsed with ochratoxin B prior to ochratoxin A exposure, the inhibitory effect of ochratoxin A on lymphocyte proliferation was markedly reduced. Ochratoxin A may bind to vital cellular parts and cause structural changes [107]. Ochratoxin B, on the other hand, could bind to the same site in a competitive manner and thus prevent the structural modification caused by ochratoxin A. The effect of ochratoxin A is not always reversed by ochratoxin B. In studies of the energy-dependent uptake of ochratoxin A by mitochondria, ochratoxin B had no effect [108] either in inhibition of respiration or competition with ochratoxin A for uptake by mitochondria.

Ochratoxin B was recently studied separately and in combination with ochratoxin A on the aminoacylation of phenylalanine tRNA catalyzed by mice liver phenylalanyl-tRNA synthetase [109]. Ochratoxin B was neither a significant inhibitor of the reaction nor an antagonist of ochratoxin A. Ochratoxin B was also assayed for its possible antagonistic effect on the in vitro protein synthesis inhibition caused by ochratoxin A in hepatoma tissue culture cells [110]. No prevention of ochratoxin A inhibition could be found for ochratoxin B.

V. IMMUNOTOXICOLOGICAL EFFECTS OF OCHRATOXIN A

A. Mice

In BALB/c mice, ochratoxin A (0.0005 mg/kg body weight) suppressed the immune response to sheep erythrocytes [111]; this effect could be prevented by the presence of phenylalanine. In Swiss mice, however, no effect was observed from the presence of ochratoxin A after daily injection of 5 mg/kg body weight for 50 days [112]. Doses of 0.001 mg/kg body weight of (4*R*)-4-hydroxyochratoxin A administered to mice caused an 80% reduction in the number of cells producing immunoglobulin M (90% with ochratoxin A) and a 93% reduction in cells producing immunoglobulin G (92% with ochratoxin A); ochratoxin α had no effect [113].

Recently, the effects of ochratoxin A on immune mechanisms associated with tumor resistance were examined in mice using dose levels similar to those that cause neoplasia [114]. Ochratoxin A specifically inhibited natural killer cell activity and increased the growth of transplantable tumor cells without altering T-cell or macrophage-mediated antitumor activity. In contrast, ochratoxin B did not influence immune function. Polyinosinic: polycytidylic acid-induced interferon was markedly reduced in mice following exposure to ochratoxin A although total serum protein levels were slightly increased. Injection of polyinosinic:polycytidylic acid enhanced natural killer cell activity in the presence of ochratoxin A, but the level of enhancement was slightly lower than the level produced by the agent in the absence of ochratoxin A. Thus, the toxin appears to suppress natural killer cell activity by inhibiting the production of basal interferon; the results also suggest a possible role for the altered natural killer cell function in the development of mycotoxin-induced carcinogenesis. Other mycotoxins such as aflatoxins [115], trichothecenes [116], and zearalenone [116] have profound effects on various aspects of the immune system. If the immune response is reduced by naturally occurring ochratoxin A, it could make animals more susceptible to disease.

B. Pigs

The effect of ochratoxin A on in vitro concanavalin-stimulated porcine blood lymphocytes has been studied [117]. The results demonstrated the capacity of ochratoxin A to bring about a dose-dependent in vitro inhibition of cell-mediated immune response in pigs. The immunosuppressive effect of ochratoxin A did not appear to be affected by increased bovine serum albumin (BSA) levels in the cell culture media; this study indicates that immunosuppression caused by ochratoxin A is not limited by the amount of free toxin. The results are in accordance with other studies [111, 113] of potent immunosuppressive effects of low doses of ochratoxin A in mice.

The pig's immune system is sensitive to doses of ochratoxin A below 1 mg/L, with a 10% inhibition of lymphocyte response being produced by approximately 0.06 mg of ochratoxin A; this level has been broadly identified in blood of slaughter pigs [56, 118]. Under practical farming conditions, a 10% inhibition of the immune response might dramatically influence swine production [117].

C. Mechanism of Ochratoxin A-Induced Immunosuppression

Incubation of 7×10^6 human T lymphocytes per milliliter was carried out in 50 μM ochratoxin A. No hydroxylated metabolites were observed in the supernates from the incubation of cells with either ochratoxin A or ochratoxin B [107]. Subsequently, peripheral blood human T lymphocytes (10^6 cells/mL) were incubated in ochratoxin A in culture medium ranging in concentrations from 100 μM to 12.5 μM for 60 min. Cells were washed three times in culture medium and then set up for mitogen-induced proliferation and interleukin-2 (IL-2) production. Ochratoxin B, ochratoxin α, and solvent-treated cells (in the same concentration range as ochratoxin A) were also studied. Ochratoxin A treatment almost completely abolished the proliferative response of the T-cell population (Table 2). Ochratoxin B did not induce a detectable effect; ochratoxin α produced significant inhibition but less than ochratoxin A. Cells pulsed with a mixture of ochratoxin A and ochratoxin B or exposed to ochratoxin A before ochratoxin B produced effects on phytohemagglutinin-induced T-cell proliferation that were about the same as those elicited by ochratoxin A alone. If cells were pulsed for 30 min with ochratoxin B prior to ochratoxin A exposure, the inhibitory effect of ochratoxin A on lymphocyte proliferation was markedly reduced (Table 3) [107].

Experiments have been carried out on highly purified subpopulations of T lymphocytes expressing the CD4 (helper/inducer) and CD8 (suppressor/cytotoxic) phenotypic markers [107]. Purity of cell populations exceeded 98% as judged by flow cytometry. Ochratoxin A treatment showed no selectivity in inhibition of the phytohemagglutinin-induced proliferative capacity of varied T-cell populations. In T-cell proliferation, the binding

TABLE 2. Effect of Ochratoxins A, B, and α on the Proliferation of Peripheral Blood T Lymphocytes Induced by Phytohemagglutinin

Ochratoxin concn.	$[^3H]$thymidine incorporation (cpm)				
	OA	OB	Oα	Control	Medium[a]
Experiment 1					
50 μM	582	28207	23584	29623	97
12.5 μM	670	29989	25316		
Experiment 2					
50 μM	957	41376	32313	42546	127
12.5 μM	1127	41319	37266		

Abbreviations: OA, ochratoxin A; OB, ochratoxin B; Oα, ochratoxin α. Results are the means of triplicate cultures. Cells were treated with toxins, washed, and stimulated with phytohemagglutinin.

[a]Medium denotes spontaneous proliferation in the absence of phytohemagglutinin.

Source: Lea et al. [107].

TABLE 3. Effect of Ochratoxin A on Proliferation of Peripheral Blood T Lymphocytes That Were Pulsed 30 Min with Varying Amounts of Ochratoxin B Prior to Ochratoxin A Exposure

Toxin concentration (nmol/100 μL)		$[^3H]$Thymidine incorporation (cpm)
Ochratoxin A	Ochratoxin B	
12.5	12.5	33,047
12.5	6.25	14,347
12.5	3.13	8,543
12.5	1.57	6,297
12.5	0.78	129
—	12.5	33,047
12.5	—	139
—	—	39,677

[a]Cells were phytohemagglutinin-induced prior to treatment.
Source: Lea et al. [107].

of IL-2 to its receptor delivers the mitogenic signal that initiates cell division. The lack of IL-2-induced cell proliferation in ochratoxin A-treated cultures could be due to low IL-2 receptor expression and/or low interleukin-2 production. Phytohemagglutinin-induced IL-2 receptor expression on T cells was investigated by flow cytometry; the procedure employed the monoclonal antibody anti-Tac (Fig. 10). Ochratoxin A-treated cells were almost negative for Tac-antigen after 24 h in culture. Interleukin-2 production was reduced to background levels in ochratoxin A-treated cultures; ochratoxin B produced almost no effect in these systems.

Ochratoxin A has been reported to influence B-cell function [113]; the results can be explained by either a direct effect upon B lymphocytes or an indirect effect on helper T lymphocytes or both. The effect of toxin treatment on the proliferative capacity of highly purified B lymphocytes was studied after activation with anti-μ antibodies in the presence of B-cell growth factor 1 (Table 4). The results demonstrate that the inhibitory effects of ochratoxin A are not restricted to T lymphocytes; negligible effects were observed for ochratoxin B in equimolar concentrations.

In an early study, ochratoxin A specifically affected natural killer cell activity [114]; the results contradicted a previous investigation that suggested that the toxin had immunosuppressive properties on B-lymphocyte function when administered to mice [113]. Subsequent results demonstrated that the toxin is taken up by human T lymphocytes [107]. After

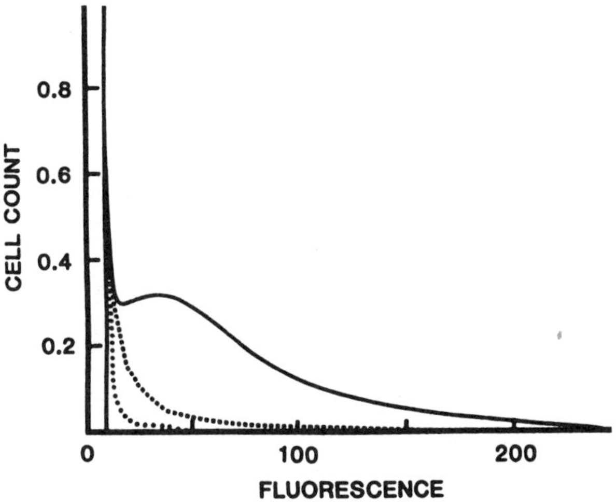

FIGURE 10. Effect of ochratoxin A on the expression of Tac antigen by T lymphocytes activated by phytohemagglutinin. (—) Histogram from untreated cells, (- - -) cells pretreated with ochratoxin A before activation. (···) Untreated cells stained with irrelevant ascitic fluid and the FITC-conjugated secondary antibody. Histograms have been normalized and superimposed for easier comparison. From Ref. 107.

washing and reincubation of toxin-loaded cells, ochratoxin A was secreted at a somewhat slower rate than uptake; the data suggested that toxin uptake is not due to simple passive diffusion through T-cell membrane. Furthermore, no hydroxylated metabolites of either ochratoxin A or ochratoxin B were released by the T lymphocytes.

The suppressive effect by ochratoxin A on B-cell function previously detected in mice [113] could also have been produced by inhibition of T-cell functions. Purified T lymphocytes were stimulated with phytohemagglutinin after being pulsed with graded concentrations of ochratoxin A, ochratoxin B, or ochratoxin α; the results demonstrated that ochratoxin A has a potent inhibitory effect on T-cell proliferation. Since phytohemagglutinin is a mitogen that is capable of inducing proliferation in the majority of human T lymphocytes, it is very unlikely that the ochratoxin-mediated inhibitory effect was produced by a functional deletion of a small subpopulation of lymphocytes in the cell suspension. Lack of T-cell proliferation can be a result of reduced production of interleukin-2 in the culture system or inhibition of pertinent receptor expression; both the parameters are negatively affected by exposure to ochratoxin A, and the toxin has a generalized inhibitory effect on the production of vital components of T-lymphocyte proliferation and effector function.

When cells are preincubated with ochratoxin B prior to ochratoxin A exposure, the inhibitory effect of ochratoxin A is strongly reduced. However, reversal of the inhibition cannot be strictly competitive because

TABLE 4. Effect of Ochratoxins A and B on the Stimulation of Human Peripheral Blood B Lymphocytes with Polyclonal Anti-μ Antibodies Coupled to Agarose Particles[a]

	[^{3}H]Thymidine incorporation (cpm)		
Stimulus	Ochratoxin A (50 μM)	Ochratoxin B (50 μM)	Medium[b]
Anti-μ	148	196	1,157
BCGF-1	179	359	1,836
Anti-μ + BCGF-1	511	9,684	11,354

BCGF-1 = B-cell growth factor.
[a]Each result is the mean of triplicate cultures.
[b]Medium denotes spontaneous proliferation in the absence of phytohemagglutinin.
Source: Lea et al. [107].

incubation with a mixture of ochratoxin A and ochratoxin B produces inhibition similar to that of ochratoxin A alone. Levels of ochratoxin B required to reduce the inhibitory effects of ochratoxin A remain unknown.

Data from in vivo experiments suggested that ochratoxin A impaired B-cell functions. A 1-h pulse with ochratoxin A abrogated the ability of purified B cells to proliferate in response to anti-μ antibodies in the presence of B-cell growth factor 1. The results provide strong evidence for ochratoxin A having potent inhibitory effects at the level of both B and T lymphocytes. There are contradictory views such as those suggesting that ochratoxin A affected only natural killer cell function [114]. Additional results support the view that ochratoxin A negatively influences the metabolism of essential components in lymphocytes with no selectivity in subpopulations; the data confirm and extend the previous observations of the potent immunosuppressive properties of ochratoxin A.

No inhibition of the phosphorylation of the two protein kinase C-specific substrates p80 and p19 in the presence of 12.5 μM ochratoxin A was obseved; this indicates that the protein kinase C-mediated phosphorylation, which is essential in activating T and B cells, is not influenced by ochratoxin A [119]. The spread of AIDS (acquired immune deficiency syndrome) has focused attention on the importance of immunosuppression and a possible connection between fungal toxins and AIDS [120].

VI. CONCLUDING REMARKS

Ochratoxin A, among other mycotoxins, may be a major cause of increased human and/or animal mortality in areas where food and/or feed is contaminated. Both historical and new information point in this direction. Many of the mycotoxins (such as ochratoxin A) act as strong immunosuppressors that may increase susceptibility to various diseases including cancer.

Available data indicate that ochratoxin A is a primary toxic agent, but not its metabolites or ochratoxin B. In some cases, ochratoxin B reverses the toxic effects of ochratoxin A, and in others it has no effect. Ochratoxin B is rarely present in contaminated commodities, and it will not play any practical role as protection against ochratoxin A toxicity.

Ochratoxin A interferes with several processes in the cell such as protein synthesis, energy metabolism, and several other enzyme activities. Despire information on ochratoxin A effects, the precise mechanism of action of the toxin is not known.

REFERENCES

1. Merwe, K. J., van der, Steyn, P. S., and Fourie, L. Mycotoxins. Part II. The constitution of ochratoxins A, B, and C, metabolites of *Aspergillus ochraceus* Willh. *J. Chem. Soc.*, *1965*:7083–7088.
2. Scott, P. M., Van-Walbeek, W., Kennedy, B., and Anyeti, D. Mycotoxins (ochratoxin A, citrinin, and sterigmatocystin) and toxigenic fungi in grains and other agricultural products. *J. Agric. Food Chem.*, *20*:1103–1109, 1972.
3. Krogh, P., Hald, B., and Pedersen, E. J. Occurrence of ochratoxin A and citrinin in cereals associated with mycotoxic porcine nephropathy. *Acta Pathol. Microbiol. Scand.* (*B*) *Microbiol Immunol.*, *81*: 689–695, 1973.
4. Ueno, Y. The toxicology of mycotoxins. *CRC Crit. Rev. Toxicol.*, *14*:99–132, 1985.
5. Krogh, P. *Mycotoxins in Food*, Academic, New York, 1987.
6. Krogh, P. Ochratoxins. In *Mycotoxins in Human and Animal Health* (J. V. Rodricks, C. W. Hesseltine, and M. A. Mehlman, Eds.), Pathotox, Forest Park South, Ill., 1977, pp. 489–498.
7. Krogh, P., Elling, F., Hald, B., Jylling, B., Petersen, V. E., Skadhauge, E., and Svendsen, C. K. Experimental avian nephropathy. Changes of renal function and structure induced by ochratoxin A-contaminated feed. *Acta Pathol. Mirobiol. Scand.* (*A*), *84*: 215–221, 1976.
8. Krogh, P., Hald, B., Plestina, R., and Ceovic, S. Balkan (endemic) nephropathy and foodborne ochratoxin A: preliminary results of a survey of foodstuffs. *Acta Pathol. Microbiol. Scand.* (*B*), *85*:238–240, 1977.
9. Visconti, A., and Bottalico, A. High levels of ochratoxins A and B in moldy bread responsible for mycotoxicosis in farm animals. *J. Agric. Food Chem.*, *31*:1122–1123, 1983.
10. Doster, R. C., Sinnhuber, R. O., and Pawlowski, N. E. Acute intraperitoneal toxicity of ochratoxin A and B derivatives in rainbow trout (*Salmo gairdneri*). *Food Cosmet. Toxicol.*, *12*:499–505, 1974.
11. Doster, R. C., Sinnhuber, R. O., and Wales, J. H. Acute intraperitoneal toxicity of ochratoxins A and B in rainbow trout (*Salmo gairdneri*). *Food Cosmet. Toxicol.*, *10*:85–92, 1972.

12. Wrigley, E. A. The growth of population in eighteenth century England: a conundrum resolved. *Past Present, 98*:121–150, 1983.
13. Wrigley, E. A., and Schofield, R. S. *The Population History of England 1541–1871. A Reconstruction*, Harvard Univ. Press, Cambridge, Mass., 1982.
14. Watkins, S. C., and van der Walle, E. Nutrition, mortality and population size: Malthus court of last resort. *J. Interdisciplinary History, 14*:205–227, 1983.
15. McKeown, T. *The Modern Rise of Population*, Academic, New York, 1976, pp. 34–40.
16. McKeown, T. Fertility, mortality, and causes of death. *Popul. Stud., 32*:537, 1978.
17. Birhaben, J. N. Alimentation et demographie historique. *Annl. Demogr. Hist., 1976*:22–40.
18. Langer, W. L. Europe's initial population explosion. *Am. Hist. Rev., 69*:1–17, 1963.
19. Langer, W. L. American foods and Europe's population growth. *J. Social Hist., 8*:51–66, 1976.
20. Matossian, M. K. Mold poisoning and population growth in England and France, 1750–1850. *J. Econ. Hist., 44*:669–686, 1984.
21. Moreau, C., and Moss, M. O. *Moulds, Toxins and Food*, Wiley, New York, 1979, pp. 13–14.
22. Parkin, J. C. Deterioration of stored grains by fungi. *Rhod. Sci. News.* 5:15–17, 1971.
23. Short, T. *A General Chronological History of the Air, Weather, Seasons, Meteors, etc. in Sundry Places and Different Times*, London, 1749, pp. 221–225.
24. Creighton, C. *History of Epidemics*, 2nd ed., New Jersey Biblio. Distribution, 1965, pp. 401–404.
25. Matossian, M. K. Death in London, 1750–1909. *J. Interdisciplinary Hist., 16*:183–197, 1985.
26. Kanai, K., and Kondo, E. Decreased resistance to mycobacterial infection in mice fed a trichothecene compound (T-2 toxin). *Jpn. J. Med. Sci. Biol., 37*:97–104, 1984.
27. Jelinek, C. F., Pohland, A. E., and Wood, G. E. Worldwide occurrence of mycotoxins in foods and feeds an update. *J. Assoc. Anal. Chem., 72*:223–230, 1989.
28. Cole, R. J., and Cox, R. H. *Handbook of Toxic Fungal Metabolites*, Academic, New York, 1981.
29. Størmer, F. C., Hansen, C. H., Pedersen, J. I., Hvistendahl, G., and Aasen, A. J. Formation of (4*R*)- and (4-*S*)-4-hydroxyochratoxin A from ochratoxin A by liver microsomes from various species. *Appl. Environ. Microbiol., 39*:971–975, 1981.
30. Størmer, F. C., Støren, O., Hansen, C. H., Pedersen, J. I., and Aasen, A. J. Formation of (4*R*)- and (4*S*)-4-hydroxyochratoxin A and 10-hydroxyochratoxin A from ochratoxin A by rabbit liver microsomes. *Appl. Environ. Microbiol., 45*:1183–1187, 1983.
31. Størmer, F. C., and Pedersen, J. I. Formation of 4-hydroxyochratoxin A from ochratoxin A by rat liver microsomes. *Appl. Environ. Microbiol., 39*:971–975, 1980.

32. Støren, O., Helgerud, P., Holm, H., and Størmer, F. C. Formation of (4*R*)-4-hydroxyochratoxin A and ochratoxin a from ochratoxin A by rats. V. Int. IUPAC Symposium on Mycotoxins and Phytotoxins, 1982, pp. 321–324.
33. Støren, O., Holm, H., and Størmer, F. C. Metabolism of ochratoxin A by rats. *Appl. Environ. Microbiol.*, *44*:785–789, 1982.
34. Chang, F. C., and Chu, F. S. The fate of ochratoxin A in rats. *Food Cosmet. Toxicol.*, *15*:199–204, 1977.
35. Galtier, P., Charpenteau, J. L., Alvinerie, M., and Labouche, C. The pharmacokinetic profile of ochratoxin A in the rat after oral and intravenous administration. *Drug Metabol. Dispos.*, *7*:429–434, 1979.
36. Nel, W., and Purchase, I. F. H. The fate of ochratoxin in rats. *J. S. Afr. Chem. Inst.*, *21*:87–88, 1968.
37. Suzuki, S., Satoh, T., and Yamazaki, M. The pharmacokinetics of ochratoxin A in rats. *Jpn. J. Pharmacol.*, *27*:735–744, 1977.
38. Hansen, C. H., Dueland, S., Drevon, C. A., and Størmer, F. C. Metabolism of ochratoxin A by primary cultures of rat hepatocytes. *Appl. Environ. Microbiol.*, *43*:1267–1271, 1982.
39. Syvertsen, C., and Størmer, F. C. Oxidation of two hydroxylated ochratoxin A metabolites by alcohol dehydrogenase. *Appl. Environ. Microbiol.*, *45*:1701–1703, 1983.
40. Hietanen, F., Bartsch, H., Castegnaro, M., Malaveille, C., Michelon, J., and Broussolle, L. Use of antibodies (Ab) against cytochrome P-450 isozymes to study genetic polymorphism in drug oxidation. *J. Pharm. Clin.*, *4*:71–78, 1985.
41. Hietanen, E., Malaveille, C., Camus, A. M., Bereziat, J. C., Brun, G., Castegnaro, M., Michelon, J., Idle, J. R., and Bartsch, H. Interstrain comparison of hepatic and renal microsomal carcinogen metabolism and liver S9-mediated mutagenicity in DA and Lewis rats phenotyped as poor and extensive metabolizers of debrisoquine. *Drug Metabol. Dispos.*, *14*:118–126, 1986.
42. Castegnaro, M., Bartsch, H., Bereziat, J. C., Arvela, P., Michelon, J., and Broussolle, L. Polymorphoc ochratoxin A hydroxylation in rat strains phenotyped as poor and extensive metabolizers of debrisoquine. *Xenobiotica*, *19*:225–230, 1989.
43. Patterson, D. S., Roberts, B. A., and Small, B. J. Metabolism of ochratoxins A and B in the pig during early pregnancy and the accumulation in body tissues of ochratoxin A only. *Food Cosmet. Toxicol.*, *14*:439–442, 1976.
44. Størmer, F. C., Kolsaker, P., Holm, H., Rogstad, S., and Elling, F. Metabolism of ochratoxin B and its possible effects upon the metabolism and toxicity of ochratoxin A in rats. *Appl. Environ. Microbiol.*, *49*:1108–1112, 1985.
45. Hagelberg, S., Hult, K., and Fuchs, R. Toxicokinetics of ochratoxin A in several species and its plasma-binding properties. *J. Appl. Toxicol.*, *9*:91–96, 1989.
46. Hamilton, P. B., Huff, W. E., Harris, J. R., and Wyatt, R. D. Natural occurrences of ochratoxicosis in poultry. *Poult. Sci.*, *61*:1832–1841, 1982.

47. Galtier, P., and Alvinerie, M. In vitro transformation of ochratoxin A by animal microbial floras. *Ann. Rech. Vet.*, *7*:91–98, 1976.
48. Fuchs, R., Hult, K., Peraica, M., Rodic, B., and Plestina, R. Conversion of ochratoxin C into ochratoxin A *in vivo*. *Appl. Environ. Microbiol.*, *48*:41–42, 1984.
49. Fogh, J., Fogh, J. M., and Orfeo, T. One hundred and twenty-seven cultured human tumor cell lines producing tumors in nude mice. *J. Natl. Cancer Inst.*, *59*:221–226, 1977.
50. Traber, M. G., Kayden, H. J., and Rindler, M. J. Polarized secretion of newly synthesized lipoproteins by the Caco-2 human intestinal cell line. *J. Lipid Res.*, *28*:1350–1363, 1987.
51. Størmer, F. C. Metabolism of ochratoxins A, B, and C by the Caco-2 human intestinal cell line. Unpublished results, 1990.
52. Galtier, P., and Alvinerie, M. The pharmacokinetic profiles of ochratoxin A in pigs, rabbits and chickens. *Food Cosmet. Toxicol.*, *19*: 735–738, 1981.
53. Kumagai, S. Ochratoxin A: plasma concentration and excretion into bile and urine in albumin-deficient rats. *Food Chem. Toxicol.*, *23*: 941–943, 1985.
54. Kumagai, S. Effects of plasma ochratoxin A and luminal pH on the jejunal absorption of ochratoxin A in rats. *Food Chem. Toxicol.*, *26*: 753–758, 1988.
55. Fuchs, R., Appelgren, L. E., and Hult, K. Distribution of ^{14}C-ochratoxin A in the rainbow trout (*Salmo gairdneri*). *Acta Pharmacol. Toxicol. (Copenhagen)*, *59*:220–227, 1986.
56. Hult, K., Hökby, E., Gatenbeck, S., and Rutqvist, L. Ochratoxin A in blood from slaughter pigs in Sweden: use in evaluation of toxin content of consumed feed. *Appl. Environ. Microbiol.*, *39*:828–830, 1980.
57. Hult, K., Plestina, R., Habazin-Novak, V., Radic, B., and Ceovic, S. Ochratoxin A in human blood and Balkan endemic nephropathy. *Arch. Toxicol.*, *52*:313–321, 1982.
58. Fuchs, R., Appelgren, L. E., Hagelberg, S., and Hult, K. Carbon-14-ochratoxin A distribution in the Japanese quail (*Coturnix coturnix japonica*) monitored by whole body autoradiography. *Poult. Sci.*, *67*: 707–714, 1988.
59. Fuchs, R., Appelgren, L. E., and Hult, K. Distribution of ^{14}C-ochratoxin A in the mouse monitored by whole-body autoradiography. *Pharmacol. Toxicol.*, *63*:355–360, 1988.
60. Fuchs, R., Radic, B., Peraica, M., Hult, K., and Plestina, R. Enterohepatic circulation of ochratoxin A in rats. *Period. Biol.*, *90*: 39–42, 1988.
61. Elling, F. Demonstration of ochratoxin A in kidneys of pigs and rats by imunofluorescence microscopy. *Acta Pathol. Microbiol. Scand.*, *85A*:151–156, 1977.
62. Elling, F., Nielsen, J. P., Lillehoj, E. B., Thomassen, M. S., and Størmer, F. C. Ochratoxin A-induced porcine nephropathy: enzyme and ultrastructure changes after short-term exposure. *Toxicon.*, *23*:247–254, 1985.

63. Berndt, W. O., and Hayes, A. W. In vivo and in vitro changes in renal function caused by ochratoxin A in the rat. *Toxicology, 12*: 5–17, 1979.
64. Mudge, G. H. Inhibitors of tubular transport of organic compounds. In *Goodman and Gilman's The Pharmacological Basis of Therapeutics* (A. G. Gilman, L. S. Goodman, and A. Gilman, Eds.), Macmillan, New York, 1980, pp. 929–934.
65. Weiner, I. M., Washington, J. A., and Mudge, G. H. On the mechanism of action of probenecid on renal tubular secretion. *Johns Hopkins Hosp. Bull., 106*:333–346, 1960.
66. Stein, A. F., Phillips, T. D., Kubena, L. F., and Harvey, R. B. Renal tubular secretion and reabsorption as factors in ochratoxicosis: effects of probenecid on nephrotoxicity. *J. Toxicol. Environ. Health, 16*:593–605, 1985.
67. Jung, K. Y., and Endou, H. Nephrotoxicity assessment by measuring cellular ATP content. II. Intranephron site of ochratoxin A nephropathy. *Toxicol. Appl. Pharmacol., 100*:383–390, 1989.
68. Meisner, H., and Selanik, P. Inhibition of renal gluconeogenesis in rats by ochratoxin. *Biochem. J., 180*:681–684, 1979.
69. Meisner, H., and Meisner, P. Ochratoxin A, an in vivo inhibitor of renal phosphoenolpyruvate carboxykinase. *Arch. Biochem. Biophys., 208*:146–153, 1981.
70. Thekkumkara, T. J., and Patel, M. S. Ochratoxin A decreases the activity of phosphoenolpyruvate carboxykinase and its mRNA content in primary cultures of rat kidney proximal convoluted tubule cells. *Biochem. Biophys. Res. Commun., 162*:916–920, 1989.
71. Brown, M. H. Szczech, G. M., and Purmalis, B. P. Teratogenic and toxic effects of ochratoxin A in rats. *Toxicol. Appl. Pharmacol., 37*:331–338, 1976.
72. Mayura, K., Reddy, R. V., Hayes, A. W., and Berndt, W. O. Embryocidal, fetotoxic and teratogenic effects of ochratoxin A in rats. *Toxicology, 25*:175–785, 1982.
73. Mayura, K., Parker, R., Berndt, W. O., and Phillips, T. D. Effect of simultaneous prenatal exposure to ochratoxin A and citrinin in the rat. *J. Toxicol. Environ. Health, 13*:553–561, 1984.
74. Mayura, K., Parker, R., Berndt, W. O., and Phillips, T. D. Ochratoxin A-induced teratogenesis in rats: partial protection by phenylalanine. *Appl. Environ. Microbiol., 48*:1186–1188, 1984.
75. Mayura, K., Stein, A. F., Berndt, W. O., and Phillips, T. D. Teratogenic effects of ochratoxin A in rats with impaired renal function. *Toxicology, 32*:277–285, 1984.
76. Hayes, A. W., Hood, R. D., and Lee, H. L. Teratogenic effects of ochratoxin A in mice. *Teratology, 9*:93–97, 1974.
77. Hood, R. D., Naughton, M. J., and Hayes, A. W. Prenatal effects of ochratoxin A in hamsters. *Teratology, 13*:11–14, 1976.
78. Gilani, S. H., Bancroft, J., and Reily, M. Teratogenicity of ochratoxin A in chick embryos. *Toxicol. Appl. Pharmacol., 46*:543–546, 1978.

79. More, J., and Galtier, P. Toxicité de l'ochratoxine A. I. Effect embryotoxique et teratogene chez le rat. *Ann. Rech. Vet.*, 5:167–178, 1974.
80. Ballinger, M. B., Phillips, T. D., and Kubena, L. F. Assessment of the distribution and elimination of ochratoxin A in the pregnant rat. *J. Food Safety*, *8*:11–24, 1986.
81. Mayura, K., Edwards, J. F., Maull, E. A., and Phillips, T. D. The effects of ochratoxin A on postimplantation rat embryos in culture. *Arch. Environ. Contam. Toxicol.*, *18*:411–415, 1989.
82. Wiger, R., and Størmer, F. C. Effects of ochratoxins A and B on prechondrogenic mesenchymal cells from chick embryo limb buds. *Toxicol. Lett.*, *54*:129–134, 1990.
83. Rahimtula, A. D., Bereziat, J. C., Bussacchini-Griot, V., and Bartsch, H. Lipid peroxidation as a possible cause of ochratoxin A toxicity. *Biochem. Pharmacol.*, *37*:4469–4477, 1988.
84. Pitout, M. J. The effect of ochratoxin A on glycogen storage in the rat liver. *Toxicol. Appl. Pharmacol.*, *13*:299–306, 1968.
85. Berndt, W. O., Hayes, A. W., and Phillips, R. D. Effects of mycotoxins on renal function: mycotoxic nephropathy. *Kidney Int.*, *18*:656–664, 1980.
86. Kane, A., Creppy, E. E., Röschenthaler, R., and Dirheimer, G. Biological changes in kidney of rats fed subchronically with low doses of ochratoxin A. *Dev. Toxicol. Environ. Sci.*, *14*:241–250, 1986.
87. Kane, A., Creppy, E. E., Röschenthaler, R., and Dirheimer, G. Changes in urinary and renal tubular enzymes caused by subchronic administration of ochratoxin A in rats. *Toxicology*, *42*:233–243, 1986.
88. Kane, A., Creppy, E. E., Roth, A., Röschenthaler, R., and Dirheimer, G. Distribution of the [^{3}H]-label from low doses of radioactive ochratoxin A ingested by rats, and evidence for DNA single-strand breaks caused in liver and kidneys. *Arch. Toxicol.*, *58*:219–224, 1986.
89. Khan, S., Martin, M., Bartsch, H., and Rahimtula, A. D. Perturbation of liver microsomal calcium homeostasis by ochratoxin A. *Biochem. Pharmacol.*, *38*:67–72, 1989.
90. Huff, W. E., Kubena, L. F., Harvey, R. B., and Doerr, J. A. Mycotoxin interactions in poultry and swine. *J. Anim. Sci.*, *66*:2351–2355, 1988.
91. Harvey, R. B., Huff, W. E., Kubena, L. F., and Phillips, T. D. Evaluation of diets contaminated with aflatoxin and ochratoxin fed to growing pigs. *Am. J. Vet. Res.*, *50*:1400–1405, 1989.
92. Kubena, L. F., Harvey, R. B., Huff, W. E., Corrier, D. E., and Phillips, T. D. Influence of ochratoxin A and T-2 toxin singly and in combination on broiler chickens. *Poult. Sci.*, *68*:867–872, 1989.
93. Engel, G., and von Milczewski, K. E. Zum Nachweis von Mykotoxinen nach Aktivierung mit Rattenleber-homogenaten mittels Histidinmangel-mutanten von *Salmonella typhimurium*. *Kiel Milchwirtsch. Forschungsber.*, *28*:359–366, 1976.
94. Bartsch, H., Malaveille, C., Camus, A. M., Martel-Planche, G., Brun, G., Hautsabadie, N., Hauterfeuille, A., Sabadie, N., Barbin,

A., Kuroki, T., Drevon, C., Piccoli, C., and Montesano, R. Validation and comparative studies on 180 chemicals with *Salmonella typhimurium* strains and V79 Chinese hamster cells in the presence of various metabolizing systems. *Mutat. Res.*, *76*:1–50, 1980.

95. Boorman, G. A. Toxicology and carcinogenesis studies of ochratoxin A. NIH Pub. No. 89-2813, Natl. Institutes of Health, Bethesda, Md., 1989.

96. Kuczuk, M. H., Benson, P. M., Heath, H., and Hayes, A. W. Evaluation of the mutagenic potential of mycotoxins using *Salmonella typhimurium* and *Saccharomyces cerevisiae*. *Mutat. Res.*, *53*:11–20, 1978.

97. Wehner, F. C., Thiel, P. G., Van Rensburg, S. J., and Demasius, I. P. C. Mutagenicity to *Salmonella typhimurium* of some *Aspergillus* and *Penicillium* toxins. *Mutat. Res.*, *58*:193–203, 1978.

98. Zeiger, E., Anderson, B., Haworth, S., Lawlor, T., and Mortelmans, K. *Salmonella* mutagenicity tests. IV. Results from testing of 300 chemicals. *Environ. Molec. Mutagen*, *11*:1–158, 1988.

99. Ueno, Y., and Kubota, K. DNA-attacking ability of carcinogenic mycotoxins in recombination-deficient mutant cells of *Bacillus subtilis*. *Cancer Res.*, *36*:445–451, 1976.

100. Mori, H., Kawai, K., Ohbayashi, F., Kuniyasu, T., Yamazaki, M., Hamasaki, T., and Williams, G. M. Genotoxicity of a variety of mycotoxins in the hepatocyte primary culture/DNA repair test using rat and mouse hepatocytes. *Cancer Res.*, *44*:2918–2923, 1984.

101. Creppy, E. E., Kane, A., Dirheimer, G., Lafarge-Frayssinet, C., Mousset, S., and Frayssinet, C. Genotoxicity of ochratoxin A in mice: DNA single-strand break evaluation in spleen, liver and kidney. *Toxicol. Lett.*, *28*:29–35, 1985.

102. Umeda, T. M., Tsutsui, T., Ithoh, M., and Saito, M. Mutagenicity and inducibility of DNA single-strand breaks and chromosome aberrations by various mycotoxins. *Gann*, *68*:619–625, 1977.

103. Kanizawa, M., and Suzuki, S. Induction of renal and hepatic tumors in mice by ochratoxin A, a mycotoxin. *Gann*, *69*:599–600, 1978.

104. Kanizawa, M. Synergistic effects of citrinin on hepatorenal carcinogenesis of ochratoxin in mice. In *Toxigenic Fungi—Their Toxins and Health Hazards* (H. Kirrata and Y. Ueno, Eds.), Elsevier, Amsterdam, 1984, pp. 245–267.

105. Bendele, A. M., Carlton, W. W., Krogh, P., and Lillehoj, E. B. Ochratoxin A carcinogenesis in the (C57BL/6JxC3H)F_1 mouse. *J. Natl. Cancer Inst.*, *75*:733–742, 1985.

106. Golinski, P., Hult, K., Grabarkiewicz-Szczesna, J., Chelkowski, J., and Szebiotko, K. Spontaneous occurrence of ochratoxin A residues in porcine kidney and serum samples in Poland. *Appl. Environ. Microbiol.*, *49*:1014–1015, 1985.

107. Lea, T., Steien, K., and Størmer, F. C. Mechanism of ochratoxin A-induced immunosuppression. *Mycopathology*, *107*:153–159, 1989.

108. Meisner, H. Energy-dependent uptake of ochratoxin A by mitochondria. *Arch. Biochem. Biophys.*, *173*:132–140, 1976.

109. Roth, A., Creppy, E. E., Kane, A., Bacha, H., Steyn, S. P., Röschenthaler, R., and Dirheimer, G. Influence of ochratoxin B

on the ochratoxin A inhibition of phenylalanyl-tRNA formation in vitro and protein synthesis in hepatoma tissue culture cells. *Toxicol. Lett.*, *45*:307–314, 1989.

110. Creppy, E. E., Størmer, F. C., Kern, D., Röschenthaler, R., and Dirheimer, G. Effects of ochratoxin A metabolites on yeast phenylalanyl-tRNA synthetase and on the growth and in vivo protein synthesis of hepatoma cells. *Chem.-Biol. Interactions*, *47*:239–247, 1983.
111. Haubeck, H. D., Lorkowski, G., Kolsch, E., and Röschenthaler, R. Immunosuppression by ochratoxin A and its prevention by phenylalanine. *Appl. Environ. Microbiol.*, *41*:1040–1042, 1981.
112. Prior, M. G., and Sisodia, C. S. The effects of ochratoxin A on the immune response of Swiss mice. *Can. J. Comp. Med.*, *46*:91–96, 1982.
113. Creppy, E., Størmer, F. C., Röschenthaler, R., and Dirheimer, G. Effects of two metabolites of ochratoxin A, (4*R*)-4-hydroxyochratoxin A and ochratoxin α, on immune response in mice. *Infect. Immun.*, *39*:1015–1018, 1983.
114. Luster, M. I., Germolec, D. R., Burleson, G. R., Jameson, C. W., Ackermann, M. F., Lamm, K. R., and Hayes, H. T. Selective immunosuppression in mice of natural killer cell activity by ochratoxin A. *Cancer Res.*, *47*:2259–2263, 1987.
115. Pier, A. C., Fichtner, R. E., and Cysewski, S. J. Effect of aflatoxin on the cellular immune system. *Ann. Nutr. Aliment.*, *31*: 781–787, 1977.
116. Cooray, R. Effects of some mycotoxins on mitogen-induced blastogenesis and SCE frequency in human lymphocytes. *Food Chem. Toxicol.*, *22*:529–534, 1984.
117. Holmberg, T., Thuvander, A., and Hult, K. Ochratoxin A as a suppressor of mitogen-induced blastogenesis of porcine blood lymphocytes. *Acta Vet. Scand.*, *29*:219–223, 1988.
118. Hult, K., Rutqvist, L., Holmberg, T., Thafvelin, B., and Gatenbeck, S. Ochratoxin A in blood of slaughter pigs. *Nord. Vet. Med.*, *36*:314–316, 1984.
119. Lea, T., and Størmer, F. C. Unpublished results, 1990.
120. Eichner, R. D., and Mullbacher, A. Hypothesis: fungal toxins are involved in aspergillosis and AIDS. *Aust. J. Exp. Biol. Med. Sci.*, *62*:479–484, 1984.

INDEX